BIOMECHANICS
of
MUSCULOSKELETAL INJURY

SECOND EDITION

William C. Whiting, PhD

California State University at Northridge

Ronald F. Zernicke, PhD

University of Michigan

Human Kinetics

Library of Congress Cataloging-in-Publication Data

Whiting, William Charles.
  Biomechanics of musculoskeletal injury / William C. Whiting, Ronald F. Zernicke. -- 2nd ed.
    p. ; cm.
  Includes bibliographical references and index.
  ISBN-13: 978-0-7360-5442-3 (hard cover)
  ISBN-10: 0-7360-5442-1 (hard cover)
  1. Musculoskeletal system--Wounds and injuries. 2. Musculoskeletal system--Mechanical properties. I. Zernicke, Ronald F. II. Title.
  [DNLM: 1. Musculoskeletal System--injuries. 2. Biomechanics. WE 140 W613b 2008]
  RD680.W47 2008
  617.4'7044--dc22

                                          2007045367

ISBN-10: 0-7360-5442-1
ISBN-13: 978-0-7360-5442-3

The Web addresses cited in this text were current as of January, 2008, unless otherwise noted.

Acquisitions Editor: Loarn D. Robertson, PhD; **Developmental Editor:** Elaine H. Mustain; **Managing Editor:** Melissa J. Zavala; **Copyeditor:** Julie Anderson; **Proofreader:** Kathy Bennett; **Indexer:** Nancy Gerth; **Permission Manager:** Carly Breeding; **Graphic Designer:** Bob Reuther; **Graphic Artist:** Angela K. Snyder; **Cover Designer:** Bob Reuther; **Photographer (cover):** Image courtesy of Primal Pictures; **Photographer (interior):** © Human Kinetics, unless otherwise noted; **Photo Asset Manager:** Jason Allen; **Art Manager:** Kelly Hendren; **Associate Art Manager:** Alan L. Wilborn; **Illustrator:** Jason M. McAlexander, MFA; **Printer:** Sheridan Books

Printed in the United States of America

10   9   8   7   6   5   4   3   2   1

Human Kinetics
Web site: www.HumanKinetics.com

United States: Human Kinetics
P.O. Box 5076
Champaign, IL 61825-5076
800-747-4457
e-mail: humank@hkusa.com

Canada: Human Kinetics
475 Devonshire Road Unit 100
Windsor, ON N8Y 2L5
800-465-7301 (in Canada only)
e-mail: info@hkcanada.com

Europe: Human Kinetics
107 Bradford Road
Stanningley
Leeds LS28 6AT, United Kingdom
+44 (0) 113 255 5665
e-mail: hk@hkeurope.com

Australia: Human Kinetics
57A Price Avenue
Lower Mitcham, South Australia 5062
08 8372 0999
e-mail: info@hkaustralia.com

New Zealand: Human Kinetics
Division of Sports Distributors NZ Ltd.
P.O. Box 300 226 Albany
North Shore City
Auckland
0064 9 448 1207
e-mail: info@humankinetics.co.nz

In loving memory of my parents, Richard and Charlotte; and to Marji, Trevor, Emmi, and Tad.

William C. Whiting

In loving memory of my parents; and to Kathy, Kristin, and Eric.

Ronald F. Zernicke

CONTENTS

Preface **vii** ▪ Acknowledgments **ix**

CHAPTER 1 Overview and Perspectives on Injury 1

Definition of Injury 2
Perspectives on Injury 3
Chapter Review and Suggested Readings 15

CHAPTER 2 Classification, Structure, and Function
of Biological Tissues 17

Embryology 18
Tissue Types 20
Arthrology 46
Chapter Review and Suggested Readings 50

CHAPTER 3 Basic Biomechanics 53

Kinematics 56
Kinetics 58
Fluid Mechanics 73
Joint Mechanics 74
Material Mechanics 80
Biomechanical Modeling and Simulation 93
Chapter Review and Suggested Readings 97

CHAPTER 4 Tissue Biomechanics and Adaptation 99

Biomechanics of Bone 101
Adaptation of Bone 107
Biomechanics and Adaptation of Other Connective Tissues 112
Biomechanics of Skeletal Muscle 118
Adaptation of Skeletal Muscle 119
Chapter Review and Suggested Readings 121

CHAPTER 5   **Concepts of Injury and Healing**. **123**

Overview of Injury Mechanisms 124
Principles of Injury 125
Inflammation and Entrapment Conditions 133
Bone Injuries. 134
Injuries in Other Connective Tissues 139
Skeletal Muscle Injuries 144
Joint Injuries 148
Nonmusculoskeletal Injuries 149
Chapter Review and Suggested Readings 151

CHAPTER 6   **Lower-Extremity Injuries** **153**

Hip Injuries 154
Thigh Injuries 161
Knee Injuries 166
Lower-Leg Injuries 184
Ankle and Foot Injuries 190
Chapter Review and Suggested Readings 201

CHAPTER 7   **Upper-Extremity Injuries** **203**

Shoulder Injuries. 204
Upper-Arm Injuries 220
Elbow Injuries 224
Forearm Injuries 232
Wrist and Hand Injuries. 237
Chapter Review and Suggested Readings 240

CHAPTER 8   **Head, Neck, and Trunk Injuries** **241**

Head Injuries. 242
Neck Injuries. 261
Trunk Injuries 272
Concluding Thoughts 286
Chapter Review and Suggested Readings 287

Glossary **289** ■ References **305** ■ Index **337** ■ About the Authors **350**

PREFACE

The purpose of this book's first edition was to explore the mechanical bases of musculoskeletal injury to better understand the mechanisms involved in causing the injury, the effect of injury on musculoskeletal tissues, and ultimately, based on our current knowledge of biomechanics, how injury might be prevented. That purpose remains unchanged, as does the fact that injury is an inevitable part of our everyday lives.

Biomechanics of Musculoskeletal Injury was written primarily for undergraduate students in the fields of exercise science, kinesiology, human movement studies, physical education, biomechanics, physical therapy, occupational therapy, and athletic training. The book may serve as a supplemental reference for practitioners in the fields of orthopedics, rheumatology, physical medicine and rehabilitation, physical therapy, occupational therapy, chiropractic medicine, ergonomics, and health and safety science.

In this second edition, the format of the inaugural volume is preserved and enhanced by new research and updated statistics, greater emphasis on lifestyle issues and a life-span approach, new topics and technologies, updated figures, and more photographs.

We begin in chapter 1 with an introduction to biomechanics as an interdiscipline and explore the mechanical aspects of injury, briefly assessing the prevalence of injury in our society and the physical, monetary, and emotional costs that result. In introducing injury, we maintain our focus on the detrimental aspects of injury while expanding on the positive aspects of injury as a stimulus for beneficial tissue adaptations.

Chapter 2 establishes the structural foundation to appreciate both the normal function of the human musculoskeletal system and how injury may affect this function. The key roles of embryology and tissue development are explained in determining the morphology and mechanical behavior of the mature human structure. We highlight the details of connective tissues (bone, cartilage, tendon, and ligament) that are most often involved in injuries of the musculoskeletal system. Chapter 2 concludes with an examination of arthrology, or joint mechanics; we emphasize this because many of the most functionally disabling injuries affect joints.

Chapter 3 presents biomechanical concepts essential for understanding injury mechanics. These mechanical parameters, such as force, stress and strain, stiffness, and elasticity are explained in the context of connective tissue injuries. The second edition is expanded to include more discussion on the application of mechanical principles to tissue mechanics and injury. Although mathematics is inextricably intertwined with biomechanics, we keep mathematical calculations to a minimum, emphasizing instead the mechanical concepts of phenomena.

Chapter 4 includes an introduction to the principle of overload and how this principle applies to tissue adaptation. The chapter integrates information from earlier chapters and explains how connective tissues respond to mechanical loading, in both normal and abnormal environments, and how these tissues are tested experimentally to quantify their mechanical behavior. A multitude of factors affect the musculoskeletal system's responses when forces are applied to tissues. Several of these factors, such as age, gender, nutrition, and exercise, are discussed with emphasis on how a person might lessen the chance or severity of injury. We emphasize the importance of lifestyle choices (e.g., nutrition, exercise) on tissue development and adaptation.

With a foundation in the scientific bases of tissue structure and function in place, we progress, in chapter 5, to the exploration of injury mechanisms. The second edition expands the link between basic mechanical properties of tissues and their clinical application and explores in greater depth applied topics such as ergonomics and osteoporosis.

The final three chapters (chapters 6, 7, and 8) present the essentials of regional injuries. We begin with the lower extremity (chapter 6), looking in detail at injuries such as inversion ankle sprains, stress fractures, compartment syndromes, and meniscal tears. Subsequent chapters examine injuries of the upper extremity (chapter 7) (e.g., rotator cuff tears, impingement syndrome, carpal tunnel syndrome) and of the head, neck, and trunk (chapter 8) (e.g., concussion, intervertebral disc injury).

In each of the final three chapters we present a detailed exploration of select injuries (in "A Closer Look") to show the depth to which any injury can be examined. We also include new or expanded sections highlighting topics of current concern such as falls in elderly populations, throwing-related rotator cuff pathologies, whiplash-related disorders, and injuries that result when youth carry heavy backpacks.

We have added a glossary at the end of the book, which includes key terms appearing in bold in the text. These terms are also found in the index with a *d* (for *definition*) following the page number. We also have added study questions at the end of each chapter to test your understanding and your ability to synthesize and apply the information presented.

In addition to these features, we also have two ancillary products available. The instructor guide presents the review questions at the ends of each chapter along with outlines for suggested student answers. The PowerPoint features the art, photos, and tables that apear in the book.

Knowledge of the biological responses of tissues to mechanical loading improves our understanding of injury and its consequences and will enable you, as a health professional, to reduce the chance that your clients, patients, or athletes will experience physical injury.

ACKNOWLEDGMENTS

A project of this scope involves the unique contributions of many more people than the two listed on the book's cover. We extend our appreciation to those friends and colleagues. We thank Rainer Martens and the staff at Human Kinetics, in particular the devoted efforts of Loarn Robertson, Elaine Mustain, and Melissa Zavala, for sharing our belief in the importance of this project. We acknowledge the hundreds of professional colleagues and thousands of students who shaped our philosophies, guided our progress, and provided inspiration for our professional work for more than 25 years. Most important, we thank our families for their support, patience, and love while we completed this project. Without them, our work and lives would have little meaning.

OVERVIEW AND PERSPECTIVES ON INJURY

All injury leaves pain in the memory except the greatest injury, that is death, which kills memory with life.

Leonardo da Vinci (1452-1519)

■ OBJECTIVES

- ■ To define and explain injury, mechanisms of injury, and biomechanics
- ■ To explain the multidisciplinary nature of injury analysis
- ■ To describe perspectives on injury, including historical, epidemiological, health professional, economic, psychological, safety professional, and scientific

Injury pervades everyday life. Although people sustain injuries of varying severity, and some people are injured more frequently than others, virtually no one is spared the pain, distraction, and incapacity caused by injury. Along with injury come the inevitable physical, emotional, and economic costs as well as loss of time and normal function.

The impact of these costs and losses is staggering worldwide. The National Safety Council (in the United States) estimated that the annual cost of injury in the United States for 2002 alone was almost $600 billion US and that about 4 of every 10 admissions to hospital emergency rooms or hospital clinics were for treatment of injury. The number of emergency room visits in the United States in 2001 represented by that proportion was nearly 40 million (National Safety Council 2004). For the same year, Health & Safety Executive Northern Ireland estimated the costs of injury to be $1 billion US, whereas Great Britain reported costs of $34 billion US. Health Canada reported that Canada spent $8.7 billion US in 1995 on preventing and treating injuries. Australia spent $11 billion US in 1986, whereas the costs of injury in New Zealand for 2003 were $1.7 billion US. The most recent figures for China (2006) put that country's costs at $25 billion US, whereas Lebanon spent approximately $13 million US in 2003.

In the United States, unintentional injury ranks immediately after heart disease, cancer, cerebrovascular disease (stroke), and chronic obstructive pulmonary disease as a leading cause of death. If intentional injuries (e.g., homicides and suicides) are added to the definition of injury, death attributable to injury vaults into fourth place, trailing only heart disease, cancer, and stroke (National Safety Council 2004). Clearly, injury is one of the most serious public health problems across the globe.

From the perspective of potential life span remaining, the impact of injury-related death is more significant than the impact of death from other causes. At the point just preceding injury, persons fatally injured in 1985 had an average remaining life span of 36 years compared with 12 and 16 years average remaining life span for those dying of cardiovascular disease and cancer, respectively (Rice and Max 1996). Estimates by the National Center for Injury Prevention and Control (NCIPC 2001-2002) confirm those earlier findings. Using years of potential life lost as a measure of impact, in 2001 the NCIPC identified unintentional injuries as the leading cause of death, accounting for more than 2 million years lost, outpacing both cancer and heart disease.

Such fatality statistics paint only a portion of the picture of the impact of injury. Nonfatal accident statistics are even more astounding. Disabling injuries affect more than 20 million people each year. National Safety Council (2004) estimates from the United States indicate that every 10 min, 2 people are killed and about 390 suffer a disabling injury, contributing to an annual cost of more than $11 billion.

On average, there are 12 deaths from unintentional injury and more than 2,300 disabling injuries every hour during the year in the United States. Despite successful efforts to reduce some types of injuries, such as those caused by automobile crashes, a substantial number of us will be victims of injury.

Definition of Injury

As will become clear in the following chapters, many injuries have a mechanical cause. Forces and force-related factors can lead to injury and may influence the severity of injury. Before delving into the multiple facets of injury, however, we establish a working definition: **Injury** is the damage sustained by tissues of the body in response to physical trauma. This definition is less encompassing than generally accepted notions of injury, but it is useful within the context of the biomechanics of musculoskeletal injury. The term *injury* is usually associated with negative consequences, and indeed these negative consequences often occur. In some situations, however, injury may be involved in events with positive consequences. In the bone remodeling process, for example, bone must first be injured to prepare it for subsequent positive adaptive changes.

Biomechanics is the area of science related to the application of mechanical principles to biology. The number of potential areas of study in biomechanics is immense. Topics as diverse as blood flow dynamics, human and animal locomotion, artificial limbs and prosthetic design, sports, and biomaterials fall under the rubric of biomechanics. The mechanical causes and effects of forces applied to the human musculoskeletal system are the primary focal points of our text, within the broader area of biomechanics.

As we explore injury biomechanics, key terms will recur, so we define them at the outset. The first, mechanics, is the branch of science that deals with the effects of forces and energy on bodies. Second, a mechanism is defined as the fundamental physical process responsible for a given action, reaction, or result. Chapters 6 through 8 examine in detail the mechanisms of many musculoskeletal injuries.

Perspectives on Injury

Exploration of the biomechanics of injury is an interdisciplinary endeavor. Among the disciplines involved are anatomy, physiology, mechanics, kinesiology, medicine, engineering, and psychology. The problem of musculoskeletal injuries cannot be addressed effectively by any single discipline examining injury in isolation. A wealth of research in this field supports the notion that "we know many facts—but we lack integrated answers" (Caine et al. 1996, p. 1). Therefore, to ensure optimal progress in addressing the problem, an interdisciplinary approach is essential.

Those with an interest in studying injury include physicians, occupational therapists, kinesiologists, prosthetists, orthotists, nurses, physical therapists, chiropractors, osteopaths, ergonomists, safety engineers, strength trainers, athletic trainers, coaches, and athletes. Each of these people will have his or her own perspective on injury.

WHAT'S IN A WORD: ACCIDENT OR INJURY?

Hear the word **accident** and most people envision an event that is unexpected, by chance, unintentional, or, as insurance companies like to say, an "act of God." The term *accident,* in the context of discussing injury, is an ambiguous and misleading descriptor. Accident implies a degree of human error or involvement, but that is not always the case. *Accident* is often used synonymously with *injury* in practical situations. This is unfortunate and inaccurate, because not all accidents involve injuries and not all injuries are accidental in nature.

Suchman (1961) provided a list of indicators that increase the likelihood that an event is accidental. These indicators are the degree of expectedness, avoidability, and intention. If an event is unexpected, unavoidable, and unintentional, it likely is accidental.

No single definition of **accident** will likely satisfy everyone. So what should be done? In some scientific circles, the word *accident* is gradually being replaced with more specific terminology. What were formerly accidental injuries are now specified as being unintentional or intentional injuries. Car accidents are now commonly termed motor vehicle crashes.

Some organizations no longer officially include the word *accident* in their professional vocabularies, but confusion may persist, as the following example shows. The U.S. National Safety Council annually publishes a comprehensive and useful volume of safety statistics that documents the prevalence of injury and its costs. The foreword of the 1994 edition begins by noting that "many organizations, including the National Safety Council, have decided to eliminate the term 'accident' from their official vocabularies." And the title of this National Safety Council publication? *Accident Facts.* That title persisted through the late 1990s, but the publication has since been renamed and now is titled *Injury Facts.*

Historical

Musculoskeletal injuries have ancient origins. Indeed, evidence of lesions in vertebrate fossils and pathologies in prehistoric bones suggests that injury is as old as life. Skeletal remains of the earliest humans reveal arthritis and fractures, suggesting that at no time were we exempt from the consequences of injury. The nature of injuries can provide insight into the history of an era. Some ancient Egyptian skeletons, for example, show a fracture of the left ulna, perhaps a result of self-defense from a blow by a club. Today, these types of fractures are sometimes called nightstick fractures. Evidence of musculoskeletal disorders is commonly seen in the art of ancient civilizations, often in the statues and wall paintings (figure 1.1).

Attempts to treat the injured are nearly as old as injury itself. Archeologists have uncovered evidence of splints and primitive surgical implements (e.g., obsidian knives). Indian surgeons circa 1000 to 600 BC, predating Hippocrates by several centuries, used instruments such as forceps, scissors, and knives.

The evolution of medicine into a specialized profession with rational tenets of practice is generally acknowledged to have begun with Hippocrates. Although their knowledge of anatomy was scant and their procedures were often crude by modern standards, Hippocrates and other Greek physicians established the foundations that are the basis for the study and treatment of injury today.

Besides the physicians of the day who studied and treated injury, some of history's great names, often heralded for other pursuits and accomplishments, highlighted injury in some form and accorded it recognition in their work. The Greek poet Homer in his classic *Iliad* wrote often of trauma and treatment, describing more than 100 specific wounds and injuries.

With the decline of the Greek empire, much of the accumulated Greek knowledge shifted

Figure 1.1 Ancient (*a*) Greek and (*b*) Egyptian depictions of injury.

(*a*) Achilles binding Patroclus wounds. Red figure drinking cup (kylix), from Vulci. Ca. 500 BCE. INV: F2278. Photo: Johannes Laurentius. © Bildarchiv Preussischer Kulturbesitz/Art Resource, NY. (*b*) New Carlsberg Glyptotek, Copenhagen to Byzantium (Asia Minor), Alexandria, and then Rome. Notable among practitioners of this era was Galen (AD 129-199). Galen's work has been credited with defining, for better or worse, the direction of medical treatment for the next 1,500 years. Among his contributions were an appreciation of the nature of muscle contraction; a fundamental understanding of anatomy (although human dissection was still centuries away); the treatment of spinal deformities such as kyphosis, scoliosis, and lordosis; and the use of pressure bandages to control limb hemorrhage (Rang 2000). Soon after Galen's death the Roman Empire declined, and with its abrupt fall in AD 476, western civilization entered the Dark Ages, virtually halting progress in medical science.

The entire world, however, did not suffer the ravages of Europe's Dark Ages. In China during the Tang dynasty (AD 619-901), for example, surgery (e.g., orthopedic treatment of fractures and dislocations) was recognized as a special branch of medicine (LeVay 1990). Later, as Europe emerged from the Dark Ages, renewed creative energies were applied to medical problems. Anatomical investigation flourished, most notably by Vesalius (1514-1564), whose anatomical drawings still inspire wonder (figure 1.2). As knowledge of human anatomy advanced, so too did understanding of how the body functions.

Leonardo da Vinci (1452-1519), perhaps the best-known figure of the Renaissance, was intrigued by the nature of pain and trauma. In his art we find exquisite depictions of physical pain and agony. In his scientific writing we also find many references to trauma, especially that caused by what he termed *percussion* (impact). From his deep interest in human anatomy, da Vinci was aware that the joints in the body serve as shock absorbers. Noticing that the pain produced by landing from a jump on the heels is much greater than when landing on the toes, he deduced "that which gives more resistance to a blow suffers most damage."

da Vinci had an abiding fascination with the body's senses and in particular with the sense of pain (figure 1.3). Although he knew that pain served an important protective function, he

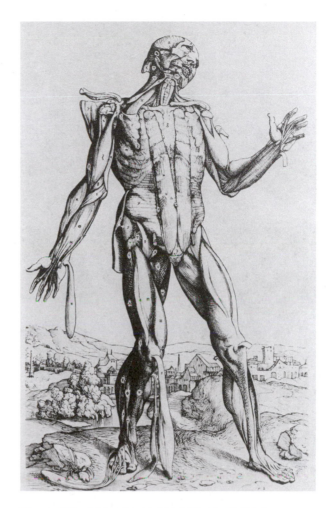

Figure 1.2 A "muscle man" from Vesalius's *Fabrica*.
Reprinted, by permission, from C.D. O'Malley, 1964, *Andreas Vesalius of Brussels, 1514-1564* (Berkeley, CA: University of California Press), 31.

also saw it as the "chief evil" in life, concluding that "the best thing is the wisdom of the soul; the worst thing is pain of the body" (Keele 1983, p. 237). The insights of da Vinci and other great thinkers of the Renaissance era may seem elementary, even naive, compared with current levels of understanding, but compared with the knowledge that was available and accepted for many centuries, their breakthroughs were extraordinary.

With the advent of the Industrial Revolution in the 19th century, medical progress accelerated. Many of the problems that were previously impossible to address were suddenly tractable. Admittedly, the industrial age created many new problems, notably injuries caused by machinery, but at the same time it brought

Figure 1.3 Leonardo da Vinci s visions of pain and trauma. *(a)* St. Jerome, a picture of pain. *(b)* The hanging of Bernardo di Bandino Baroncelli.

(*a*) Reprinted, by permission, from K.D. Keele, 1983, *Leonardo Da Vinci s elements of the science of man* (New York, NY: Academic Press), 10

(*b*) Reprinted, by permission, from K.D. Keele, 1983, *Leonardo Da Vinci s elements of the science of man* (New York, NY: Academic Press), 236.

a welcomed prospect for rapid developments in medicine. With the discovery of anesthesia and antiseptics, surgical success improved dramatically. Advances such as clinical arthroscopy, pioneered by Bircher in the early 1900s, showed the promise of rapidly developing technologies.

Progress continues today, and advancements in the diagnosis and treatment of injury show no sign of slowing. Even a few decades ago the suggestions of routine joint replacement, laser surgery, advanced imaging techniques (e.g., magnetic resonance imaging), microsurgery, and computer- or robot-assisted surgery were viewed as futuristic speculation. Continuing advancements in materials science, computer technology, nanotechnology, robotics, tissue engineering, and genetics promise even more spectacular advances. Although technological progress extends great promise, we must not forget that it can be a double-edged sword. The technological saber swung in one direction has the potential to prevent injury or aid in its diagnosis and treatment, but wielded in the opposite direction it has the potential for creating or exacerbating injury as well. As long as injury remains an unfortunate fact of everyday life, challenges will undoubtedly change but will not disappear.

HIPPOCRATES AND INJURY

Hippocrates (460-377 BC), called the "father of medicine," treated numerous injuries in his role as physician and described in detail many of the orthopedic conditions he encountered. Although some of his descriptions were flawed in light of our current understanding, he successfully treated injuries on a regular basis and related his techniques and results in documentary form. His descriptions of treating shoulder dislocations, for example, gave numerous artists the material to depict the procedures. Hippocrates, with biomechanical insight, noted that even an old dislocated shoulder could be reduced (i.e., shoulder bones brought back to their normal position), "for what could not correct leverage move?" (LeVay 1990, p. 24).

Among the many other injuries Hippocrates described were acromioclavicular dislocation ("I know many otherwise excellent practitioners who have done much damage in attempting to reduce shoulders of this kind"), spinal deformities (with vertebrae "drawn into a hump by diseases"), and leg fractures ("All bones unite more slowly if not placed in their natural position and immobilized in the same position, and the callus is weaker") (LeVay 1990, pp. 26-37).

Hippocrates exhibited great insight in this summary observation:

> All parts of the body which have a function, if used in moderation and exercised in labours to which each is accustomed, thereby become healthy and well-developed: but if unused and left idle, they become liable to disease, defective in growth, and age quickly. This is especially the case with joints and ligaments, if one does not use them. (LeVay 1990, p. 30)

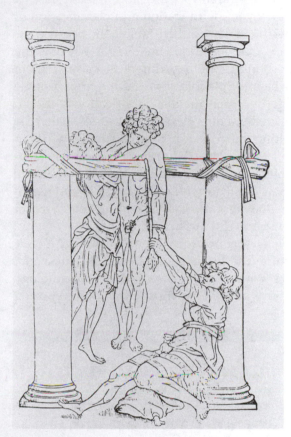

Depiction of historical technique of reducing a shoulder dislocation using a large wooden beam.

Epidemiological

Questions about injuries, such as how many, how often, what kind, and to whom, are central to epidemiology. **Epidemiology** is the study of the distribution and determinants of disease and injury frequency within a given human population. In most cases, the distinction between disease (e.g., measles) and injury (e.g., torn ligament) is clear. In other cases the picture is less clear, and deciding whether a malady is a disease or an injury may not be as obvious. In our text we focus on musculoskeletal injuries, but disease can be a factor, because certain diseases predispose an individual to injury (e.g., osteoporosis can lead to bone fractures).

Epidemiological studies are typically either descriptive or analytical in nature. The first of these, **descriptive epidemiology,** is the most common form of epidemiological research. Types of descriptive epidemiological designs include case reports or case series, cross-sectional surveys, and correlational studies. The purpose of such approaches is to quantify the *distribution* of disease or injury and address questions pertaining to occurrence (how many injuries occur?), person (who is getting injured?), place (where are the injuries occurring?), and time (when are the injuries happening?). On the surface, this should be straightforward, and in many cases it is. However, identification and classification of a specific injury can be problematic because the clinical manifestations are similar whereas the underlying pathology differs or because there may be multiple injuries resulting from a single incident, which makes classification difficult. Care is essential in classifying injuries so that the resulting categories are mutually exclusive (is an injury suffered in a delivery truck crash a vehicular injury or a work-related injury?), exhaustive (is there a category for every injury?), and useful (does the classification system have a practical and meaningful application?).

Clear terminology is essential when we examine the biomechanics of musculoskeletal injury. With respect to descriptive epidemiology of injury, results are most commonly reported as either incidence or prevalence rates. Many people use the terms *incidence* and *prevalence* interchangeably and synonymously. In fact, incidence and prevalence are distinctly different terms and, when one is analyzing injuries, provide very different estimates. A **prevalence** rate describes the number of cases (e.g., injuries), both new and old, that exist in a given population at a specific point in time divided by the total population number. An **incidence** rate describes the number of new injuries that occur within a given population at risk over a specified time period.

Analytical epidemiology involves complex research strategies to reveal the determinants or underlying causes of disease and injury. Questions such as why or how injuries are taking place are addressed by examining or identifying factors that may contribute to the occurrence of injury. These contributing factors are known as **risk factors** and are classified as either intrinsic or extrinsic. Intrinsic risk factors are characteristics of a biological or psychological nature that may predispose an individual to injury. Examples of intrinsic risk factors include physical characteristics such as gender, age, a family

PUBLIC HEALTH APPROACH TO INJURY PREVENTION

One valuable use of descriptive epidemiology in the community is public health surveillance. Surveillance is a systematic and ongoing collection, analysis, interpretation, and dissemination of public health information to assess public health status, define public health priorities, and evaluate programs set in place to improve the health of a community. Surveillance is carried out by health agencies for a number of reasons, such as estimating the magnitude of a problem, detecting epidemics, generating hypotheses, stimulating further research, evaluating present levels of health care, determining the geographic distribution of a disease, and facilitating planning and resource allocation. Who, what, where, and when are addressed, resulting in valuable information applicable and accessible to the public.

There are four steps to gathering this information, termed the **public health approach:**

1. Define the problem—this is the purpose of Public Health Surveillance, the who, what, where, and when of injury occurrence

   ↓

2. Risk factor identification—what factors place people at risk for injury?

   ↓

3. Evaluation—developing and testing strategies based on gathered public data to test the efficacy of injury prevention programs

   ↓

4. Implementation—replicating successful strategies within the community

history of injury or disease, and percent body fat; performance characteristics such as muscular strength, balance, flexibility, or endurance; and cognitive characteristics such as level of anxiety, self-esteem, and self-efficacy.

In contrast, extrinsic risk factors are external or environmental characteristics that influence a person's injury risk. Examples of extrinsic risk factors include safety conditions, programs, or the use or misuse of protective equipment within a workplace or, with respect to sporting events, the level of competition, training schedule, or weather conditions during the event. Difficulties arise in the identification of risk factors because in most situations, factors act in concert, and injury or disease is a result of this interaction. Before a **causal association** between a risk factor and a disease outcome (or injury) can be established, the factor under investigation must be examined through a multifactorial model of causation. Intrinsic risk factors are thought to predispose an individual to injury, and once a person is susceptible, extrinsic or "enabling" factors may interact with predisposing factors to increase the likelihood of injury (Meeuwisse 1994) (figure 1.4). Investigators must exercise caution in assigning causal relations to injury by ruling out the possibility of mere correlation or coincidence.

Relative risk is an epidemiological measure used to quantify the likelihood of an injury occurrence in one group versus another group. Relative risk is calculated as the injury incidence in group A divided by the injury incidence in group

B. You could calculate the relative risk of stress fractures, for example, if you knew the incidence data of stress fractures in a group of female long-distance runners versus a group of age-matched sedentary females.

When using a statistical measure, investigators must ensure that injury rates are calculated from reliable data and that conclusions based on rates are valid. Caution, care, and clear thinking are warranted before using or accepting any statistical measures, including those for rates of injury.

Many organizations in the United States, including the National Safety Council, insurance companies, law enforcement agencies, the Occupational Safety and Health Administration (OSHA), and traffic safety boards, routinely collect and publish accident and injury data. Health Canada collects similar data annually for Canada. In 2002, the Consumer Safety Institute from Amsterdam, The Netherlands, undertook a surveillance-based assessment of the medical costs of injury (EUROCOST) in Europe from data obtained in 1999. China undertook an assessment in 2006 of workplace injuries under the auspices of the Chinese Center for Disease Control and Prevention. Similar accident and injury data are collected and published by the Department of Labour in New Zealand and the Commonwealth Department of Health and Ageing in Australia. The World Health Organization tracks and publishes data on the costs of injury for many European countries and others (e.g., Lebanon).

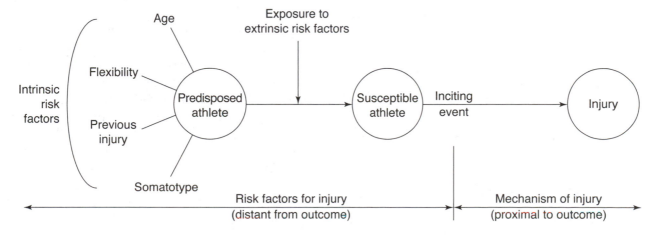

Figure 1.4 Assessing causation in sport injury: A multifactorial model.

Adapted, by permission, from W.H. Meeuwisse, 1994, "Assessing causation in sport injury: A multifactorial model," *Clin J Sport Med* 4(3):168.

For two primary reasons, much of the available injury data are for injury-related deaths. First, the catastrophic nature of fatalities makes them prominent, and second, death statistics are easy to compile. Less attention is paid to the documentation of nonfatal injuries, especially those of a minor nature that may never be reported at all. This raises a question: What percentage of all injuries are actually reported? The answer to that question is unknown. Certainly many injuries are never officially recorded, and therefore the published statistics undoubtedly underestimate the true injury toll.

In the United States, the 2003 unintentional-injury deaths totaled an estimated 101,500, which translates to a rate of 34.9 injury deaths per 100,000 population (National Safety Council 2004). In 1998, the World Health Organization estimated that on average, 6.5% of all mortality occurring in its member countries was attributable to unintentional injury, and that percentage varied substantially between high-income (6.2%) and low- and middle-income countries (11.5%). Variation in rates around the world is attributed to the influence of many economic, social, and cultural factors.

Deaths attributable to injury are proportionately higher in the young. According to the U.S. National Safety Council, injury is the number one cause of death among individuals aged 1 to 34. Statistics for nonfatal injuries are equally daunting. In the United States, for example, the National Safety Council (2004) estimated that more than 20.7 million disabling injuries occurred in 2003. Virtually all nonfatal injury statistics are approximations based on hospital or commission records or extrapolations from interview surveys.

Much of the epidemiological information about the biomechanics of musculoskeletal injury cannot be used directly because it is described according to circumstance (e.g., injury attributable to automobile collision) rather than by the specific causal agent or mechanism.

Health Professional

Many health professionals and organizations are involved in strategies to reduce the incidence and severity of injuries. At an organizational level, in the United States, groups such as the National Safety Council, the Centers for Disease Control and Prevention, and the Consumer Product Safety Commission perform injury prevention analyses. Both Canada and New Zealand have injury prevention strategy organizations that function in much the same way as the American organizations. In Europe, the World Health Organization (WHO) is actively involved in similar analyses. At an individual level, safety engineers, ergonomists, safety consultants, job supervisors, health professionals, parents, teachers, coaches, and many others are in positions to stress safety and injury prevention.

Severe injuries often require immediate medical attention. Emergency medical personnel, such as paramedics, emergency room physicians, and support staff, provide often life-saving emergency diagnosis and treatment. Less severe injuries may require nonemergency treatment. Physicians, athletic trainers, and other allied health professionals perform these less urgent diagnostic and treatment tasks. Many injuries, especially those requiring surgical treatment, require rehabilitation to ensure a return to preinjury performance levels. Rehabilitation personnel (e.g., physical and occupational therapists) perform these essential services.

Economic

In addition to the physical and emotional costs associated with injury, the financial costs of injury are enormous. Because public policy decisions are often based on fiscal considerations, the economic perspective of injury requires comment.

In comparison to long-recognized public health hazards such as cardiovascular disease and cancer, only in the last several decades has injury been recognized as a true public health hazard. A watershed study, *Injury in America* (Committee on Trauma Research 1985), helped bring injury into the public health spotlight and prompted the U.S. Congress to commission a study on the economic and noneconomic impact of injury. Results of that study showed that injury had a tremendous effect on both individuals and society as a whole (Rice and Max 1996).

The National Safety Council (2004) estimated the total cost of unintentional injuries for 2003 at nearly $608 billion. That total included estimates of economic costs of fatal and nonfatal unintentional injuries together with employer costs, vehicle damage costs, fire losses, wage and productivity losses, medical expenses, and administrative expenses. If to that $608 billion you add the more than $1.3 trillion estimated cost for lost quality of life from those injuries, the resulting comprehensive cost of injury for 2003 alone is almost $2 trillion.

In summarizing the cost of injury, Runge (1993) issued a challenge to health professionals to become more involved as advocates for injury prevention and control, thereby contributing to efforts at limiting the present cost of injury. His challenge should be heeded by everyone, because we all pay the price that injury exacts.

Psychological

The most obvious consequence of injury is the direct, physical damage to bodily tissues. Often overlooked are the psychological factors that may be involved before, during, and after the injury. These factors can influence the likelihood and severity of injury and the course of healing and rehabilitation. Aspects of injury in which psychological factors may be integrally involved include risk behaviors and predisposition to injury, human error and accidents, theories of causation, risk evaluation, and emotional response to injury.

A person's likelihood of being injured depends largely on the task in which he or she is engaged, the environment in which the injury occurs, and the person's psychological state. Some activities such as playing football or occupations such as oil drilling are inherently riskier than others. Certain environments, such as rugged outdoor terrain or construction sites, are more risk laden than others. In addition, certain psychological states, such as inattention, distraction, fatigue, or stress, may predispose a person to injury. Andersen and Williams (1999) proposed a theory of stress responsivity that suggests a strong relation between the likelihood of injury and stress factors such as stress history, personality factors, and coping resources.

Key points from a discussion of the role of human error, accident causation, and risk evaluation in injury prevention and control by Sanders and McCormick (1993) highlight the importance of including psychological factors in the overall context of injury analysis:

▪ Human error (defined as an inappropriate or undesirable human decision or behavior that reduces or has the potential to reduce effectiveness, safety, or system performance) is responsible for most, if not all, events leading to injury. Human error leading to injury typically results from the direct action of the injured person but also may be an indirect human error, such as a poor decision made by an engineer in designing a particular product or device. Human error may be reduced by (1) selection of people with appropriate skills and capabilities to perform a particular task, (2) proper training, and (3) effective design of equipment, procedures, and environments.

▪ Theories have been proposed to explain accident causation, including accident-proneness theory (some people are more prone to accidents than others); accident-liability theory (people are prone to accidents in given situations and this proneness is not permanent); capability–demand theory (accidents increase when job demands exceed the capability of workers); adjustment-to-stress theory (accidents increase in situations with stress levels that exceed an individual's coping capabilities); arousal–alertness theory (accidents are more likely when arousal is too low or too high); and goals–freedom–alertness theory (freedom of workers to set their own goals results in high-quality performance, which reduces accidents). No single theory adequately explains all accidents and their resulting injuries; a more likely scenario is that a unique combination of factors is involved in each injury.

▪ Many factors contribute to accidents and injuries. Sanders and Shaw (1988) proposed a comprehensive model, termed **contributing factors in accident causation,** which provides a framework for categorizing the many contributing factors to injury occurrence. More recently, Ghosh and colleagues (2004) examined relations of working conditions and individual characteristics to occupational injuries. Both studies identified the following categories of factors that can be important to accident causation: management, physical environment, equipment design, the work itself, social and psychological environment, and workers and coworkers.

▪ **Risk** refers to the likelihood of injury or death associated with a particular object, task, or environment. The perception and evaluation of risk are important for determining whether an injury will occur and, if it does, the severity of the injury. Interestingly, studies indicate that whereas most people are quite capable of discerning the relative risk between various products (e.g., using a computer is less risky than riding a bicycle), their ability to estimate the absolute risk is not nearly as accurate. Perception of risk may be distorted by overestimating the value of one's own expertise and experience, overemphasizing situations receiving media attention, and adopting a philosophy that "it can't happen to me."

Psychological factors are important influences before, during, and immediately following the injury and in the postinjury period, which may last for weeks, months, or even years. Although the psychological factors summarized by Heil (1993) are specific to athletes, many of the factors are applicable to general injury situations as well (table 1.1). Other factors that could be added to the list include family support structures, need to work, and malingering, and there is little doubt that psychological considerations play an influential role in a comprehensive assessment of injury.

Many injuries, and certainly those that draw the most media attention, occur among athletes. The psychological profiles of highly competitive, elite athletes are in some ways different from those of the general population. The differences can be both beneficial and deleterious in dealing with the injury and the recovery process. As noted by Heil (1993), the positive psychological attributes found in many athletes are high levels of motivation, pain tolerance, goal orientation, and good physical training habits. On the negative side, athletes may experience a higher sense of loss, greater threat to their self-image, unrealistic expectations, and desire for a quick recovery, and they may have higher, sport-specific demands to meet than do those in the general population. Whether in specific populations, such as athletes, or in the general public, psychological factors play a critical role in injury and should be neither underestimated nor ignored.

Safety Professional

The prevention and control of injuries, although not the primary focus of this text, are integral to a broad discussion of injury, and we would be remiss not to mention the role of safety profes-

TABLE 1.1

PSYCHOLOGICAL FACTORS IN INJURY

Factors preceding injury	Factors associated with injury	Factors following injury
Medical history	Emotional distress	Culpability
Psychological history	Injury site	Compliance with treatment
Somatization	Pain	Perceived effectiveness
Life stress and change	Timeliness	Treatment complications
Sport stress and change	Unexpectedness	Pain
Approach of major competition		Medication use
Marginal player status		Social support
Overtraining		Personality conflicts
Sport-related health risk factors		Fans and the media
		Litigation

Adapted, by permission, from J. Heil, 1993, *Psychology of sport injury* (Champaign, IL: Human Kinetics), 75.

sionals, such as safety consultants, ergonomists, safety engineers, and health and safety educators, in dealing with injury and its prevention.

Injury prevention programs are typically of two types: injury control programs and health and safety education programs. One commonly referenced injury control program lists strategies that include prevention of hazard creation; modification of the hazard's rate of release from its source; protection from the hazard source; and stabilization, repair, and rehabilitation of the damaged object (Haddon and Baker 1981). The second approach seeks to reduce the incidence and severity of injury through health and safety education. A third injury prevention program combines community programming with education and control programs. The establishment of injury prevention foundations (e.g., Think First National Injury Prevention Foundation) that act through a multilevel approach to injury prevention might make a difference (Rosenberg et al. 2005).

None of these approaches has been entirely successful in achieving injury prevention. The lack of success is likely attributable to complex political, economic, sociological, or behavioral influences on injury prevention and control. Recent efforts to improve safety have concentrated on providing frameworks that combine elements of injury control and health education. For example, Gielen (1992) noted that a comprehensive injury control program must be both effective and feasible and proposed a four-step process: (1) epidemiological diagnosis, (2) environmental and behavioral diagnosis, (3) influencing factors diagnosis, and (4) formulation of an intervention plan. Although many theories and models exist to guide program design and implementation and the development of evaluation measures, a systematic 20-year review of published literature found few examples of theory testing (Trifiletti et al. 2005). That suggests that greater collaborative research could enhance behavioral approaches to injury control.

At least three strategies are available to control or prevent injuries (Committee on Trauma Research [CTR] for the U.S. National Academy of Science 1985): (1) persuade (educate) those at risk of injury to alter their behavior to increase self-protection (e.g., to use helmets while cycling or to use seat belts while driving cars or flying in planes), (2) require changes in individual behavior by law or rules (e.g., enforce laws for mandatory seat belt use in cars, penalize football players who spear-tackle an opponent with the top of the helmet, require protective eyewear while playing squash), and (3) provide automatic protection by product or environmental design (e.g., air bag passive restraints in cars, multi-directional release mechanisms for ski bindings, padding for fixed goal posts, enhanced rear foot stabilizers with shock-absorbing heel materials for running shoes).

Of these three injury prevention strategies, the third (automatic protection) is the most effective, followed by the second strategy (requiring behavioral change). Persuading is the least effective of the three. Although education about injuries is important, many injuries result less from a lack of knowledge than from failure to apply what is known. Most people will acknowledge that it is safer to wear a mask as a baseball catcher or hockey goalie, but sometimes a mask is not available or the player chooses not to wear one. Health behavior research has shown that as the amount of individual effort required to adopt a safer behavior increases, the proportion of the population that will respond by adopting the behavior decreases. For example, the more difficult or cumbersome the protective equipment is to put on, the less likely players are to use it.

Education alone has rarely proven to be an adequate preventive strategy (CTR 1985). The most successful attempts at changing individual behavior to prevent injuries have involved behaviors that were easily observable and required by law. For example, when laws required helmet use for motorcyclists, almost all complied. In Thailand, after enforcement of the helmet act, the number of helmet wearers increased five-fold (Ichikawa et al. 2003).

Interestingly, more than two decades ago the CTR (1985) reported that injury prevention for most types of recreational activities remained nearly unresearched. That statement is still accurate. More information is needed to assess the effectiveness and use of protective sports equipment and modifications, such as energy-absorbing gymnastics mats, playground surfaces, running surfaces, or gymnasium walls.

A fourth important strategy for preventing musculoskeletal injury that should be mentioned is maintenance of strength, flexibility, and good physical condition. Whether in the home, in the workplace, or in sports, people with better physical conditioning and flexibility are less likely to be injured and are more likely to recover faster after being injured than are those who are in poor shape. Indeed, one of the most important benefits of regular stretching and flexibility training may be the prevention of musculoskeletal injury. Proper flexibility training and pre-exercise stretching can reduce joint stiffness, muscle and tendon tightness, and exercise-related muscle soreness. For example, the probability of reinjuring calf muscles could be reduced to less than 1%

as a result of stretching exercises (Miller 1979), whereas tightness, or lack of flexibility, tends to increase the likelihood of musculoskeletal injuries (Witvrouw et al. 2003).

Considering the enormous numbers and types of injuries that occur, many challenges remain for safety professionals worldwide. The obstacles cross educational, legal, scientific, political, and economic disciplines, suggesting that the most effective solutions will likely be interdisciplinary or multidisciplinary in scope.

Scientific

Among all the perspectives on injury, the one that predominates in the following chapters is a scien-

WARNING: HAZARDOUS TO YOUR HEALTH

Warning signs, it seems, are everywhere. Their purpose is to inform product users or people in a certain environment of potential dangers posed by the product or place. For any warning to be effective, it must be designed to ensure that the person at risk senses the warning (e.g., with bright colors or flashing lights), receives the message of the warning (sensing a warning does not ensure that it will be read), and understands the warning (e.g., the message must be short, simple, and unambiguous). At a minimum, an effective warning should include the following four elements (Sanders and McCormick 1993):

- Signal word: to convey the level of risk (e.g., *danger, caution*)

- Hazard: the nature of the hazard

- Consequences: what will likely happen if the warning is not heeded

- Instructions: appropriate behavior to reduce or eliminate the hazard

Effective warnings can go a long way toward reducing the incidence of injury and death in both work and recreational situations.

Sign warning of potential dangers: lives lost, posted alongside the Kern River on Rte. 155, heading north. Sequoia National Forest, Southern Sierra, California.

tific perspective. As stated earlier, many scientific disciplines have a role to play in a comprehensive understanding of injury. Anatomists, for example, study which structures and tissues are actually injured, physiologists examine the biological processes involved in tissue repair, psychologists are interested in the behavioral aspects of injury, and engineers design equipment and structures to prevent and minimize injury.

Of all the scientific disciplines, physics and its subdiscipline mechanics are arguably most central to the study of injury. The common denominator of this area of science is energy. Indeed, energy is called the agent of injury. Although thermal, electrical, magnetic, and chemical energy can cause injuries, most injuries involve mechanical energy. The fundamental relation between mechanical energy and injury highlights biomechanics as the logical discipline to study the causes and effects of human musculoskeletal injury.

In *Injury in America*, the Committee on Trauma Research (1985) reinforced the important role of biomechanics research in the prevention of injury with the following conclusions—which remain highly relevant more than two decades later:

- High priority should be given to research that can provide a clearer understanding of injury mechanisms.
- Quantification of the injury-related responses of critical body areas (such as the nervous system, thoracic and abdominal viscera, joints, and muscles) to mechanical forces is needed.
- High priority should be given to defining limits of human tolerance to injury, particularly with regard to segments of the population for whom data are extremely limited, including children, women, and elderly people.
- Improvement in injury assessment technology is needed, including the development of methods for assessing important debilitating injuries and causes of fatality, improvement of anthropomorphic dummies, and development of valid computer simulation models to predict injury in complex crash conditions.
- Organizations are needed to administer research on injury mechanisms and injury biomechanics and ensure a supply of scientists trained in injury biomechanics.

▪ CHAPTER REVIEW

Key Points

- Statistics emphasize that injury is a serious public health problem that deserves our full attention, should be given greater priority, and should be addressed with combined approaches to prevention and control. Injury is a multifaceted problem, requiring a multidisciplinary approach to find and implement effective solutions.
- An accurate and comprehensive awareness of injury can be developed only by examining it from numerous perspectives. Historically, the ancient origins of injury, prevention, and treatment procedures began with Hippocrates, and these practices have evolved over time. The historical perspective also highlights the achievements of many individuals who advanced our knowledge

in anatomy, physiology, injury, and trauma, and our current level of knowledge would not exist without the keen investigation, curiosity, and observations of these historical figures.

- An epidemiological perspective offers a chance to answer health questions both observationally (e.g., who, what, where, and when, with respect to injury) and analytically (why and how). Incidence, prevalence, and risk factors are key aspects to consider from an epidemiological perspective.
- Health and safety professionals are intimately involved with injury prevention and treatment. Injury control and safety education programs can reduce the incidence and severity of injury. Nonetheless, injuries will still occur, and thus the health professionals,

whether emergency medical personnel, athletic trainers, or physicians, are needed to perform essential services critical to injury diagnosis and treatment.

- The cost of injury is enormous: One must consider not only the direct costs of injury from medical and nonmedical treatments but also the indirect costs, such as morbidity and mortality costs. Estimated costs of injury would exceed trillions of dollars per year.

- Often overlooked, but nonetheless extremely important to both prevention and recovery from injury, are the psychological factors that may affect a person before, during, or after injury. Many theories have been proposed identifying psychological state as a predisposing risk factor to injury. Additionally, psychological state after injury greatly affects rehabilitation and recovery.

- Many scientific disciplines collaborate to address etiology, affected tissue, and the biological processes underlying injury. Arguably, the scientific discipline of physics, and more specifically the subdiscipline of mechanics, is most pertinent to understanding musculoskeletal injury and its prevention.

Can we eliminate injury? The answer is no. Can we reduce the incidence and severity of injury? The answer is assuredly yes. Thus, we embark on our exploration of the biomechanical aspects of injury, with the goals of increasing awareness of its importance to individuals and to society as a whole and identifying ways in which biomechanics can contribute to the problem's solution.

Questions to Consider

1. Explain why effective consideration of musculoskeletal injury requires a multidisciplinary approach.

2. Our understanding of injury mechanisms, diagnosis, and treatment has increased rapidly in recent decades. Looking into the future, if you were to write a chapter on the history of injury research from the present until the year 2030, what would you cover?

3. What are the limitations of examining injury from a single perspective (e.g., only from a biomechanical viewpoint)?

4. As an injury epidemiologist, you have been hired by a manufacturing company to investigate a recent increase in the rate of work-related injuries. What steps would you take in conducting a comprehensive assessment of the problem?

5. Consider a sport psychologist working with an Olympic-level athlete who has recently suffered a potentially career-ending injury. What factors should the psychologist keep in mind while assisting this athlete?

6. What do you see as the most important future areas of injury-related research?

Suggested Readings

Alberta Injury Control Strategy Report. Available: www.med.ualberta.ca/acicr.

Baker, S.P., B. O'Neill, M.J. Ginsburg, and G. Li. 1992. *The Injury Fact Book*. New York: Oxford University Press.

Caine, D.J., C.G. Caine, and K.J. Lindner. 1996. *Epidemiology of Sports Injuries*. Champaign, IL: Human Kinetics.

Haddon, W., Jr., E.A. Suchman, and D. Klein. 1964. *Accident Research: Methods and Approaches*. New York: Harper & Row.

Meeuwisse, W.H. 1994. Assessing causation in sport injury: A multifactorial model. *Clinical Journal of Sport Medicine* 4: 166-170.

Nahum, A.M., and J.W. Melvin, J.W., eds. 2002. *Accidental Injury: Biomechanics and Prevention*. New York: Springer-Verlag.

National Safety Council. Available: www.nsc.org.

National Safety Council. 2004. *Injury Facts*. Itasca, IL: National Safety Council.

Praemer, A., S. Furner, and D.P. Rice. 1999. *Musculoskeletal Conditions in the United States*. Park Ridge, IL: American Academy of Orthopaedic Surgeons.

Rivara, F.P., P. Cummings, T.D. Koepsell, D.C. Grossman, and R.V. Maier. 2001. *Injury Control: A Guide to Research and Program Evaluation*. Cambridge, UK: Cambridge University Press.

Robertson, L.R. 1983. *Injuries: Causes, Control Strategies and Public Policy*. Lexington, MA: Lexington Books.

U.S. Centers for Disease Control and Prevention. 1993. Years of potential life lost before age 65—United States, 1990 and 1991. *Morbidity and Mortality Weekly Report* 42: 251-253.

U.S. Department of Health and Human Services. 1989. *Promoting Health/Preventing Disease, Year 2000 Objective for the Nation*. Washington, DC: U.S. Government Printing Office.

CHAPTER 2

CLASSIFICATION, STRUCTURE, AND FUNCTION OF BIOLOGICAL TISSUES

Form follows function.

Louis Henri Sullivan (1856-1924)

■ OBJECTIVES

- ■ To learn the origins of the body's tissues, focusing on the connective tissues that form the key elements of the musculoskeletal system
- ■ To understand the common and uniquely distinct constituents and features of musculoskeletal tissues, including bone, cartilage, tendons, ligaments, skeletal muscle, and joints
- ■ To be able to describe the unique roles that connective tissues and skeletal muscles play during normal function and after injury

For centuries, load-transmitting connective tissues, such as bone, ligament, tendon, and articular cartilage, were considered inert structures. In reality, these tissues are dynamic and respond acutely to many physiological and mechanical stimuli—including injury. This chapter provides background on the derivation, structure, and mutability of these primary tissues of the musculoskeletal system so we can better understand the underlying mechanisms of their responses.

Embryology

To understand normal musculoskeletal system function and injury sequelae, we need to know the organization of body tissues. As threads are woven together to form a fabric, so too are cells, fibers, and other matrix components blended to form tissues. A tissue is an aggregation of cells and intercellular substances that performs a specialized function.

Each tissue in the body has a specialized function and a distinctive organization. To better understand how tissues are organized and how they function, it is useful to review where they come from and how they were differentiated and formed developmentally. The following brief review of tissue embryology highlights commonalities in tissue organization and lays the groundwork for later discussions of the role of cells (e.g., mesenchymal stem cells) in the repair and healing processes of tissues of the musculoskeletal system.

The following descriptions are arranged in developmental, chronological order. The embryo stage of human development comprises the time from fertilization to week 8, and the fetal stage denotes the time from week 9 until birth. Our emphasis is on the development of the musculoskeletal system.

Fertilization of the egg (or *oocyte)* by the spermatozoa produces the zygote. The zygote begins to mitotically divide 1 day after fertilization, and that division continues as the zygote travels down the uterine tube (figure 2.1). Between days 1 and 4, the zygote undergoes rapid cell division with no growth in size by a process called embryonic cleavage to produce a solid ball called the morula, which hollows out, fills with fluid, and "hatches" from the zona pellucida—the original thick transparent membrane around the oocyte. By day 5 after fertilization, the morula, now referred to as the embryo, consists of a mass of about 64 cells arranged in a hollow ball shape called a blastocyst, which is composed of an outer sphere of trophoblast cells and an eccentric cell cluster called the *inner cell mass*. By day 8, the blastocyst is partially implanted in the uterine wall, and by day 10 two cavities have developed within the cell mass. The primitive amniotic cavity (sac) is associated with a layer of cells called the ectoderm, whereas the second cavity, the primitive yolk sac, is associated with a layer of cells called the endoderm. The embryo proper is the region where the endoderm and ectoderm are in contact (figure 2.2).

The contiguous layers of cells form the bilaminar embryonic disk, which is the fundamental cellular mass that develops into the fetus. Figures 2.1 and 2.2 illustrate this stage of development and the structures of the embryo.

The first vestiges of the primitive spinal cord (notochord) are apparent by day 16 of embryonic development. Further specialization and differentiation of cells in the embryo occur during the latter second week to early third week of

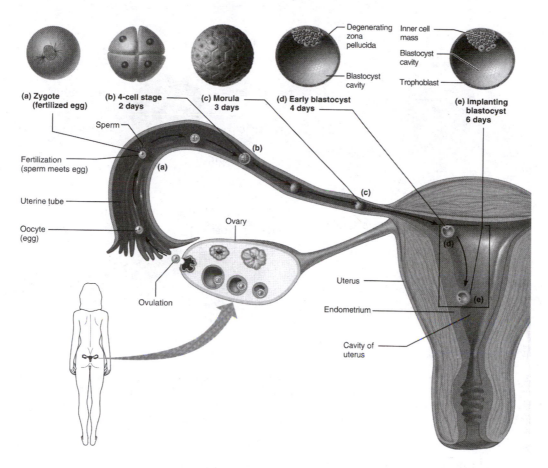

Figure 2.1 The zygote begins to divide about 24 hours after fertilization and continues the rapid mitotic divisions of cleavage as it travels down the uterine tube. Three to four days after ovulation, the pre-embryo reaches the uterus and floats freely for 2 to 3 days, nourished by secretions of the endometrial glands. At the late blastocyst stage, the embryo is implanting into the endometrium; this begins at about day 7 after ovulation.

Fig. 28.4, p. 1114 from HUMAN ANATOMY AND PHYSIOLOGY, 6th ed. by Elaine N. Marieb. Copyright © 2004 by Pearson Education, Inc. Reprinted by permission.

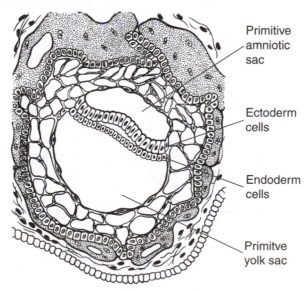

Figure 2.2 The bilaminar germ cell layers are evident, as well as the primitive amniotic sac (associated with the ectoderm) and the primitive yolk sac (associated with the endoderm).

Reprinted, by permission, from J. Langman, 1969, *Medical embryology*, 2nd ed. (Baltimore, MD: Lippincott, Williams & Wilkins), 41.

development. At this time a third layer of cells, called the *intra-embryonic mesodermal layer*, is produced. The developing mesodermal cells invaginate and spread between the endodermal and ectodermal layers. This transformation of the bilaminar embryonic disc into a three-layered disc containing the three primary germ layers—ectoderm, **mesoderm**, and endoderm—is referred to as **gastrulation.**

By day 20 there is evidence of the formation of distinct neural structures (plate, groove, and folds). Dorsal and frontal views of the embryo at this stage are shown in figure 2.3. Along with the neural structures, there is evidence of the first somites in figure 2.3*b*.

The **somites** are cuboidal bodies that form distinct surface elevations and influence the external contours of the embryo. By the beginning of week 4, the ventral and medial walls of the somites show highly proliferative activity,

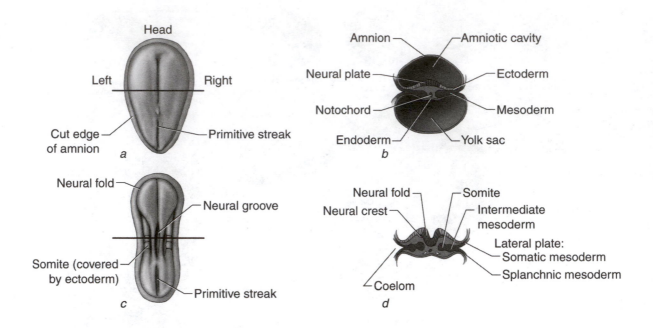

Figure 2.3 *(a)* Dorsal and *(b)* frontal views of a late presomite embryo (approximately 2 mm in length, at 17 days). The amniotic sac has been removed, and the neural plate is visible. *(c)* Dorsal and *(d)* frontal view of the embryo at 20 days, with the emergence of the somites and the formation of a distinct neural groove and fold.

Figs. 28.9 a&c, p. 1121 from HUMAN ANATOMY AND PHYSIOLOGY, 6th ed. by Elaine N. Marieb. Copyright © 2004 by Pearson Education, Inc. Reprinted by permission.

become polymorphous in shape, coalesce, and migrate toward the notochord. Collectively, the migrating cells are known as the **sclerotome.** After the sclerotome has condensed near the notochord, the remaining wall of the somite, the dorsal aspect, gives rise to a new layer of cells called the **dermatome.** Cells arising from the dermatome form a tissue known as the **myotome,** which gives rise to the musculature (figure 2.4). The chronology of key embryological events is presented in table 2.1.

The undifferentiated cells of the sclerotome form a loosely woven tissue known as **mesenchyme** (or primitive connective tissue). Mesenchyme is the progenitor tissue of adult connective tissues, such as cartilage, ligaments, fascia, tendons, blood cells, blood vessels, skin, bone, and muscle. One of the primary attributes of mesenchymal cells (stem cells) is their ability to differentiate into a variety of more specialized cells; thus, they are **pluripotent.** They may become fibroblasts (associated with the formation of elastic or collagen fibers), chondroblasts (involved in the formation of cartilage matrix), or osteoblasts (associated with bone extracellular matrix).

Tissue Types

Tissue is classified as one of four types: epithelial, nervous, muscle, and connective (figure 2.5). The ectoderm and endoderm are primarily epithelial tissues. Most of the epithelial tissues of the body are derived from these two embryonic layers. **Epithelial tissue** is a covering (lining) tissue. It can be specialized to absorb, secrete, transport, excrete, or protect the underlying organ or tissue. Epithelial membranes consist entirely of cells and have no capillaries but are nourished via tissue fluid from the capillaries of connective tissues. These membranes are not strong and typically are bound firmly to connective tissue separated by a thin layer of material called a **basement membrane.**

Epithelial tissue is subject to wear, and its cells are constantly being lost and regenerated. Structurally, the number of cells and the arrangement of the cellular layer provide the generic names of epithelial tissues. A single layer of cells is described as simple. Tissue with two or more layers of cells is *stratified*. For cell shape, the usual categories include squamous, cuboidal, and columnar. For example, an epithelial tissue

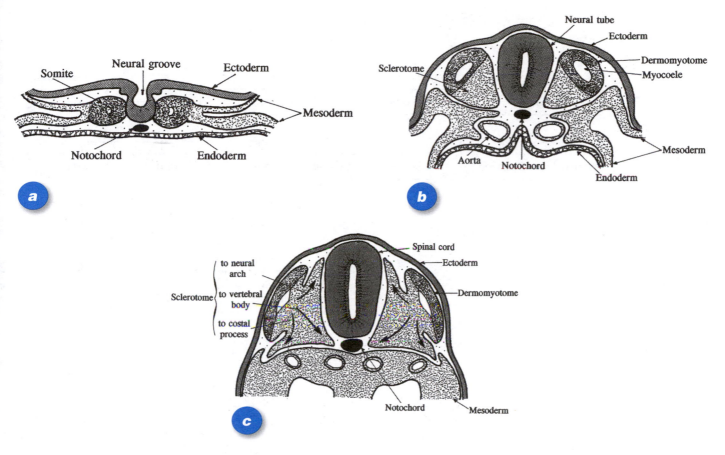

Figure 2.4   *(a)* Stages in development of a somite. *(b)* Mesoderm cells are arranged around a small cavity. *(c)* Cells of the ventral and medial walls of the somite lose their epithelial arrangement and migrate in direction of the notochord. These cells collectively constitute the sclerotome. Cells at the dorsolateral portion of the somite migrate as precursors to limb and body wall musculature. Dorsomedial cells migrate beneath the remaining dorsal epithelium of the somite to form the myotome. After ventral extension of the myotome, dermatome cells lose their epithelial configuration and spread out under the overlying ectoderm to form dermis.

Reprinted, by permission, from B.R. MacIntosh, P.F. Gardiner and A.J. McComas, 2005, *Skeletal muscle: Form and function*, 2nd ed. (Champaign, IL: Human Kinetics), 55.

TABLE 2.1

CHRONOLOGY OF KEY EVENTS IN EARLY EMBRYOLOGY

Days after fertilization	Event
5	Cells arranged into blastocyst
8	Blastocyst partially implanted into uterine wall
10	Two cavities form within cell mass
16	Notochord forms
20	Evidence of distinct neural structures and somites

could be classified as simple cuboidal or stratified squamous. Figure 2.6 illustrates four types of epithelial cells.

The layered or columnar arrangement of epithelial cells is mechanically weak. Epithelial tissues, however, have a prominent role in the diffusion of tissue fluid and heat and in bioelectric conduction.

Nervous tissue, a second type of tissue, develops from the ectoderm. It comprises the main parts of the nervous system, including the brain, spinal cord, peripheral nerves, nerve endings,

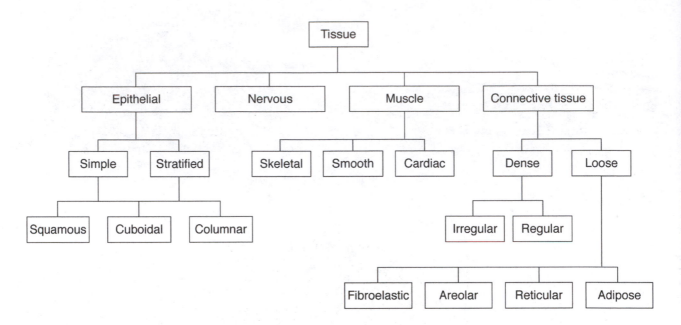

Figure 2.5 Organizational relations of tissue types.

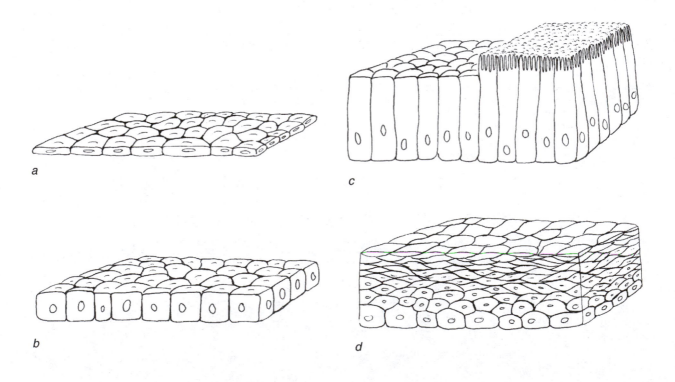

Figure 2.6 Examples of cellular shapes and arrangements of various types of epithelial tissue. *(a)* Simple squamous epithelium contains platelike cells organized in a single layer that adhere closely to each other by their edges. *(b)* Simple cuboidal epithelium and *(c)* simple columnar epithelium appears similar on the surface (which may have hairlike cilia attached) as a continuous mosaic of polygons. From the side view, however, it is evident that the heights of the cells are substantially different, one being essentially a cube arrangement and the other being a much taller column. *(d)* Stratified squamous epithelium consists of multiple layers, ranging from cuboidal or columnar cells to irregularly shaped polyhedral cells to superficial layers consisting of thin squamous cells.

Adapted, by permission, from D.W. Fawcett, 1994, *Bloom and Fawcett: A textbook of histology*, 12th ed. (New York, NY: Chapman & Hall), 58.

and sense organs. The basic unit of nervous tissue is the **neuron,** or nerve cell. Communication is a vital feature of nervous tissue. Inherent characteristics of nervous tissue include **irritability** (capacity to react to chemical or physical agents) and **conductivity** (ability to transmit impulses from one location to another). Nerve impulses are conducted toward the cell body by dendrites and away from the cell body by an axon. There is only one axon per cell, but there may be many dendrites. Nerve tissue can be injured by excessive tension or compression, but its prime physiological function is not load bearing.

The third type of tissue, **muscle,** can be divided into three categories: skeletal, smooth, and cardiac. Muscle tissue is derived from the mesoderm. All three types of muscle cells perform the specialized tasks of conductivity and **contractility. Skeletal muscle** is also called striated muscle because its fibers exhibit cross striations. Skeletal muscle fibers have multinucleated cells. Skeletal muscle is under voluntary control and is enveloped in a sheath of connective tissue that blends with its tendon. Because of the contractility of skeletal muscle cells, they have a prime function of generating force to maintain posture and produce body movements.

Smooth muscle does not appear striated and is not considered under voluntary control. It is found in the walls of tubes in the arterial, intestinal, and respiratory systems. Smooth muscle is innervated by both the sympathetic and parasympathetic nerves.

Cardiac muscle displays structural and functional characteristics of both skeletal and smooth muscle. Cardiac muscle is striated in appearance but is considered involuntary. Cardiac muscle cells form a functional **syncytium,** as the cardiac tissue acts electrically as though it were a single cell.

Connective tissue is the fourth type of tissue. It too is derived from the mesoderm and differs primarily from the other three types of tissue in the amount of extracellular substance. Cells are soft, easily deformable structures and by themselves would be unable to transmit substantial loads. The **extracellular matrix** that holds the connective tissues together gives it form and allows the tissue to transmit load. The ratio of cells to extracellular matrix and the composition of the

matrix establish the physical characteristics of the connective tissue. The composition of the matrix can range from a relatively soft, gel-like substance (as in skin or ligament) to the rigid extracellular matrix found in bone. A primary role of the cells in bone, cartilage, tendons, and ligaments is to produce and maintain the extracellular matrix.

Connective Tissues

Connective tissues are aggregated materials consisting of cells, fibers, and other macromolecules embedded in a matrix that can also contain tissue fluid. The principal fibers in connective tissues are **collagenous, reticular,** and **elastic** fibers, although collagenous and reticular fibers are basically different forms of collagen (Fawcett and Raviola 1994). The arrangement and packing density of fibers in connective tissues distinguish loose connective tissue and dense connective tissue. In addition, the term **dense irregular connective tissue** is used to describe fibrous connective tissue with loosely and randomly interwoven fibers, such as fascia, and the term **dense regular connective tissue** refers to organized fibrous tissues, such as tendons, ligaments, or aponeuroses (fibrous ribbonlike membranes similar in composition to tendons).

Specialized load-bearing tissues such as bone, cartilage, tendon, and ligament (dense connective tissue) are detailed in later sections of this chapter, and here we only describe loose connective tissues. More prevalent than dense connective tissues, loose connective tissues have four basic types in the adult: fibroelastic, areolar, reticular, and adipose tissues. All of these tissues contain some elastic fibers that provide extensibility to the tissues. Collagen fibers are also evident in the tissues, along with a liquid extracellular matrix that bathes the cells and fibers.

Fibroelastic tissue is a loose, woven network of fibers that encapsulates most organs. The extensibility of loose connective tissue is attributable to the organization of the collagen fibers. The tissue as a whole can be stretched without initially deforming the fibers. Because the collagen in loose connective tissue is configured as a mesh, the mesh is deformed first, and the fibers are aligned before the individual collagen fibers are stretched. This contrasts with dense fibrous connective tissues, such as tendon, in which the collagen is arranged in parallel rows. Because the tendon collagen is already aligned with the tensile load, the fibers quickly resist the applied tensile load. After a load deforms fibroelastic tissue, elastic fibers help return the stretched connective tissue back to its original position when the load is released.

Areolar tissue saturates almost every area of the body. The tissue is called areolar because there are spaces or holes where only fluid extracellular matrix exists. Fibroblasts and macrophages are abundant and collagenous, and some elastic and reticular fibers give limited structural strength to areolar tissue. Nonetheless, areolar tissue is a weak connective tissue and can be easily pulled apart. Reticular network fibers act as a boundary between areolar connective tissues and other structures.

Reticular tissue contains reticular fibers and some primitive cells and thus resembles early mesenchymal tissues. The primitive cells within reticular tissue can differentiate into fibroblasts, macrophages, and even some plasma cells. Reticular tissue is found near lymph nodes and in bone marrow, liver, and spleen. Reticular fibrils are also found in many other areas of the body, such as around nerves, muscles, and blood vessels.

Adipose tissue is the fourth type of loose connective tissue. Microscopically, this tissue appears as an aggregate of fat cells surrounded by areolar tissue. Each adipose cell has a fat droplet. Any loose connective tissue can accumulate fat, and when it predominates, the term *adipose tissue* is used. Reticular fibers enclose each fat cell, and capillaries are found between the cells. The rich vascularity is consistent with the elevated metabolism of adipose tissue. It can be mobilized for use in the body when carbohydrates are not immediately available and is readily stored when not needed. Adipose tissue is commonly found around the organs in the abdominal cavity, under the skin, and in bone marrow. Adipose tissue, when present under the skin, may prevent heat dissipation and act as a cushion for the skeleton during external impacts.

Constituents of Connective Tissues

Cells, extracellular matrix (including fibers and matrix glycoproteins), and tissue fluid are the structural elements of connective tissues. The specific constituents of bone, cartilage, tendon, and ligament are discussed later in this chapter; here we present only the generic components of these tissues.

Cells

Several cell types exist within connective tissues. They are classified as either **resident** (fixed) or **migratory** (wandering) **cells** (Fawcett and Raviola 1994). Resident cells are relatively stable within a tissue, and their role is to produce and maintain the extracellular matrix.

Undifferentiated mesenchymal cells (a type of **stem cell**) are resident cells that can differentiate into a variety of connective tissue cells, including fat cells. An important characteristic of mesenchymal cells is that they can differentiate

FIBROBLAST VERSUS FIBROCYTE

Mesenchymal cells are the undifferentiated progenitors of connective tissue cells. Fawcett and Raviola (1994) explained that the suffix *–blast* (derived from the Greek *blastos*, meaning germ) is frequently used to refer to the immature stages of some cell types. The term *fibroblast* has been used to describe the undifferentiated stage of a fibrocyte. In turn, the term *fibrocyte* was used to indicate the relatively quiescent and mature phase of the cell's development. This is a misnomer, however. Fibroblast already means a "fiber-forming" cell. Because the mature fibroblast is the principal site of collagen and elastin biosynthesis, it is not necessary to change the name to fibrocyte when the cell becomes mature. Thus *fibroblast* is the preferred term, but *fibrocyte* is also allowable.

into fibroblasts, chondroblasts, or osteoblasts. Subsequently, chondroblasts and osteoblasts mature into chondrocytes and osteocytes. **Fibroblasts** are the principal cells in many fibrous connective tissues. Their function includes the formation of fibers as well as other components of extracellular matrix. Although there is some difference of opinion, generally it is agreed that the terms *fibroblast* and **fibrocyte** are synonymous. In the remainder of this text, we use only the term *fibroblast* to refer to this mature cell.

The migratory cells that enter connective tissue (e.g., macrophages, monocytes, basophils, neutrophils, eosinophils, mast cells, lymphocytes, and plasma cells) travel to the tissue via the bloodstream. These cells are usually associated with the tissue's reaction to injury through the initiation and regulation of an immune response and inflammation. The numbers of these cells in connective tissues are quite variable, but two of the cells warrant further mention.

A **macrophage** contains empty reservoirs or vacuoles that can accumulate foreign material, old red blood cells, and bacteria. Because of this ability, the macrophage is part of a larger phagocytic system, the reticuloendothelial system, a major defense system in the body. **Mast cells** are relatively large cells because of their substantial amount of cytoplasm. The many granules in their cytoplasm are thought to contain heparin, which acts as a blood anticoagulant. Histamine (a vasodilator) and serotonin (a vasoconstrictor) may also be present in mast cells.

Extracellular Matrix

The extracellular matrix in connective tissues is a blend of components, including protein fibers **(collagen** and **elastin),** simple and complex matrix glycoproteins, and tissue fluid, all of which interact and contribute to the mechanical properties (e.g., stiffness and strength) of connective tissues. The extracellular matrix can be attached or linked with cells via transmembrane **integrins,** and those interactions may be involved in **mechanotransduction**—the conversion of mechanical stimuli into biological responses.

Collagen is the most abundant protein in the animal world and constitutes more than 30% of the total protein in the human body (Eyre 2004; Parry and Squire 2005). Collagen is an umbrella

term; there are many forms of collagen. Collagen fibers are present in varying amounts in all types of connective tissue. The organization of collagen fibers is tissue specific and can range from a relatively random arrangement of fibers in loose connective tissue to a very organized and parallel arrangement in dense regular connective tissues such as tendon. All of the key cells of connective tissue (fibroblasts, chondroblasts, chondrocytes, osteoblasts, and osteocytes) are able to produce collagen.

Collagen fibers are proteins (long chains of amino acids with peptide linkages). The fundamental unit of collagen is the tropocollagen molecule. The molecule is made from three spiraled polypeptide chains of about 1,000 amino acids that are intertwined to form a helix. Parallel rows of tropocollagen form microfibrils, and these microfibrils aggregate in a parallel fashion into fibrils. Collagen fibrils are aligned into bundles to form collagen fibers. The stability of the collagen fibers can be enhanced with the formation of collagen cross-links both within and between the collagen molecules. Excessive cross-linking, however, makes the collagen stiff and inextensible. Collagen is classified according to its molecular organization as type I, type II, type III, and so forth. At least 27 types of collagen have been reported (Eyre 2004). Type I collagen is found in skin, bone, tendon, ligament, and cornea and is the most abundant type of collagen in the body. Type II collagen is primarily found in cartilage, and type III is most abundant in loose connective tissue, the dermis of the skin, and blood vessel walls.

Elastic fibers are more slender and extensible than collagen fibers. They can be stretched to about 150% of their original length before the fiber breaks (Fawcett and Raviola 1994). **Microfibrils** and elastin are the two components of elastic fibers. Microfibrils are aggregated into small bundles and embedded within a relatively amorphous elastin. The chemical composition of elastin has some components that are similar to collagen.

Besides the collagen and elastic fibers, another major protein fraction of the extracellular matrix is **complex glycoproteins.** A **proteoglycan** is a protein to which are attached one or more specialized carbohydrate side chains, called

glycosaminoglycans (Lo et al. 2003; Silver and Bradica 2002) (figure 2.7). Glycoproteins occupy the spaces between fibers and constitute the so-called ground substance of connective tissues. These complex matrix glycoproteins are negatively charged and hydrophilic (attract water molecules), attributes that have significant effects on the mechanical behaviors of connective tissue.

In addition to proteoglycans, other specialized glycoproteins can also be found in the connective tissue matrix. These are called *cell-associated glycoproteins,* because they are important for the adhesion of cells. One type, *fibronectin,* plays an important role in cell migration. Also, cells involved with tissue repair can use fibronectin as a stable attachment site during the repair process. Other types of cell-associated glycoproteins include *chondronectin,* which helps to stabilize the chondrocyte in its matrix, and anchorin CII, which may be important in mediating the binding of type II collagen to chondrocytes (Lo et al. 2003).

Tissue Fluid

Tissue fluid is a filtrate of the blood and resides in the intercellular (interstitial) spaces. It aids in the transport of materials (e.g., nutrients, trophic factors, and wastes) between the capillaries and cells in the extracellular matrix. The tissue fluid carries food and oxygen to the cells by diffusing through the arterial end of the capillary. Moving in the opposite direction, the fluid has two paths to remove wastes from the cell. Either it returns the wastes to the venous end of the capillary for removal by the blood, or wastes are disposed of by the lymphatic system. The latter provides channels and filtering points (lymph nodes and spleen) to cleanse the tissue fluid before it is returned to the bloodstream. If blockage occurs in the lymphatics, the tissue fluid is trapped in the intercellular spaces, and **edema** (tissue swelling) results.

Tissue fluid is retained in the interstitial spaces of the intertwined proteoglycans and glycosaminoglycans. The interaction of the fluid with these macromolecules gives the extracellular matrix its gel characteristics and contributes to the mechanical behavior of the tissue.

Bone

The specialized connective tissue known as bone is one of the hardest and strongest tissues in the

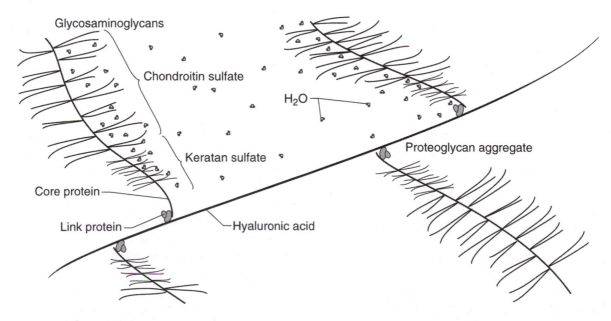

Figure 2.7 Proteoglycan aggregate. Glycosaminoglycans are composed of negatively charged sulfate groups attached to a core protein. The abundance of negative charges in close proximity to one another causes the sulfate groups to repel one another and gives rise to the characteristic bottle-brush appearance of the glycosaminoglycan groups. Many glycosaminoglycan groups can be attached to a long central protein (hyaluronic acid) via link proteins.

Adapted, by permission, from J.A. Buckwalter and J. Martin, 1995, "Degenerative joint disease," *Clinical Symposia* 47(2): 7.

body. The skeleton of humans and other vertebrates protects vital organs, serves as a mineral storehouse, houses bone marrow hematopoietic cells, and provides levers that contribute to muscles' ability to generate and control movements. Bone is a dynamic structure that perpetually remodels and responds to alterations in mechanical loads, systemic hormones, and serum calcium levels. Each of these factors is synergistically interrelated, and the anatomy of a bone reflects the interaction of these factors.

Bone can be studied as an organ, as a tissue, or in terms of its cells, because bone is a functional unit at each of these organizational levels. As an organ, bone accounts for a substantial percentage of the total body mass and is involved with metabolic processes such as **hematopoiesis** (formation of blood cells). Bone tissue can be classified either as **cortical** (also called **compact) bone** or **trabecular** (also called **cancellous** or **spongy)**

bone. Although cortical bone and trabecular bone have the same cells, their mechanical behavior and adaptive responses are different. Many types of cells are found in bone tissue, and these cells function interactively to maintain bone as a tissue and an organ.

Bone Development

Skeletal development begins when mesenchymal cells from the mesodermal germ layer condense. In a few bones (cranium and facial bones and, in part, the ribs, clavicle, and mandible), the cellular condensations form fibrous matrices that subsequently ossify directly **(intramembranous ossification).** Figure 2.8 provides a schematic representation of the stages involved in intramembranous ossification.

In most limb and axial bones, mesenchymal condensations form a cartilaginous model **(anlage)** of the bones rather than proceeding

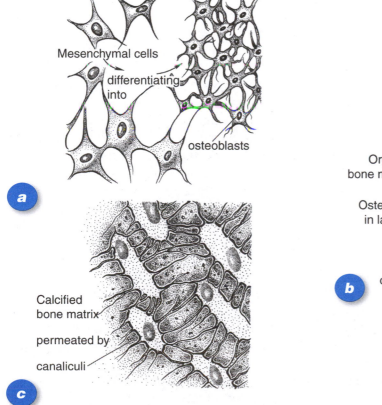

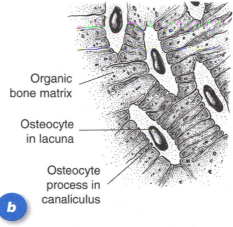

Figure 2.8   Intramembranous ossification. The sequence illustrates the temporal stages of ossification, starting with *(a)* differentiation of mesenchymal cells into osteoblasts. *(b)* Osteoblasts then secrete bone matrix, including type I collagen. *(c)* When the bone extracellular matrix calcifies, distinct **canaliculi** (channels) are left intact to permit communication between osteocytes.
Adapted, by permission, from D.H. Cormack, 1987, *Ham s histology*, 9th ed. (Philadelphia, PA: Lippincott-Raven Publishers), 276.

directly to calcification and ossification. **Endochondral ossification** proceeds during long-bone development through a cartilaginous **growth plate** (or **physis).** Long bones typically have two physes (one at each end), although some bones (e.g., metacarpals) may only have one functional physis.

Longitudinal growth occurs when tissue is added to the **metaphyseal** side of the physis and also occurs through activity of the chondrocytes within three functionally distinct regions of the growth plate: the regions of proliferation and growth, maturation, and transformation. The region of growth contains two types of chondrocytes. Resting cells lie close to the secondary ossification center *(zone 1)*. These cells are associated with the small arterioles and capillaries from the epiphyseal vessels. These vessels are important in transporting undifferentiated cells to add to the pool of resting cells. Away from the resting cells is an area of active cell division *(zone 2)*. In this area, the cells are organized in longitudinal columns, and during a period of rapid growth the columns may account for more than half the height of the growth plate (Ogden et al. 1987).

The region of maturation *(zone 3)* is associated with hypertrophic chondrocytes that actively synthesize and secrete cartilaginous extracellular matrix. Cells adjacent to the region of growth are large and actively produce matrix components, whereas the cells near the ossification front become trapped in the rapidly calcifying cartilaginous matrix and, therefore, are not as active in matrix production.

The third region has an area of transformation where the cartilage matrix becomes increasingly calcified *(zone 4)*. The calcification of the matrix leads to the death of the chondrocytes, and only large empty spaces (lacunae) remain where the chondrocytes once resided *(zone 5)*. That lattice of calcified cartilage is the **primary spongiosa.** Blood vessels invade that lattice and deliver cells of monocytic origin (septoclasts and osteoclasts) that degrade the calcified cartilage and cells that secrete bone matrix (osteoblasts) to replace the degraded calcified cartilage. At that time, metaphyseal trabeculae are referred to as **secondary spongiosa.** Eventually, all of the cartilaginous trabeculae are replaced by bone.

Concurrently, bone grows in circumference as the perichondrium surrounding zone 5 thickens and lays down a thin layer of osteoid tissue that subsequently mineralizes, forming a bony collar **(periosteal collar)** at the midshaft level. Vascular channels penetrate the central region and bony collar, ultimately forming the **primary ossification center.** Ossification proceeds quickly toward the ends to form the bone diaphysis and metaphysis. Figure 2.9 illustrates the process of endochondral ossification.

An important anatomical region within the developing long bone is the **zone of Ranvier,** found at the cortical margins of the growth plate toward the primary ossification center (Ogden and Grogan 1987). This complex zone is where the increase in metaphyseal diameter occurs during growth. Therefore, if trauma damages the zone of Ranvier, the normal circumferential growth of the long-bone metaphysis can be disrupted.

In the epiphyseal regions, the vascular channels directly invade the cartilage, which subsequently ossifies and forms **secondary ossification centers.** Vascular ingrowth (irruption) is an integral step in the formation of the primary and secondary ossific centers, because the blood supply ensures the arrival and subsequent differentiation of osteogenic precursor cells (Zelzer et al. 2004).

Between the bone formed by the primary and secondary ossification centers, the cartilage anlage persists as a physis between the shaft and ends of the long bone. As growth proceeds from infancy to maturity, the physes typically change from a relatively flat plate dividing the epiphysis and metaphysis to a complex series of curves and interdigitating epiphyseal and metaphyseal ridges and valleys. That change in geometry has significant implications for resistance to fracture. Cartilage, such as the cartilage in the physis, can withstand large compressive loads but cannot withstand large shear and tensile loads. Because of that, physeal injuries in which the epiphysis "shears" off the metaphysis are common in young children (e.g., slipped capital femoral epiphysis). The tortuous series of interdigitations in the physes that comes about with growth may help prevent those shearing injuries by locking the epiphysis and metaphysis together.

Eventually, chondrocyte differentiation and proliferation slow in the regions of growth and maturation, allowing the bone mineralization (encroaching from the diaphyseal edge of the

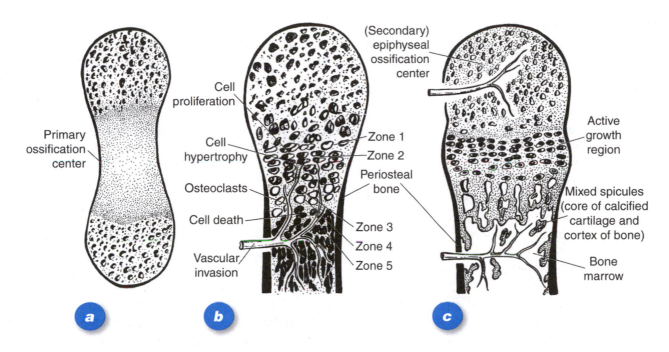

Figure 2.9 Schematic representation of endochondral ossification. *(a)* The hyaline cartilage anlage with its primary ossification center. *(b)* The transitional zones (1-5), ranging from a site of cell proliferation, through hypertrophy, to cell death and ossification. *(c)* The mixed spicules, which form in the midshaft region, around the primary ossification center. The active growth region is the primary site of bone accretion and, thus, is the primary site for long-bone growth.
Reprinted, by permission, from J. Langman, 1969, *Medical embryology*, 2nd ed. (Baltimore, MD: Lippincott, Williams & Wilkins), 32.

plate) to catch up. This unites the bone formed by the primary and secondary centers, epiphyseal and metaphyseal vascularity, and marks the culmination of long-bone growth. This process is known as epiphyseal plate **closure.** Physeal closure typically occurs 2 to 3 years earlier in girls than it does in boys, which can contribute to the shorter stature of women relative to men.

Both intramembranous and endochondral ossification can occur in the same bone. For example, the shaft of the clavicle is formed by intramembranous ossification, but a secondary ossification center develops within a cartilaginous epiphysis to form the sternal end of the bone. A primary ossification center is present in most bones at birth, but the secondary ossification center of the distal femur is the only secondary center present at birth and is often used to identify a full-term fetus. Both the endochondral and intramembranous ossification processes persist postnatally and are similar to those during fracture repair (endochondral) and periosteal bone deposition (intramembranous).

Movement and its related forces during skeletal development are among the stimuli that can influ-

ence the final skeletal form. Carter and colleagues (2004) proposed that the regulation of skeletal biology by mechanical forces is accomplished by the transfer of strain energy. To this end, they suggested that cyclic shear stresses generated during movement accelerate the rate of chondrocytic proliferation, maturation, degeneration, and ossification that occur during endochondral ossification, whereas compressive stresses tend to retard the same sequence. Carter and colleagues proposed that some of the energy imparted to the skeletal structures during movement is stored within the tissue and later released during unloading. The remaining energy is transferred to the tissue in the form of heat or a change in internal energy. This latter form of energy transfer may be an important factor in a bone's ability to recognize and respond to mechanical cues.

Extrinsic factors, such as hormones, influence the rate and extent of long-bone growth. Thyroxine, growth hormone, and testosterone can all stimulate cartilage cell differentiation in the growth plate. Estrogen exerts a greater stimulatory influence on the bony tissue while suppressing cartilage growth. The distinct influences of testosterone and estrogen may account for the differences in the timing of physeal closure between boys and girls.

Normal skeletal growth can also be interrupted by trauma or fracture. Physeal injuries account for approximately 15% of all fractures in children. Girls are more prone to physeal injury from 9 to 12 years of age, whereas boys are more prone between the ages of 12 to 15 years (Ogden 2000a). The periods of increased incidence of fracture parallel the times of rapid growth during which hormone-mediated changes in the growth plate cartilage may alter the response of the cartilage to mechanical stress (Sands et al. 2003). Most pediatric fractures are classified according to a system developed by Salter (Ogden 2000b). The system considers the location of the fracture, whether the fracture disrupts the growth plate, and the extent of the growth plate damage. Growth disturbances may result if the fracture and subsequent callus formation stimulate the premature closure of the growth plate, thereby preventing the normal longitudinal growth of the bone. Angular deformities may result if only one portion of the growth plate sustains damage while normal growth

occurs in the remaining portion of the growth plate. An example of an angular deformity of the proximal radius is demonstrated in the X ray in figure 2.10.

Bone Tissue Components

The four primary bone cells are osteoblasts, osteocytes, bone-lining cells, and osteoclasts. Osteoblasts, mononuclear cells of mesenchymal origin, are located on the bony surface and are the primary bone-forming cells. Once osteoblasts have produced a sufficient amount of unmineralized matrix (osteoid) and become relatively quiescent, one of three things can happen. Osteoblasts can (1) undergo cell death, (2) persist on the bone surface (i.e., become bone-lining cells), or (3) become trapped and surrounded by osteoid that mineralizes shortly after deposition, at which point they have become osteocytes.

Osteocytes pervade the entire bone cortex (25,000 cells/mm^3 of tissue). When the active osteoblast begins the transition to osteocyte, cell volume decreases by 30% initially, and as the metabolic activity of the osteocyte gradually decreases, cell volume also continues to decrease.

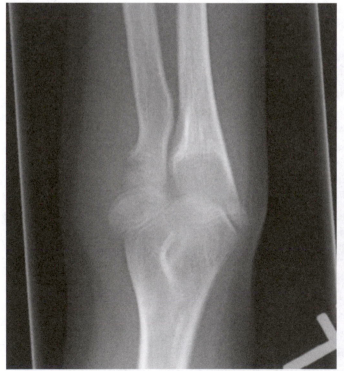 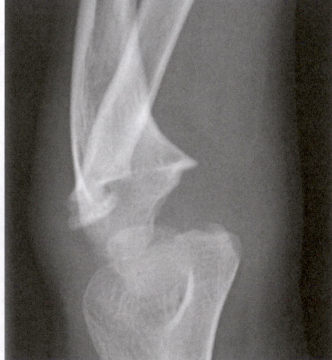

Figure 2.10 An X ray of a bone deformity resulting from a damaged growth plate.
Courtesy of Dr. Gerhard Kiefer.

The osteocyte slowly fills in its surrounding lacuna with matrix, and thus both cell and lacunar size decrease.

Osteocytes communicate with one another, and the deeper osteocytes communicate with the surface-covering osteoblasts by a network of interconnecting processes (gap junctions) housed in canaliculi within the extracellular matrix (figure 2.8). The presence of connections between adjacent processes between bone cells suggests that osteoblasts, osteocytes, and bone-lining cells form a functional *syncytium* that may play an integral role in many physiological functions, including the conversion of mechanical signals into remodeling activity and mineral movement into and out of the bone (Currey 2002).

Osteoclasts are multinuclear cells of hematogenous (from blood) origin that are located on the bony surface and are the primary bone-resorbing cells. The most distinguishing feature of the osteoclast is the extensive in-folding of the cell plasma membrane that gives rise to a ruffled border. This border is functionally significant because it greatly increases the surface area along which the cell can interact with the surrounding bony matrix (Rosier et al. 2000). Osteoclasts resorb bone by anchoring to the bone surface and secreting hydrogen (H^+) ions and proteolytic enzymes across their plasma membrane. The enzymes are released into the extracellular matrix by lysosomes. Once released, the enzymes digest the organic components of the matrix. Osteoclasts move along the bone surface and leave behind a trail of resorbed bone that has the appearance of an etched surface.

The extracellular bone matrix has inorganic (mineral), organic, and fluid components. Minerals, primarily in the form of calcium hydroxyapatite crystals, contribute about 50% of the total bone volume. Organic components constitute 39% of the volume (95% type I collagen and 5% proteoglycans), whereas fluid-filled vascular channels and cellular spaces constitute the remaining volume. Bone mechanical behavior reflects a balance between the mineral and organic phases, with minerals contributing stiffness and the organic matrix adding strength to bone.

Mineral content distinguishes bone from other connective tissues and provides bone with its characteristic rigidity; bone serves as a mineral storehouse. The mechanism responsible for calcification of the extracellular matrix of bone and not of other connective tissues is not completely understood, but apparently the ability of type I collagen to bind mineral crystallites (hydroxyapatite) is unique among the collagen molecules. Neither type II nor type III collagen can bind to minerals. More than 200 noncollagenous proteins are also found within bone's extracellular matrix. In terms of concentration, however, collagen occupies the greatest portion of the matrix. Because collagen provides major structural support in connective tissues, abnormalities in collagen production can have far-reaching consequences in the ability of the skeleton to resist mechanical stresses. For example, if a genetic defect affects bone collagen formation (e.g., osteogenesis imperfecta), the bone can become very fragile and easily fractured.

Bone, including the marrow, periosteum, metaphysis, diaphysis, and epiphysis, is richly supplied with blood vessels (figure 2.11). Studies report that approximately 7% of the cardiac output is sent to the skeleton (Shim et al. 1967; Tothill and MacPherson 1986). Blood reaches each area of the bone via extensive arterial interconnections (anastomoses) that feed a network of sinusoids (dilated venous channels). The sinusoids in turn empty into central venous channels deep within the medullary canal in long bones or a central canal in flat bones. The primary nutrient artery

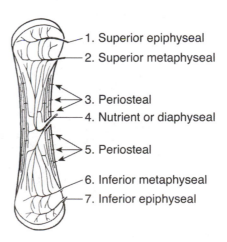

1. Superior epiphyseal
2. Superior metaphyseal

3. Periosteal
4. Nutrient or diaphyseal

5. Periosteal

6. Inferior metaphyseal
7. Inferior epiphyseal

Figure 2.11 Diagram of the arterial supply of a long bone with its six groups of vessels.

Adapted, by permission, from R.P. Ficat, J. Arlet and D.S. Hungerford, 1980, *Ischemia and necroses of bone* (Baltimore, MD: Lippincott, Williams & Wilkins), 2.

enters the medullary canal via an obliquely oriented nutrient foramen. Once within the marrow cavity, the artery divides into two longitudinal branches, one toward each of the epiphyses. Each branch gives rise to many parallel branches that can pierce through the cortex and anastomose with periosteal vessels that supply the outer third of the cortex. Within compact bone, primary arteries and veins travel relatively parallel to the osteonal longitudinal axes within structures called Haversian canals. Transversely oriented vessels are contained within structures called Volkmann canals (figure 2.12).

The terminal ends of the nutrient artery branches anastomose with branches from the metaphyseal system. Preceding closure of the physes, the physes can be seen as a boundary largely separating the epiphyseal and metaphyseal systems; only a few metaphyseal branches perforate the physes and anastomose with epiphyseal branches. After closure of the physes, the metaphyseal and epiphyseal branches are richly interconnected. Epiphyseal vessels branch into extensive arcades that supply the bony ends.

Bone Macrostructure

Despite differences in size and mechanical properties, bone tissue is basically similar in all bones. As described earlier, bone can be divided at a gross structural level into cortical (compact) and trabecular (cancellous or spongy) bone. At a tissue level, bone may be divided into three broad categories: woven, primary, and secondary bone. These categories are described in table 2.2.

Woven bone is laid down rapidly as a disorganized arrangement of collagen fibers and osteocytes. Although the mineral content of woven bone may be even higher than that of primary and secondary bone, the disorganized pattern and generally lower proportions of noncollagenous proteins decrease the mechanical strength of woven bone compared with primary or secondary bone. Developmentally, woven bone is unique because it can be deposited de novo (without a preexisting membrane, bone, or cartilaginous model; Martin et al. 1998). The cell-to-bone volume ratio is high in woven bone, confirming its role in providing temporary, rapid mechanical support, such as following traumatic injury.

In the adult skeleton, woven bone is not usually present but can be found in a fracture callus, in areas undergoing active endochondral ossification, and in some skeletal pathologies. During maturation, primary bone systematically replaces woven bone, providing the mature skeleton with the appropriate functional stiffness.

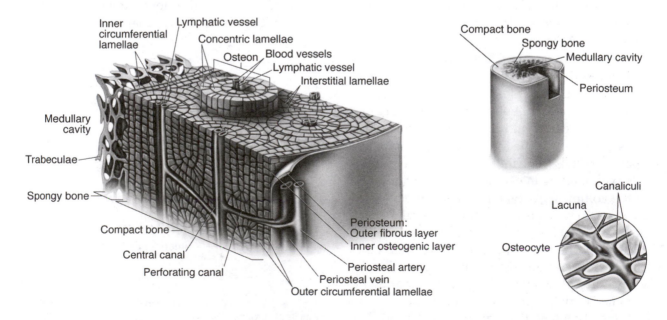

Figure 2.12   Ultrastructure of cortical bone, emphasizing the organization of intracortical Haversian and Volkmann canals containing blood vessels, as well as the canalicular connections within osteons and between osteocytes.

Reprinted, by permission, from W.C. Whiting and S. Rugg, 2006, *Dynatomy: Dynamic human anatomy* (Champaign, IL: Human Kinetics), 27.

TABLE 2.2

BONE MACROSTRUCTURE

Type of bone tissue	Deposition	Example
Woven bone	Without preexisting membrane, bone, or cartilage model	Fracture callus
Primary bone (lamellar)	Membrane or cartilage foundation	Trabecular bone
Primary bone (primary osteons)	Vascular channels	Young bone
Secondary bone	Replacement of preexisting bone	Human cortical bone

Adapted, by permission, from R.B. Martin and D.B. Burr, 1989, *Structure, function, and adaptation of compact bone* (New York, NY: Raven Press), 19.

Primary bone comprises several types of bone, each with unique morphology and function. A common factor among the types of primary bone, however, is that unlike woven bone, primary bone must replace a preexisting structure, either a cartilaginous model or previously deposited woven bone. Primary bone is composed of multiple thin layers (lamellae) of bone matrix and cells organized in parallel with the bone surface. That bone is referred to as **lamellar bone** and can exist in cortical and trabecular bone. Vascular channels are sparse in primary bone, and therefore this bone can be very dense. Where vascular channels are present, they are associated with several lamellae circumferentially surrounding the vascular channels. The outer most lamella has a smooth surface. The units of vascular channels and lamellae are referred to as **primary osteons** or principal structures of compact bone.

Primary bone is also found in cancellous bone. For example, the trabeculae (small rods) found in the vertebral bodies and in long-bone epiphyses are mostly primary bone in young people. In this case, although vascular channels are not enclosed within the lamellar structure, the individual struts or trabeculae of cancellous bone are in intimate contact with a rich vascular supply. Because of this close proximity, cancellous bone has a very important role in mineral homeostasis because calcium stores can be mobilized quickly in response to decreased serum calcium.

Secondary bone is deposited only during remodeling and replaces preexisting primary cortical or trabecular bone. Remodeling can create osteons, but unlike primary osteons, the outer-most lamellae of secondary bone have a scalloped surface. Differences between the developmental process of primary and secondary bone imply that a different controlling mechanism may be responsible for the endosteal or periosteal deposition of primary bone versus the intracortical deposition of secondary bone during remodeling.

Cartilage

Cartilage contains the basic elements of a connective tissue, namely cells and extracellular matrix (tissue fluid and macromolecules). The relative amounts and types of matrix constituents distinguish three kinds of cartilage: hyaline, elastic, and fibrocartilage. Hyaline cartilage is the most abundant of the three. None of the cartilage types has intrinsic blood vessels, nerves, or lymph vessels. The absence of vascular structures in cartilage makes it necessary for cartilage cells **(chondrocytes)** to receive their nutrients and remove metabolic waste by diffusion.

All cartilage develops from mesenchyme (primitive connective tissue). Mesenchymal cells produce the extracellular matrix (including collagenous fibrils) and differentiate into chondroblasts, which are the precursors of chondrocytes. The chondrocytes, when formed, are encapsulated in caves **(lacunae)** in the cartilage matrix. The unit including a chondrocyte in its lacunae is termed a **chondron.** A connective tissue membrane **(perichondrium)** surrounds the new cartilage. Within the perichondrium are blood and lymph vessels and nerves; nutrients leave this area to enrich the chondrocytes within the cartilage. The perichondrium also contains fibroblasts, collagen fibers, and elastic fibers.

Chondrocytes can multiply by two mechanisms. **Interstitial growth** occurs in young cartilage as the chondrocytes divide within the lacunae, forming cell nests. Younger cartilage is more flexible than mature cartilage, and the matrix can accommodate the interstitial expansion. **Appositional growth** proceeds in the cartilage layers immediately beneath the perichondrium. The mesenchymal cells in this superficial zone develop into new cartilage cells. These, in turn, are laid between the older cells and perichondrium, and new cartilage cells produce new matrix components.

Figure 2.13a shows the distribution of cells in joint-surface cartilage from the surface of the perichondrium, through the chondrogenic layer immediately beneath the perichondrium (where appositional growth occurs), and on to the middle region where interstitial growth occurs within the chondrocytes. The distribution of the collagen fibers throughout this same region is illustrated in figure 2.13b. In the superficial tangential zone (perichondrium), the collagen fibers are arranged tangentially to the joint surface. In the middle 40% to 60% of the cartilage, the collagen fibers appear more randomly organized. In the deepest layer

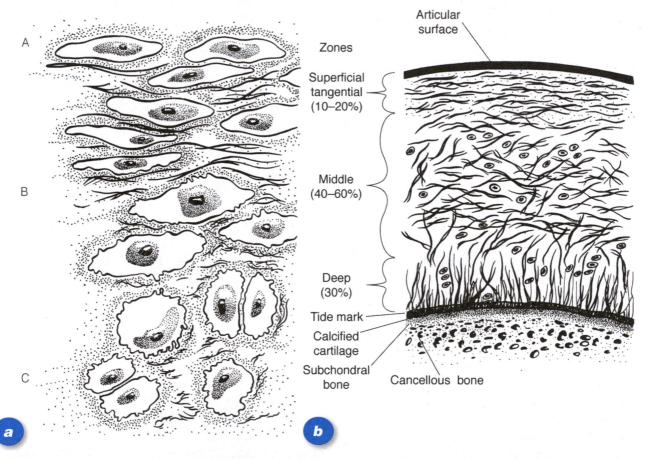

Figure 2.13 (a) Cellular organization of hyaline cartilage. In A, cells are flattened, chondrogenic, and adjacent to the perichondrium. B represents a continuation of the perichondrium level as cartilage cells produce extracellular matrix to distance themselves from their neighboring cells. In C, columns of cartilage cells are seen in lacunae. In the orientation represented in the figure, the joint surface is at the top, and the deepest part of the cartilage is at the bottom near the attachment to bone. (b) Collagen fiber organization in articular (hyaline) cartilage. The superficial tangential zone is at the region closest to the joint surface, and the collagen fibers are aligned tangential to the articular surface. In the middle zone, the collagen fibers are relatively random in orientation, and in the deep zone the collagen fibers are in a radial direction (with respect to the surface of the joint). The collagen fibers in the deep zone penetrate through the **tide mark** (the transition zone between calcified and noncalcified articular cartilage) and into the calcified cartilage overlying the subchondral bone.

(a) Reprinted, by permission, from D.W. Fawcett, 1994, *Bloom and Fawcett: A textbook of histology*, 12th ed. (New York, NY: Chapman & Hall), 191.
(b) Reprinted, by permission, from V.C. Mow, Ch.S. Proctor and M.A. Kelly, 1989, Biomechanics of articular cartilage. In *Basic biomechanics of the musculoskeletal system*, 2nd ed., edited by M. Nordin and V.H. Frankel (Philadelphia, PA: Lea & Febiger), 34.

of the articular cartilage, the collagen fibers are oriented radially and penetrate into the underlying calcified cartilage to maintain a solid adhesion to the underlying bone.

Hyaline Cartilage

Hyaline cartilage gets its name from its glassy appearance. The fetal skeletal anlage is basically hyaline cartilage before being replaced by bone later in life. The surfaces of most of the joints, the anterior portions of the ribs, and areas of the respiratory system (e.g., trachea, nose, bronchi) are composed of hyaline cartilage throughout life. Hyaline cartilage within joints is known as articular cartilage and provides a suitable surface for joint lubrication. The hyaline matrix appears blue and homogeneous in the fresh state, is firm and resilient in texture, and contains collagen fibers. The collagen fibers give cartilage its tensile strength (Lo et al. 2003; Silver and Bradica 2002). The mechanical stiffness and strength throughout the cartilage depth vary with the changes in collagen fiber organization (Morel et al. 2005; Silver and Bradica 2002). The principal collagen in hyaline cartilage is type II (90% of total collagen). The perichondrium provides the nutrients for the cartilage and surrounds all hyaline cartilage except articular cartilage, which receives nutrients by diffusion via the joint synovial fluid.

Cells constitute less than 10% of hyaline cartilage volume, with the principal components being macromolecules (about 20% of volume) and tissue fluid (about 70% of volume; Lo et al. 2003). The main structural macromolecule (besides type II collagen) in hyaline cartilage is proteoglycan. The integrated structure–function relations among the collagen fibers, proteoglycans, and fluid contribute to the unique mechanical behavior of articular cartilage (figures 2.7 and 2.13).

Hydrophilic proteoglycans tend to draw water into the matrix, and the negatively charged proteoglycans tend to repel each other. Articular cartilage wants to swell, but this expansion is resisted by the tensile restraint provided by the collagen fibrils. A dynamic interaction occurs between these matrix constituents during the loading of normal articular cartilage. Figure 2.14 illustrates the basics of this dynamic interaction. Figure 2.14a illustrates articular cartilage without any external load being applied. The cartilage is swollen with water that has been attracted by

the proteoglycans, and the negative charges of the proteoglycans are repelling each other while the collagen fibers provide the restraining tensile forces to maintain the structure. In figure 2.14b, an external compressive load has been applied. Now the fluid exudes from the articular cartilage, and the proteoglycan monomers and aggrecan are forced closer together. If a constant load is applied, the cartilage will creep (slowly deform) until a new equilibrium has been reached.

In adults, chondrocytes slowly continue to produce and renew proteoglycan macromolecules. With age, however, the turnover rate of proteoglycan diminishes, and some of the proteoglycan monomers or individual glycosaminoglycans can become disconnected, which diminishes the resilience of the articular cartilage to external loads.

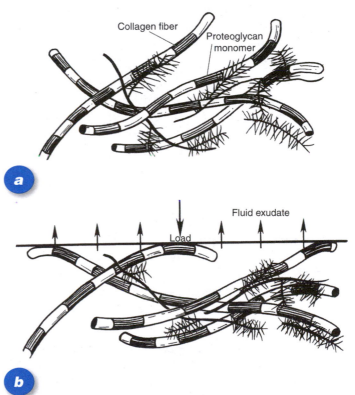

Figure 2.14 Unloaded and loaded extracellular matrix of articular cartilage. *(a)* Collagen provides the tensile restraint to the swelling pressures generated by the influx of tissue fluid (assisted by the hydrophilic, negatively charged proteoglycans). In resting articular cartilage, there is a normal swelling pressure. *(b)* An external load has been applied to the surface of the articular cartilage, and the matrix is being compressed. Tissue fluid exudes from the cartilage matrix, and the negatively charged proteoglycan monomers and aggrecan are being pushed closer together. For a given (constant) load, eventually a new equilibrium will be reached with the matrix in a compressed state.

Reprinted, by permission, from H.M. Mankin et al., 1994, Form and function of articular cartilage. In *Orthopaedic basic science*, edited by S.R. Simon (Park Ridge, IL: American Academy of Orthopaedic Surgeons), 21.

Other Types of Cartilage

Although hyaline cartilage is the most common type of cartilage, several other types exist that have unique compositions and functions. The other types are broadly separated into elastic cartilage and several specialized fibrocartilages, described in more detail next.

■ **Elastic cartilage** is found in the external ear, the epiglottis, portions of the larynx, and the eustachian tube. Consistent with its name, elastic cartilage possesses a great deal of flexibility. The matrix contains elastic fibers as well as collagen fibers. The matrix appears more yellow because of the higher percentage of elastic fibers and is not translucent like hyaline cartilage. Elastic cartilage is able to develop both interstitially and through appositional growth. The perichondrium is, again, a dense type of connective tissue with more elastic fibers, and the layer of cells immediately beneath the perichondrium is chondrogenic.

■ **Fibrocartilage** is strong and flexible because of its endogenous collagen fibers and is resilient because of its matrix. Fibrocartilage is found in many areas of the body, especially at stress points where friction could be problematic. It is distinct from hyaline and elastic cartilage because it contains no perichondrium. The fibrocartilage develops much like other ordinary connective tissue: Fibroblasts produce matrix and then differentiate into chondrocytes. Fibrocartilage is essentially a filler material between hyaline cartilage and other connective tissues and is found near joints, ligaments, and tendons and in the intervertebral discs. Four categories have been labeled, each with a specific function: interarticular, connecting, stratiform, and circumferential.

■ **Interarticular fibrocartilage** (meniscus) is found in the wrist and knee joints as well as in the temporomandibular joint and at the junction of the sternum and clavicle. In these joints, in which frequent movement and potential impact occur, the fibrocartilage provides a buffer. Interarticular fibrocartilages are flattened plates that are interposed between the joint surfaces and held in position by ligaments and tendons that connect to the edges of the fibrocartilage. The surfaces of these interarticular fibrocartilages, however, are free of connections and help to prevent friction between the moving joints. Furthermore, the interarticular fibrocartilages act as spacers to fill the gap between the joints, improve joint geometry, and protect the surfaces of the underlying articular cartilage.

■ **Connecting fibrocartilage** occurs at limited motion joints, such as intervertebral discs. These fibrocartilage plates allow the surfaces of the adjacent vertebral bodies to move slightly with respect to each other—read more about intervertebral discs in chapter 8.

■ **Stratiform fibrocartilage** forms layers over bone where tendons may act and can also be an integral part of the tendon surface. When a muscle contracts and a tendon is forced to slide over a bony surface, friction is minimized by interposing stratiform fibrocartilage between the bone and tendon.

■ **Circumferential fibrocartilage** acts as a spacer in the joints of the hip and the shoulder (e.g., glenoid labrum). Circumferential fibrocartilage is a circular ring without a center. Thus, it protects only the edge of the joints and improves the bony fit.

Tendons and Ligaments

Fibrous connective tissue can be classified into dense or loose fibrous varieties, depending on the proximity and the packing of the fibers. In turn, dense fibrous tissue can be either organized or unorganized, depending again on the fiber organization (figure 2.5).

In organized tissue, collagen fibers run in parallel bundles. These regularly arranged tissue types include tendons, ligaments, and aponeuroses. In each of these, the tissue is primarily composed of fibers and extracellular matrix components. Fibroblasts are the principal cells in these tissues. These tissues have great tensile strength but are able to resist stretching primarily in one direction—along a tensile force generated parallel to the fiber line.

Tendons are white, collagenous, flexible bands that connect muscle to bone. Figure 2.15 shows the generic structure of a tendon. The building blocks of a tendon are the tropocollagen molecules. Tropocollagen molecules generally are aligned in parallel rows to form a microfibril. Subsequently, the microfibrils aggregate into parallel bundles to form subfibrils and then fibrils. Fibrils are gathered into fascicles bound together by a loose connective tissue (endotendineum), which permits relative motion of the collagen

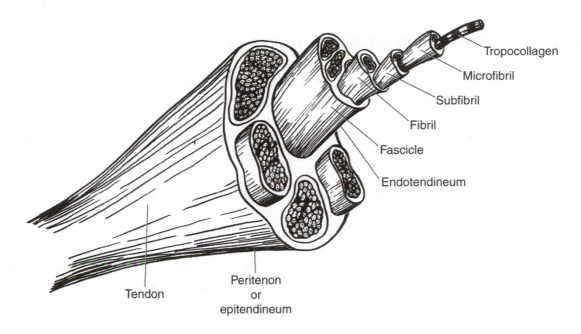

Tropocollagen
Microfibril
Subfibril
Fibril
Fascicle
Endotendineum

Tendon

Peritenon
or
epitendineum

Figure 2.15 Schematic representation of the hierarchy of a tendon.

fascicles and supports blood vessels, nerves, and lymphatics (Woo et al. 2000). Tendon fascicles are grouped into the tendon proper. When a tendon is relaxed (no tensile load), it takes on a crimped or wavy appearance. As a load is applied, the wavy pattern is straightened.

As is evident from the description of the collagen fiber organization in tendon, the major component of tendon is type I collagen, which accounts for about 86% of the dry weight of a tendon (Woo et al. 2000). Elastic fibers are present in small quantities in the matrix of tendons.

The surface of the tendon can be covered with an **epitendineum,** usually seen as a tendon sheath, to act as a pulley and direct the path around sharp corners as in the flexor tendons of the hand. When tendons are not enclosed with the epitendineum and move in a relatively straight line, a loose areolar connective tissue **(peritenon)** envelopes the tendon. The peritenon contains blood vessels and nerves to serve the tendon.

The insertion of tendon into bone **(osteotendinous junction)** has been classified as direct or indirect. The direct insertion is characterized by four layers (zones) with a gradual transition from tendon to unmineralized fibrocartilage, to mineralized fibrocartilage, and finally to bone. The indirect insertion is characterized by an interface made up of three layers: tendon, **Sharpey's fibers,** and bone.

At the opposite end of the tendon, the **myotendinous junction** (also **musculotendinous junction)** is a specialized region of longitudinal membranous invaginations that increase the surface area and reduce stress on the junction during the contractile force transmission. The strength of an adhesive junction such as the myotendinous junction depends both on the properties of the adjoining structures and on the orientation of forces across the junction. Junctions that are loaded in shear, with the force being parallel to the membrane surface, are stronger than junctions with a large tensile component perpendicular to the membrane (Tidball 1991).

Aponeuroses are fibrous, ribbonlike membranes similar in composition to tendons. These structures are sometimes called flattened tendons. For example, the palmar aponeurosis encloses the muscles of the palm of the hand. Aponeuroses are whitish in appearance because of the presence of collagen. The fibers of aponeuroses run in a single direction and thus differ from unorganized (or irregular, dense fibrous) connective tissue (fascia).

Ligaments are dense, regular connective tissue structures that join bone to bone. The primary function of ligaments, like tendons, is to resist tensile forces along the line of the collagen fibers. Ligaments are classified and named by several criteria. Examples of criteria are attachment sites (coracoacromial), shape (deltoid), function (capsular), position or orientation (collateral, cruciate), position relative to the joint capsule (extrinsic, intrinsic), and composition (elastic).

CASE STUDIES IN ANKLE AND KNEE SPRAINS

Sprains are injuries to ligaments usually caused by sudden overstretching. Sprains can occur in many joints, most commonly the ankle, knee, fingers, and wrist. Typically the symptoms associated with sprains are related to those of inflammation and include pain, swelling, and loss of function.

Ankle Orthotics and Sprains

Inversion sprains of the ankle are common in running and jumping sports. If a person has had a previous ankle sprain, a subsequent sprain is much more likely to happen.

To reduce the number of ankle sprain injuries, **taping** or semirigid **orthotics** have been advocated. Taping has advantages and disadvantages. Although prophylactic taping can effectively reduce excessive ankle inversion before exercise, in many cases its restraint is lost during an exercise bout. Taping can also be expensive and requires a skilled individual to apply the tape. Semirigid ankle orthotics have been proposed as a substitute for taping. Few prospective studies, however, have evaluated the effectiveness of ankle orthotics for reducing or preventing ankle sprains.

Notably, researchers in South Africa completed a prospective study over a 1-year soccer season (Surve et al. 1994). Adult male soccer players with a previous history of ankle sprain and those with no previous history of ankle sprain were identified. Each player was then randomly assigned to an orthotic group or a control group (no orthotic or taping). Thus, four groups of players were studied, comprising more than 500 players. Injury was defined as any sprain that occurred during a scheduled match or practice that caused the soccer player to miss the next game or practice. Marked differences occurred among the groups during the 1 year of play. The principal finding was that the application of a semirigid ankle orthotic resulted in a fivefold reduction in the incidence of ankle sprains in players with a previous history of ankle sprains. Ankle orthosis, however, did not significantly alter the incidence of ankle sprains in soccer players who had never sprained their ankles before.

Why this difference? Many people suggest that the positive effects of an external support orthotic are primarily attributable to mechanical support that limits excessive inversion and eversion of the ankle. Nevertheless, the previously injured and previously uninjured athletes responded differently from one another. Only the athletes who had previously sprained their ankles received positive benefits from wearing the orthotic. Surve and colleagues (1994) suggested that proprioceptive defects can happen after an ankle sprain because of damage to sensory receptors in the ligaments of the ankle. This may impair the reflex stabilization of the ankle. The application of an external orthotic may have stimulated mechanoreceptors to improve proprioceptive function of the previously injured ankle, rather than just provide mechanical support.

Loss of ACL Can Alter Neuromotor Control Patterns

When people tear the anterior cruciate ligament (ACL) in the knee joint so that it no longer restrains the forward motion of the tibia (with respect to the femur), the majority (75%) may alter their patterns of neuromotor control of muscles surrounding the knee to accommodate changes in function. Berchuck and colleagues (1990) found that when an activity such as walking requires the quadriceps to be active while the knee is flexed between 0° and 45°, the contraction of the quadriceps tends to move the proximal end of the tibia anteriorly, thereby straining the anterior cruciate ligament. If a person does not have an anterior cruciate ligament, however, what can he or she do to avoid the forward motion of the tibia on the end of the femur?

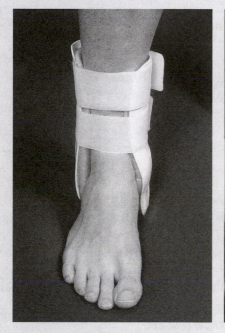
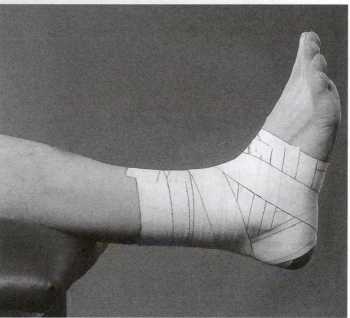

(a) Ankle brace. *(b)* Closed basket weave taping for the ankle.

In analyzing the gait of patients with ACL-deficient knees, Berchuck and colleagues (1990) reported that patients reduced contraction of their quadriceps during the stance phase of walking, using a so-called quadriceps-avoidance gait. The patients also may have increased the action of the hamstring muscles to pull back on the tibia during stance, but that was not measured.

But if these patients avoided using the quadriceps to prevent collapse of the knee during midstance, then how did they maintain an extended knee? Why didn't their knees collapse? Apparently the patients learned to increase the amount of hip extensor activity to compensate for the reduction in knee extensor activity. Interestingly, the patients walked with quadriceps-avoidance gait on both their ACL-deficient side and on the other (normal) knee.

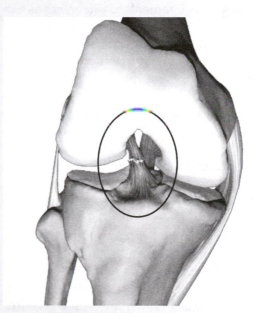

Complete rupture of the anterior cruciate ligament.
Copyright Primal Pictures Ltd.

The researchers suggested that after ligament injury there is a reprogramming of the locomotor process so that excessive anterior displacement of the tibia is prevented. How this abnormal function affects the long-term outcome and the likelihood of posttraumatic osteoarthritis remains unknown.

Joint ligaments have a structure that is similar to tendon, but whereas collagen fibril bundles in tendons are typically aligned parallel to each other (in line with the pull of the muscle), the collagen fibril bundles in ligaments may be oriented in parallel, obliquely, or even in a spiral arrangement. The geometry of the collagen fibril bundles in ligaments is specific to a ligament's function. The color of collagenous ligaments is a duller white than tendon because of the slightly greater percentage of elastic and reticular fibers found between the collagen fiber bundles.

The ligament insertion to bone is either direct or indirect (Lo et al. 2003; Woo et al. 2000). The direct attachment is comparable to the specialized collagen fibers (Sharpey's fibers) that attach tendon to bone. The indirect route is one in which the collagen fibers from the ligament blend with the fibrous periosteum of the bone.

Fibroblasts are the principal cells in ligaments, whereas the main fibrous component of the extracellular matrix is type I collagen (36% of wet weight). Several other types of collagen are also found in ligaments. Proteoglycans are present, although fewer than in articular cartilage. Because almost two-thirds of a ligament is composed of water, the proteoglycans (which are hydrophilic) may play a role in the mechanical behavior of a ligament.

Joint ligaments, such as those in the knee, contain several sensory receptors (Ruffini corpuscles, Pacinian corpuscles, Golgi tendon organs, and free-nerve endings) that are capable of providing the nervous system with information about movement, position, and pain. Nevertheless, the exact neurosensory role of ligaments and receptors in joint proprioception is controversial and continues to be studied. After synthesizing anatomical, neurophysiological, and mechanical data, with a particular focus on the sensory receptors of the knee joint ligaments, Kim and colleagues (1995) and more recently Solomonow (2004) concluded that ligaments may provide sensory information about changes in the stiffness of muscles around the knee joint. In this way, ligaments can have an important function in regulating the stability of the knee joint.

Yellow elastic ligaments are less common in the body than collagenous ligaments. Parallel elastic fibers, which predominate in elastic ligaments, are surrounded by loose connective tissue.

Elastic ligaments in humans include the vocal cords and the ligamenta flava of the vertebrae. A classic example of an elastic ligament in animals is the ligamentum nuchae of cattle, which helps the animal hold up its head while grazing.

Fascia is a catch-all category that includes dense, fibrous, unorganized tissues that do not logically fall into the categories of tendon, aponeurosis, or ligament. The principal fibers in fascia are collagenous, although some elastic and reticular elements also exist. Fascia contains interwoven, meshlike, nonparallel fibers and is usually found in layers or sheaths around organs, blood vessels, bones, and cartilage as well as in the dermis of the skin. It provides firm support for muscles. The fibers in fascia traverse in different directions and, in some cases, in different planes (as in the dermis). Because of this organization, fascia withstands stretching in many directions.

Skeletal Muscle

Skeletal muscles are the prime executors of the peripheral nervous system's motor division. Contractile proteins and a network of connective tissue are the two basic elements of muscles. Fibrous connective tissues within the muscle belly and those that blend with the tendon provide important functional stiffness, which enhances the transmission of tension. Significant cellular interactions direct a muscle's physiological response, but muscle adaptation and injury are best described by considering the mechanics of a muscle's functional units.

Microstructure and Function

The structure of skeletal muscle is diagrammatically presented in figure 2.16. From the whole-muscle level, individual muscle fibers (muscle cells) are subdivided into myofibrils, sarcomeres, and actin and myosin. The connective tissue surrounding the entire muscle is called the epimysium, and the bundles of muscle fibers (fascicles) are surrounded by the perimysium. Each individual muscle fiber is surrounded by endomysium.

A skeletal muscle fiber is composed of contractile proteins called myofibrils. The myofibril is the contractile unit of muscle, with hundreds of myofibrils combining to form a single muscle

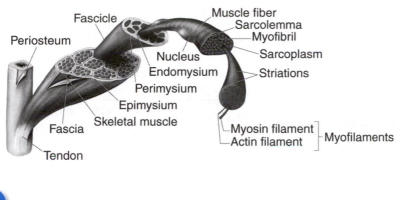

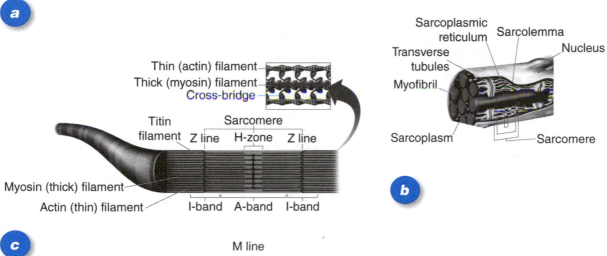

Figure 2.16 The structural composition and organization of *(a)* skeletal muscle tissue, *(b)* a muscle fiber, and *(c)* a sarcomere. Reprinted, by permission, from W.C. Whiting and S. Rugg, 2006, *Dynatomy: Dynamic human anatomy* (Champaign, IL: Human Kinetics), 69.

fiber. Each myofibril is composed of two contractile **myofilaments: actin** (thin filament) and **myosin** (thick filament). The myofibril has a striated appearance with transverse bands of repeated units called **sarcomeres.** The striations are created by the overlapping of myosin and actin filaments.

An array of other proteins (e.g., titin) are present within a sarcomere and can contribute to the structure and passive properties of the sarcomere. **Titin** is a large protein that spans from the Z disc to the M band of a sarcomere (figure 2.17). It is generally accepted that titin acts as a spring to develop tension as the sarcomere is stretched and may act to center the thick filament within the sarcomere when forces on each side of the sarcomere are unequal.

In the **sliding filament theory** of muscle contraction, muscle shortens when the sarcomere structure changes. The cross-bridge cycle inter-

actions between the actin and myosin filaments allow for an increase in the overlap between the actin and myosin fibers and thus a decrease in overall sarcomere length. Figure 2.17 depicts the sliding filament theory of muscle contraction and shows a schematic of a contracted and extended sarcomere. During muscle contraction, sarcomere length decreases, which results in coiling of the I-band part of titin. Simultaneously, interfilament spacings increase, which is likely to reorient or stretch Z- and M-line proteins that cross-link thin and thick filaments and myofibrils. During extension, sarcomere length increases and interfilament spacings decrease. This extends titin and releases tension on the Z- and M-line proteins. Overextension of muscle results in the unraveling of titin polypeptide, which starts from the least mechanically stable PEVK region. This is accompanied by compression and reorientation of Z- and M-line proteins.

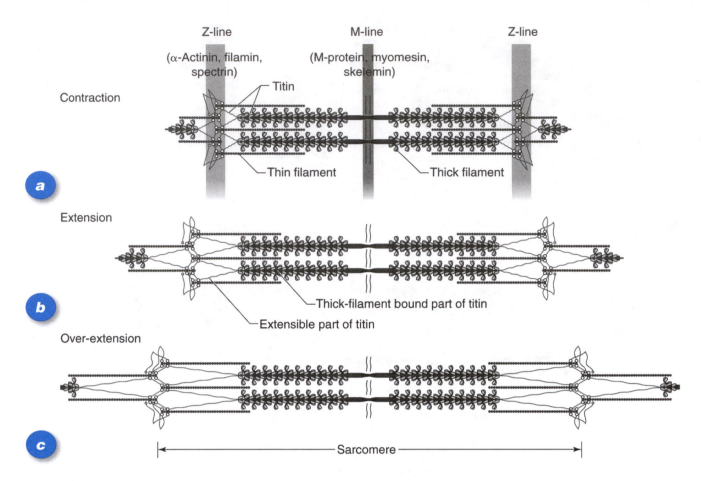

Figure 2.17 Sarcomere structure at various degrees of extension: *(a)* contracted, *(b)* extended, and *(c)* overextended.

Adapted, by permission, from L. Tskhovrebova and J. Trinick, 2003, "Titin: Properties and family relationships," *Nature Reviews: Molecular Cell Biology* 4(9): 679-689.

The process of muscular contraction commences with a neural **action potential.** Action potentials travel down motor neurons that are connected to muscle fiber membranes (neuromuscular junction or **synapse).** At that synapse, acetylcholine is released from the presynaptic terminal and binds to receptors on the postsynaptic terminal. The binding of acetylcholine to the postsynaptic membrane increases that membrane's permeability to sodium (Na^+). If the depolarization of that membrane attributable to Na^+ exceeds a certain threshold, the action potential will propagate down the length of the muscle fiber. That action potential is transmitted to the interior of the muscle fiber by specialized cell membrane invaginations called **T tubules**. Depolarization of those T tubules leads to the release of Ca^{2+} from intracellular stores into the sarcomere. Ca^{2+} binds to specialized sites that normally inhibit filament sliding in the relaxed state. That binding removes the inhibition, and the sarcomeres are ready to contract.

Figure 2.18 illustrates the four major states of interactions between myosin and actin during the cross-bridge cycle. The myosin head attaches to a binding site on the actin filament, forming a **cross-bridge.** The breakdown of bound adenosine triphosphate (ATP) into adenosine diphosphate (ADP) and phosphate (P) with adenosine triphosphatase (ATPase) (an enzyme) provides the energy required to create a configurational change (the *power stroke)* in the myosin head that slides the actin filament past the myosin filament. The myosin portion of the cross-bridge binds to a fresh ATP molecule that facilitates detachment of the cross-bridge and thus prepares the cross-bridge for another cycle.

Muscle fibers vary in length and can shorten to approximately one-half of their resting length. Human skeletal muscle contains several fiber

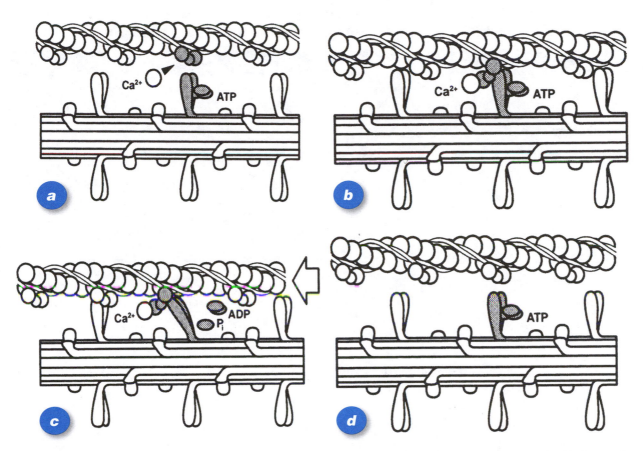

Figure 2.18 Schematic illustration of the cross-bridge cycle. *(a)* The muscle is at rest. The attachment site on the thin filament is covered by the tropomyosin–troponin complex. Adenosine triphosphate (ATP) is bound to the myosin cross-bridge. *(b)* On activation, calcium concentration increases in the sarcoplasm and calcium (Ca^{2+}) binds to troponin C, thereby causing a configurational change that exposes the actin binding site. *(c)* The cross-bridge attaches to the actin and goes through a configurational change. The splitting of ATP into adenosine diphosphate (ADP) and inorganic phosphate (Pi) provides the energy that results in contraction (i.e., movement of the thin past the thick filaments). *(d)* A new ATP attaches to the cross-bridge and the cross-bridge can detach from the thin filament, is ready for a new interaction with another attachment site on the thin filament.

From *Biomechanics of the musculoskeletal system*, 2nd ed., B.M. Nigg and W. Herzog. Copyright 1999, John Wiley & Sons Limited. Adapted with permission.

types that have different functional characteristics. Garrett and Best (2000) provided a summary of the three types of muscle fibers and their physiological, metabolic, and structural characteristics (table 2.3). Most muscles in the body are a mixed variety containing a combination of muscle fiber types.

Type I muscle fibers tend to have slower contraction and relaxation times and are very fatigue resistant. Their motor unit size is typically small with a high capillary density. Type II muscle fibers can be subdivided into types IIA and IIB. Type IIA fibers are fast-twitch, fast oxidative glycolytic fibers that have high contraction speeds but are relatively fatigable, with a larger motor unit size and relatively high capillary density. Type IIB muscle fibers are fast-twitch fibers and use glyco-

lytic metabolic processes. Type IIB fibers are the most fatigable but have the highest contraction speeds and the greatest contraction strength. Their motor unit size is the largest of the three, but their capillary density is relatively low.

Muscle fibers with the same biochemical profiles tend to have similar force-producing characteristics. A muscle fiber shortening to one-half of its length will have the same force characteristics whether it is long or short because the sarcomeres are in series. Increasing the number of muscle fibers in parallel, however, increases the effective force of the muscle.

One important aspect of muscle architecture is the angle of pennation of the muscle fibers. Longitudinal or fusiform muscles have muscle fibers lying parallel to the line of pull of the

TABLE 2.3

CHARACTERISTICS OF HUMAN SKELETAL MUSCLE FIBER TYPES

	Type I	Type IIA	Type IIB
Other names	Red, slow twitch (ST) Slow oxidative (SO)	White, fast twitch (FT) Fast oxidative glycolytic (FOG)	Fast glycolytic (FG)
Speed of contraction	Slow	Fast	Fast
Strength of contraction	Low	High	High
Fatigability	Fatigue-resistant	Fatigable	Most fatigable
Aerobic capacity	High	Medium	Low
Anaerobic capacity	Low	Medium	High
Motor unit size	Small	Larger	Largest
Capillary density	High	High	Low

Adapted, by permission, from W.E. Garrett, Jr. and T.M. Best, 1994, Anatomy, physiology, and mechanics of skeletal muscle. In *Orthopaedic basic science*, edited by S.R. Simon (Park Ridge, IL: American Academy of Orthopaedic Surgeons), 100.

tendon. These fibers pull in a straight line, and the full magnitude of the force is directed along the tendon's line of action. The fibers of pennate muscle (unipennate, bipennate, or multipennate) arise at an oblique angle to the line of pull (usually considered as a straight line along the tendon). Thus, only a portion of the force generated by the contracting fiber is transmitted along the tendon. Pennation of the fibers allows the number of fibers to increase without significantly increasing the muscle's diameter. Although only one component of the muscle fiber force is used effectively to move the tendon, the advantage of the unipennate muscle system is that an increased number of sarcomeres can be placed in parallel to increase the effective force of the muscle (figure 2.19).

Motor Units

The fundamental neuromuscular unit is the **motor unit** (figure 2.20). Motor units consist of a motoneuronal cell body located within the spinal cord, its axon, and the muscle fibers that it innervates at the motor end plates. When the neuron depolarizes, all the muscle fibers in the motor unit contract as one **(all-or-none principle).** Muscle tension can be increased by sending action potentials to the muscle more frequently (increasing stimulation rate) and by contracting more muscle fibers (recruiting additional motor

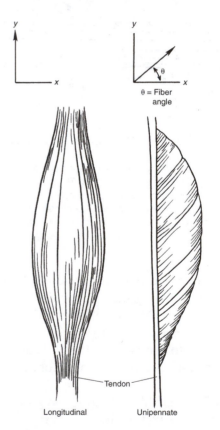

Figure 2.19 Effect of muscle fiber pennation. In the longitudinal (fusiform) muscle (left), all the force generated within the muscle fiber is directed through the tendon (no horizontal *x* component). The advantage of this arrangement is an increased range of motion (excursion of the tendon end with respect to muscle fiber excursion). By comparison, the unipennate muscle fibers (right) are directed at an angle with respect to the tendon. Thus, the force generated within the muscle has an *x* component that pulls perpendicular to the tendon, whereas the *y* component moves the tendon in the intended direction.

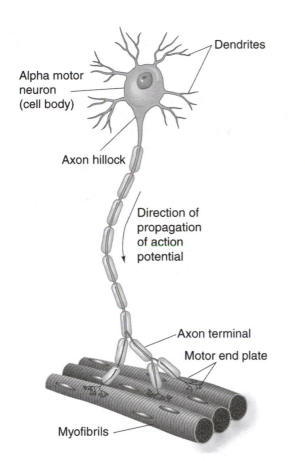

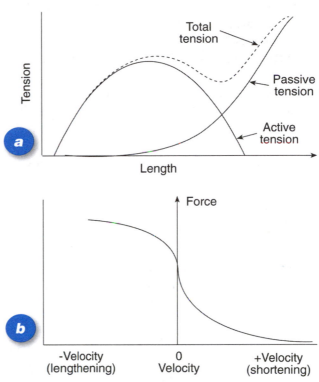

Figure 2.21 *(a)* Skeletal muscle length–tension curve and *(b)* skeletal muscle force–velocity curve.

Figure 2.20 Single motor unit. A motor unit consists of the motoneuron (cell body) with its accompanying axon and the total number of muscle fibers innervated by that neuron.

Reprinted, by permission, from J.H. Wilmore, D.L. Costill and W.L. Kenney, 2007, *Physiology of sport and exercise*, 4th ed. (Champaign, IL: Human Kinetics), 31.

units). Recruiting additional motor units is the more potent mechanism for initial force development. Only at higher force levels does increasing the firing frequency assume a prominent role.

In addition to the neural determinants of force, a muscle's force output also is modulated by the length of the muscle when contraction is initiated and by the velocity of contraction. Force, velocity, and length are interrelated variables that affect a muscle's mechanical response. These relations are usually summarized as **force–velocity relation** and **length–tension relation** curves (figure 2.21). Length and velocity are not independent of each other; they are both related to force. The maximal tension can be generated when a muscle is forcibly lengthened while it attempts to shorten (eccentric action), and the tension declines as an active muscle shortens (concentric action). Maximal strength in rapid eccentric

muscle action exceeds the maximum in isometric work, and the strength is even less in concentric muscle action.

Skeletal muscle–tendon units also have inherent passive properties that affect force output. The tension developed in a muscle–tendon unit is transmitted to the skeleton from an integrated blend of muscle cells and fibrous connective tissues. The fibrous connective tissue components within muscles include sarcoplasm, sarcolemma, and endomysium. The other major load-carrying connective tissues in the muscle–tendon unit include the tendon and the collagen fibers that permeate the muscle belly. Some of these passive structures function in series with the active muscle cells, whereas others function in parallel. The terms **series-elastic component** and **parallel-elastic component** are derived from these functions. Together, the two components account for the passive tension properties of muscle, which can be important in muscle mechanics. As Åstrand and colleagues (2003) noted, a given tension in a muscle–tendon unit can be produced at a lower metabolic energy cost in eccentric work than in concentric muscle actions because of the mechanical energy that can be stored in the elastic components.

Activated cross-bridges within the myofibrils exhibit a resistance to stretching, thus generating an internal force that is often termed *muscle stiffness.* Measured as change in force per change in length, stiffness is a property of muscle believed to operate over length changes and to have functional significance during locomotion and other movements.

Arthrology

The classification of joints and joint motion (arthrology) focuses on the classes, types, and examples of various joints in the human body. The words articulation and joint are used synony-

mously to describe the junction of two or more bones at their sites of contact. Some joints allow free movement (e.g., hip and knee joints), whereas others allow little or no movement between the connecting bones (e.g., sutures of the skull). A useful classification that organizes joints by structure and action is described in table 2.4.

Joints are divided into those with and those without a joint cavity. Synarthrodial (immovable) and amphiarthrodial (slightly movable) joints do not have a cavity. Diarthrodial (movable) joints typically have a joint cavity. This is the type of joint usually analyzed in movement-related injuries. For example, the knee joint is a frequently injured diarthrodial joint. A cross section of a knee joint, illustrating its many complex components, is provided in figure 2.22.

TABLE 2.4

SUMMARY OF JOINT STRUCTURE AND MOVEMENT

(ALL MOVEMENTS BEGIN FROM ANATOMICAL POSITION)				
PELVIS AND LOWER EXTREMITY				
Joint	**Structural classification**	**Movement**	**Plane**	**Axial and planar characteristics**
Sacroiliac	Synovial (plane)	Gliding		Nonaxial and nonplanar
Pubic symphysis	Symphysis	Distraction, separation during birth		
Pelvic girdle (movement of pelvis relative to femur)	Synovial (ball and socket)	Anterior tilt	Sagittal	Triaxial and triplanar
		Posterior tilt		
		Lateral tilt right	Frontal	
		Lateral tilt left		
		Rotation right	Transverse	
		Rotation left		
Hip (movement of femur relative to pelvis)	Synovial (ball and socket)	Flexion	Sagittal	Triaxial and triplanar
		Extension		
		Hyperextension		
		Abduction	Frontal	
		Adduction		
		Internal (medial) rotation	Transverse	
		External (lateral) rotation		
	(starting with hip flexed 90°)	Horizontal abduction (extension)	Transverse	
		Horizontal adduction (flexion)		

Joint	Structural classification	Movement	Plane	Axial and planar characteristics
Patellofemoral	Synovial (plane)	Gliding		Nonaxial and nonplanar
Tibiofemoral (knee)	Synovial (bicondyloid)	Flexion Extension	Sagittal	Biaxial and biplanar
		Internal (medial) rotation External (lateral) rotation (with knee flexed)	Frontal	
Ankle	Synovial (hinge)	Dorsiflexion Plantar flexion	Sagittal	Uniaxial and uniplanar
Subtalar	Synovial (plane)	Inversion Eversion	Frontal	Uniaxial and uniplanar
Intertarsal	Synovial (plane)	Gliding		Uniaxial and uniplanar
Tarsometatarsal	Synovial (plane)	Gliding		Uniaxial and uniplanar
Metatarsophalangeal	Synovial (condyloid)	Flexion Extension Hyperextension	Sagittal	Biaxial and biplanar
		Abduction Adduction	Transverse	
Interphalangeal	Synovial (hinge)	Flexion Extension	Sagittal	Uniaxial and uniplanar

UPPER EXTREMITY				
Joint	**Structural classification**	**Movement**	**Plane**	**Axial and planar characteristics**
Sternoclavicular (shoulder girdle)	Synovial (ball and socket)	Anterior rotation Posterior rotation	Sagittal	Triaxial and triplanar
		Upward rotation Downward rotation	Frontal	
		Abduction Adduction	Transverse	
Acromioclavicular	Synovial (plane)	Gliding		Nonaxial and nonplanar
Glenohumeral (shoulder)	Synovial (ball and socket)	Flexion Extension Hyperextension	Sagittal	Triaxial and triplanar
		Abduction Adduction	Frontal	
		Internal (medial) rotation	Transverse	
		External (lateral) rotation	Transverse	
	(starting with shoulder flexed 90°)	Horizontal abduction (extension)		
		Horizontal adduction (flexion)		

UPPER EXTREMITY				
Joint	Structural classification	Movement	Plane	Axial and planar characteristics
Elbow	Synovial (hinge)	Flexion Extension	Sagittal	Uniaxial and uniplanar
Radioulnar	Proximal: synovial (pivot) Middle: syndesmosis Distal: synovial (pivot)	Pronation Supination	Transverse	Uniaxial and uniplanar
Radiocarpal (twist)	Synovial (condyloid)	Flexion Extension Hyperextension Radial deviation (abduction) Ulnar deviation (adduction)	Sagittal Frontal	Biaxial and biplanar
Intercarpal	Synovial (place)	Gliding		Nonaxial and nonplanar
Carpometacarpal	Synovial (plane)	Gliding		Nonaxial and nonplanar
Metacarpophalangeal	(1) Thumb: synovial (saddle) (2-5): Synovial (condyloid)	Flexion Extension Hyperextension Abduction Adduction	(1) Frontal (2-5) Sagittal (1) Sagittal (2-5) Frontal	Biaxial and biplanar
Interphalangeal	Synovial (hinge)	Flexion Extension	Sagittal	Uniaxial and uniplanar

HEAD, NECK, AND TRUNK				
Joint	Structural classification	Movement	Plane	Axial and planar characteristics
Intercranial	Suture	None		
Temporomandibular	Synovial (condyloid)	Elevation Depression Protraction Retraction	Sagittal Transverse	Biaxial and biplanar
Atlanto-occipital	Synovial (hinge)	Flexion Extension	Sagittal	Uniaxial and uniplanar
Vertebral column: atlantoaxial	Synovial (pivot)	Rotation right Rotation left	Transverse	Uniaxial and uniplanar

C2-L5: (Vertebral bodies: symphysis) (Articular processes: synovial [plane])		Flexion Extension Hyperextension	Sagittal	Triaxial and triplanar
		Lateral flexion (right)	Frontal	
		Lateral flexion (left)		
		Rotation right	Transverse	
		Rotation left		
Costovertebral	Synovial (plane)	Gliding		Nonaxial and nonplanar
Sternomanubrial	Symphysis	Sternal angle increase		Nonaxial and nonplanar
		Sternal angle decrease		

Adapted, by permission, from W.C. Whiting and S. Rugg, 2006, *Dynatomy: Dynamic human anatomy* (Champaign, IL: Human Kinetics), 50-52.

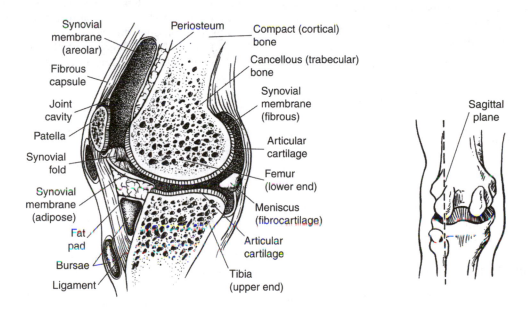

Figure 2.22 A sagittal plane cross section of an adult human knee.

How does such a complex diarthrodial joint develop? To answer this question, we return to the description of the somite-stage embryo provided in the embryology section earlier in this chapter. After the appearance of the limb buds (26- to 28-day embryo), a group of mesenchymal cells coalesces within the developing limb to form a **blastema** (Lo et al. 2003). The blastema is the foundation material that produces the capsule, ligaments, synovial lining, and menisci of the joint. The adjoining bones at this stage are cartilage models that are undergoing endochondral ossification. At the juncture between two of these cartilage–bone models, the interzonal mesenchyme condenses and forms an articular disc (primitive joint plate). The central cavity within the developing joint emerges at about 10 weeks. This cavity ultimately becomes the synovial cavity, which will contain the synovial fluid that assists with joint lubrication.

The development from blastema to definable skeletal elements happens between weeks 4 and 10 in the developing human embryo. Many stimuli can influence the development of joints. Movement appears to be one of the important factors and may result from the extrinsic hydrodynamic forces that act in utero or from the nascent actions of the developing skeletal muscle tissues in the limb.

■ CHAPTER REVIEW

Key Points

- Musculoskeletal tissues allow bodily motion. Motion is created through articulations that can include complex interactions with all musculoskeletal tissues that begin to take form very early during embryological development.

- Muscle is the force-producing tissue and acts on bones through tendons to control movement. Collagenous bands that tether bones together (ligaments) and fibrocartilaginous discs (menisci) can passively control the relative motion of articulating bones.

- Cartilage covering those bone segments within joints provides suitable surfaces for joint lubrication.

- All of those musculoskeletal tissues have structural subcategories.

- This chapter emphasized the complexity of these tissues and sets the stage for understanding the potential of these tissues to respond to stimuli, including injury. This information was presented with broad strokes. There are excellent and extensive sources available to provide the fine details of tissue anatomy and histology, and you are encouraged to explore these details.

Questions to Consider

1. You're giving a lecture to a lay audience on the dynamic nature of biological tissues, and someone makes the comment, "I always thought that bone was hard and lifeless." Provide arguments to convince the audience member to the contrary.

2. Consider a musculotendinous unit (e.g., calf muscles and the calcaneal [Achilles] tendon). Describe the embryological processes for each component of the unit (i.e., muscle and tendon).

3. Biological tissues work in concert to complete the many tasks and satisfy the functional needs of the human organism. Each tissue has unique characteristics that distinguish it from other tissues. Describe the unique characteristics of muscle, nervous, epithelial, and connective tissues.

4. Describe the steps involved in bone growth and development, and describe several potential impediments to normal, healthy bone growth.

5. Describe what is meant by the *structure-function relation* of biological tissues. Provide specific examples that illustrate this concept.

6. List and explain the factors that affect the amount of force a muscle can produce.

Suggested Readings

EMBRYOLOGY AND DEVELOPMENT

Iannotti, J.P., S. Goldstein, J. Kuhn, L. Lipiello, F.S. Kaplan, and D.J. Zaleske. 2000. The formation and growth of skeletal tissues. In *Orthopaedic Basic Science* (2nd ed.), edited by S.R Simon. Park Ridge, IL: American Academy of Orthopaedic Surgeons.

Sadler, T.W. 2003. *Langman's Medical Embryology* (9th ed.). Philadelphia: Lippincott Williams & Wilkins.

Pettifor, J.M., and H. Juppner. 2003. *Pediatric Bone: Biology and Diseases*. London: Academic Press.

HISTOLOGY

Cormack, D.H. 1987. *Ham's Histology* (9th ed.). Philadelphia: Lippincott.

Fawcett, D.W., and E. Raviola. 1994. *Bloom and Fawcett: A Textbook of Histology* (12th ed.). New York: Chapman and Hall.

Fawcett, D.W., and R.P. Jensh. 2002. *Bloom and Fawcett: Concise Histology* (2nd ed.). London: Hoddar Arnold.

Garner, L.P., J.L. Hiatt, J.M. Strum, T.A. Swanson, S.I. Kim, and A.S. Schneider. 2002. *Cell Biology and Histology* (4th ed.). Philadelphia: Lippincott Williams & Wilkins.

BONE

Currey, J.D. 2002. *Bones: Structure and Mechanics*. Princeton, NJ: Princeton University Press.

Hall, B.K. ed. 2005. *Bones and Cartilage: Developmental and Evolutionary Skeletal Biology*. London: Elsevier/Academic Press.

Bostrom, M.P.G., A. Boskey, J.J. Kaufman, and T.A. Einhorn. 2000. Form and function of bone. In *Orthopaedic Basic Science* (2nd ed.), edited by S.R. Simon. Park Ridge, IL: American Academy of Orthopaedic Surgeons.

Martin, R.B., D.B. Burr, and N.A. Sharkey. 1998. *Skeletal Tissue Mechanics*. New York: Springer.

CARTILAGE

Mankin, H.J., V.C. Mow, J.A. Buckwalter, J.P. Iannotti, and A. Ratcliffe. 2000. Articular cartilage structure, composition and function. In *Orthopaedic Basic Science* (2nd ed.), edited by S.R. Simon. Park Ridge, IL: American Academy of Orthopaedic Surgeons.

Mow, V.C., A. Ratcliffe, and A.R. Poole. 1992. Cartilage and diarthrodial joints as paradigms for hierarchical materials and structures. *Biomaterials* 13: 67-97.

TENDON AND LIGAMENT

Woo, S.L.-Y., K.-N. An, C.B. Frank, G.A. Livesay, C.B. Ma, J. Zeminski, J.S. Wayne, and B.S. Myers. 2000. Anatomy, biology, and biomechanics of tendon and ligament. In *Orthopaedic Basic Science* (2nd ed.), edited by S.R. Simon. Park Ridge, IL: American Academy of Orthopaedic Surgeons.

SKELETAL MUSCLE

Garrett, W.E., Jr., and T.M. Best. 2000. Anatomy, physiology, and mechanics of skeletal muscle. In *Orthopaedic Basic Science* (2nd ed.), edited by S.R. Simon. Park Ridge, IL: American Academy of Orthopaedic Surgeons.

Jones, D.A., J. Round, and A. de Haan. *Skeletal Muscle, from Molecules to Movement*. Edinburgh, UK: Churchill Livingstone.

Lieber, R.L. 2002. *Skeletal Muscle Structure, Function, & Plasticity* (2nd ed.). Philadelphia: Lippincott Williams & Wilkins.

MacIntosh, B.R., P. Gardiner, and A.J. McComas. 2006. *Skeletal Muscle: Form and Function* (2nd ed.). Champaign, IL: Human Kinetics.

GENERAL

Martin, R.B., D.B. Burr, and N.A. Sharkey. 1998. *Skeletal Tissue Mechanics*. New York: Springer-Verlag.

Simon, S.R., ed. 2000. *Orthopaedic Basic Science* (2nd ed.). Park Ridge, IL: American Academy of Orthopaedic Surgeons.

CLASSIC REFERENCE

Thompson, D.W. 1992. *On Growth and Form*, edited by J.T. Bonner (abridged ed.). Cambridge: Cambridge University Press. (Originally published 1917)

BASIC BIOMECHANICS

Mechanics is the paradise of the
mathematical sciences because by means of
it one comes to the fruits of mathematics.

Leonardo da Vinci (1452-1519)

■ OBJECTIVES

- To identify the major areas of biomechanics relevant to human movement: movement mechanics, fluid mechanics, joint mechanics, and material mechanics
- To explain biomechanics concepts and measures, including linear and angular motion, center of gravity, stability, mobility, and movement equilibrium
- To explain concepts of movement mechanics, including kinematics, kinetics, force, pressure, lever systems, torque (moment of force), Newton's laws of motion, work, power, energy, momentum, and friction
- To explain concepts of fluid mechanics, including fluid flow and resistance
- To explain concepts of joint mechanics, including range of motion, joint stability, joint mobility, lever systems, and joint reaction force
- To explain concepts of material mechanics, including stress, strain, stiffness, bending, torsion, viscoelasticity, and material fatigue and failure
- To describe various model types and model selection criteria
- To describe rheological, finite element, and musculoskeletal models

Movement is essential to life. Not only life processes such as blood circulation, respiration, and muscle contraction require motion, but activities such as walking, bending, and grasping inherently involve movement. Consider how the human organism seeks, consciously or not, to move. When you sit in a chair, for example, do you remain motionless? Hardly. You cross and uncross your legs, slouch, squirm, and slide to create some degree of movement. Children provide perhaps the best evidence of the inherent nature of humans to move. They never seem to stop. Even as we age and slow down, movement remains a quintessential element of our being.

In times past, movement meant survival. Those not able to move—or to move rapidly enough—often met with injury or death. Although we no longer have to escape from predators (except on rare occasions), our ability to move can serve us well in avoiding problems that confront us (e.g., dodging an oncoming vehicle). Limited movement, such as when a person is bedridden or elects a sedentary lifestyle, can contribute to deleterious health effects such as cardiovascular disease, diabetes, and cancer. Thus, our inability to move or the choice to limit movement may contribute, either directly or indirectly, to our susceptibility to musculoskeletal injury.

In mechanical terms there are two basic forms of movement: (1) **translational** or **linear motion,** in which a body moves along either a straight line **(rectilinear motion)** or a curved line **(curvilinear motion),** and (2) **angular** or **rotational motion,** in which the body rotates about an **axis of rotation** (figure 3.1). Although there are theoretically an infinite number of axes about which a body can rotate, only a few are of practical interest in discussing movement of human body segments (figure 3.2).

Many of the movements performed by living organisms are a combination of both *linear* and *angular motion*. Simultaneous linear and angular motion is termed **general motion.** Consider, for example, the movement of a person's thigh during walking: linear motion of the entire thigh in a forward direction combined with angular motion of the thigh as it rotates about the hip joint axis in alternating phases of flexion and extension.

The movement of an inanimate object can also exhibit combined motion. The flight of a basketball shot toward the rim demonstrates both linear motion (the curved path, or arc, of the ball) and angular motion (the backspin of the ball). As we explore the mechanical concepts in this chapter, the notions of linear and angular motion recur often. Combinations of these two simple movement forms result in the wide and nearly endless variety of human movement patterns.

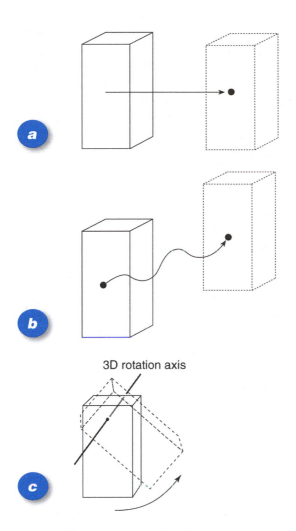

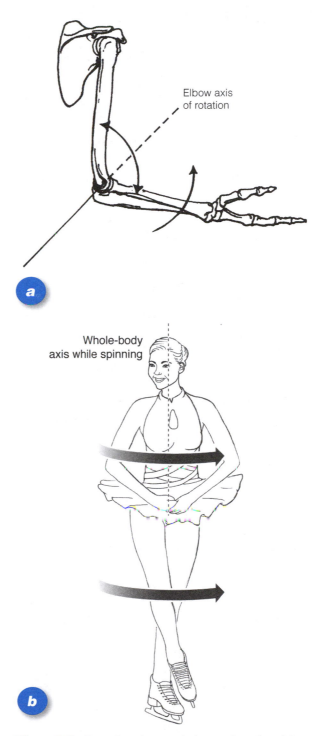

Figure 3.1 Linear and angular motion. *(a)* Rectilinear (straight line) motion. *(b)* Curvilinear motion. *(c)* Angular (rotational) motion.

Human movement can be viewed from several perspectives. One considers whether mechanical factors that produce and control movement work inside the body **(internal mechanics)** or affect the body from without **(external mechanics).** Examples of internal mechanical factors include the forces produced by muscle action and the stability provided by ligaments surrounding joints. External mechanical factors include gravity and other external forces, such as a foot striking the ground or a falling brick hitting the top of one's head.

Another important perspective on movement involves the difference between describing a movement versus identifying the forces involved in producing or controlling the movement. The description of movement without regard to the

Figure 3.2 Examples of anatomical axes of rotation. *(a)* Elbow flexion and extension about the elbow s axis of rotation. *(b)* Ice skater s whole body rotating about a longitudinal (vertical) axis.

forces involved is known as **kinematics.** The assessment of movement with respect to the forces involved is called **kinetics.**

Kinematics

Kinematics involves five primary variables:

1. Temporal (timing) characteristics of movement
2. Position or location
3. Displacement (describing what movement has occurred)
4. Velocity (a measure of how fast something has moved)
5. Acceleration (an indicator of how quickly the velocity has changed)

The last four variables (position, displacement, velocity, and acceleration) can be expressed in linear or angular form, giving rise to the general descriptors of linear kinematics and angular kinematics. Keep in mind that displacement, velocity, and acceleration are all **vector** measurements, which have both magnitude and direction.

We can further assess kinematics according to whether the motion is viewed two-dimensionally *(planar kinematics)* or three-dimensionally *(spatial kinematics)*. The essential terminology and formulations for planar kinematics are described in the following sections and summarized in figure 3.3.

▪ **Time:** The first variable, *time,* provides a measure of the duration of a particular event. Noting that during a single step a person's right foot is in contact with the ground for 450 m/s would be an example of a temporal kinematic measure. The duration *(Δt)* of force application associated with acute musculoskeletal injuries is typically quite short and may last only a fraction of a second. This short time interval necessarily results in high loading rates. As we see later, loading rate is an important factor in determining a tissue's mechanical response to applied forces.

▪ **Position:** The position of a person's whole body, or a segment of the body, plays a critical role in determining the likelihood of injury. Forces applied to an arm that is hyperextended and externally rotated will cause a different injury pattern than the same forces applied to an arm that is flexed and internally rotated at the moment of force application. Similarly, a force applied to the top of one's head when the neck is flexed will

result in different injuries than if the same force is applied to the head while the neck is hyperextended. The position of a body segment can be described qualitatively (e.g., arm is abducted) or quantitatively (e.g., forearm is positioned with the elbow flexed 45°). The position of a specific point, or landmark, on the body can be specified quantitatively using, for example, either Cartesian *(x,y)* coordinates or polar *(r,θ)* coordinates (see figure 3.3).

▪ **Displacement:** When a body moves from one location to another we measure the **linear displacement** *(Δd)* in a straight line from the starting position (A) to the ending position (B), regardless of the path taken. The **distance** measures how far the body has moved along any given path in getting from A to B. A body rotating about an axis experiences **angular displacement** *(Δθ)*, which is measured as the number of degrees (or radians) of rotation (e.g., the knee flexed through an angular displacement of 35°). A direct relation exists between the linear and angular measures of distance and displacement, as shown in figure 3.4*a*.

▪ **Velocity:** Velocity is a measure of the time rate of displacement. The average **linear velocity** *(v)* is given by the quotient of linear displacement *(Δd)* divided by *Δt*. **Angular velocity** *(ω)* is calculated by dividing the angular displacement *(Δθ)* by the change in time *(Δt)*. A direct relation exists between the linear and angular measures of velocity (figure 3.4*b*). In common usage, the terms *velocity* and *speed* often are used interchangeably. In mechanical terms, however, they have distinct—although related—meanings. Velocity is a vector quantity (magnitude and direction), whereas **speed** is a **scalar** (magnitude only) measure. The speed of a runner might be 5 m/s. To transform the movement measure to velocity, we must indicate the running direction, for example, 5 m/s due north.

▪ **Acceleration:** Acceleration measures the time rate of change in a body's velocity. **Linear acceleration** *(a)* is measured as the change in linear velocity *(Δv)* divided by the change in time *(Δt)*. Similarly, **angular acceleration** *(α)* is the change in angular velocity *(Δω)* divided by change in time *(Δt)*. As was the case with linear and angular velocity, a direct relation exists

	Linear		Angular	
	Symbol	Formula or relation	Symbol	Formula or relation
Time	t	$t_2 - t_1 = \Delta t$	t	$t_2 - t_1 = \Delta t$
Position	(x, y)	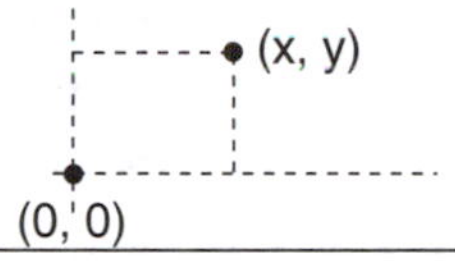 (x, y) at $(0, 0)$	(r, θ)	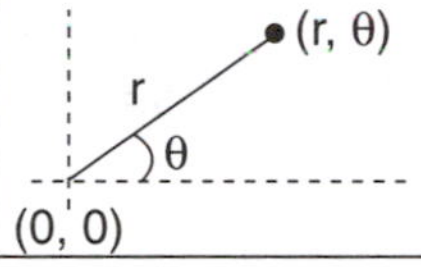 (r, θ) at $(0, 0)$
Displacement (x-direction only)	d	$d = x_2 - x_1$ $x_1 \bullet \xrightarrow{d} \bullet x_2$	θ	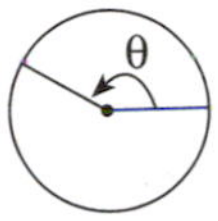
Average velocity	$\overline{v}$	$\overline{v} = \dfrac{d_2 - d_1}{t_2 - t_1} = \dfrac{\Delta d}{\Delta t}$	$\overline{\omega}$	$\overline{\omega} = \dfrac{\theta_2 - \theta_1}{t_2 - t_1} = \dfrac{\Delta \theta}{\Delta t}$
Instantaneous velocity	v	$v = \dfrac{dx}{dt} = \dot{x}$	ω	$\omega = \dfrac{d\theta}{dt} = \dot{\theta}$
Average acceleration	$\overline{a}$	$\overline{a} = \dfrac{v_2 - v_1}{t_2 - t_1} = \dfrac{\Delta v}{\Delta t}$	$\overline{\alpha}$	$\overline{\alpha} = \dfrac{\omega_2 - \omega_1}{t_2 - t_1} = \dfrac{\Delta \omega}{\Delta t}$
Instantaneous acceleration	a	$a = \dfrac{dv}{dt} = \ddot{x}$	α	$\alpha = \dfrac{d\omega}{dt} = \ddot{\theta}$

Figure 3.3 Terminology and formulas for planar kinematics. Standard units: s (time), m (linear displacement), rad or deg (angular displacement), m/s (linear velocity), rad/s or deg/s (angular velocity), m\s² (linear acceleration), rad/s² or deg/s² (angular acceleration).

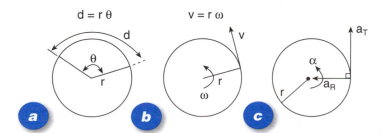

Figure 3.4 Relation between linear and angular measures. (a) Linear distance (d) moved along the circumference of a circle (radius = r) equals $r \cdot \theta$. (b) Linear velocity (v) of a point on the circumference of a circle equals $r \cdot \omega$. (c) Linear tangential acceleration (a_T) of a point on the circumference of a circle equals $r \cdot \alpha$. Radial acceleration (a_R) equals v^2/r. (Note: Angular measurements of τ, ω, and α in these equations must be expressed in units of radians, rad/s, and rad/s², respectively).

between the linear and angular measures of acceleration (figure 3.4c). Many musculoskeletal injuries are acceleration related. Rapid acceleration or deceleration of the head, for example, can result in concussive injury to the brain. Linear acceleration often is expressed in units of g, where 1 g is the acceleration created by the earth's gravitational pull (~9.81 m/s^2). Thus, a boxer's head hit with a force of 5 g would be accelerated at five times the acceleration caused by gravity.

Kinetics

Description is an important first step in analyzing any movement. Kinematic analyses, however, are limited to describing the spatial geometry of movement without investigating the forces involved. Because force is a causal agent in movement, *kinetics* (the study of forces and their effects) is an area worthy of our consideration. Keep in mind that the following force-related concepts are interrelated, and to consider each of them in isolation limits their applicability and our ability to analyze injury biomechanics.

Linear Kinetics

If the applied forces are large enough to overcome a body's resistance to movement, the body moves linearly. *Linear kinetics* examines the relation between a body's resistance to a change in its linear state of motion and the effect of applied forces.

Mass, Inertia, and Force

The quantity of matter in a body is its **mass.** Mass is measured in SI units (Système international d'unités); that is, in kilograms (kg). Common sense suggests that the greater an object's mass, the more difficult it is to move. **Inertia** is the resistance to being moved linearly and is the property of matter by which it remains at rest or in uniform motion in a straight line (i.e., constant velocity). To move an object that is at rest, we must overcome its inertia, or its tendency to remain stationary.

An analogous resistance concept applies to angular movement. Resistance to a change in a body's state of angular motion is termed *moment of inertia*. Discussion of moment of inertia requires an understanding of several other concepts and so is discussed later.

Force is the most fundamental element in injury. **Force** is defined as the mechanical action or effect applied to a body that tends to produce acceleration. The standard SI unit of force is the **newton (N),** defined as the force required to accelerate a 1 kg mass at 1 meter per second in the direction of the force (1 N = 1 kg · m · s^{-2}). In the British system of measurement, the unit of force is the pound (lb). One pound equals 4.45 N.

In preparation for a more general discussion of force, we introduce the concept of an **idealized force vector.** Consider, for example, the forces acting on the head of the femur during the process of standing up: An infinite number of force vectors could be distributed over the articular surface. We can, however, create a single force vector (idealized force vector) that represents the net effect of all the other vectors, essentially idealizing the situation through simplification. What is lost in information describing the distribution of forces is gained by creating a model with a single vector from which calculations and evaluations can be made. This notion of an idealized force vector is useful in many situations, as noted shortly.

Forces inherent to injury analysis are those that act in or upon the human body. Among these are gravity (which accelerates objects at ~9.81 m/s^2); the impact of the feet, hands, or body on the ground; the impact of objects colliding with the body (e.g., thrown ball or bullet); musculotendinous forces; ligament forces acting at joints; and compressive forces exerted on long bones of the lower extremities.

In injury-causing situations, seven factors combine to determine the nature of the injury, the tissues injured, and the severity of the injury:

1. Magnitude (How much force is applied?)
2. Location (Where on the body or structure is the force applied?)
3. Direction (Where is the force directed?)
4. Duration (Over what time interval is the force applied?)
5. Frequency (How often is the force applied?)
6. Variability (Is the magnitude of the force constant or variable over the application interval?)
7. Rate (How quickly is the force applied?)

In the human body, rarely does a single force act in isolation. Much more common are cases involving multiple forces. To aid in analysis, it is useful to categorize multiple forces as force systems. Types of force systems include **linear, parallel, concurrent,** and **general force systems** (figure 3.5, *a-d*). A special case of force application is a **force couple,** which is composed of two oppositely directed parallel forces that tend to create rotation about an axis (figure 3.*5e*).

The engineering approach uses a free-body diagram for biomechanical analysis of a force system. A **free-body diagram (FBD)** is simply a graphic representation of all the forces acting in the system. Figure 3.6 depicts an FBD for a simple biomechanical application. Note that the effect of gravity is represented as a single vector, another example of an idealized force vector. In actuality, gravity acts on each small element of body mass.

Center of Mass and Center of Gravity

When an idealized force vector is developed, many vectors are reduced to a single vector. A similar process can be applied to the mass of a body by reducing its distributed mass to a single point **(point mass)** that represents the entire body. Again, this type of simplification will facilitate analysis but with loss of information. The word

body is taken to mean any collection of matter and may refer to the entire human body or any collected mass (e.g., a body segment such as the thigh or upper arm or a block of wood).

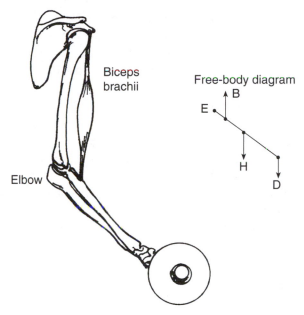

Figure 3.6 The free-body diagram (FBD) *(right)* represents the forces acting on the upper extremity while a subject is holding a dumbbell in the hand. Gravity creates force (weight) vectors for the dumbbell *(D)* and forearm and hand *(H)*. The biceps brachii creates a muscle force represented by the vector *(B)*. All force vectors tend to cause rotation about the elbow joint axis *(E)*.

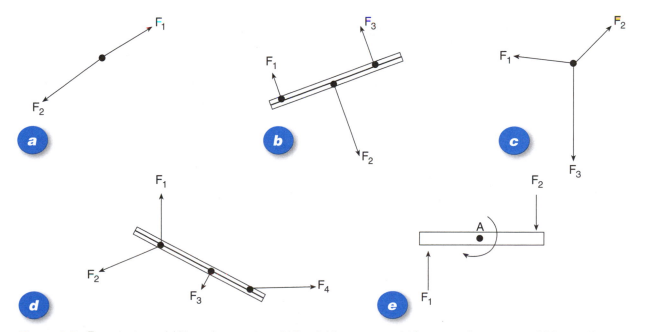

Figure 3.5 Force systems. *(a)* Linear force system. *(b)* Parallel force system. *(c)* Concurrent force system. *(d)* General force system, the designation given to a force system that does not fall under one of the classifications *a* through *c*. *(e)* Force couple; parallel and oppositely directed forces F_1 and F_2 cause rotation about axis *A*.

For any body, there exists a point, known as the **center of mass** or **center of gravity,** at which, if we concentrated the body's mass into a point mass, this point mass would move exactly the same as the body would in its distributed state. Even though there is a technical distinction between the center of gravity and the center of mass of a body, in practical terms they are located at the same point, and therefore we use the terms interchangeably.

The center of mass alternatively may be defined as the point about which a body's mass is equally distributed. The center of gravity also acts as a balance point, such as when a food server places his hand at the center of gravity of a tray full of dishes and balances the tray over his head.

The human body's center of mass typically is located within the body's boundaries (figure 3.7a), but this may not always be the case (figure 3.7b).

Pressure

Because many injuries occur as a result of one object impacting another, it is important to know how the force of impact is distributed across the surface being contacted. A sharp object contacting the skin with 300 N of force will likely have a different effect than a blunt object impacting the skin with a similar force. A principle of injury mechanics suggests that as the area of force application increases, the likelihood of injury decreases.

The measure of force and its distribution is **pressure,** defined as the total applied force divided by the area over which the force is applied. In equation form,

$$p = F/A \tag{3.1}$$

where p = pressure, F = applied force, and A = area of contact. The standard unit of pressure, the **pascal (Pa),** is equal to a 1 N force applied to an

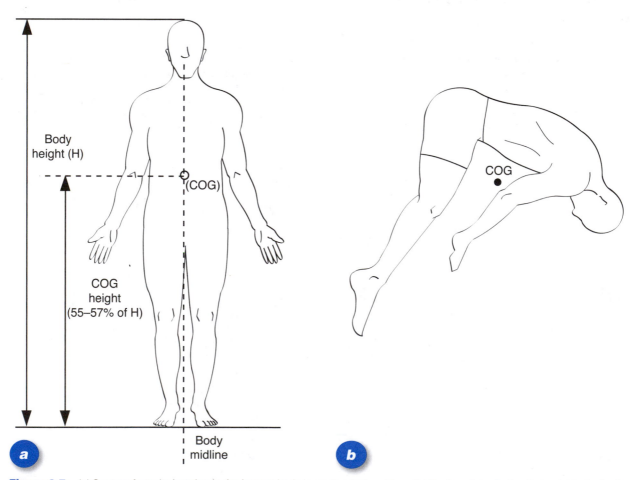

Figure 3.7 *(a)* Center of gravity location in the human body in anatomical position. *(b)* Center of gravity located outside the body when the body is in a bent-over, or pike, position, as during a gymnastic maneuver or dive.

Reprinted, by permission, from W.C. Whiting and S. Rugg, 2006, *Dynatomy: Dynamic human anatomy* (Champaign, IL: Human Kinetics), 10.

area 1 meter square ($1 Pa = 1 N/m^2$). In injury situations, the pressures exerted on body structures can be quite high and are often expressed with a unit of **mega-pascal (MPa),** which is equal to a 1 N force applied to an area 1 millimeter square ($1 MPa = 1 N/mm^2$).

Angular Kinetics

Angular kinetics examines the relation between a body's resistance to a change in its angular state of motion and the effect of applied torques. Its major concepts include moment of force and moment of inertia.

Moment of Force (Torque)

In the case of linear motion, force is the mechanical agent creating and controlling movement. For angular motion the agent is known as a *moment of force*, or **moment** *(M)*, or **torque** *(T)*, and is generally defined as the effect of a force that tends to cause a change in a body's state of angular position or motion (figure 3.8). More specifically, *torque* typically refers to the twisting action (torsion) created by a force, as seen in turning a screwdriver, or torsional loading of the lower leg (tibia) in a skiing fall. *Moment* relates to the rotational (e.g., knee extension) or bending action (e.g., pole vault) of a force. Despite this technical distinction, the two terms often are used interchangeably.

The mathematical definitions of moment and torque are the same. The magnitude of a moment or torque is equal to the applied force times the shortest (perpendicular) distance from the axis of rotation to the line of force action. This perpendicular distance is known as the **moment arm, torque arm,** or **lever arm.** The standard unit of moment (torque) measurement comes from the product of the two terms: *force* (N) times *moment arm* (m). The resulting unit is a newton-meter (N·m).

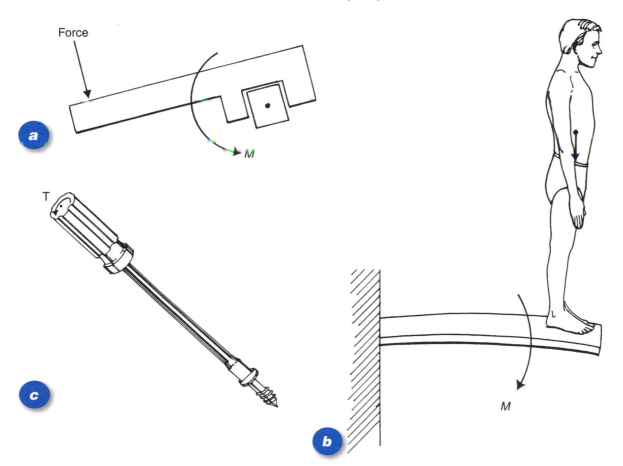

Figure 3.8 Applied examples of moment of force *(M)*, or torque *(T)*. *(a)* Force applied to a wrench creates a moment to turn a nut on a bolt. *(b)* The body weight of a diver creates a moment that bends the diving board. *(c)* Torque shown by the twisting acting of a screwdriver.

For a force acting at a right angle to the body being rotated, the moment arm is the distance d (figure 3.9a), and the magnitude of the moment (M) is given by this equation:

$$M = F \cdot d \qquad (3.2)$$

If the force F in figure 3.9a was 175 N, for example, and it was acting at a distance d of 1.2 m from the axis, the moment (torque) created would equal $F \cdot d$, or $M = 175\ N \cdot 1.2\ m = 210\ N \cdot m$.

In cases where the force is not acting perpendicularly to the segment, the moment arm is smaller and is calculated using the appropriate trigonometric function as shown in figure 3.9b. The magnitude of the moment (M) in this case is

$$M = F \cdot d \cdot \sin(\beta) \text{ or } M = F \cdot d \cdot \cos(\theta) \qquad (3.3)$$

If the same 175 N force as in the preceding example was acting at an angle $\beta = 35°$ (as shown in figure 3.9b), the moment arm would be $d' = d \cdot \sin(\beta) = 1.2\ m \cdot 0.574 = 0.688\ m$. The moment created now is $M = F \cdot d' = 175\ N \cdot 0.688\ m = 120.4\ N \cdot m$.

Closer examination of the moment equation 3.2 reveals several principles that are important when applying torque concepts to injury biomechanics. First, there is an obvious interaction between the force and the moment arm that directly affects the magnitude of the applied torque. To increase the moment, we have the following options:

- Increasing the force while holding the moment arm constant
- Increasing the moment arm while holding the force constant

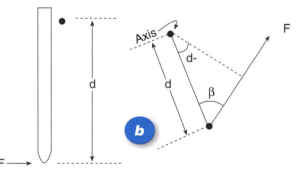

Figure 3.9 Moment (torque) arm. *(a)* When the force acts perpendicularly to the segment, the moment arm is the distance *d*. The moment is given by equation 3.2. *(b)* When the force acts at an angle (β) to the segment of length *d*, the moment arm *d'* = *d* · sin(β). The moment in this case is given by equation 3.3.

- Increasing both the force and the moment arm
- Decreasing the force while increasing the moment arm more than proportionally so that the net effect is an increase in moment
- Decreasing the moment arm while increasing the force more than proportionally so that the net effect is again an increase in moment

To decrease the moment, we have only to reverse the logic in each of these five cases.

A second moment-related concept, although simple in statement, is powerful in its application. That is, when a force is applied through the axis of rotation, no moment is produced. This concept follows directly from the moment equation $M = F \cdot d$, where d = moment arm. If the force passes through the axis, the moment arm is zero, and hence no moment is produced. This creates the potential for a situation in which body tissues are exposed to extremely high forces but with no moment created. Compressive forces acting through the center of a vertebral body, for example, will cause no vertebral rotation but will increase the likelihood of a compressive fracture.

A third moment concept arises from the fact that in many instances, only a portion of the applied force is involved in producing a moment, as can be seen in the two examples in figure 3.10. In the first situation (figure 3.10a), the weight attached to the foot (F_w) can be broken into two force components: F_r, which causes rotation about the knee joint axis and is termed the **rotatory component** of force, and F_d, whose line of action passes through the joint axis and contributes nothing to the moment about the knee. F_d acts to pull the segment away from the joint axis and is thus referred to as a *distracting, dislocating,* or **destabilizing component** of force. Similarly, in figure 3.10b, the biceps force (F_b) has a rotatory component (F_r). In contrast to the previous example, however, the component (F_s) passing through the axis is directed toward the axis and is called a **stabilizing component** of force.

A fourth moment concept arises from many real situations in which more than one moment is applied to the system. The system's response is based on the **net moment** (also called **net torque**) or the result of adding together all the moments

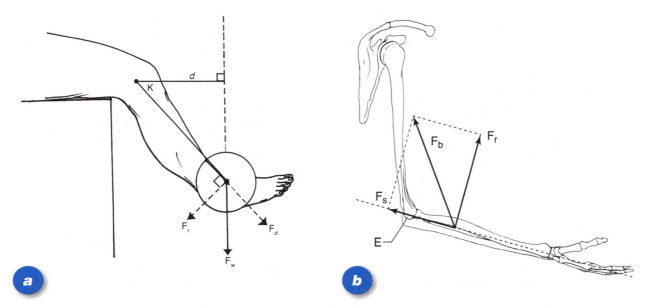

Figure 3.10 Components of force. (a) Rotational (F_r) and destabilizing (F_d) force components created by a weight (F_w) secured to the foot of a person performing a leg extension exercise about the knee joint axis (K). (b) Rotational (F_r) and stabilizing (F_s) force components created by the biceps brachii (F_b) during elbow flexion about the elbow axis (E).

acting about the axis. A simple glenohumeral abduction exercise provides an example (figure 3.11). Gravity, acting on the arm and the dumbbell, creates a moment about the glenohumeral axis of rotation that tends to adduct the arm. The magnitude of this moment (M_1) is given by

$$M_1 = W_a \cdot d_a + W_b \cdot d_b \qquad (3.4)$$

where W_a = weight of the arm and hand, d_a = moment arm (distance from the glenohumeral axis of rotation to the arm's center of gravity), W_b = weight of the dumbbell, and d_b = moment arm (distance from axis to dumbbell's center of gravity). By convention, moments tending to create clockwise rotation are designated as negative (–) moments. Moments in the counterclockwise direction are positive (+). M_1 is, therefore, a negative moment.

If this were the only moment, the arm would immediately adduct under the effect of gravity. However, the abductor muscles acting about the glenohumeral joint create a moment that acts in the opposite direction and is termed a **counter-moment** (M_2) or **countertorque.** The counter-moment in this example is a positive moment and will tend to abduct the arm. The movement that results depends on the relative magnitudes of M_1

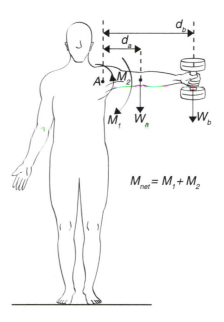

Figure 3.11 Net joint moment. The net moment (M_{net}) of the combined moments created by $M_1 + M_2$. The weight of the arm (W_a) and dumbbell (W_b) combine to create an adductor moment (M_1) acting about the glenohumeral joint axis A. The force of the glenohumeral abductor muscles creates an abductor moment (M_2) to counteract M_1. (M_1 acts in a clockwise direction and thus creates a negative moment. Conversely, M_2 acts counterclockwise and is positive).

Reprinted, by permission, from W.C. Whiting and S. Rugg, 2006, *Dynatomy: Dynamic human anatomy* (Champaign, IL: Human Kinetics), 109.

and M_2. By adding the two moments together, we create a net moment (M_{net}):

$$M_{net} = M_1 + M_2 \qquad (3.5)$$

In this example we have three possible scenarios: (1) If M_1 is equal to M_2, then $M_{net} = 0$ and the arm remains in its horizontal position; (2) if $M_1 > M_2$, then $M_{net} < 0$, and the arm will adduct; (3) if $M_1 < M_2$, then $M_{net} > 0$ and the arm will abduct. The resulting movement thus depends on the net moment acting at the joint about which the movement occurs.

Moment of Inertia

Just as bodies resist change in their state of linear motion, they also tend to resist forms of angular movement. The term used to describe this resistance to change in angular state of motion or position is **moment of inertia.** There are three types of moment of inertia corresponding to the three forms of angular movement: rotation, bending, and torsion.

A body at rest and with a fixed axis (e.g., a pendulum) will resist being moved rotationally, just as a body that is already rotating at a constant angular velocity (ω) will tend to maintain that angular velocity and will resist a change in its velocity. The measure of this resistance

to change in a body's state of angular motion (rotation) is termed **mass moment of inertia** (*I*). Recall that the magnitude of resistance in the case of linear movement was determined by the mass of the object. In the case of angular movement, the magnitude of the resistance is determined by the mass and the mass's distribution with respect to the specified axis of rotation. For a point mass, the mass moment of inertia is defined as

$$I = m \cdot r^2 \qquad (3.6)$$

where m = mass of the body and r = distance from the axis of rotation to the point mass (figure 3.12). For a distributed mass, such as a limb segment, the mass moment of inertia is

$$I = \int m_i \cdot r_i^2 \qquad (3.7)$$

where m_i = mass of the ith point mass and r_i = distance of the ith point mass from the axis of rotation. As the mass is moved farther from the axis, the resistance, or mass moment of inertia, increases as a function of the square of the distance moved.

The other two types of moment of inertia (*area moment of inertia* and *polar moment of inertia*) are explained in later sections.

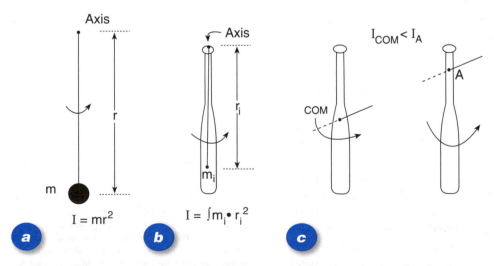

Figure 3.12 Moment of inertia. *(a)* Moment of inertia *(I)* for a point mass equals the product of the mass *(m)* and the square of the distance *(r)* from the axis to the mass (equation 3.6). *(b)* A baseball bat illustrates a distributed mass with an axis shown in the handle end of the bat. The moment of inertia for a distributed mass is given by equation 3.7. *(c)* The moment of inertia *(I_{com})* of any body (e.g., a baseball bat as shown) about an axis through its center of mass is less than the moment of inertia *(I_A)* for the same body about an axis at point A because more of the mass is farther away from the axis in the second case.

HIP JOINT MOMENTS

Hip joint injuries are, unfortunately, a common injury, especially in elderly people. Structural features of the hip combined with osteoporosis (low-density bone) provide a potent tandem that increases the risk of hip fracture. Among the structural features is the relation of the femoral neck and head to the long axis of the femur. Normally, the angle between the long axis and the neck is about 120°. Abnormal angles can alter the mechanical loading of the proximal femur. In **coxa vara,** the angle is less than normal. The contrasting condition, **coxa valga,** has an angle greater than normal. When the bone is loaded, as during the stance phase of gait, compressive forces act on the femoral head. Because these forces are offset from the long axis of the femoral diaphysis, **cantilever bending** occurs. This produces a moment about an axis denoted as *A* in the figure. In coxa vara, the moment arm *(d$_v$)* is longer than in the normal condition *(d$_n$)*. For a given force *(F)*, the coxa vara moment *(M$_v$)* is greater than normal *(M$_n$)*. In contrast, the deviation in coxa valga results in a shorter moment arm *(d$_l$ < d$_n$)* and, hence, a smaller moment *(M$_v$ > M$_n$ > M$_l$)*. Larger moments create greater stresses in the bony tissue. If there is an area of relative weakness, as in an osteoporotic patient, the likelihood of bone fracture increases. With this potential mechanism of injury, an older woman, for example, on occasion may break her hip and fall, instead of the more common case in which she falls and breaks her hip.

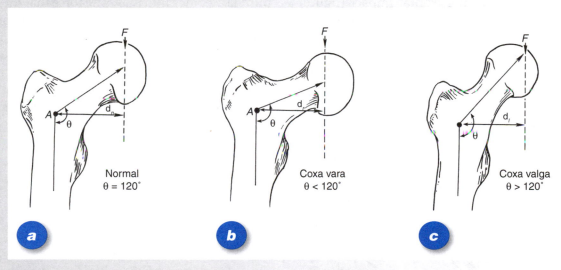

Structural geometry of the femur: (*a*) normal, (*b*) coxa vara, and (*c*) coxa valga.

Kinetic Concepts and Measures

Several kinetic concepts are important in our discussion of injury biomechanics. These include Newton's laws of motion, equilibrium, work, power, energy, momentum, collisions, and friction. These concepts are discussed in the following sections.

Newton's Laws of Motion

Among Sir Isaac Newton's (1642-1727) many scientific contributions, perhaps most profound and enduring are his laws of motion, which form the basis for classical (Newtonian) mechanics:

- **First law of motion:** A body at rest or in uniform linear motion (moving in a straight line at a constant velocity) will tend to remain at rest or in uniform motion, unless acted upon by an external force. A body at rest or in uniform angular motion (moving about an axis at a constant angular velocity) will tend to remain at rest or in uniform motion, unless acted upon by an external torque.

- **Second law of motion:** A force *(F)* acting on a body with mass *(m)* will produce an acceleration *(a)* proportional to the force, or mathematically $F = m \cdot a$. Angularly, a torque *(T)* applied to a body will produce an angular acceleration (α) proportional to the torque, or mathematically $T = I \cdot \alpha$ (where *I* = mass moment of inertia).

- **Third law of motion:** For every action there is an equal and opposite reaction.

Several examples show how these laws play a role in determining the mechanisms of performance and injury (figure 3.13). A cervical whiplash mechanism during frontal impact (figure 3.13*a)* is simply a consequence of the first law of motion. Just before impact, the automobile and its seat-belted and shoulder-harnessed driver are moving at a constant velocity. At impact, the outside force abruptly decelerates the vehicle and the belted occupant's body. However, for a brief interval the head obeys the first law of motion and continues in its uniform motion (straight ahead). Resistance forces provided by neck structures rapidly decelerate the head, causing a violent flexion motion at the cervical spine. The head then rebounds into hyperextension. This flexion–extension pattern is typical of many whiplash-related injuries and is explained by Newton's first law.

In the second example (figure 3.13*b*), the weightlifter must exert considerable force to accelerate the bar upward. Newton's second law of motion determines the magnitude of the acceleration in response to the applied force, $F = m \cdot a$. More detailed application of the laws of motion allows us to estimate the forces acting at various joints throughout the body. If these forces exceed the ability of the body's structures to tolerate load, injury occurs.

In the third example (figure 3.13*c*), the marathon runner's feet contact the ground many thousands of times. At each contact, Newton's third law comes into play. The force that the foot exerts on the ground is equally and oppositely resisted by the ground, giving rise to the term **ground reaction force (GRF)** to describe the forces acting on the foot by the ground. Increasing magnitude and frequency of force application increase the chances of injury, perhaps as a stress fracture.

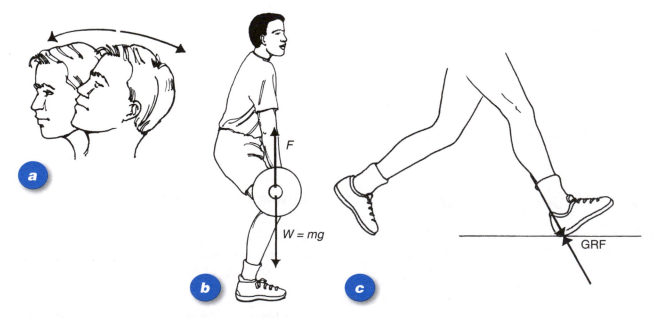

Figure 3.13 Newton's laws of motion. *(a)* First law of motion applied to a whiplash mechanism. *(b)* Second law of motion involved during a weightlifting movement. *F* = force applied by the weight lifter; *W* = weight; *m* = mass; *g* = acceleration attributable to gravity. *(c)* Third law of motion exemplified by the ground reaction force created when a runner's foot strikes the ground.

Equilibrium

The word *equilibrium* implies a balanced condition. From a mechanical standpoint, **equilibrium** exists when forces and moments are each balanced. Equilibrium exists for a body at rest or for one moving with constant linear and angular velocities; the net force (ΣF, where the Greek letter sigma means "sum of") and net moment (ΣM) acting on the body are both equal to zero. A body at rest is in a state called **static equilibrium.** In spatial terms (three-dimensional space), for a body in static equilibrium the following equations must be satisfied:

$$\Sigma F_x = 0 \quad \Sigma F_y = 0 \quad \Sigma F_z = 0 \tag{3.8}$$

$$\Sigma M_x = 0 \quad \Sigma M_y = 0 \quad \Sigma M_z = 0 \tag{3.9}$$

where F_x, F_y, and F_z are the forces in the x, y, and z directions, respectively, and M_x, M_y, and M_z are the moments about the x, y, and z axes, respectively.

Bodies in motion and experiencing external forces and moments are in **dynamic equilibrium** and must adhere to the following equations:

$$\Sigma F_x = m \cdot a_x \quad \Sigma F_y = m \cdot a_y \quad \Sigma F_z = m \cdot a_z \tag{3.10}$$

$$\Sigma M_x = I_x \cdot \alpha_x \quad \Sigma M_y = I_y \cdot \alpha_y \quad \Sigma M_z = I_z \cdot \alpha_z \tag{3.11}$$

where a_x, a_y, and a_z are the linear accelerations of the center of mass in the x, y, and z directions, respectively; α_x, α_y, and α_z are the angular accelerations about the x, y, and z axes, respectively; and I_x, I_y, and I_z are the mass moments of inertia about the x, y, and z axes, respectively.

Work and Power

The term **work** is used in many ways, including reference to physical labor ("I'm working hard"), physiological energy expenditure ("I worked off 100 calories"), or an occupation ("I went to work"). In mechanical terms, work has a specific meaning. **Mechanical work** is performed by a force acting through a displacement in the direction of the force. By definition, **linear work** (W) is a scalar measure equal to the product of force (F) and the displacement (d) through which the body is moved (figure 3.14a):

$$W = F \cdot d \tag{3.12}$$

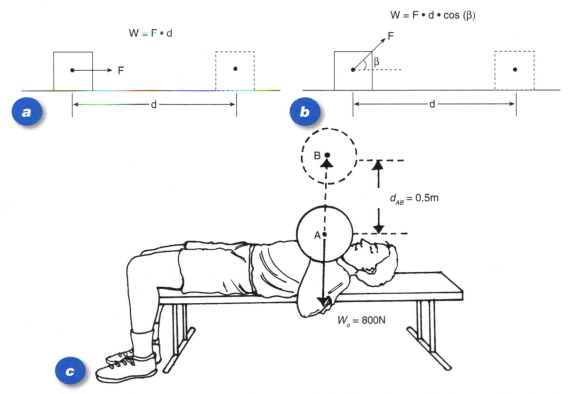

Figure 3.14 Mechanical work. *(a)* Linear work *(W)* as the product of force *(F)* and distance *(d)* in the case where the entire force acts in the direction of movement (equation 3.12). *(b)* Linear work performed when only part of the force acts in the direction of movement (equation 3.13). In the case shown, the force *(F)* is directed at an angle β above the horizontal. *(c)* A weightlifter who bench-presses 800 N through a distance of 0.5 m performs 400 J of work.

The standard (SI) unit of work is the **joule** (1 J = 1 N·m). If the entire force is not acting in the direction of motion (figure 3.14b), then only the component of force in that direction is used in calculating the work done. Figure 3.14b shows a force (F) at an angle (β) above the horizontal. In this case, the work performed is

$$W = F \cdot d \cdot \cos(\beta) \tag{3.13}$$

where F = applied force, d = displacement, and β = angle of force above the horizontal.

In the example depicted in figure 3.14c, the work performed in lifting the barbell from point A to point B is equal to the product of the barbell's weight (W_b) and the distance from A to B (d_{AB}). If, for example, the barbell weighed 800 N and was lifted 0.5 m, the work done would be 400 J.

In the previous example, the force was assumed constant. In real-world situations, that often is not the case. Determination of work done by a varying force is more involved, because it must be calculated at successive intervals and requires the use of calculus. The equation becomes

$$W = \int F_x \cdot dx \tag{3.14}$$

In similar fashion, **angular work** ($W_\angle$) is defined as the product of torque (T) times the angle (θ) through which a body rotates. In equation form,

$$W_\angle = T \cdot \theta \tag{3.15}$$

The calculation of work alone often is insufficient to completely describe the mechanics of a body's movement. In many cases, the *rate* of work also is important. The rate of work is termed **power. Linear power** (P) is defined as the rate at which linear work is done:

$$P = W / \Delta t \tag{3.16}$$

where W = work performed and Δt = change in time, or the time interval over which the work was done.

Angular power ($P_\angle$) is defined as the rate at which angular work is done:

$$P_\angle = W_\angle / \Delta t \tag{3.17}$$

where $W_\angle$ = angular work performed and Δt = change in time.

Power is expressed in units of **watts** (1 W = 1 J/s). In the previous barbell example, a person lifting the 800 N barbell 0.5 m in 2 s would perform 400 J of work, with a power of 200 W, whereas a lift done in 0.5 s would have the same work (400 J) but a power of 800 W. A given amount of work performed in a shorter time usually will have greater power.

Alternatively, linear power may be expressed as the product of force (F) and linear velocity (v):

$$P = F \cdot v \tag{3.18}$$

and angular power as the product of torque (T) and angular velocity (ω):

$$P_\angle = T \cdot \omega \tag{3.19}$$

Equations 3.18 and 3.19 highlight the fact that to be powerful, from a mechanical perspective, one has to be able to generate large forces at high velocities. In other words, a powerful athlete must be both strong and fast. Strength (force) alone is not sufficient; to be powerful, one must be able to generate force and do so quickly.

Energy

In discussing the epidemiology of injury, Robertson (1998) concluded, "The leading source of injury by far is **mechanical energy,** the characteristics of which have been known since Sir Isaac Newton's work on the laws of motion in the seventeenth century. Although Newton's laws of motion do not apply near the speed of light, they are applicable to moving motor vehicles and bullets or to falling human beings (p. 7)."

As the primary agent of injury, energy is critical to an understanding of injury biomechanics. **Energy,** defined as the capacity or ability to perform work, can assume many forms, including thermal, chemical, nuclear, electromagnetic, and mechanical. Although each form of energy has the potential to cause injury, mechanical energy is the one most frequently involved. Mechanical energy is measured in joules (J).

The mechanical energy of a body can be classified according to its **kinetic energy** (energy of motion) or its **potential energy** (energy of position or deformation). Kinetic energy can be either linear (E_k) or angular ($E_{\angle k}$). These two types of kinetic energy are defined, respectively, as

$$E_k = 1/2 \cdot m \cdot v^2 \quad\quad (3.20)$$

$$E_{\angle k} = 1/2 \cdot I \cdot \omega^2 \quad\quad (3.21)$$

where m = mass, v = linear velocity of the center of mass, I = mass moment of inertia, and ω = angular velocity.

Potential energy can take two forms. The gravitational form (*positional potential energy*) measures the potential to perform work as a function of the height a body is elevated above some reference level, most typically the ground. The equation describing gravitational potential energy is

$$E_p = m \cdot g \cdot h \quad\quad (3.22)$$

where m = mass, g = gravitational acceleration (~ 9.81 m/s^2), and h = height in meters above the reference level.

The second form of potential energy is **deformational energy,** or **strain energy,** which is stored in a body by virtue of its deformation. Common examples of strain energy include a stretched rubber band, a pole vaulter's bent pole, and a drawn bow prior to arrow release. The equation describing the amount of stored energy depends on the **material properties** of the deformable body. No single equation describes the strain energy of all bodies.

Biomechanists studying whole-body or limb segment movement dynamics often assume that each body segment is a rigid (nondeformable) body. When this simplifying assumption is made, there is no strain energy component in the system. In these cases, the total mechanical energy is simply the sum of the linear kinetic, angular kinetic, and positional potential energies.

The **total mechanical energy (TME)** = linear kinetic energy + angular kinetic energy + positional potential energy. If we use equations 3.20 to 3.22, this TME equation becomes

$$\text{TME} = E_k + E_{\angle k} + E_p = (1/2 \cdot m \cdot v^2) + (1/2 \cdot I \cdot \omega^2) + (m \cdot g \cdot h) \quad (3.23)$$

where m = mass of the shank, v = linear velocity of the shank's center of mass (COM), I = mass moment of inertia, ω = angular velocity, $g = -9.81$ m/s^2, and h = height of the relevant body part's COM.

Consider, for example, a soccer player swinging her leg to kick the ball. Each of the lower-limb segments (thigh, lower leg, and foot) possesses a continuously changing total mechanical energy. Let's focus on the shank (lower leg) to illustrate how we would calculate the total mechanical energy of a segment at a given instant. Assume that the shank has a mass of 2.6 kg and a mass moment of inertia (I) of 0.04 kg/m^2, at one point in time is rotating with an angular velocity of 7.0 rad/s, has a COM that is moving at a linear velocity of 3.5 m/s and is positioned at a height of 0.38 m above the ground.

By substituting the values given into equation 3.23, we can determine that the TME for the shank at this instant is 26.6 J. Similar calculations done at successive points in time create an energy profile for the shank as a function of time and show how the energy increases and decreases (i.e., flows) throughout the kick. The same analysis can be done for the thigh and foot to create a complete mechanical energy profile of the lower extremity during a kick or any other activity.

Two principles governing energy and its effect are important in assessing the effects of energy on injury. These are **conservation of energy** and **transfer of energy.** Each of these principles applies to both linear and angular energy. *Conservation of energy* indicates how much of a system's energy is conserved and how much is gained or lost during a given time period. The more energy conserved, the greater the potential for injury. True conservation results in no net gain or loss in system energy.

The corollary principle, *transfer of energy*, is the mechanism by which energy is transferred from one body to another. This can take many forms in the course of human movement. Transfer during a throwing motion can happen as energy moves from a proximal segment (e.g., upper arm) to a more distal segment (e.g., forearm, hand) as the throw progresses. Transfer of energy can also happen between different bodies, as in the case of an American football player blocking or tackling an opponent or automobiles colliding in an automobile crash. Transfer of energy in these cases often results in injury, when the energy transferred exceeds the tolerance of the tissues in one or both of the bodies.

Momentum

Momentum measures quantity of motion. An old adage suggests that the bigger they are, the harder they fall. With respect to injury, this maxim can be accurately modified to state that the bigger and faster they are, the harder they hit. This revised maxim embodies the concept of momentum. In mechanical terms, **linear momentum** *(p)* is defined as

$$p = m \cdot v \tag{3.24}$$

where *m* = mass and *v* = velocity of the body's center of mass. Increasing either a body's size (mass) or speed (velocity) will increase its linear momentum.

Similarly, **angular momentum** *(L)*, or the quantity of angular motion, is defined as

$$L = I \cdot \omega \tag{3.25}$$

where *I* = mass moment of inertia about the center of mass and ω = angular velocity.

The principles of *conservation* and *transfer* apply to momentum in the same way they do for energy. **Conservation of momentum** measures how much of a system's momentum, or quantity of motion, is conserved and how much is gained or lost during a given time period. True conservation would imply no net gain or loss in system momentum. **Transfer of momentum** is the mechanism by which momentum is transferred from one body to another, either from one body segment to another (e.g., momentum transfer from the upper arm to the forearm during a throw) or between different bodies (e.g., from the foot to the ball in a soccer kick).

There is a direct relation between an applied force and the change in momentum it creates. Consider a force applied to a particular body over a very short time interval, as is often the case in force-related injuries. Such a force is referred to as an **impulsive force.**

This impulsive force *(F)* relates to momentum in what is termed the **impulse–momentum principle,** which states that the **impulse** equals the change in momentum. In the linear case, the **linear impulse** *(F · Δt)* equals the change in linear momentum (Δ *m · v*).

$$F \cdot \Delta t = \Delta m \cdot v \tag{3.26}$$

where *F* is the impulsive force, Δ*t* is the time period of force application, *m* = mass, and *v* = linear velocity.

In the angular form, **angular impulse** *(T · Δt)* equals the change in angular momentum (Δ*I* · ω).

$$T \cdot \Delta t = \Delta I \cdot \omega \tag{3.27}$$

where *T* is the impulsive torque (moment), Δ*t* is the time period of torque application, *I* = mass moment of inertia, and ω = angular velocity.

The importance of impulsive force in our exploration of injury can be seen in the following example. Consider a person landing on the ground after jumping from an elevated surface. Would the person rather land on a concrete surface or a padded one? At the instant before ground contact, the person will have a certain momentum (*m* · *v*), calculated as the product of the falling velocity *(v)* times the person's body mass *(m)*. At the end of the landing, the person's body comes to a stop and, hence, has no momentum. The impulse (*F* · Δt) created between the ground and the jumper caused the momentum to change from *m* · *v* to zero. If the jumper landed on a concrete surface, the impulse time (Δ*t*) would be very short and the impulsive force *(F)* would be high. In contrast, if the jumper landed on a padded surface, the impulse time would be longer and as a consequence the impulsive force would be lower. The lower impulse force answers the question of which would be the preferred landing surface. Falling from an extreme height (which creates considerable momentum) onto an unyielding surface (which causes high impulsive forces) is a recipe for injury.

Collisions

In many cases, musculoskeletal injuries occur as a result of one object impacting another. In athletic contests, body–body and body–ground impacts are common. In automobile crashes, multiple impacts occur between the various parts of both vehicles and occupants. In slips and falls, an impact occurs between the person and the ground. Because of their impact characteristics, all of these situations have injury potential. Injury happens when the forces applied during an impact exceed the body tissues' ability to withstand the force.

A forceful impact between two or more bodies is known as a **collision.** Collisions have relatively large impact forces acting over a relatively short time interval. In every collision the contacting bodies undergo deformation; that is, their shape (configuration) changes. In some instances, the deformation is negligible (e.g., a collision between two billiard balls), whereas in others the deformation can be considerable (e.g., a forceful blow to a person's abdomen). The deformed body may experience a plastic deformation, an elastic deformation, or a combination of both. In a **plastic deformation,** the body's change in physical configuration is permanent. In an **elastic deformation,** the body recovers from the deformation and returns to its original configuration when the force is removed. The ability of a material to return to its original shape is termed **elasticity** and is an essential characteristic of body tissues.

The nature of the collision between two bodies depends on their relative masses and velocities (both magnitude and direction) and on the material properties of the respective bodies. In theory, collisions occur along a continuum ranging from **perfectly plastic (inelastic) collisions** at one extreme to **perfectly elastic collisions** at the other. In a perfectly plastic collision, the bodies stick together and move together with a common velocity after impact with no loss of energy or momentum. In contrast, a perfectly elastic collision involves bodies that rebound away from each other following the collision with no energy or momentum loss.

In real terms, most collisions involving the human musculoskeletal system are **elastoplastic** in nature. The bodies deform, sometimes permanently, and energy is transferred and lost in the collision. The greater the energy involved, the more likely injury will occur and the more severe it will be. In elastoplastic collisions energy is lost, and the relative postcollision velocity between the two bodies decreases. To measure this loss of separation velocity, we introduce the concept of **coefficient of restitution** (e), defined as the ratio between the relative postcollision velocity (RV_{post}) of two bodies and their relative precollision velocity (RV_{pre}):

$$e = RV_{post}/RV_{pre} \qquad (3.28)$$

The coefficient of restitution can range between 0 and 1. A hard rubber ball dropped onto a hard surface would have an *e* value near 1, indicating little energy loss. A partially deflated basketball, in contrast, would bounce very little and would have an *e* value close to 0. The material properties of the colliding bodies will determine where on the collision continuum each impact falls.

Friction

Newton's first law of motion tells us that bodies in motion tend to remain in motion unless acted upon by an outside force. The force may be an abrupt one, such as a collision, or may be a force of lower magnitude and greater duration, such as the force of friction. **Friction** is defined as the resistance created at the interface of two bodies in contact with one another and acting in a direction opposite impending or actual movement. Frictional resistance results from microscopic irregularities, known as *asperities,* on the opposing surfaces. Asperities tend to adhere to each other, and efforts to move the bodies result in very small resistive (shear) forces that oppose the motion.

In the simple case of a body at rest on a surface, **static friction** resists movement until a force sufficient to overcome the frictional resistance is applied. The magnitude of this static friction (f_s) is given by

$$f_s \le \mu_s \cdot N \qquad (3.29)$$

where μ_s = the coefficient of static friction and N = the component of the contact force that is normal, or perpendicular, to the surface. This **normal force** is also referred to as the **reaction force** (R). In the case of a horizontal surface, N equals the weight of the body. For bodies on an inclined surface, N changes as a function of the angle of inclination, becoming less as the incline becomes steeper.

As the force applied to a body at rest increases, a level is reached at which the static resistance is overcome, and the body begins to slide along the surface. Once the body begins moving, the friction decreases slightly and then is known as **kinetic friction** or **dynamic friction** (f_d) with a dynamic coefficient of friction (μ_d):

$$f_d = \mu_d \cdot N \tag{3.30}$$

The relation between static and dynamic friction is depicted in figure 3.15. Coefficients of sliding friction generally are between 0 and 1, where $\mu = 0$ indicates a frictionless surface.

Sliding friction plays a critical role in many injuries. A person walking on a wet, slippery surface, for example, is more likely to slip, fall, and become injured because the wet surface has a lower coefficient of friction than a dry one. Similarly, an individual navigating an icy stretch of sidewalk needs to be careful because of the very low frictional coefficient that ice provides.

Another example of the prominent role that friction plays is seen in the sporting world. Artificial turf was developed in the 1960s in response to inadequate outdoor playing fields in urban areas and the construction of domed stadiums. This type of playing surface withstands the harsh use and environmental conditions of outdoor fields and fulfills the need for an indoor surface where

natural turf is not feasible. Considerable research has characterized the biomechanical aspects of artificial surfaces (Andreasson et al. 1986; Nigg and Yeadon 1987; Skovron et al. 1990). The differences in frictional characteristics between natural turf and artificial surfaces have been implicated as a major factor in injuries in sports such as soccer, tennis, and American football (Allen et al. 2004; Ekstrand and Nigg 1989; Nigg and Segesser 1988; Skovron et al. 1990). As new artificial surfaces are developed, ongoing research is warranted to assess their injury potential (Meyers and Barnhill 2004).

Surfaces with low frictional resistance are associated with fewer injuries. The typically higher friction on artificial surfaces is assumed to cause more injuries, particularly to the knee and ankle. Caution is warranted in making broad generalizations, however, because the frictional characteristics in a given situation are determined by the interactive effect of many factors such as shoe type, surface wear, type of sport, weather conditions, and the athlete's individual anthropometrics (e.g., height and weight), experience, and skill level.

When an automobile or bicycle is moving, the tires are in contact with the road surface. If the tires are prevented from rotating (e.g., when one applies the brakes), the vehicle slides along the road, resisted by sliding friction. Most of the time, however, the wheels are free to rotate, and the vehicle rolls along. Even in rolling, friction is present. This rolling resistance is not as obvious as sliding resistance, because rolling resistance is much lower, often by a factor of 100 to 1,000 times. The actual value of resistance depends on the material properties of the body and surface and on the normal force acting between them.

In some cases friction works to our advantage. In fact, we would be unable to walk or run without friction acting between our shoes and the ground. Too much friction, however, may contribute to injury. High levels of friction lead to abrupt deceleration, which causes high forces and extreme loading of body tissues.

Our examples so far have focused on friction acting on the body. Friction also plays an important role within the human body. During normal

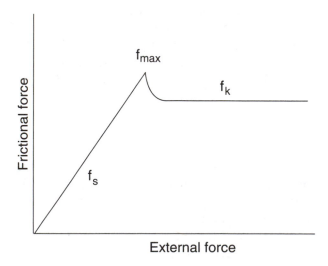

Figure 3.15 Relation between static friction and kinetic friction. For an object at rest, static friction (f_s) increases linearly to oppose the externally applied force. When maximum static friction (f_{max}) is reached, the object breaks loose and begins sliding. While sliding, kinetic friction (f_k) resists the movement.

limb movements, for instance, the friction in joints is extremely low, allowing for freedom of movement with minimal resistance. Details of this low joint friction are presented in the section on joint mechanics to follow.

Internal friction can also lead to injury, as in the case of a tendon sliding along its tendon sheath. Excessive friction between the tendon and sheath can hasten an inflammatory response and conditions such as *tendinitis* (see chapter 5).

Fluid Mechanics

Fluid mechanics, the branch of mechanics dealing with the properties and behavior of gases and liquids, assumes an important role in our framework for considering human biomechanics. Areas as diverse as *performance biomechanics* (study of human mechanical function), **biotribology** (study of the friction, lubrication, and wear of diarthrodial joints), *tissue biomechanics* (study of mechanical response of tissues), and **hemodynamics** (study of blood circulation) all rely on the principles of fluid mechanics.

We live and operate in various fluid environments, with air as the principal gas and water as the predominant liquid. The temperature, density, and composition of each fluid contribute to its mechanical properties. We consider these mechanical properties in two broad categories: fluid flow and fluid resistance. Another important term in fluid mechanics is *viscosity*. Flow, resistance, and viscosity are all critical to our understanding of body function and tissue response and are intrinsically related to the biomechanics of injury:

■ **Fluid flow.** This term refers to the characteristics of a fluid, whether liquid or gas, that allow it to move and that govern the nature of this movement. Blood circulating through a coronary artery provides a biomechanical example of fluid flow. Fluid flow can exhibit many movement patterns. **Laminar flow** is characterized by a smooth, essentially parallel pattern of movement. **Turbulent flow** exhibits a more chaotic pattern of flow, characterized by areas of turbulence (eddies) and multidirectional movement. Arterial blood flow provides us with a good example of these differences. Factors contributing to turbulent flow include the roughness (degree of irregularity) of the surface over which the fluid flows, the diameter of the vessel through which the fluid flows, obstructions, and the speed of flow.

■ **Fluid resistance.** Fluids also provide resistance, such as the resistance we might experience while running into a head wind or swimming in a pool. Fluid resistance takes many forms, some of which are advantageous and others that may be detrimental. Examples of the positive effects of fluid resistance include **buoyant force,** which allows a person or object to float in water (according to **Archimedes' principle,** which states that the magnitude of the buoyant force equals the weight of the displaced liquid); **lift** and **drag forces,** which assist in keeping an object in flight (aerodynamics) or allow a person to swim (hydrodynamics); and **magnus forces,** which affect the trajectories of objects spinning through the air. Negative effects of fluid resistance are evident in the extra physiological work expenditure required of a cyclist riding into the wind or by the severe and unpredictable forces acting on an airplane during a storm.

■ **Viscosity.** The resistance to flow produced by fluids may be considered "fluid friction" and is termed viscosity. *Viscosity* is the property of a fluid that enables it to develop and maintain a resistance to flow dependent on the flow's velocity (rate of flow). This viscous effect and its dependence on velocity can be seen in a familiar example. When you move your hand slowly through water, the resistance is minimal. Increasing the speed of movement markedly increases the resistance. Because all biological tissues have a fluid component, it is logical that tissue response to mechanical loading will include a viscous component. For example, the response of tendon and ligament to stretch will vary depending on the rate of stretch created by the applied load. The details of this rate-dependent response will be considered later.

FLUID MECHANICS OF ATHEROSCLEROSIS

Normal physiological function depends on efficient transport provided by the cardiovascular system. Compromise of this system's efficiency can have harmful, even fatal, consequences. Normal, healthy vessels allow for smooth, unobstructed blood flow with minimal resistance. Fatty (plaque) buildup on vessel walls signals the onset of **atherosclerosis.** As the amount of atherosclerotic plaque increases, the vessel wall becomes rougher and more irregular, and the vessel cavity (lumen) is occluded and, therefore, narrows.

These changes can have serious physiological consequences as a result of the mechanical changes that accompany plaque accumulation. Roughened arterial walls increase the turbulence of the blood flow and increase resistance. Narrowing of the arterial cavity (lumen) also increases resistance to flow. The increase in resistance can be dramatic for even small degrees of narrowing. A reduction in radius of one half normal, for example, produces a 16-fold increase in resistance. The heart, which must pump much harder to force blood through the narrowed lumen, is at risk for serious and deleterious long-term effects.

The injury that atherosclerosis inflicts on vessel walls can have catastrophic consequences if left untreated. For example, when a piece of plaque becomes dislodged from an arterial wall and blocks blood flow, the situation can become life-threatening. Such a blockage in a coronary artery is termed a *myocardial infarction,* or heart attack.

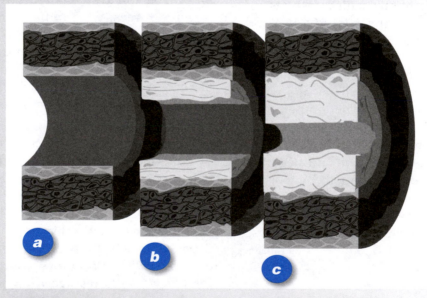

Stages of atherosclerosis: *(a)* normal artery, *(b)* partially blocked artery, and *(c)* significantly blocked artery.

Reprinted, by permission, from E.T. Howley and B.D. Franks, 2007, *Fitness professional s handbook,* 5th ed. (Champaign, IL: Human Kinetics), 298.

Joint Mechanics

Hundreds of articulations (joints) in the human body allow us to move. Because many injuries occur to joint structures, a study of their mechanical characteristics is essential to our discussion of injury biomechanics. No two joints are structurally the same; each has its own distinct combination of tissues, tissue configuration, and movement potential. This variety of joint structure and function results in many complex injuries, which we explore in subsequent chapters.

The major articulations (e.g., hip, knee, and elbow) are diarthrodial (synovial) joints containing synovial fluid that serves lubricative, shock absorptive, and nutritive functions. In its lubricative role, synovial fluid reduces friction in the joint to extremely low levels. This minimal friction plays an important part in the durability (i.e., lack of wear) of normal articular cartilage. The lubrication mechanisms acting in synovial joints are complex and not completely understood. These mechanisms are detailed in chapter 4.

Movement analysis depends on proper description of the joint motions that constitute each movement pattern, and joint motions are defined with respect to anatomical position. In this position, the body is referenced according to three mutually perpendicular planes: sagittal, frontal, and transverse (figure 3.16). Primary joint motion typically occurs in one of these movement planes. For example, knee flexion from anatomical position occurs in the sagittal plane, glenohumeral abduction happens in the frontal plane, and hip internal (medial) rotation occurs in the transverse plane (table 3.1).

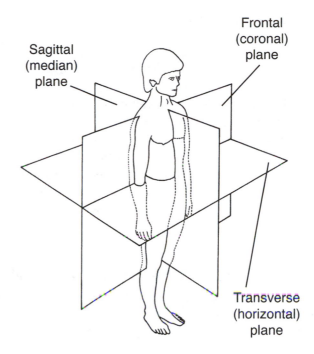

Figure 3.16 Three primary movement planes shown for a person standing in anatomical position. These planes are the transverse (horizontal), frontal (coronal), and sagittal planes. These names are given to any plane parallel to the ones shown. The sagittal plane that bisects the body is sometimes referred to as the midsagittal or median plane.

Adapted, by permission, from J. Watkins, 1999, *Structure and function of the musculoskeletal system* (Champaign, IL: Human Kinetics), 55.

TABLE 3.1

Summary of Primary Joint Motions and Planes of Action

Joint	Joint motion	Plane of action[a]
Hip	Flexion–extension	Sagittal
	Abduction–adduction	Frontal
	Internal–external rotation	Transverse
Knee	Flexion–extension	Sagittal
Ankle	Plantar flexion–dorsiflexion	Sagittal
Shoulder	Flexion–extension	Sagittal
	Abduction–adduction	Frontal
	Internal–external rotation	Transverse
	Horizontal flexion–extension	Transverse
Elbow	Flexion–extension	Sagittal
Radioulnar	Forearm pronation–supination	Transverse
Wrist	Flexion–extension	Sagittal
	Ulnar–radial deviation	Frontal
Intervertebral (spine)	Flexion–extension	Sagittal
	Lateral flexion	Frontal
	Rotation	Transverse

[a]Planes of action for movements begun from anatomical position.

Mobility and Stability

Each joint in the body has a **range of motion (ROM)** throughout which the joint normally operates. This ROM determines a joint's mobility. The magnitude of allowable ROM is specific to both the joint and the person. Joints with an ability to move in more than one plane have ROMs specific to each particular plane of movement. ROMs vary considerably from one person to another, and thus individual measurement is the surest method of determining accurate joint ROM (table 3.2). Intrinsically related to ROM is the notion of **joint stability,** defined as "the ability of a joint to maintain an appropriate functional position throughout its range of motion" (Burstein and Wright 1994, p. 63).

Injuries often occur when a joint exceeds its normal ROM, as when an elbow hyperextends, which raises the question of what determines normal ROM. Joint ROM is determined by the combined effects of

- the shape of the articular surfaces and their geometric interaction (degree of bony fit);

TABLE 3.2

AVERAGE RANGES OF MOTION (ROM) OF JOINTS[a]

Joint	Joint motion	ROM, degrees
Hip	Flexion	90-125
	Extension	10-30
	Abduction	40-45
	Adduction	10-30
	Internal (medial) rotation	35-45
	External (lateral) rotation	45-50
Knee	Flexion	120-150
Ankle	Plantar flexion	20-45
	Dorsiflexion	15-30
Shoulder	Flexion	130-180
	Extension	30-80
	Abduction	170-180
	Adduction	50
	Internal (medial) rotation[b]	60-90
	External (lateral) rotation[b]	70-90
	Horizontal flexion (adduction)[b]	135
	Horizontal extension (abduction)[b]	45
Elbow	Flexion	140-160
Radioulnar	Forearm pronation (from midposition)	80-90
	Forearm supination (from midposition)	80-90
Cervical spine	Flexion	40-60
	Hyperextension	40-75
	Lateral flexion	40-45
	Rotation	50-80
Thoracolumbar spine	Flexion	45-75
	Hyperextension	20-35
	Lateral flexion	25-35
	Rotation	30-45

[a]ROM for movements made from anatomical position (unless otherwise noted). Averages reported in the literature vary, sometimes considerably, depending on method of measurement and population measured. [b]Movement from 90° abducted position.

- the restraint provided by ligaments, joint capsule, and other periarticular structures; and
- the action of muscles around the joint.

When the limits imposed by these stabilizing factors are exceeded, normal ROM is violated, and the tissues may experience injury-producing forces.

One way of viewing joint stability is the joint's ability to resist dislocation. Stable joints have a high resistance to dislocation. Unstable joints tend to dislocate more easily. Joints can be classified along a mobility–stability continuum, which specifies that joints that have a tight bony fit or numerous ligamentous and other supporting structures or that are surrounded by large muscle groups will be very stable and relatively immobile. Joints with a loose bony fit, limited extrinsic support, or minimal surrounding musculature tend to be very mobile and unstable. One exception to this categorization is the hip joint, which is both very mobile—with large ROM potential across all three primary planes—and very stable, as seen by the rarity of its dislocation.

Lever Systems

Most motion at the major joints results from the body's structures acting as a system of levers. A **lever** is a rigid structure, fixed at a single point, to which two forces are applied at two points. One of the forces is commonly referred to as the **resistance force** (R); the other is termed the **applied force** or **effort force** (F). The fixed point, known as the **axis, pivot,** or **fulcrum,** is the point about which the lever rotates. In the human body, these three components are typically an external force (R), a muscle force (F), and a joint axis of rotation (A), such as when one is performing an elbow curl exercise to lift a dumbbell (R) using the biceps brachii (F) about the elbow joint axis (A).

These lever system components may be spatially related to one another in three configurations, giving rise to three classes of levers. Distinctions among the classes are determined by the location of each component relative to the other two (figure 3.17, a-c). In a **first-class lever,** the pivot-point axis (A) is located between the resistance (R) and the

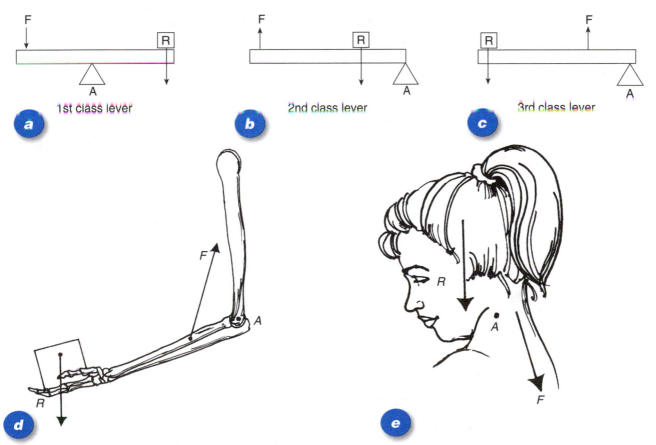

Figure 3.17 Lever systems. Simple lever systems comprise three elements: an axis *(A)*, effort force *(F)*, and resistance force *(R)*. *(a)* First-class lever system. *(b)* Second-class lever system. *(c)* Third-class lever system. *(d)* Biceps brachii acting about the elbow joint as a third-class lever system. *(e)* Extensors of the head and neck as a first-class lever system to counteract the tendency of the head s weight to flex the cervical spine.

effort force (*F*). A **second-class lever** has *R* located between *F* and *A*, whereas a **third-class lever** has *F* between *R* and *A*. Joints in the human body are predominantly third-class levers (figure 3.17*d*), with some first-class levers (figure 3.17*e*) and few second-class lever systems.

Lever systems in the human body perform two important functions. First, they increase the effect of an applied force, because the applied force and the resisting force have different moment arms, in the same way that a leverage advantage is gained by using a bar to pry a large rock loose from the ground. In a first-class lever, for example, increasing the moment arm on the side of the applied force increases the effective force seen on the other side of the pivot point.

The second function of levers is to increase the effective speed (or velocity) of movement. During knee extension (figure 3.18), a given angular displacement (Δθ) produces different linear displacements of points *x* and *x'* on the lower leg. Similarly, if the knee is extended at a constant angular velocity (ω), the linear velocity of point *x'* will be greater than that of *x*. Thus, by increasing the lever arm distance from *x* to *x'*, we have increased the linear velocity of movement. The human body effectively uses both the force and speed advantages provided by lever systems in accomplishing the many tasks it performs daily. As expected, these force and speed enhancements

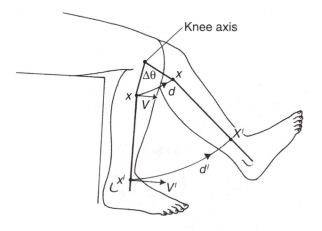

Figure 3.18 Knee extensor mechanism acting as a lever system. For a given angular displacement (Δθ), the curvilinear displacement (*d'*) for point *x'* will be greater than the displacement (*d*) for point *x*. Similarly, for a given angular velocity (ω), the linear velocity (*v'*) for point *x'* exceeds the velocity (*v*) of point *x*.

play a role in injury occurrence and prevention as well.

Moment of Force (Torque) and Joint Motion

We defined *moment of force*, or *torque*, as the effect of a force tending to cause rotation about an axis. With respect to joint function, moments created by the action of skeletal muscles are the essential element in controlling joint motion. Figure 3.19*a* depicts moment production at a joint. The muscle force *F* acts at a perpendicular distance (moment arm = *d*) from the elbow joint axis *A*, producing a moment *M* = *F* · *d*. In static situations such as this, calculation of the moment is straightforward.

In real-life cases involving joint motion, the calculation becomes much more complex. Consider each component of the moment calculation. The magnitude of muscle force *F* typically varies and is determined by a combination of factors including the muscle's length, velocity, level of neural activation, and fatigue, along with the external resistance that the system is experiencing. Instantaneous changes in any or all of these factors directly influence the amount of muscle force produced.

As the joint moves throughout its range, the musculotendinous line of action (i.e., direction of pull) changes continuously, thus affecting the moment arm distance. Because joints in the human body are not perfect hinge joints, the location of the axis of rotation relative to the bony structures at any instant in time **(instantaneous joint center)** changes as well (figure 3.19*b*). These changes in muscle force, line of action, and moment arm result in a continuously varying moment of force.

The asymmetry of joint motion that accounts for movement of the instantaneous joint center motion (figure 3.19) is caused by a combination of three movements: rotation, sliding, and rolling (figure 3.20). In rotation, the motion is purely angular, with rotation about a fixed axis (figure 3.20*a*). Sliding joint motion occurs when one articulating surface moves linearly relative to the other (figure 3.20*b*). Rolling results in angular joint movement combined with linear displacement of the axis of rotation (figure 3.20*c*).

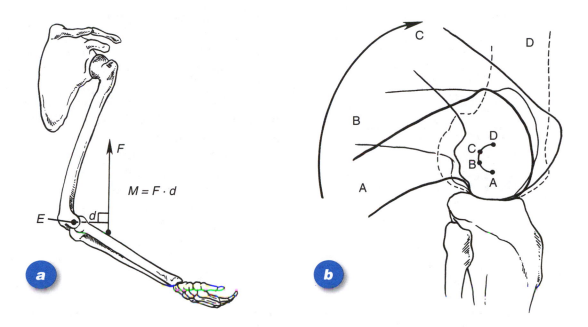

Figure 3.19 Moment production at a joint and instantaneous joint center of rotation. *(a)* Biceps brachii muscle force *(F)* with a moment arm *(d)* producing a moment of force *(M)* about the elbow joint axis *(E)*. *(b)* Structural asymmetries result in movement of the instantaneous joint center with respect to the bones constituting the joint. Movement of the instantaneous joint center is shown for the knee as the joint extends from a flexed position (A) to full extension (D).

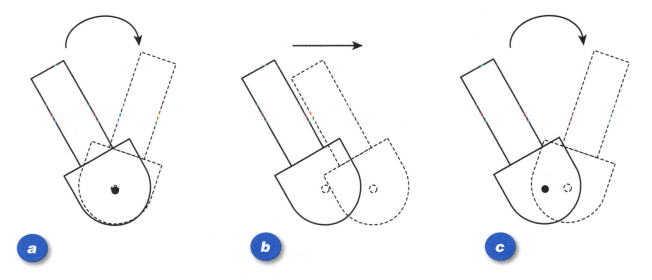

Figure 3.20 Components of joint motion: *(a)* rotation, *(b)* sliding, and *(c)* rolling.

Joint Reaction Force Versus Bone-on-Bone Forces

Articular surfaces can experience high forces such as those occurring at the knee during running or jumping. Repeated high-force loading may lead to joint injuries (e.g., meniscal tears, articular cartilage degeneration). Actual measurement of these forces can be a complex undertaking, and thus mathematical models (discussed later in this chapter) are often used to estimate these joint loads. The net effect of muscle and other forces acting across a joint is called the **joint reaction force (JRF).** Examples of JRFs are depicted in figure 3.21. The JRF is not the same as the actual bone-on-bone force. Calculation of actual

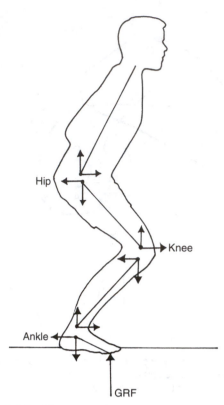

Figure 3.21 Joint reaction forces. Equal and opposite joint reaction forces (in accordance with Newton's third law) act at the ankle, knee, and hip joints in a squatting position with each joint in a flexed position. These joint reaction forces develop in response to the ground reaction force (GRF).

bone-on-bone forces is complex (Winter 2005), and valuable information can be gained from in vivo force-sensing transducers implanted in human joints, although a discussion of this is beyond the scope of this text (e.g., Kirking et al. 2006; Stansfield et al. 2003).

Material Mechanics

Our discussion so far has focused on the movement of bodies and the causal agents affecting those movements. We have concentrated on external mechanics or the effect of external forces on the movement of bodies. In this section we shift our attention to the internal mechanics of structures, focusing on the internal response of materials to externally applied loads.

In previous sections we considered bodies as if they were rigid structures whose size and shape do not change when loads are applied. This approach, known as **rigid-body mechanics,** is useful when we examine movement charac-

teristics. Rigid-body formulations make certain assumptions about the body, including non-deformability, fixed center of mass, and homogeneity of the composite material. Although biological tissues are deformable, viewing body segments as rigid bodies is a reasonable approximation in fields of inquiry such as movement mechanics. In examining the biomechanics of injury, however, we also need to explore the mechanics of deformable solids, because injury mechanisms often result in considerable tissue deformation.

Biological materials such as tissues exhibit many properties that influence each material's response to loading and, hence, the likelihood and severity of injury. Among these *material properties* are size, shape, area, volume, and mass. Additional properties derived from these fundamental ones include **density** (ratio of mass to volume) and *center of mass* (also *centroid).* The tissue's **structural properties,** discussed in the previous chapter, are also important factors in describing tissue mechanical response.

Tissue Response to Loading

In an earlier section we described *internal* and *external* mechanics. Consistent with that distinction, we can describe forces as being either internal forces (i.e., forces acting within the body) or external forces (i.e., forces acting on the body from without). An externally applied force is called a mechanical **load.** There are three load types: **compression** (compressive load), **tension** (tensile load), and **shear** (shear load), as depicted in figure 3.22. Tensile loads tend to pull the ends of a body apart, compressive loads tend to push the ends together, and shear loads tend to produce horizontal, or parallel, sliding of one layer over another.

Load Deformation and Stiffness

When a load is applied to a body, the body changes shape or configuration, although sometimes imperceptibly. This change in shape is termed **deformation.** Tensile loads create material elongation; compressive loads shorten, or compress, the material; and shear loads cause a sliding or angulation in the material. Deformation typically is measured in absolute units (e.g., a tendon was stretched, or elongated, by 10 mm).

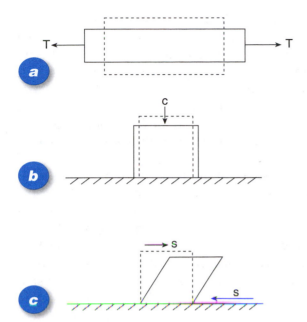

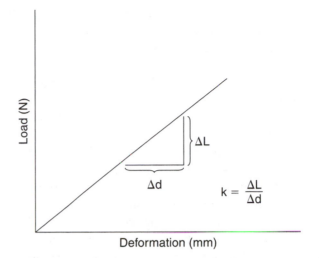

Figure 3.23 Relation between the load and deformation when a force is applied to a tissue or structure. Stiffness *(k)* is measured by the slope (ΔL/Δd) of the linear portion of the load–deformation curve.

Figure 3.22 Mechanical loading. External loads change the shape of a material. This deformation can manifest as *(a)* elongation, or tensile deformation; *(b)* compression, or compressive deformation; or *(c)* angulation, or shear deformation. Dashed lines indicate unloaded condition. Solid lines indicate loaded condition.

Each material or tissue has a characteristic relation between load and deformation that is graphically depicted by a **load–deformation curve** (figure 3.23). The slope of the linear portion of the load–deformation curve measures the tissue's **stiffness** (often represented by *k*). Curves with a steeper slope represent stiffer materials. The opposite, or inverse, of stiffness is known as **compliance** (1/*k*). The terms *stiffness* and *compliance* are used as relative terms, meaning that a material is not absolutely stiff or compliant; rather one material is stiffer, or more compliant, than another.

Stress and Strain

Any tissue, when loaded, develops an internal resistance to the external load. In the case of a small rubber band, this resistance is minimal. In contrast, a steel bar provides considerable resistance. This internal resistance to an axial load is common to all materials and in mechanics is called **stress** (σ). Axial, or normal, stresses are categorized as either **compressive stress** (i.e., resistance to being pushed together) or **tensile stress** (i.e., resistance to being pulled apart). The magnitude of the normal stress is

$$\sigma = F/A \qquad (3.31)$$

where F = the magnitude of the axial load and A = the cross-sectional area over which the load is distributed.

Those forces acting parallel, or tangential, to the applied load create **shear stress** (τ):

$$\tau = F/A \qquad (3.32)$$

The standard (SI) unit of stress is the *pascal* (Pa), defined as 1 N distributed over 1 m² (1 Pa = 1 N/m²). To avoid possible confusion, note that although both pressure $(p = F/A)$ and stress $(\sigma = F/A)$ are defined as force divided by area, the area (A) is different between the two. Pressure is an *external* measure of the force divided by the surface area over which the force is distributed (see equation 3.1). Stress is an *internal* measure of the force divided by the cross-sectional area of the tissue. As with pressure, the megapascal (1 MPa = 1 N/mm²) is a commonly used unit for stress when applied to tissue mechanics

Earlier, we defined *deformation* as the change in shape or configuration of a body. Mechanical **strain** (ε) provides another measure of shape change, but in contrast to deformation, which is measured in absolute units (e.g., mm), strain is measured in relative terms. Strain is measured as the change in dimension divided by the unloaded dimension and, therefore, is technically dimensionless. Strain typically is reported as mm/mm (or inch/inch) or as a percent (%) strain:

strain (%) = (dimension change)/(unloaded dimension) × 100   (3.33)

For example, an unloaded tissue of length 50 mm, which is elongated by 3 mm, has a percent strain of (3 mm)/(50 mm) × (100) = 6% strain.

Deformations in biological tissues such as bone are so small that their strain is measured in units of **microstrain** ($\mu\varepsilon$), which is (10^{-6}). A measured strain of 0.001 (0.1%) would be equal to 1000 $\mu\varepsilon$.

As with load deformation, a direct relation exists between stress and strain, and the consequences of this relation in a tissue determine its susceptibility to injury. Compressively loaded bone, for example, develops high resistance while deforming very little. Skin, in contrast, deforms considerably more at substantially lower forces. The strain responses of tendon, ligament, and cartilage fall somewhere between these two. Plotting stress as a function of strain (figure 3.24a)

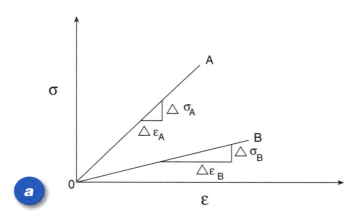

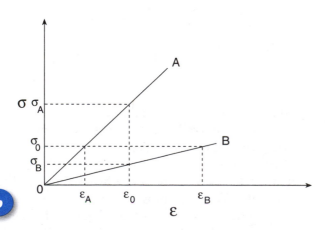

Figure 3.24 Relation between mechanical stress (σ) and strain (ε). *(a)* Linear stress–strain (σ–ε) curves for two materials. The slope ($\Delta\sigma$–$\Delta\varepsilon$) of each line measures each material s modulus of elasticity *(E)*. *(b)* The relative *E* of the materials determines their response to loading. For a given stress level (σ_o), material B shows more strain than material A.

allows us to visualize the σ–ε relation. The figure shows linear σ–ε curves for two materials labeled A and B. A closer look at the curves reveals several important relations. For a given stress (σ_o), material B exhibits more strain than material A, which is evident as $\varepsilon_B > \varepsilon_A$. Conversely, for a given strain (ε_o), material A develops a greater stress than material B, as demonstrated by $\sigma_A > \sigma_B$ (figure 3.24b).

The stress–strain (σ–ε) relation can be summarized in a single measure as the ratio of the two values. This ratio is termed the **modulus of elasticity,** or **elastic modulus** *(E)*, also known as **Young's modulus** *(Y)*. Stiff materials such as bone have a steeply sloped σ–ε curve and a high *E* value. More compliant materials such as skin have flatter σ–ε slopes and lower *E* values.

Thus far we have considered only the linear σ–ε relation. Linear materials are said to operate according to **Hooke's law,** which posits that stress and strain are linearly related; that is, the resulting strain is directly proportional to the developed stress. Mathematically, Hooke's law is expressed by

$$\sigma = E \cdot \varepsilon \qquad (3.34)$$

The mechanical response of biological tissues typically is not linear throughout its physiological range, attributable largely to nonlinear characteristics created by the tissue's fluid component.

Loading Types

Recall from earlier discussion the seven factors involved in force application (page 58). These factors (magnitude, location, direction, duration, frequency, variability, and rate of force application) are fundamental determinants of loading response. The type of loading also plays a primary role in the response of biological tissues. We focus now on various types of loading.

Uniaxial Loading

The simplest form of force application, **uniaxial loading,** refers to forces applied along a single line, typically along a primary axis of the structure. For any uniaxially loaded material, the location and direction of the force will determine how each of the three loading types (tension, compression, and shear) presents.

As the magnitude of applied load increases, the tissue eventually is unable to withstand the loading and fails (i.e., tears apart or ruptures). The level of force **(ultimate load)** at which **failure** occurs defines the tissue's **structural strength.** The concept of structural strength has obvious implications as we examine failure characteristics of tissue in injury situations.

We noted that the direction of force application is an essential factor. A homogeneous material will respond the same irrespective of the direction of loading. Biological tissues, however, are generally not homogeneous (i.e., their structure varies throughout the tissue), and as a result the loading response depends on the direction of load. A material exhibiting this direction-dependent response is termed **anisotropic.** As an example, consider a long bone undergoing uniaxial compression. The structural strength for a compressive load along the longitudinal axis is much greater than the structural strength for a force directed perpendicular to the long axis. That anisotropic effect largely results from the bone's structure, wherein the osteons in the cortical bone of the diaphysis are aligned with the long axis of the bone to accept compressive loads.

Biological tissues can exhibit linear behavior through certain loading ranges but typically are nonlinear through other parts of their physiological range. Such nonlinearities result in a generalized σ–ε curve (figure 3.25). We apply this curve to specific tissues in the next chapter, explaining at this point only general aspects of the curve.

Consider a tissue experiencing a gradually increasing tensile load. At low loads, with commensurate low stress levels, the σ–ε response is linear (Hookean). The proportional response continues until point B in figure 3.25. At stresses greater than σ_B, the response becomes nonlinear. Point B is, therefore, known as the **proportional limit** (also **linear limit).** As the stress continues to increase, we reach point C, known as the **elastic limit.** At stresses less than σ_C, the material is elastic (i.e., returns to its original shape when the load is removed). At stresses greater than σ_C, the material is no longer elastic and experiences permanent *plastic deformation* (also *plastic strain).* This plastic deformation is shown in figure 3.26. When the load is removed, stress decreases, and the material shortens. Because the tissue

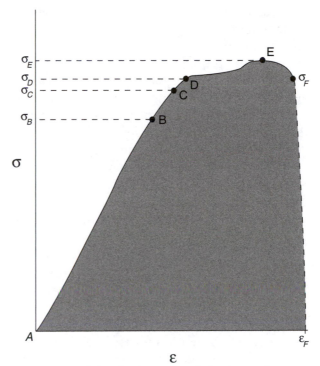

Figure 3.25 Generalized stress–strain (σ–ε) curve for biological tissues.

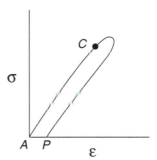

Figure 3.26 Permanent set *(AP)* created by stress exceeding the elastic limit at point *C.*

has exceeded its elastic limit, the tissue is no longer able to return to its original shape (point *A).* Instead, it deforms to an unloaded length at point *P.* The difference between points *A* and *P* is the amount of plastic deformation, or **permanent set.**

Point *D* in figure 3.25 is approached as stress continues to increase. This is known as the **yield point,** at which there begins a brief region of relatively large strain for little increase in stress. This yielding phenomenon is characteristic of many biological tissues. Further increase in stress eventually brings the material to its ultimate stress (σ_E), or **material strength,** where failure begins to occur. Because the failure of some tissues is not instantaneous, the actual completion of failure may occur at a stress level below σ_E, at what is termed the **rupture point** *(F)* at a stress level of σ_F. The actual σ–ε curves for specific tissues vary in their response characteristics and may not exactly mirror the curve presented in figure 3.25. Chapter 4 presents detailed discussion of the differences for bone, cartilage, tendon, and ligament.

Two other important mechanical parameters shown in figure 3.25 are measures of energy stored by the tissue and the tissue's dimensional change. On the load–deformation curve, the area under the curve represents the **energy to failure** (measured in joules), whereas the dimensional change is termed the **deformation to failure** (measured in absolute units, e.g., millimeters) Comparable measures are found on the σ–ε curve in figure 3.27. The area under the curve is the **strain energy density,** a measure of the relative strain energy stored by the tissue prior to failure, and the dimensional change is termed **strain to**

failure *(ε_F)*, a relative measure of how much the tissue has deformed to the point of failure.

When a compressive load is applied to an object (e.g., a ball), a deformation results, which compresses the object in the direction of the load (figure 3.28*a*). At the same time, an accompanying deformation happens perpendicular to the axial load. This distortion is a tensile deformation, and the ball in this example gets wider. This simple case is an example of **Poisson's effect,** which says that when a body is subjected to a uniaxial load and its dimension decreases in the axial direction, its perpendicular, or transverse, dimension increases. Poisson's effect applies in the opposite sense as well. A body experiencing a tensile load shows an increase in its axial dimension and a decrease in its transverse dimension (figure 3.28*b*). The quantitative measure of this effect is given by **Poisson's ratio** *(v):*

$$v = -(\varepsilon_t / \varepsilon_a) \qquad (3.35)$$

where ε_t = transverse strain and ε_a = axial strain.

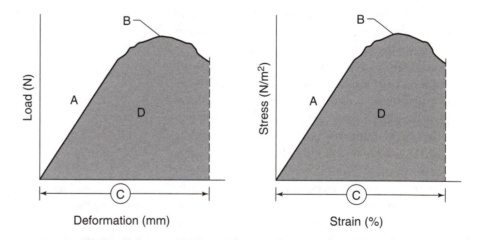

	Load-deformation	Stress-strain
Units	Absolute	Relative
A-Slope of linear portion of curve	Stiffness	Elastic modulus (E) or Young's modulus (Y)
B-Highest point on curve	Ultimate (maximum) load or structural strength	Ultimate (maximum) stress or material strength
C-Change in dimension	Deformation at failure	Strain at failure
D-Area under curve	Energy at failure	Strain energy density

Figure 3.27 Comparison of a load–deformation curve versus a stress–strain curve and related measures.

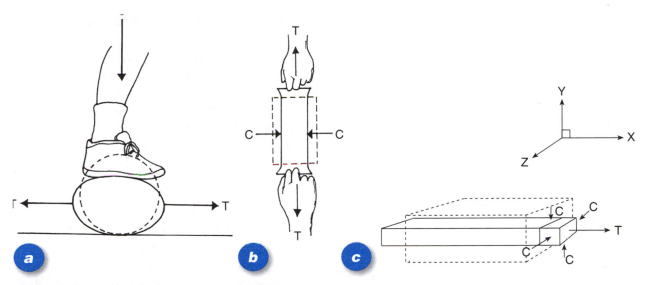

Figure 3.28 Poisson s effect. *(a)* An applied compressive load causes tensile stress and strain perpendicular to the imposed load. *(b)* An applied tensile load results in compressive stress and strain perpendicular to the applied load. *(c)* Poisson s effect shown for a three-dimensional case in which tension *(T)* applied in the *x* direction causes contraction *(C)* in the *y* and *z* directions. Dashed lines indicate conformation prior to loading. Solid lines indicate conformation while loaded.

Similar to the two-dimensional case just presented, Poisson's effect also occurs in three-dimensional space, with transverse strains occurring in two dimensions in response to a uniaxial load in the third dimension (figure 3.28*c*).

Multiaxial Loading

We have so far considered only the simple case of uniaxial loading. In most real-life situations, however, the forces applied to a body are multidimensional, and hence an understanding of **multiaxial loading** and its effects is essential. An analysis of multiaxial loading uses the same stress and strain concepts just discussed and extends them into two- and three-dimensional space. Although the biaxial and triaxial responses are illustrated for tensile loading only, the concepts are equally applicable to compressive loading and to force vectors with reversed orientation, and the following formulations are for linearly elastic materials.

Biaxial (Two-Dimensional) Loading Responses

Consider a three-dimensional body (figure 3.29) with sides of length x', y', and z' that is subjected to axial forces F_x and F_y. The stresses produced in the x and y directions are as follows:

$$\sigma_x = F_x/A_x = F_x/(y' \cdot z') \tag{3.36}$$

$$\sigma_y = F_y/A_y = F_y/(x' \cdot z') \tag{3.37}$$

The x-direction stress (σ_x) will, according to Poisson's effect, produce deformation in all three directions. The elongation in the x direction and contraction in the y and z directions are shown in figure 3.29*b*. By applying equations 3.34 and 3.35, we obtain x and y direction strains attributable to σ_x:

$$x \text{ direction: } \varepsilon_{x\sigma x} = \sigma_x/E \tag{3.38}$$

$$y \text{ direction: } \varepsilon_{y\sigma x} = -\nu \cdot \varepsilon_{x\sigma x} = -\nu \cdot (\sigma_x/E) \tag{3.39}$$

We similarly obtain y and x direction strains attributable to σ_y:

$$y \text{ direction: } \varepsilon_{y\sigma y} = \sigma_y/E \tag{3.40}$$

$$x \text{ direction: } \varepsilon_{x\sigma y} = -\nu \cdot \varepsilon_{y\sigma y} = -\nu \cdot (\sigma_y/E) \tag{3.41}$$

To obtain the combined effect of σ_x and σ_y, we add the strain effects just calculated. The net strains in the x and y directions are then

$$\varepsilon_x = \varepsilon_{x\sigma x} + \varepsilon_{x\sigma y} = \sigma_x/E - \nu \cdot (\sigma_y/E) \tag{3.42}$$

$$\varepsilon_y = \varepsilon_{y\sigma y} + \varepsilon_{y\sigma x} = \sigma_y/E - \nu \cdot (\sigma_x/E) \tag{3.43}$$

We have presented two normal stresses, σ_x and σ_y. As previously described, a tangential, or shear, stress *(t)* is also created. In the case of **biaxial loading,** shear stress (τ_{xy}) is created as shown in figure 3.30*a*.

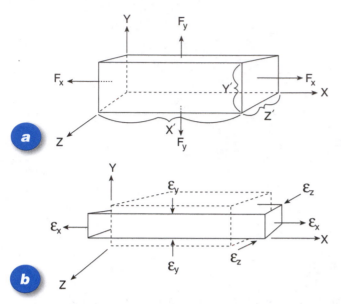

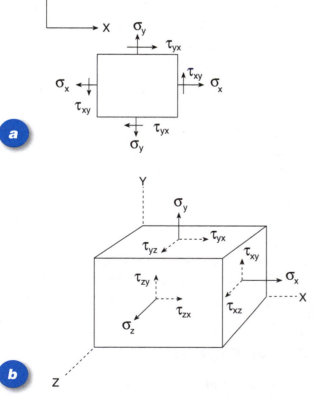

Figure 3.29 Biaxial loading. *(a)* Applied forces in the *x* and *y* direction (*F_x* and *F_y*, respectively) load the material biaxially. *(b)* Elongation caused by *F_x* and the resulting perpendicular contraction in the *y* and *z* directions. Dashed lines indicate conformation prior to loading. Solid lines indicate conformation while loaded.

Figure 3.30 Shear stresses in response to biaxial and triaxial loading. *(a)* Shear stress (τ_{xy}) created by biaxial loading. Because of equilibrium conditions, $\tau_{xy} = \tau_{yx}$. *(b)* Triaxial tensile stresses $(\sigma_x, \sigma_y, \sigma_z)$, shown as solid vectors, result in shear stresses $(\tau_{xy}, \tau_{yz}, \tau_{zx})$, depicted as broken-line vectors. Equilibrium constraints dictate that $\tau_{xy} = \tau_{yx}$, $\tau_{yz} = \tau_{zy}$, and $\tau_{zx} = \tau_{xz}$.

Triaxial (Three-Dimensional) Loading Responses

The addition of a third axial force in the *z* direction complicates the conceptual model only slightly, whereas the mathematical aspects of the model become quite involved. Focusing on the conceptual application (figure 3.31), we now have

- F_x, which creates $\sigma_x = F_x/(y' \cdot z')$ and produces elongation in the *x* direction and contraction in the *y* and *z* directions
- F_y, which creates $\sigma_y = F_y/(x' \cdot z')$ and produces elongation in the *y* direction and contraction in the *x* and *z* directions
- F_z, which creates $\sigma_z = F_z/(x' \cdot y')$ and produces elongation in the *z* direction and contraction in the *x* and *y* directions

The equations for resultant strains seen in triaxial loading are

$$\varepsilon_x = (1/E) \cdot [\Sigma_x - \nu \cdot (\Sigma_y + \Sigma_z)] \qquad (3.44)$$

$$\varepsilon_y = (1/E) \cdot [\Sigma_y - \nu \cdot (\Sigma_x + \Sigma_z)] \qquad (3.45)$$

$$\varepsilon_z = (1/E) \cdot [\Sigma_z - \nu \cdot (\Sigma_x + \Sigma_y)] \qquad (3.46)$$

The shear stresses $(\tau_{xy}, \tau_{yz}, \tau_{zx})$ produced in triaxial loading are shown in figure 3.30*b*.

Bending

So far we have discussed axial loading. Two other types of loading happen frequently in the human body: **bending** and **torsion.**

Any structure that is relatively long and slender (e.g., long bone) may be considered in mechanical terms as a **beam.** Any force, force

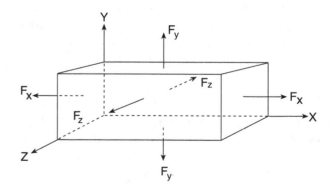

Figure 3.31 Triaxial loading. Applied forces in the *x*, *y*, and *z* directions (*F_x*, *F_y*, and *F_z*, respectively) load the material triaxially.

component, or moment acting perpendicular to the longitudinal axis of such a beam will tend to deflect, or bend, the beam.

In bending, the material on the **concave** (inner) surface of the structure experiences compressive stress, whereas that on the **convex** (outer) surface is subject to tensile stress (figure 3.32*a*). These tensile and compressive stresses are maximal at the outer surfaces of the beam, with the material closer to the middle experiencing less stress than at the surfaces. The line along which neither compressive nor tensile stress exists is known as the **neutral axis.**

Each force acting on the beam may create a moment in the beam. The sum of these moments is referred to as the **bending moment** (M_b), which creates different stress levels in the beam that vary with the distance from the neutral axis. At a distance *y* from the neutral axis (figure 3.32*b*), a normal stress (σ_x) is created with a value given by

$$\sigma_x = (M_b \cdot y)/I \qquad (3.47)$$

where M_b = bending moment, *y* = distance from neutral axis, and *I* = area moment of inertia of

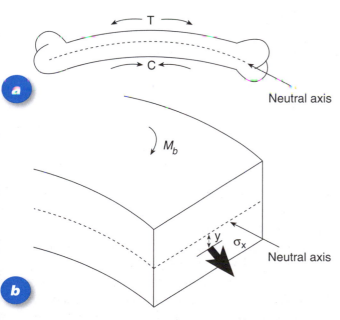

Figure 3.32 *(a)* Material stresses in response to bending. Bending creates compressive stress on the concave (inner) surface and tensile stress on the convex (outer) surface. Maximal stress happens at the surfaces, with lower stress levels toward the center of the bent object. A neutral axis exists along which there are no tensile or compressive stresses present. *(b)* Bending moment. A beam subjected to a bending moment (M_b) develops a normal stress (σ_x) as given by equation 3.47.

the cross-sectional area about the neutral axis. The **area moment of inertia** (*I*) measures the resistance to bending, and its value depends on the cross-sectional shape of the structure. The area moment of inertia for common shapes is depicted in figure 3.33.

The area moment of inertia (*I*) for a solid cylinder is measured by

$$I = (\pi \, r^4)/4 \qquad (3.48)$$

where *r* = the radius of the cylinder.

The area moment of inertia (*I*) for a hollow cylinder is measured by:

$$I = [\pi \, (r_o^4 - r_i^4)]/4 \qquad (3.49)$$

where r_o = outer radius and r_i = inner radius of the cylinder.

In instances when shear forces are acting on a beam, shear stresses are created that are maximal at the neutral axis and zero at the surfaces (figure 3.34). The magnitude of this shear stress (τ) is given by

$$\tau = (Q \cdot V)/(I \cdot b) \qquad (3.50)$$

where *Q* = the first moment of the area about the neutral axis, *V* = shear force, *I* = area moment of inertia, and *b* = width of the cross section.

Two common bending modes seen in biomechanical cases are **three-point bending** and **four-point bending** (figure 3.35, *a* and *b*). A boot-top skiing injury mechanism illustrates three-point bending (figure 3.35*c*), whereas the forces acting on a barbell produce four-point bending (figure 3.35*d*).

The failure differences between these two modes are important when considering injury mechanisms. In three-point bending, failure occurs at the middle point of force application (figure 3.36*a*). In contrast, failure in four-point bending occurs at the weakest point between the two inner forces but not necessarily at the midpoint (figure 3.36*b*).

An important combined loading mode is *cantilever bending,* in which a force offset from the longitudinal axis creates both compression and bending. Cantilever bending (figure 3.37) occurs in a loaded femur when compressive forces are applied to the femoral head, creating a bending moment in the diaphyseal bone shaft combined

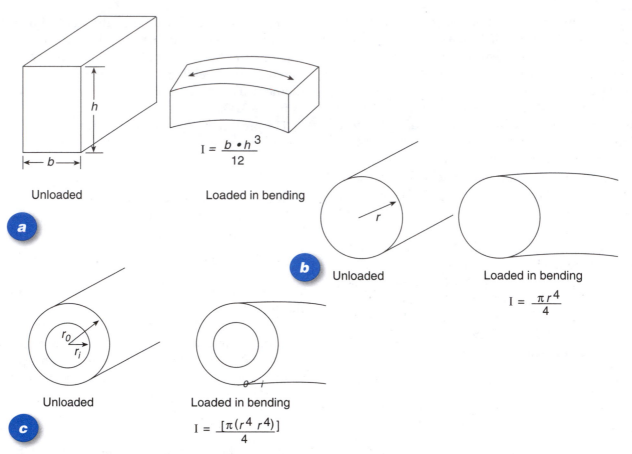

Figure 3.33 *(a)* The area moment of inertia for a rectangular cross section depends on the height *(h)* and base *(b)* dimensions of the cross section. *(b)* The area moment of inertia for a solid cylinder (equation 3.48) depends on the radius *(r)*. *(c)* For a hollowed cylinder (tube), the area moment of inertia (equation 3.49) depends on the outer radius *(r_o)* and the inner radius *(r_i)*.

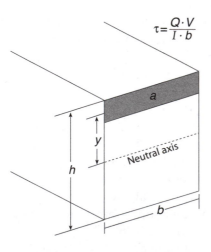

Figure 3.34 Shear stress in response to bending. The magnitude of the shear stress (τ) is given by equation 3.50. *V* = vertical shear force (obtained from a shear force diagram or from static analysis), *Q* = area moment (calculated as the product of the shaded area *a* and the distance *y* from the neutral axis to the centroid of area *a*), *I* = moment of inertia of the total cross section (see figure 3.33*a*), *b* = width of cross section.

with axial compressive effect (e.g., Poisson's effect).

Torsion

Any twisting action applied to a structure results in *torsion*, as seen in the simple example of unscrewing a lid from its jar. Although torsional concepts are applicable to both cylindrical and noncylindrical structures, we limit our mathematical formulations to solid, circular shafts because of the complex mathematics necessary to explain torsional loading of noncylindrical structures. The following torsion formulations are based on assumptions of tissue isotropy, linear elasticity, and structural homogeneity.

Earlier we presented two types of moment of inertia: *mass moment of inertia* (resistance to rotation about a fixed axis) and *area moment of inertia* (resistance to bending of a beam about its neutral axis). A third form of angular resistance is involved when torsional loads are applied to a

Figure 3.35 Three- and four-point bending. *(a)* Three-point bending caused by the action of three parallel forces. The middle force is in the direction opposite to the outer two forces. *(b)* Four-point bending caused by two pairs of parallel forces. The inner pair is in the direction opposite the outer pair. *(c)* Skiing boot-top fracture exemplifies three-point bending. *(d)* An athlete lifting a barbell creates a four-point bending system.

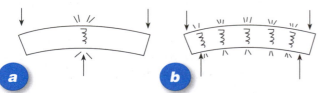

Figure 3.36 Failure attributable to bending. *(a)* Failure (fracture) cause by three-point bending happens at the point of middle force application. *(b)* In four-point bending, failure (fracture) happens at the weakest point between the two inner forces, not necessarily at the midpoint.

body. The internal stresses developed in response to the torsional loading produce resistance to the applied torque. This resistance to torsional loading about the longitudinal axis is termed **polar moment of inertia** *(J)*, and its magnitude for a solid shaft (figure 3.38*a*) is

Figure 3.37 Cantilever bending. Compressive loading offset from the longitudinal axis creates a combined loading situation (compression and bending) known as cantilever bending. Solid lines indicate conformation prior to loading. Dashed lines indicate conformation while loaded.

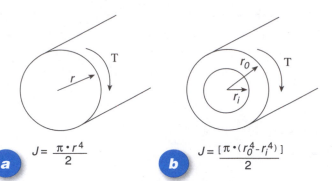

a $J = \dfrac{\pi \cdot r^4}{2}$

b $J = \dfrac{[\pi \cdot (r_0^4 - r_i^4)]}{2}$

Figure 3.38 Polar moment of inertia. Resistance to an applied torque, measured by the polar moment of inertia, for (a) a solid shaft and (b) a hollow cylindrical shaft.

$$J = (\pi \cdot r^4)/2 \tag{3.51}$$

where r = radius of the shaft. For a hollow cylindrical shaft (figure 3.38b), such as a long bone, the polar moment of inertia (J) is

$$J = [\pi \cdot (r_o^4 - r_i^4)]/2 \tag{3.52}$$

where r_o = outer radius of the shaft and r_i = inner radius.

Torsion creates stresses throughout the shaft with the magnitude of shear stress (τ) being a function of shaft radius (r), applied torque (T), and polar moment of inertia (J) (figure 3.39a), expressed as

$$\tau = (T \cdot r)/J \tag{3.53}$$

The resulting shear strain (γ) is shown in figure 3.39b. The ratio of shear stress (τ) to shear strain (γ) is called the **shear modulus of elasticity** (G):

$$G = \tau/\gamma \tag{3.54}$$

The angle of twist (θ) shown in figure 3.39c is given by

$$\theta = (T \cdot l)/(G \cdot J) \tag{3.55}$$

where T = applied torque, l = shaft length, G = shear modulus of elasticity, and J = polar moment of inertia.

Several important generalizations emerge from an examination of torsional loading:

- The larger the radius of the shaft, the more resistance it creates and the more difficult it is to deform.

- The stiffer the material being loaded, the harder it is to deform in torsion.

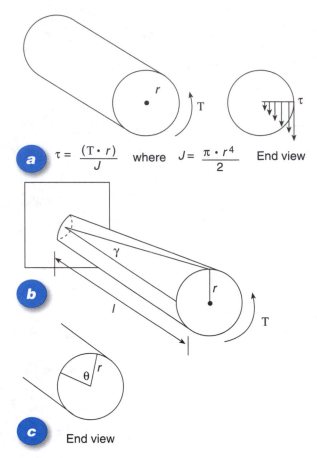

a $\tau = \dfrac{(T \cdot r)}{J}$ where $J = \dfrac{\pi \cdot r^4}{2}$ End view

b

c End view

Figure 3.39 Torsional loading. (a) Shear stress (τ) developed in response to torsional loading (T) (equation 3.53), where r is the radius of the cylinder and J is the polar moment of inertia. (b) Shear strain (γ) created by torsional loading (T). (c) Angle of twist (θ).

- In addition to shear stress, torsion produces normal stresses (tensile and compressive) in the form of helical stress trajectories (figure 3.40a). These stresses are maximal at the outer surfaces and may result in spiral failure lines (figure 3.40b), as seen in a spiral fracture of a tibia loaded in torsion.

Viscoelasticity

As noted in the discussion of fluid mechanics, the mechanical response of a material depends on its constituent matter, which in the case of biological tissues usually has a fluid component. This viscous element provides resistance to flow and affects the stress–strain (σ–ε) relation. The stress response is a function of both the strain and the strain rate ($\dot{\varepsilon}$). Such tissues are said to be **strain-rate dependent.** Tissues with conjoint properties

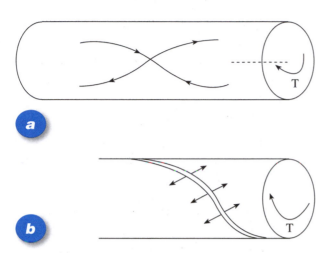

Figure 3.40 Helical stress trajectories. *(a)* Applied torque *(T)* creates helical (spiral) stress lines. *(b)* When the stresses exceed the material s threshold, tensile failure happens along the helical stress trajectories. This fracture pattern is seen in spiral fractures of long bones.

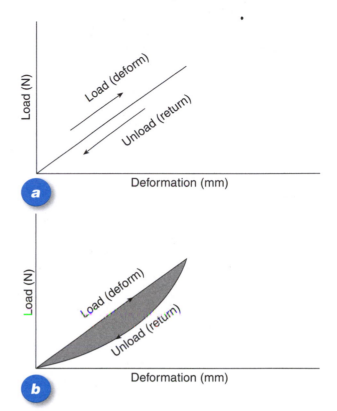

Figure 3.41 *(a)* Perfectly elastic materials return along the same load–deformation path and therefore lose no energy. *(b)* Viscoelastic materials exhibit a delayed return response (hysteresis) and lose energy (heat) during the deformation–return cycle. The shaded area within the hysteresis loop measures the energy loss.

of viscosity (i.e., strain–rate dependency) and elasticity (i.e., ability to return to original shape when load is removed) are termed **viscoelastic.** In viscoelastic tissues, an increasing strain rate steepens the slope of the σ–ε curve and increases the tissue's stiffness.

Purely elastic (i.e., nonviscous) materials subjected to load will deform according to their particular load–deformation (L–d) or stress–strain (σ–ε) relation and store energy in the process. When the load is removed, the stored *strain energy* is returned, and the tissue returns to its original shape with no energy loss by retracing the L–d path traversed during loading (figure 3.41*a*). No biological tissue, however, is purely elastic.

Viscoelastic tissues, in contrast, lose energy to heat during deformation, and the return following unloading is retarded, or delayed, resulting in a return path different from the initial path during loading (figure 3.41*b*). This retarded return is termed **hysteresis.** The **rate of elastic return** is determined by the material properties, in particular the amount of viscous resistance. Materials that quickly return to their original shape are termed **resilient;** those that return more slowly exhibit a **dampened response,** or **damping.** The path of a loading–unloading cycle graphed on a load–deformation curve creates a characteristic pattern known as a **hysteresis loop.** The area within the hysteresis loop (shaded area in figure 3.41*b*) represents the energy lost during the loading–unloading cycle.

Viscous effects also are responsible for a characteristic **biphasic response** in biological tissues. When loaded, viscoelastic tissues exhibit an immediate mechanical response (first phase), followed by a delayed second phase. We explore two common biphasic, time-dependent phenomena associated with biological tissues. The first of these, **creep response,** is seen when a tissue is subjected to a *constant load.* At initial loading, the tissue deforms rapidly (first phase) until the specified constant force level is reached. Instead of maintaining this deformation under the constant load, the tissue continues to deform (second phase), or **creep,** as it approaches an asymptotic deformation plateau (figure 3.42).

The second phenomenon, seen in viscoelastic tissues subjected to *constant deformation,* is the **stress-relaxation response** (also **force-relaxation response).** A tissue stretched (or compressed) to a given length (first phase) and then held at that length develops an initial resistance, or stress. While being maintained at the constant deformation (second phase), the stress decreases, or relaxes, as shown in figure 3.43*a*.

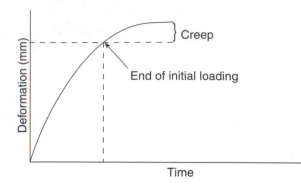

Figure 3.42 Mechanical creep in response to constant load. After an initial deformation while the load is initially applied, the material further deforms (creeps) to an asymptotic value while maintained at a constant load.

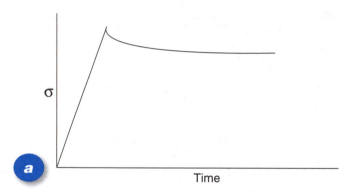

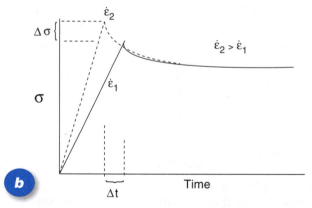

Figure 3.43 *(a)* Stress relaxation in response to constant deformation. The initial deformation elicits a stress (Σ) response. Once the material reaches and maintains its constant deformation, the stress decreases (relaxes) to a constant level. *(b)* The effect of strain rate on the stress-relaxation response. At the higher strain rate ($\dot{\varepsilon}_2$), the maximum stress is greater (by $\Delta\sigma$) and reaches this peak stress earlier (by Δt).

The creep and stress-relaxation responses are *strain-rate dependent* (i.e., the mechanical response depends on the rate of deformation). This is shown for the stress-relaxation curve in figure

3.43*b*. With increased strain rate ($\dot{\varepsilon}_2$), the tissue is stiffer (i.e., steeper σ–ε slope), and the peak stress (σ_{max}) is higher and occurs sooner than with the slower strain rate (ε_1).

Material Fatigue and Failure

Materials, including biological tissues, that are subjected to repeated loads above a certain threshold experience material **fatigue** and exhibit a decreased ability to withstand applied forces. Continued loading of a fatiguing material leads to eventual material failure. The number of loading cycles required before failure may range from a few, as in the case of repeatedly bending a paper clip, to many millions.

An important fatigue-related concept, known as the **initial-cycles** or **first-cycle effect,** implies that the mechanical response seen in initial loading cycles may differ from the response seen during later loading cycles. This effect is shown in the σ–ε relations depicted in figure 3.44. We see a gradual shift in tissue behavior from the initial cycles to later ones. Reasons for this shift include temperature fluctuation, fluid shift, and viscous response characteristics.

The susceptibility to tissue failure is determined in large part by how the stresses generated in response to loads are distributed throughout the material. If the stress is equally distributed, as in a smooth, homogenous solid, there is less chance of failure. If the stress is concentrated at a specific location, the likelihood of failure at that point increases. Stress concentration tends to occur at locations of material discontinuity within the tissue. These discontinuities create **stress**

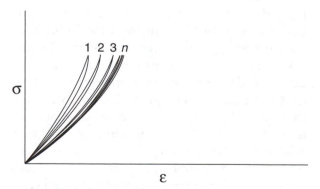

Figure 3.44 First-cycle or initial-cycles effect. Schematic representation of the different σ–ε responses seen in early loading cycles compared with later cycles when the response settles into a steady-state pattern.

risers (also *stress raisers)*, or points of focused stress. Examples of stress risers include abrupt tissue interfaces (e.g., osteotendinous junction), fracture sites, and bone screw insertion points.

The material response prior to failure can vary considerably. Some materials (e.g., glass or bone) deform very little before failure and are classified as **brittle.** Other materials (e.g., putty or elastic ligament) undergo considerable deformation before failing and are called **ductile.** Caution must be taken not to confuse the strength of a material with its brittleness or ductility. Brittle materials, for example, may possess high strength (e.g., steel) or may fail quite easily (e.g., chalk).

Various theories have been advanced to explain material failure, but the critical point is that biological tissue failure is inextricably linked to severe injury. Greater insight into a tissue's mechanical behavior may lead to more specific strategies to control or reduce injury.

Biomechanical Modeling and Simulation

A **model** is defined as a representation of one or more of an object's or system's characteristics. One of the primary goals of *modeling* is to improve, often through idealization and simplification, our understanding of the system or phenomenon being studied. Virtually every field of endeavor uses models in some way. Engineers and architects construct miniature versions of machines and buildings before constructing the real structure; economists create elaborate models to represent how the financial markets behave; psychologists construct models of human behavior. Simulations are intrinsically related to models, so much so that the two terms often and erroneously are used interchangeably. The distinction is that a model uses the equations describing the system, whereas a **simulation** uses the validated model to perform experiments to address questions related to a system and its operation.

Biomechanical models typically exist in one of two forms: a physical model or a mathematical (or computer) model. Many of these models have the potential to provide valuable insights into the mechanisms of injury. Physical models in biomechanics are perhaps best exemplified by crash-test dummies, which have generated valuable data to improve vehicle occupant safety.

CRASH-TEST DUMMIES ARE NO DUMMIES

Anthropometric models, commonly known as crash-test dummies, are much more than friendly reminders for us to buckle our seat belts. The data they provide about the body's response in collision have proved invaluable in making automobiles safer. Countless lives have been saved through the innovations these dummies have pioneered. State-of-the-art versions are much more than lifeless masses; current models use sophisticated instrumentations that allow measurement of body velocities, accelerations, impact forces, and much more. The effective designs of current passive restraint systems, airbags, crumple zones, side-impact reinforcements, collapsing steering columns, and safety windshields are in large part products of information generated from crash-test dummies.

Crash-test dummy.

The biomechanics literature is replete with examples of computer models. These models use mathematical equations as the language of expression to characterize aspects of the system being modeled. In biomechanics, mathematical models have addressed movements such as walking, running, jumping, and throwing, along with more sport-specific movement patterns in swimming, diving, track-and-field events, gymnastics, ice skating, golf, and other sports. Recent biomechanical models have probed areas as diverse as muscle force production, meniscal dysfunction, arterial hemodynamics, prosthetic design, and foot placement during gait.

Why choose to develop a model over other means of investigation, such as direct experimentation? First, mathematical models prove useful in situations that are not easily duplicated in real life. For obvious reasons, studies involving collisions or injuries are not tenable using human subjects. Computer models provide a means of manipulating potential injury conditions without risk. Second, models allow investigators to make changes in a system that could not be accomplished readily by an organism operating in a real environment. A human running, for example, would be unable to modulate his or her performance to produce specific changes in ground reaction forces at each step. A model of the runner, however, could easily produce these forces through appropriate inputs and can be used to predict the response of the system across a range of values.

The third major reason is time. Time-consuming direct experimental paradigms can be simulated in a fraction of the time required by direct experimentation. The continuing development of sophisticated mathematical models in recent decades parallels the development of computing power. The speed and calculation power afforded by computers allow the implementation of complex models that in the past would have been computationally intractable.

Model Selection Criteria

Once the decision is made to create a model of a device, object, or system, the next step is to select the most appropriate type or class of model. Proper selection of model type depends largely on the questions being posed. An important caveat is that model complexity is directly related to the difficulty of model formulation and interpretation.

Biomechanical model selection relies on many considerations. These criteria are not mutually exclusive but rather are complementary. The human body can be studied at many levels, ranging from molecular to whole body. The questions being addressed by the model dictate whether the model is a molecular, cellular, tissue, organ, segmental, or whole-body formulation.

Body tissues are subjected to external loads and will deform to varying extents. Thus, different structural types can be modeled. Tissues experiencing measurable deformation are best modeled using a **deformable-body model.** Tissues exhibiting negligible deformation or bodies assumed to be nondeformable (e.g., limb segments) can be represented by a **rigid-body model.** Structures considered without regard to their molecular characteristics are best examined using a **continuum mechanical model.** A contrasting view of a structure according to its component parts is best considered with a *discrete-element,* or **finite-element model.**

The level of system motion determines whether the appropriate model is a **static model** (e.g., assessing the loads on the low back while in a bent-over position), a **quasi-static model** (e.g., patellofemoral joint loads during a slow squat), or a **dynamic model** (e.g., ankle joint moments during a vertical jump).

Most human body functions are inherently nonlinear across their physiological ranges. Thus, even though linear models are easier to formulate and manipulate, more complex nonlinear models often are the models of choice. The complexity of the process being modeled dictates the level of mathematical sophistication required for its formulation. Some simple systems may only require algebraic calculations for their solution, but complex systems may be assessable only through advanced mathematics. Another aspect affecting model selection is whether the system is fully determined (i.e., **deterministic model)** or comprises functions based on certain probability behaviors (i.e., **stochastic model).**

Activities involving movement in a single plane, or primarily in a single plane (e.g., walking), can be represented by a two-dimensional **planar model.** Most human movements, however, occur in multiple planes and require a more complex three-dimensional **spatial model.**

If actual forces and moments (i.e., kinetics) are measured and used in a model to predict the details of movement (kinematics), we use what is termed a **forward solution,** or **direct solution,** approach. In contrast, using measured kinematics (velocities and accelerations) to predict the kinetics (forces and moments) is referred to as an **inverse solution** approach. This second approach, sometimes called **inverse dynamics,** is useful when the movement characteristics are measurable but measuring the actual forces or moments is either very difficult or impossible. Inverse dynamics models are commonly used in biomechanics to estimate internal forces and moments at joints. Direct measurements of muscle forces and joint torques are difficult at best, and inverse dynamics techniques provide a noninvasive means of calculating these values.

All models are simplifications of the actual situation being modeled. Simplification, however, does not necessarily imply simplicity. Even with simplification, models can be quite complex, and with available computer processing power increasing exponentially, there is a temptation to create extremely complex models. How complex a model is needed? Hubbard provided sage advice: "Always begin with the simplest possible model, which captures the essence of the task being studied" (1993, p. 55).

Elements of a model are often idealized representations of real-life variables, in much the same way as an idealized force vector is representative of many force vectors. When a system component is idealized, however, some information is lost about its actual functional properties. The ability of a model or simulation to achieve its purpose is only as good as the data that it receives as input.

A stable model can maintain its validity over an appropriate range of values and conditions. The predictive ability and, therefore, the usefulness of any model are related to the accuracy with which variables can be specified over a range of values. The critical issue becomes how well the model predicts values between known data points **(interpolation)** and beyond the ranges of known values **(extrapolation).**

Biomechanical modeling is a useful tool for exploring many areas of human function, particularly in describing and evaluating human movement. Modeling has yet to reach its full potential in assessing the biomechanics of injury and holds much promise. As technological capabilities increase, we must maintain our focus on the physiological processes of interest and not become distracted by mathematical sophistication.

Tissue Models

Our discussions of tissue structure (chapter 2) and material mechanics (this chapter) have highlighted the viscoelastic nature of tissues such as bone, tendon, ligament, and cartilage. The details of these tissues' mechanical characteristics are presented in chapter 4, but it is instructive at this point to examine the viscoelastic properties of tissue in some modeling applications.

Rheological Models

Rheology is the study of the deformation and flow of matter. Given that body tissues all have a fluid component and, therefore, have flowlike characteristics, we introduce the concept of a **rheological model,** which has been used extensively to examine the mechanical behavior of human tissues.

Rheological models of tissue interrelate stress (σ), strain (ε), and strain rate ($\dot{\varepsilon}$) of biological tissues. They use three model components that, although having no direct association with actual tissue structural elements, allow us to examine tissue response to loading. These three model components are the linear spring, the dashpot, and the frictional element (figure 3.45). In normal situations, internal friction is negligible compared with other forces and so is often omitted from rheological models of biological tissues.

The **linear spring** represents the elastic properties of the tissue, assuming that the material deforms and returns to its original shape linearly with respect to the applied force and the deformation. The relation between the spring's stress (σ) and strain (ε) is given by equation 3.34, where E is the modulus of elasticity, or Young's modulus.

The fluid component of biological tissues dictates a loading response that is strain-rate

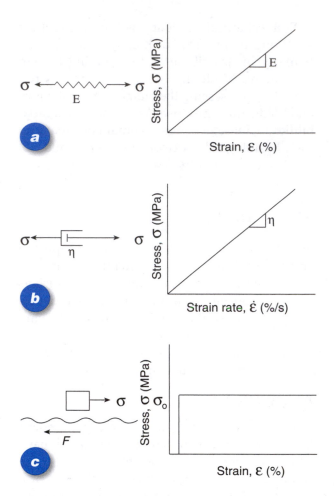

Figure 3.45 Rheological model components: *(a)* spring, *(b)* dashpot, and *(c)* frictional element.

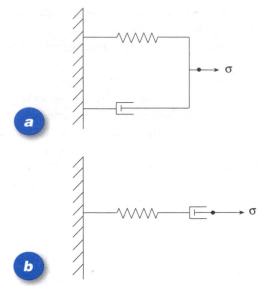

Figure 3.46 Rheological models. *(a)* Kelvin–Voight model, with spring and dashpot in parallel. *(b)* Maxwell model, with spring and dashpot in series.

dependent. The **dashpot** models this viscous contribution to the overall response. If the fluid's stress–strain rate response is linear, the fluid is termed a **Newtonian fluid.** The following equation indicates the dashpot's linear relation:

$$\sigma = \eta \cdot \dot{\varepsilon} \qquad (3.56)$$

where η = the proportionality coefficient relating stress and strain rate.

Researchers can use linear springs (or nonlinear springs) and dashpots as building blocks in constructing composite models with the goal of accurately predicting the response of real tissues. Two standard combinations of linear spring and dashpot are the **Kelvin–Voight model** (spring and dashpot in parallel) and the **Maxwell model** (spring and dashpot in series). These models and their loading responses are shown in figure 3.46. Neither of these models is directly applicable for

modeling tissue behavior, but combining them with other components produces models with good predictive abilities. A simple, standard solid model and a more complex model are depicted in figure 3.47. If we vary the coefficients E_i and n, the models can be fine-tuned to provide an accurate representation of a tissue's response to loading.

Finite-Element Models

The finite-element (FE) method originated in the mid-1950s as a tool to assist engineers in the design of structures. The FE approach often requires lengthy and complex calculations and, therefore, became tractable only with advances in computer technology. Originally, its use was restricted to experts specializing in FE methods who used large, mainframe computers to solve problems. As computer size decreased and speed increased, FE methods became more accessible to nonspecialists via commercially available FE program packages. Originally used in aerospace engineering (primarily by NASA), FE methods gradually found their way into other areas of structural analysis in mechanical and civil engineering. In recent years, biomechanists have found finite-element modeling to be a valuable tool for investigating a wide range of biological problems, such as designing artificial hip joints.

Finite-element modeling uses simple shapes, known as elements (building blocks), which

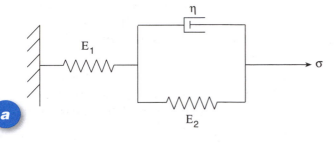

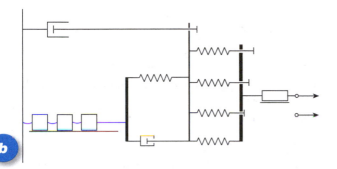

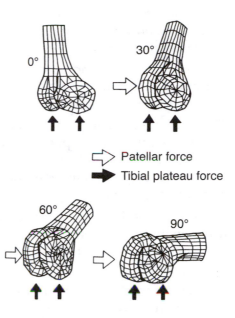

⇨ Patellar force
➡ Tibial plateau force

Figure 3.48 Finite element model of bone.

Reprinted from *Journal of Biomechanics* Vol. 24, T. Nambu et al., Deformation of the distal femur: A contribution towards the pathogenesis of osteochondrosis dissecans in the knee joint, p. 426, Copyright 1991, with permission from Elsevier.

Figure 3.47 Examples of rheological models. *(a)* Standard solid model. *(b)* Complex model.

Reprinted from *Journal of Biomechanics*, Vol. 1, A. Viidik, "A rheological model for uncalcified parallel-fibered collagenous tissue," pp. 3-11, Copyright 1968, with permission from Elsevier.

are assembled to form complex geometrical structures. The elements are connected at points known as nodes. In a model, a finite number of elements (or shapes) is connected at nodes to form a mathematical representation of a structure such as bone (figure 3.48). As forces are applied to the model or as the model is deformed, elaborate equations predict the structure's stress and strain responses to loading. The complexity of a finite-element model is determined by the imagination of its creator, the level of mathematical sophistication, and computing power.

Current research combines finite-element modeling with imaging techniques to create sophisticated models to characterize complex musculoskeletal systems. For example, Blemker and Delp (2005) developed a three-dimensional finite-element model using magnetic resonance images to describe moment arms of fibers within specific muscles (i.e., psoas, iliacus, gluteus maximus, and gluteus medius). Such efforts improve the accuracy of musculoskeletal computer models and allow researchers to answer emerging questions about how the human body functions.

CHAPTER REVIEW

Key Points

- Four major areas of biomechanics form the foundation for an understanding of musculoskeletal injury: movement mechanics, fluid mechanics, joint mechanics, and material mechanics.

- From a mechanical perspective, human movement depends on effective integration of linear and angular motion, center of gravity position and control, stability, mobility, and equilibrium.

- The primary mechanical agents involved in injury are force and energy. These two mechanical variables and related measures (e.g., work, power, and torque) explain the mechanics of injury.

- Human movement and susceptibility to musculoskeletal injury are governed by Newton's three laws of motion.

- The fluid component in musculoskeletal tissues governs tissue responses to loading (e.g., viscoelasticity and biphasic response) and, hence, plays an important role in the mechanics of injury.

- Biological tissue responses can be described using engineering variables such as load–deformation, stress–strain, and related measures.

- Mechanical models are used to represent the human body at all levels (i.e., from molecular to whole body) and are useful in explaining tissue response to loading and mechanisms of injury.

Questions to Consider

1. Many engineering texts present linear mechanics and angular mechanics in separate chapters. From a biomechanical perspective, why is it important to consider linear and angular mechanics simultaneously?

2. Explain the relation between muscle force (chapter 2) and muscle torque (chapter 3).

3. Describe the importance of Sir Isaac Newton's contributions to our understanding of mechanics, and specifically of the biomechanics of musculoskeletal injury.

4. Explain how the principles of fluid mechanics apply to cardiovascular disease.

5. Consider a person jumping down and landing from a platform 1 m above the ground. What movement strategies might the jumper employ to reduce injury risk at landing? What mechanical principles are being applied in devising an effective landing strategy?

6. Describe three detailed examples that illustrate the role of friction in the biomechanics of injury.

7. Describe the utility of using both absolute and relative measures to explain tissue response to loading and injury potential.

8. List and briefly describe what factors should be considered in selecting the type of model to be used for assessing a biomechanical problem.

9. When a long bone is bent until it fractures, what factors will determine the bone's resistance to bending and the location of eventual fracture?

Suggested Readings

Enoka, R.M. 2002. *Neuromechanics of Human Movement* (3rd ed.). Champaign, IL: Human Kinetics.

Green, M., and L.D.M. Nokes. 1988. *Engineering Theory in Orthopaedics.* New York: Halsted Press.

Hall, S.J. 2006. *Basic Biomechanics* (5th ed.). New York: McGraw-Hill.

Hamill, J., and K. Knutzen. 2003. *Biomechanical Basis of Human Movement* (2nd ed.). Baltimore: Lippincott Williams & Wilkins.

Hay, J.G. 1993. *The Biomechanics of Sports Techniques* (4th ed.). Englewood Cliffs, NJ: Prentice Hall.

Jackson, J.J., and H.G. Wirtz. 1983. *Statics and Strength of Materials.* New York: McGraw-Hill.

Levangie, P.K., and C.C. Norkin. 2000. *Joint Structure and Function: A Comprehensive Analysis* (3rd ed.). Philadelphia: Davis.

Luttgens, K., and N. Hamilton. 2000. *Kinesiology: Scientific Basis of Human Motion* (9th ed.). New York: McGraw-Hill.

McGinnis, P.M. 2004. *Biomechanics of Sport and Exercise* (2nd ed.). Champaign, IL: Human Kinetics.

Morecki, A., ed. 1987. *Biomechanics of Engineering: Modelling, Simulation, Control.* New York: Springer-Verlag.

Mow, V.C., E.L. Flatow, and R.J. Foster. 1994. Biomechanics. In *Orthopaedic Basic Science,* edited by S.R. Simon. Park Ridge, IL: American Academy of Orthopaedic Surgeons.

Mow, V.C., and R. Huiskes, ed. 2005. *Basic Orthopaedic Biomechanics & Mechano-Biology.* New York: Lippincott Williams & Wilkins.

Nahum, A.M., and J. Melvin, eds. 2001. *Accidental Injury: Biomechanics and Prevention* (2nd ed.). New York: Springer-Verlag.

Nigg, B.M., and W. Herzog, eds. 1999. *Biomechanics of the Musculo-Skeletal System* (2nd ed.). New York: Wiley.

Nordin, M., and V.H. Frankel. 2001. *Basic Biomechanics of the Musculoskeletal System* (3rd ed.). New York: Lippincott Williams & Wilkins.

Spiegel, L., and G.F. Limbrunner. 2003. *Applied Statics and Strength of Materials* (4th ed.). Englewood Cliffs, NJ: Prentice Hall.

Winter, D.A. 2005. *Biomechanics and Motor Control of Human Movement* (3rd ed.). New York: Wiley.

CHAPTER 4

TISSUE BIOMECHANICS AND ADAPTATION

The form . . . of matter . . . and the changes
of form which are apparent in . . . its growth
. . . are due to the action of force(s).

D'Arcy Thompson (1917)

■ OBJECTIVES

- ■ To explain the mechanical behaviors of the human musculoskeletal system
- ■ To highlight form and function of load-bearing tissues such as bone, tendon, ligament, cartilage, and muscle
- ■ To link the comprehensive knowledge of tissue adaptation to its environment with tissue biomechanical function and injury

Building on the anatomy presented in chapter 2 and the biomechanical principles detailed in chapter 3, here we review the viscoelastic behaviors and adaptive responses of the tissues of the human musculoskeletal system (i.e., bone, articular cartilage, tendon, ligament, and skeletal muscle). These tissues exhibit complex mechanical behaviors, such as stress relaxation at constant strain, creep at constant stress, hysteresis under cyclic loading, strain-rate dependency, and cyclic stress fatigue. Skeletal muscle, in addition to these passive phenomena, also has active force, length, and velocity properties that are unique and synergistic. Information about the mechanical properties and behaviors of musculoskeletal tissues has been gathered using in vitro, in situ, and in vivo methods.

The adaptive capabilities of musculoskeletal tissues are as important to understanding injury as is tissue biomechanical function. Although homeostasis is a tenet of physiology, when we examine that tenet more closely we see that the actual situation is closer to a continually changing equilibrium—tissues in the body are constantly adapting. Adaptation is a natural, form–function interaction and can be defined as the "modification of an organism or its parts that makes it more fit for existence under the conditions of its environment" (Merriam-Webster 2003).

Throughout the life span, dramatic changes and adaptations happen in bone, cartilage, tendon, ligament, and muscle. Factors such as physical activity, immobilization with a cast or brace, or changes in diet can profoundly affect the quality and quantity of load-bearing connective tissues. For each musculoskeletal tissue, we first review the biomechanical properties and then summarize each tissue's adaptive capabilities.

IN VITRO, IN SITU, OR IN VIVO?

Accurate testing and measurement of mechanical properties are essential to understand tissue function and responses, but the act of measuring, in itself, can change a tissue's behavior. The dilemma facing scientists who study the properties of biological tissues is highlighted by these two quotes:

The most likely way, therefore, to get any insight into the nature of those parts of creation which come within our observation, must in all reason be to number, weigh and measure.

S. Hales, 1727 (p. 17)

Error is all around us and creeps in at the least opportunity. Every method is imperfect.

C.J.H. Nicolle in Beveridge, 1957 (p. 115)

The methodological approaches for studying the full range of joints, tissues, or tissue constituents fall broadly into three categories: in vitro, in situ, and in vivo. Each successive category gets closer to measuring tissue behaviors as they exist in the body—although each method has advantages and disadvantages.

In vitro literally means "within a glass," but in the generic sense in vitro connotes testing done in an artificial environment. The specimen (from whole bones to cells) is usually immersed in a physiological buffer solution maintained at body temperature, but the environment is artificial, and the test is therefore in vitro.

One advantage of in vitro tests is that *direct measurements* can be taken. One disadvantage, however, is that the in vitro method is invasive—a tissue is removed from the body and its normal environment. The cells no longer have their native chemical and physical connections to the surrounding tissues and fluids.

In situ means "in its normal place," or confined to the site of origin, and one advantage of in situ preparations is that some elements of the natural environment are preserved in the testing. Notably, much of the information available on skeletal muscle properties has been gathered through in situ techniques. During in situ experiments to record skeletal muscle contractions, a muscle's natural blood supply can be maintained as well as the terminal nerve–muscle interface. The orientation of the muscle with respect to its bony attachments can be maintained, and the muscle temperature can be kept within the physiological range. Although this method is closer to natural, components of the test environment are still artificial. The properties are determined under constrained conditions and not in the freely moving organism.

In vivo indicates that the testing is done within the living body, and this approach might appear to be the best. However, obtaining accurate in vivo data is technically very challenging. Even if the data are recorded successfully, the transducers that are implanted can affect the measurements being taken. Furthermore, ethical concerns must be addressed, especially when trying to measure responses of musculoskeletal tissues in humans. A few investigators have recorded in vivo muscle–tendon forces (Gregor et al. 1991) or bone strains (Burr et al. 1996) in humans, but the great majority of in vivo experiments involving musculoskeletal tissues have been conducted with animal models.

Biomechanics of Bone

There are two major types of bone—compact or cortical bone and trabecular or cancellous bone. Eighty percent of the weight of the human skeleton comes from compact bone, which is dense and forms the surface of bones. Trabecular bone is spongy and makes up the interior of most bones. Its role is to support the body, protect the internal organs, provide levers for movement, and store minerals.

Compact (Cortical) Bone

The mechanical behavior of compact (cortical) bone can be assessed in a variety of ways. The method of testing should be selected to approximate most closely the loading situation that the structure experiences in the body. Because bone is usually loaded multiaxially, it is difficult to test each condition in which bone is loaded. To simplify assumptions, compact bone is treated as an elastic beam of uniform dimension and tested in three- or four-point bending, or a sample of bone of known dimension is machined out of a larger piece and tested in uniaxial compression or tension.

In a typical bending test (figure 4.1), as the bone is initially loaded, the load–deformation curve is concave toward the load axis. As the load increases, load and deformation increase in a relatively linear fashion, obeying Hooke's law. The

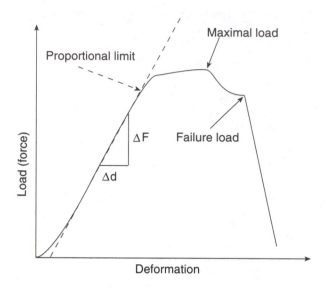

Figure 4.1 Example of a load–deformation curve from a sample of cortical bone that is being tested in three-point bending. In three-point bending, the flexural rigidity of the bone is calculated as $(L^3/48) \cdot (F/d)$, where L is equal to the distance between the two supporting points on which the bone rests and (F/d) is the slope of the force–deformation curve in the linear region. Flexural rigidity itself is the product of the elastic modulus (E) and the cross-sectional area moment of inertia (I) of the sample.

slope of this linear region is related to the bone's **flexural rigidity,** a measure of bending stiffness. The *proportional limit* marks the end of the linear region. In compact bone, the proportional limit and the elastic limit are closely related. The *elastic limit* demarcates the transition from bone's elastic behavior into its plastic region. As the response of bone moves into the region of *plastic deformation,* smaller and smaller increases in load will produce greater and greater increases in deformation. If the applied load is removed just after the transition to plastic deformation (but before maximal and failure loads are reached), the bone does not return to its original (preloaded) configuration. Instead, the bone takes on a permanent set—it remains bent.

Flexural rigidity, load behaviors, and energy (area under a load–deformation curve) are *structural properties* of the bone. If the geometry (shape) of the bone sample is known, then the *material properties* of the bone can be calculated. **Material** refers to the mechanical *quality* of the bone. The elastic modulus and stresses (force per unit area) at the proportional limit and at the maximal and failure points are examples of the bone's material properties.

The distinction between structural and material properties is easily recognized by considering the diaphyses of a femur and a phalanx. The obviously larger femoral diaphysis is able to carry substantially greater loads than the smaller phalanx, and thus the structural properties (e.g., maximal load and bending stiffness) of the femur will be much greater than the phalanx. The compact bone within the femur and phalanx, however, could have very similar composition, and thus, if their differences in size were removed (normalized), the material properties of the two bones may be very similar (e.g., maximal stress, calculated as force per unit area).

Understanding the importance of bone shape and geometry leads us to consider how the *area moment of inertia (I)* influences the structural properties of bone during bending. But before using a bone example, try the following thought experiment. Imagine that you are standing on a 2 m long oak plank that is 5 cm thick and 30 cm wide. The board is supported only at its two ends; if you stand at midlength, the board is being loaded in three-point bending. When you stand on the flat (30 cm) side of the board, you will easily bend the board as it sags under your weight. If, however, you rotate the board 90° so you balance on the 5 cm edge, the sag will be much less, and the board will be much stiffer.

This potent effect of bone cross-sectional geometry can be seen in the three scenarios illustrated in figure 4.2. Assume that you are testing tubular long bones of three different cross-sectional shapes in three-point bending. The examples in figure 4.2, *a* and *b,* have the same periosteal diameter, but the example in figure 4.2*a* has marrow core, whereas the example in figure 4.2*b* has no marrow core (a solid bone). Figure 4.2*c* shows a slightly larger periosteal diameter of 2.5 cm and a relatively large marrow core (a thin-walled, tubular bone). As seen in table 4.1, these shape differences generate pronounced differences in area moment of inertia and bending behaviors. The bone cross-sectional areas in figure 4.2*a* (2.95 cm²) and figure 4.2*b* (3.14 cm²) are more than 66% greater than the cross-sectional area of the large tubular bone shown in figure 4.2*c* (1.77 cm²). At the same time, because the area moment of inertia (figure 3.33) is related to the amount of bone and the distribution of the bone about the bending axis, the thin-walled, tubular bone (figure 4.2c) has an area moment of inertia that is substantially greater (>40%) than either example a or b. In figure 4.2, a and c, area = $\pi \cdot (r_o^2 - r_i^2)$, and area moment of inertia $(I) = \pi \cdot (r_o^4 - r_i^4)/4$ (see equation 3.49). In figure 4.2b, which has no endosteal (marrow) cavity, area = $\pi \cdot r_o^2$, and area moment of inertia $(I) = \pi \cdot r_o^4/4$ (see equation 3.48). Thus, if the same bending load is applied to each of the bones, approximately 13% more stress would be developed within both the smaller tubular bone

(figure 4.2a) and the solid bone (figure 4.2b) than in the large tubular bone (figure 4.2c).

Changing the ratio of the periosteal-to-endosteal diameters increases the bending stiffness of a long bone without necessarily adding large amounts of bone mass. The theoretical implications and potential optimization of the structural features of the wall thickness of tubular bones were explored in a classic paper by Currey and Alexander (1985) (see *What if Bones Were Solid Beams...*, p. 104).

TABLE 4.1

CROSS-SECTIONAL GEOMETRY AND BONE BENDING BEHAVIOR

	Example a	Example b	Example c
Outer (periosteal) radius (r_o)	1.00	1.00	1.25
Inner (endosteal) radius (r_i)	0.25	0.00	1.00
Bone area (A), cm²	2.95	3.14	1.77
Area moment of inertia (I), cm⁴	0.78	0.79	1.13
Force (F), N	20.0	20.0	20.0
Stress $(\sigma = MC/I)$, N/cm²	256	253	221

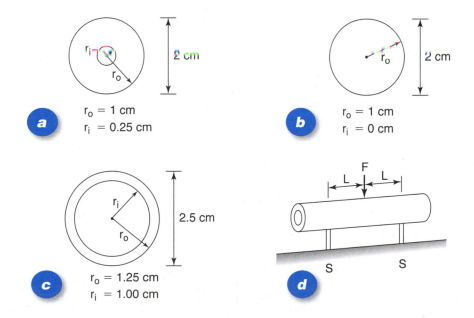

Figure 4.2 Cross-sectional geometry affects bending behavior of tubular (a and c) and solid (c) bone of different dimensions (a-c), where r_o = outer (periosteal) radius and r_i = the inner (endosteal) radius of long bone. (d) Three-point-bending loads being applied to the bone samples. F = applied force; S = positions of supports under bone sample; L = distance between applied force F and each support S. For examples a through c, during three-point bending (d), the stress (σ) developed in the bone is σ = MC/I, where M = moment, C = distance from center of cross section to periosteal surface (outer radius), and I = area moment of inertia of the cross section. M = (F/2) · L, where F = 20N, and L = 20 cm.

WHAT IF BONES WERE SOLID BEAMS RATHER THAN THIN-WALLED TUBES?

Two renowned biologists, John Currey and McNeill Alexander (1985), hypothesized that the mechanical design of tubular long bones suggests that there may be an optimal ratio of the diameter of thin-walled bones to their cortical thickness. It is accepted that the tubular structure of most limb bones ensures that they will be lighter without sacrificing strength or stiffness. But does this notion hold across a wide range of species?

In assessing the concept of an optimal ratio, Currey and Alexander considered the counter-balancing of relative bone mass against structural properties, such as yield (or fatigue) strength, ultimate (or impact) strength, and stiffness. By measuring the limb bones of a substantial array (56 species) of extant and some extinct mammals, birds, and reptiles, these researchers came up with intriguing results showing why tubular, marrow-filled bones appear to be an effective design for bones that undergo substantial bending loads—such as the bones of limbs.

Currey and Alexander calculated the ratio of cross-sectional radius (C) to the thickness of the bone cortical shell (t). They predicted that if marrow-filled bones were basically tubular structures of minimal mass and optimal strength, then the C/t ratio would be 2:3. By comparison, if stiffness was the criterion for bone design, then C/t would be 3:9. Using the measurements of actual limb bones, they found that the median C/t ratio for terrestrial mammals was about 2:0—close to the predicted optimal values for impact loading and ultimate strength. Nonetheless, a spectrum of experimental results emerged from their analysis. For example, the manatee (*Trichechus manatus,* an aquatic mammal) has a solid humerus ($C/t = 1.0$), and the elephant *(Loxodonta africana)* has solid metacarpals. On the other hand, the humerus of a bat (*Pipistrellus* sp.) had thinner walls than most mammals ($C/t = 3.7$).

The extinct *Pteranadon,* with its huge wing span—presumably used for soaring and gliding on wind currents—had astonishingly thin-walled but very large diameter wing bones that would be particularly well designed if minimal mass was vital. The *Pteranadon* humerus, for example, had a C/t value of 20:0. It was the only animal with C/t values greater than 7:7. At the other end of the spectrum, alligators have a femoral C/t value of 1:3—the walls of their femurs are very thick with a small marrow cavity.

To get another view of the effects of the different-sized tubular bones, compare the three long bones (examples *a, b,* and *c* in figure 4.2) for which we contrasted cross-sectional geometry and stresses in three-point bending. In the following table, we provide values for their cross-sectional radius *(C),* cortical thickness *(t),* and the ratio C/t. The larger diameter and thinner walled bone (example *c*) has the much larger C/t ratio.

Interestingly, chickens trained by running on a treadmill also demonstrated a shift toward Currey and Alexander's predicted ultimate (and impact) strength optimal C/t for a weight-bearing limb bone (Loitz and Zernicke 1992; Matsuda et al. 1986). The adult roosters were exercised on a treadmill for 1 h/day, 5 days/week for 9 weeks, at 70% and 75% predicted maximal aerobic capacity (Loitz and Zernicke 1992) and had an average C/t of 2:3 in their tarsometatarsal bones—precisely the values predicted by Currey and Alexander (1985).

Why is this important? Consider the following. Currey calculated that when a horse gallops at 54 km/h, about 50% of its power is used to accelerate and decelerate the bones of its limbs. Thus, a 10% decrease in limb bone mass would generate a 5% power savings. This savings could be significant in the context of natural selection (escape from a predator) or even in a thoroughbred horse race.

	Example a*	Example b*	Example c*
Radius *(C)*, cm	1.00	1.00	1.25
Cortex thickness *(t)*, cm	0.75	1.00	0.25
Radius/thickness *(C/t)*	1.33	1.00	5.00

*See figure 4.2 for sketches and dimensions of these long bones.

Extensive data on the material properties of cortical bones of many species (including human) were reported by Yamada (1973). Besides a wide assortment of human cortical bones (e.g., femur, tibia, fibula, humerus, radius, and ulna), bone samples from a range of animals were also tested (e.g., horses, cattle, wild boars, deer, and ostriches). Yamada noted that the average ultimate tensile stress for the long bones of limbs from humans 20 to 39 years of age ranged between 120 and 150 MPa. Later studies (Martin et al. 1998) of the ultimate tensile stress in cortical bone reported the same range (e.g., 108-130 MPa). In terms of ultimate compressive stress, the human femur had the greatest value (up to 166 MPa in persons 20-39 years of age), followed by the tibia and then the humerus, with the lowest

values in the fibula, ulna, and radius—although even those three bones had an average ultimate compressive stress greater than 115 MPa. Compared with human cortical bone, compact bone from a host of species have been studied extensively in terms of mechanical properties (Currey 1988, 2002, 2003a, 2003b, 2005).

Currey's comparative osteological results have shown that a bone's calcium content and porosity can explain about 80% of the variation seen in a bone's elastic modulus (stiffness). Sampling 28 bones from 17 species of mammals, birds, and reptiles, Currey (1988) reported that bone's maximal strain and the mechanical work under the stress–strain curve are sharply decreased with excessive mineralization.

COMPARATIVE PROPERTIES OF BONE

From his perspective as a biologist, John Currey viewed the structure–function relation of bones as it relates to natural selection. For example, in 1979 he used machined specimens to assess the mechanical properties of three types of bones with very different functions and consequently very different mechanical properties—a deer's *(Cervus elaphus)* antler, a cow's *(Bos taurus)* femur, and a fin whale's *(Balaenoptera physalus)* tympanic bulla.

The principal functions of the red deer's antlers are display and dueling, and **impact strength** is probably an important property for winning the fight. Currey found a high impact resistance in the antlers but a concomitantly low elastic modulus and relatively low bending strength.

The fin whale's tympanic bulla, in contrast, had the highest mineral content and highest elastic modulus of the three bones, and Currey described the bulla as "quite rocklike to handle" and about the size of your fist. Like the auditory ossicles in your own ear, the tympanic bulla of the fin whale is securely protected in the skull from the outside world. Stiffness (high modulus) would be a very important property to ensure that sound waves are transmitted with fidelity for accurate hearing. The fracture strength of the fin whale's bulla therefore is less important, and there is a natural blend of form and function in this bone. The cow's femur, on the other hand, has an elastic modulus, impact strength, and bending strength that are all rather high but not extreme.

Compact bone is an *anisotropic* material, and thus bone's elastic modulus and strength depend on the orientation of the collagen–mineral matrix with respect to the loading direction. As with a piece of wood, bone has a grain associated with its structure. If you apply a compressive load along the long axis of a piece of wood or bone (in line with its grain), the piece of wood or bone has a much greater elastic modulus and strength than

if you applied the load at right angles to the long axis (perpendicular to the grain). In contrast, if you apply loads from any direction to an **isotropic** metal (e.g., stainless steel), the elastic modulus and strength are the same in all directions.

Poisson's ratio (v) is the other parameter used to quantify a bone's elastic behavior. This effect was illustrated in chapter 3, figure 3.28, and mathematically defined in equation 3.35. The ratio v is the negative of the strain transverse to the load divided by the strain of bone in line with the applied load. Bone has a relatively high Poisson's ratio ($v \leq 0.6$), much higher than found in metals.

Toughness is a measure of a tissue's ability to absorb mechanical energy. Cortical bone is considered to be tough because it is able to absorb a great deal of mechanical energy before it fractures. Bone is also a relatively *ductile* material. If a metal is ductile, it can be hammered into thin sheets without breaking. A gold nugget, for example, can be flattened into a wafer-thin foil and thus is considered to be very ductile. Bone, of course, is not as ductile, but it can be deformed (to some extent) without fracture. The opposite of ductile is *brittle*. With increasing age, bones have a tendency to become less ductile and more brittle, less stiff, and more fragile.

Finally, compact bone is viscoelastic. That is, it exhibits strain-rate sensitivity, creep behavior, hysteresis, and fatigue. Indeed, some of the mechanical properties of cortical bone are very sensitive to differing strain rates. As a bone is loaded more and more rapidly, its ultimate strength increases at a faster rate than does its elastic modulus. *Fatigue* is the loss of strength and stiffness that occurs in materials subjected to repeated cyclic loads. Although bone can withstand substantial stresses when loaded only once, as the number of cyclic loads increases, the ability of a bone to withstand the stress decreases exponentially. In bone, fatigue has been attributed to microscopic cracks that develop within and between the osteons (Martin et al. 1998). In healthy bone, if damage is not excessive, remodeling resorbs the material around microcracks, and new bone is deposited. If the damage is excessive, however, and the normal remodeling process cannot keep up with the repair, macroscopic failure (fractures) can happen.

When mechanical properties of a material such as bone are compromised, it can be said that the material is damaged. Using a new type of microscopy, Zioupos and Currey (1994) visualized the three-dimensional distributions of microcracks in cortical bone. Because microcracks usually develop within the interior of compact bone, they are hard to see in situ. But Zioupos and Currey used a laser-scanning confocal microscope to take optical sections through a translucent piece of bone. By focusing the microscope at a given depth within the bone, one can digitize an image. Then through step-by-step refocusing, images can be made at successive layers of the tissue. The thin, sequential sections are then digitally reconstructed to give a three-dimensional picture of the tissue and its components. Zioupos and Currey loaded small strips of bovine compact bone and examined the microcracks that developed. They were satisfied that the microcracks were not caused by machining artifacts and that the cracks were associated with regions of high strain (stress). The microstructure (grain) of the bone affected the propagation of the microcracks within the bone, and microcracks were most likely to occur in the most highly mineralized parts of the bone.

Why are these microcracks significant? As we know, bone has both elastic and plastic responses. In the plastic region, even with decreasing load levels during continuous and cyclic loading, there is an increasing amount of deformation—the bone becomes damaged and increasingly compliant. Dispersed microcracks may weaken bone and decrease its structural stiffness. Clinically, the accumulation of fatigue microdamage may produce a stress fracture—an injury not uncommon for athletes such as ballet dancers, gymnasts, basketball players, and long-distance runners, who place highly repetitive, cyclic loads on their weight-bearing bones.

Trabecular (Cancellous) Bone

The latticework organization of trabecular bone is diverse, and the apparent density and architecture of trabecular bone have potent, nonlinear effects on the elastic modulus and strength of bones. The elastic modulus of trabecular bone can vary from 10 to 2,000 MPa, in contrast to cortical bone, which has an elastic modulus around 13 to 17 GPa.

The spaces in trabecular bone are typically filled with marrow, which can play an important part in the load-bearing capabilities of trabecular bone. If the bone marrow is left in specimens of trabecular bone during impact-speed tests, the strength and elastic modulus are dramatically greater than if the marrow is removed before the test. The enhancing effect of the marrow is minimal, however, at physiological rates of loading. Nevertheless, with or without the marrow, trabecular bone exhibits viscoelastic effects because its fundamental building block is viscoelastic lamellar bone (as exists in compact bone).

Adaptation of Bone

Bone is a dynamic tissue that is exquisitely adapted to the multiple internal factors (e.g., systemic calcium or hormone levels) and external factors (e.g., mechanical loads) that can affect the structure, composition, and quantity of bone. The capacity of bone to adapt its structure to imposed loads has become known as **Wolff's law.** Although Julius Wolff, a 19th-century surgeon, is given credit for this idea, he was not the first to observe bone's adaptive abilities, did not define many of the concepts now associated with his law, and was mistaken in some of his observations. Nonetheless, he did recognize the adaptive capacity of bone, and his name has become inseparably associated with bone adaptation. Wolff was quoted by Keith (1918) as saying that "Every change in the form and function of . . . bone[s] or of their function alone is followed by certain definite changes in their internal architecture, and equally definite secondary alterations in their external conformation, in accordance with mathematical laws" (Martin et al. 1998, p. 225).

Among the concepts that arose in the 19th century and that are now incorporated into Wolff's law are the optimization of bone strength with respect to bone weight, trabecular alignment with the lines of principal stress, and the self-regulation of bone structure by cells responding to mechanical stimuli (Martin et al. 1998).

Events that signal change in bone are classified as either modeling or remodeling. **Modeling** is the addition (formation) of new bone, whereas **remodeling** involves resorption and formation or reformation of existing bone. Differences between modeling and remodeling of bone are summarized in table 4.2.

Modeling can happen at various rates and is a continuous process that can occur on any bony surface to produce a net gain in bone. During modeling, osteoclasts and osteoblasts are not active along the same surface; resorption can occur along one cortex while deposition occurs along another. The specific stimulus that initiates modeling remains unclear. Modeling happens mainly during the growing years. The ability of bone to adapt to mechanical loading is greater during growth than after maturity, but limited modeling can still happen after skeletal maturation (Suva et al. 2005).

Remodeling is the resorption and replacement of existing bone. Skeletal remodeling triggers a release of mineral stored in the bone in response to a low serum calcium level, to repair

TABLE 4.2

BONE MODELING AND REMODELING

	Modeling	Remodeling
Timing	Continuous	Cyclical (ARF)
Resorption and formation surfaces	Different	Same
Surfaces affected	100%	20%
Activation	Not required	Required
Balance	Net gain	Net loss
Coupling of formation and resorption	Systemic? (No ARF)	Local

ARF = activation–resorption–formation.

Adapted, by permission, from A.M. Parfitt, 1984, "The cellular basis of bone remodeling: The quantum concept reexamined in light of recent advances in the cell biology of bone," *Calcified Tissue International* 36(Suppl 1): S38.

skeletal microdamage or to balance mechanical and mass needs of the skeleton (Pogoda et al. 2005). The sequence of remodeling events can be remembered as **ARF (activation–resorption–formation).** The first step in remodeling is activation of osteoclasts to resorb existing bone. New bone is deposited by osteoblasts that follow the resorptive front of osteoclasts. Deposition of new bone takes three times longer than resorption. This translates into a 1-week time lapse between resorption and formation (Martin et al. 1998).

Lanyon (1987) described functional remodeling as an "interpretation and purposeful reaction" to a bone's strain state, allowing for adaptation to both increased and decreased strains. "Functional strains are both the objective and the stimulus for the process of adaptive modeling and remodeling" (Lanyon 1987, pp. 1084-1085). Rubin and Lanyon (1985) proposed that if functional strains are too high, the incidence of damage and probability of failure increase. If strains are too low, the metabolic cost of maintaining the unnecessary bone mass is high, and bone is resorbed. Thus, functional strain appears to be a relevant parameter to control. A question remains, however: To which attribute or combination of attributes of strain (magnitude, rate of application, frequency, distribution, or gradient) does bone have the greatest sensitivity?

Another remodeling theory related to mechanical use of the bone suggests that remodeling is stimulated by fatigue damage that happens during physical activity. In bone, fatigue (loss of strength and stiffness that occurs in materials subjected to repeated cyclic loads) has been attributed to the microscopic cracks that develop within and between the osteons (Martin et al. 1998). In healthy bone, if the damage is not excessive, remodeling resorbs the material around the crack, and new bone is deposited. If the damage is excessive and normal remodeling cannot keep up with the repair demands, macroscopic failure and fracture may result.

Development and Maturation

Much of the accumulated information about the relation between bone growth and bone mineral acquisition is summarized in figure 4.3. Although undoubtedly a wealth of bone model-

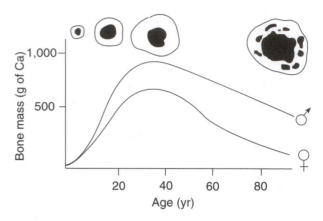

Figure 4.3 Relations among bone mass, age, and sex. The ordinate, bone mass, is represented by the total grams of calcium (Ca) in the skeleton.

Adapted, by permission, from F.S. Kaplan et al., 1994, Form and function of bone. In *Orthopaedic basic science*, edited by S.R. Simon (Park Ridge, IL: American Academy of Orthopaedic Surgeons), 167.

ing and growth happens in both the axial and appendicular (limb) skeleton during childhood until the pubertal growth period, the differences between sexes in **bone mineral content (BMC)** are negligible. At about 13 years of age, the bone mineral content for boys and girls diverges, with boys having a greater rate of gain than girls (figure 4.3).

Usually about 1 year after the peak rate of longitudinal growth **(peak height velocity, PHV),** the peak rate of gain in BMC occurs (figure 4.4). Peak gains in BMC occur at about 13 years for girls and about 14.4 years for boys. Thus, in the year between the time of PHV and the rapid gain in BMC, there is a period of relative bone weakness with a greater chance of fracture. Growth and sex hormones are mainly responsible for the rapid increases in BMC (Suva et al. 2005).

As longitudinal growth begins to slow toward late adolescence, about 90% of adult bone mineral content has already been deposited. Maximal BMC is attained between 20 and 30 years of age. Furthermore, it seems that neither *bone mineral content* nor **bone mineral density (BMD)** increases after the age of 30 for men or women. At skeletal maturity, men have greater BMC than women, and most of that difference is because of men's thicker cortical bone.

By the fifth decade, bone mass begins to decline. Both men and women lose cortical bone at about the same rate, but women lose trabecular bone much more rapidly than men—especially

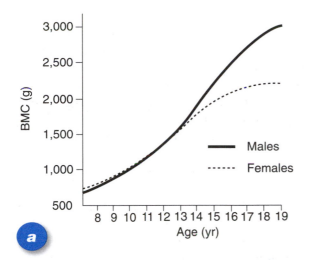

a

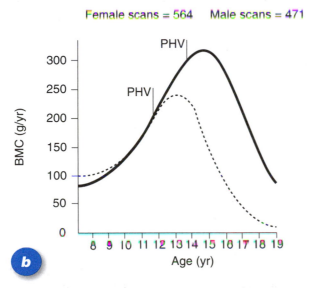

b

Figure 4.4 Bone mineral content (BMC) and peak rate of change in height (peak height velocity, PHV) are shown for both the boys and girls in this cross-sectional sample. *(a)* BMC is given in total grams of mineral in the body, whereas *(b)* the BMC rate-of-gain curve is presented in grams per year.

Reprinted, by permission, from D.A. Bailey, R.A. Faulkner and H.A. McKay, 1996, "Growth, physical activity, and bone mineral acquisition," *Exercise and Sport Sciences Reviews* 24: 240.

when women enter menopause. Between the ages of 40 and 50, men will lose up to 0.75% of total bone mass yearly, whereas women can lose bone at more than twice that rate. In the few years immediately following menopause, the yearly rate of bone loss in some women can be as much as 3% (Bostrom et al. 1994).

Reaching maximal bone mass in adolescence and early adulthood, therefore, is crucial for reducing the effects of bone loss and mitigating

fracture risk later in life. Suggestions for optimizing bone mineral acquisition during the growing years include making a lifelong commitment to weight-bearing physical activity at an early age; engaging in a variety of vigorous daily activities of short duration as opposed to prolonged repetitive activity; doing activities that increase muscle strength and work all large muscle groups; and avoiding periods of immobility (Suva et al. 2005).

Nutrition

Nutrition has potent effects on bone growth and remodeling and therefore on bone quality and mechanical properties. Next we discuss a few of the nutritional factors that can influence bone quality and quantity: calcium, vitamin D, protein, fats, and sugar.

Normally, the body's bone–mineral balance is regulated by the synergistic actions of vitamin D metabolites, parathyroid hormone, and calcitonin—substances that influence the dietary absorption of calcium, bone mineral resorption and deposition, and the renal secretion and resorption of calcium and phosphorus. Of the total-body calcium, about 99% is found in the skeleton, with the remaining 1% circulating in the extracellular fluid (Bostrom et al. 1994). Calcium compounds constitute more than half of the mass of bone. Because calcium is excreted throughout the day, adequate calcium intake is vital for bone health. Bones with diminished bone mineral content are less stiff and consequently may be more prone to fractures.

Vitamin D, dietary protein, phosphorus, fiber, and fats all can affect calcium absorption. Vitamin D is a fat-soluble molecule that can be stored in body fat. Stores of vitamin D primarily depend on the amount of time the skin is exposed to the sun and on the size of the exposed area. A relatively small fraction of the body's vitamin D stores comes from the diet. Vitamin D helps to increase calcium absorption from the intestinal tract; thus, a person deficient in vitamin D poorly absorbs dietary calcium (Staud 2005).

Dietary protein has a significant effect on urinary calcium handling (Ashizawa et al. 1997). Protein deficiency can lead to decreased levels of calcium in the urine (hypocalciuria) and reduced calcium absorption in the intestine (Hengsberger et al. 2005). Conversely, excessive dietary protein can result in greater renal calcium loss and the development of negative calcium balance (Giannini et al. 2005). Protein deficiency has been implicated in the genesis of osteopenia (reduced bone mass) in malnourished humans (Deprez and Fardellone 2003) and animals (Bourrin et al. 2000).

For optimal bone health, excessive ingestion of saturated fats and refined sugar should be avoided because both can have negative effects on the body's ability to absorb calcium. High levels of sugar cause excess acidity, which causes the calcium to leach out of the bones to balance the pH level. High levels of dietary fatty acids reduce the amount of dietary calcium absorbed in the intestine and thereby lead to lower calcium levels in bones (Liu et al. 2004).

Use Versus Disuse

Exercise and physical activity can stimulate bone remodeling, but how exercise affects the skeleton is profoundly complex. Exercise intensity, skeletal maturity, type of bones (trabecular or cortical), and anatomical location (axial or extremity bones) can all influence the response of specific bones to exercise (Lieberman et al. 2003).

Use

The following five points summarize current knowledge of the relations between physical activity and bone mass: (1) Growing bone responds to low or moderate exercise through significant addition of new cortical and trabecular bone, with periosteal expansion and endocortical contraction. (2) A threshold of activity exists above which some bones respond negatively by suppressing normal growth and modeling activity. (3) Moderate to intense physical training can generate modest increases (1-3%) in BMC in men and premenopausal women; in young adults very strenuous training may increase BMC of the tibia by up to 11% and its bone density by 7%. Some evidence shows that exercise can also add bone mass to the postmenopausal skeleton, although the amounts are modest and site specific. After 1 to 2 years of intensive exercise, increases as high as 5% to 8% can be found, but usually less

than 2%. (4) The long-term benefits of exercise are retained only by continuing to exercise. (5) The amount of bone mass that can be achieved appears to depend primarily on the initial bone mass, suggesting that individuals with extremely low initial bone mass may have more to gain from exercise than those with moderately reduced bone mass (Lorentzon et al. 2005).

In the adult skeleton, regular prolonged exercise can increase the skeletal mass (Bass et al. 2005; Lorentzon et al. 2005; Suva et al. 2005), cortical thickness (Hiney et al. 2004; Specker et al. 2004), and bone mineral content (Engelke et al. 2006; Iwamoto et al. 2004; Korpelainen et al. 2006). If, for example, a person participated in a sport such as tennis for many years, greater bone density would be expected in the radius and the ulna of the dominant arm.

The Question of Diminishing Returns

Are there thresholds above which there is a diminishing return in adding new bone in response to greater and greater exercise? One facet of that question was examined by MacDougall and colleagues (1992) at McMaster University.

These researchers investigated the relation between the amount of running and bone mineral mass in adult male runners. Using dual-photon absorptiometry, they examined the bone density of the trunk, spine, pelvis, thighs, and lower legs of 22 sedentary controls and 53 runners who were selected according to their running mileage. The runners were grouped by their weekly mileage as follows: 5 to 10, 15 to 20, 25 to 30, 40 to 55, and 60 to 75 miles per week. The ages (20-45 years) and dietary habits of the runners were similar. In this cross-sectional study, the researchers found no significant differences in bone density measurements, except in the lower legs. The bone mineral density of the lower legs of runners in the 15 to 20 miles per week group was significantly greater than that of the control or 5 to 10 miles per week groups. Interestingly, the researchers found no further increase in bone mineral density in the lower legs of runners who covered more than 20 miles per week. Indeed, there was a tendency for decreased bone mineral density for runners who covered 60 to 75 miles per week. These high-mileage runners had bone mineral density that was no different from that of the sedentary controls.

These data suggest that the amount of running may influence bone mineral density and bone thickness in weight-bearing bones but that there also may be a threshold effect for these adaptations—both at the high and low ends of the loading spectrum. Other investigators have found similar lower bone mineral densities in the vertebrae (Bilanin et al. 1989) and the legs (Ormerod et al. 1988) of very high-mileage male runners compared with normal subjects and weightlifters. In the aggregate, these studies suggest that there is an upper limit for exercise beyond which bone mechanical integrity may stop increasing and actually decrease.

Female Athlete Triad

Whether it is during puberty with its rapid bone growth or during menopause with its precipitous loss of bone, sex hormones have a potent effect on bone health. For females, estrogen is important in building or maintaining the skeleton, and it is apparent that highly intensive training by young female athletes can lead to deleterious consequences on the skeleton as a result of disturbances in normal menstrual cycles.

In sports such as gymnastics, ballet, running, and figure skating, there are numerous examples of young female athletes who experience no menstruation (amenorrhea) or only intermittent

---

IS EXERCISE ENOUGH?

Snow-Harter and Marcus (1991) stated that the efficacy of exercise in preventing and treating osteoporosis still remains unclear. Nevertheless, they summarized what is known by answering the following questions (p. 381):

- Can exercise maximize peak bone mass?
- Can exercise forestall or reduce age-related bone losses?
- Does exercise enhance bone mineral density in people with existing osteoporosis?
- Can exercise supersede estrogen replacement therapy during the postmenopausal years?

With reasonable qualifications, the answer is yes to the first three questions, but the answer to the fourth question is no. To date, there is no basis to state that exercise alone is as effective as estrogen replacement for maintaining bone mass and reducing the fracture risk for post-menopausal women (Maddalozzo et al. 2007). Current thinking on hormone replacement therapy (HRT) still supports the view that HRT should not be used to prevent or treat chronic diseases such as osteoporosis, due to deleterious side effects (Dull 2006).

---

and irregular menstruation (oligomenorrhea) as a consequence of altered hormone levels. With the lower estrogen levels that accompany amenorrhea and oligomenorrhea, these young athletes may not achieve a peak bone mass as great as they would have if they had experienced normal levels of estrogen. When this is coupled with calorie-restricted diets to reduce body weight, there is a further danger of slipping into what is termed the female athlete triad, namely disordered eating, disrupted hormone levels (and accompanying

menstrual dysfunction), and increased risk of poor bone quantity and quality. A comparison of the vertebral mineral mass of amenorrheic elite athletes and cohorts with normal menstrual function found that the amenorrheic athletes had up to 25% less mineral mass than the normal women, suggesting that estrogen deprivation has a powerful effect on trabecular mineral mass (Marcus et al. 1985).

Not only are these young female athletes at potentially greater risk of osteoporosis in later years, but they also have a likelihood of developing stress fractures in their late teens and early 20s (Zernicke et al. 1994). Among the multiple factors that could contribute to lower-extremity stress fractures in elite intercollegiate runners, one of the most significant predictors of who did and who did not sustain a stress fracture was how much time elapsed between beginning serious, high-intensity training and the runner's first menses. Those runners who had been training in earnest longer before their first menses (and with the heavy training, that first menses may also have been delayed) had a significantly greater chance of having a stress fracture while they were running at the intercollegiate level.

Interestingly, high-intensity and high-impact loading activities (even though they may disrupt normal hormonal balance in young females) may partially counterbalance the full effects of the menstrual disturbances (Bailey et al. 1996). For example, Robinson and colleagues (1995) reported that although 47% of the young female gymnasts they studied were either oligomenorrheic or amenorrheic, these gymnasts had greater bone mineral density than either a cohort of runners or control females with normal menses. According to Bailey and colleagues (1996, p. 256), the data suggest that high-impact loading activities in female athletes with disrupted menstrual function have a "sparing effect at weight-bearing sites but not to the same extent at non-weight-bearing sites."

Disuse

Disuse-related changes in bone are commonly associated with bed rest, immobilization, or space flight. Without normal loading, resorption substantially increases and deposition of bone decreases. In some cases, bone resorption can increase dramatically in a very short period of time (Loitz-Ramage and Zernicke 1996).

In humans, the adverse effects of reduced loading on bone have been highlighted by the substantial skeletal degeneration and calcium loss that can occur in space flight. The loss of bone density may not be as dramatic in the non-weight-bearing bones (e.g., radius and ulna), but weight-bearing trabecular bones (e.g., calcaneus) are particularly sensitive to the lack of normal loads experienced during microgravity (Doty 2004; Oganov 2004).

Because disuse changes are dramatic and immobilization may be necessary in some instances, the question of whether disuse-related changes are reversed by remobilization is important. To examine this issue, a study by a group from Finland (Tuukkanen et al. 1991) reported the effects of 1 or 3 weeks of immobilization followed by 3 weeks of remobilization in rats. The researchers examined the tibia and the femur; after 3 weeks of immobilization, the bone ash weights decreased as much as 12% compared with nonimmobilized controls. After a period of remobilization, the tibia recovered 62% of its mineral mass, whereas the femur regained only 38% of the lost mineral mass. This research shows that mineral loss caused by immobilization can be reversed to some extent but that the recovery does not occur as rapidly as the loss of bone. Furthermore, the degree of recovery is related to the length of immobilization (Loitz-Ramage and Zernicke 1996). These potent immobilization and remobilization effects have fueled the use of fracture braces and early mobilization in the management of orthopedic injuries.

Biomechanics and Adaptation of Other Connective Tissues

Whereas articular cartilage functions primarily under compressive loads, ligaments and tendons are usually under tensile loads. The extracellular matrix composition of ligaments and tendons is specially formulated for these specific tasks. However, there are special cases where specialized structures must be able to function under both compressive and tensile loads. A classic example of this is the Achilles tendon, which connects the calf muscles to the heel and has both a region capable of bearing tension and another that is subjected to compressive forces.

Articular Cartilage Biomechanics

Understanding the synergy among collagen fibers, proteoglycans, and synovial fluid is essential for understanding the mechanical behavior of articular cartilage—the thin layer of hydrated soft tissue covering the ends of the bones of diarthrodial joints. Normally, these components "perform their functions so well that we are often not even aware of their existence nor the functions they provide until injury strikes or arthritis develops" (Mow et al. 1992).

Load Response

Type II collagen is the principal fibrous protein in hyaline cartilage. Figure 4.5 shows how collagen fibers are recruited as a tensile load is applied. Initially **(toe region),** the fibers are partially relaxed and wavy in appearance. As the tensile load continues to increase (linear region), the fibers straighten and become taut. If the load is increased even further, individual fibers begin to tear, and finally large groups of fibers fail in tension (failure).

We usually think of a tensile load applied to tissue as tending to pull the tissue apart. With a straight tendon, a simple tensile load is easy to envision. With articular cartilage, however, recall that collagen changes its orientation throughout the depth of the cartilage (figure 2.13b). The fibers of the superficial tangential zone are oriented parallel to the surface of the joint, whereas in the middle region the fibers are more randomly arranged. In the deepest layer, the radially directed fibers penetrate into the underlying bone.

Tension is applied to the collagen fibers in the cartilage matrix in many ways. Shearing forces can be generated along the surface of the articular cartilage as the two bone ends move past each other. The surface collagen fibers can be deformed in this way. Also, recall that the negatively charged and hydrophilic proteoglycan aggregates (aggrecan) repel each other and draw water into the extracellular matrix. The cartilage has a natural tendency to swell, and this tendency to expand is resisted by the matrix collagen. Thus, a homeostatic balance, or equilibrium, develops because of the cartilage's swelling pressure and the tensile restraint of the collagen fibers.

Creep refers to the deformation of a tissue as a constant load is applied, and the creep response is a flow-dependent mechanical behavior of the cartilage. That is, a constant load causes extracellular fluid to exude from the matrix. After time, the cartilage reaches a stable compression deformation, or equilibrium strain. If the compressive load is released, the fluid is drawn into the matrix by the hydrophilic proteoglycans (Mow et al. 1992). Cartilage replete with water is easier to compress, and cartilage densely packed with charged glycosaminoglycan is more resistant to compression.

The cyclic loading and unloading of cartilage allow for a dynamic flow of fluid (with accompanying nutrients and waste products) into and out of the tissue. Thus, cyclic loading of cartilage can be beneficial for normal matrix and cell health. Excessive loading, however, can be destructive to the matrix.

Lubrication Mechanisms

Diarthrodial joint surfaces have remarkably low coefficients of friction attributable in part to the synovial fluid found in the cavities of synovial joints. Synovial fluid lubricates and cushions the tissues of the joint during movement. Articular cartilage contributes significantly to joint lubrication by means of cartilage's intrinsic properties and fluid flux. The coefficient of friction for articular cartilage in human diarthrodial joints has been reported to range from 0.01 to 0.04. By comparison, the coefficient of friction for ice sliding on ice at 0 °C ranges between 0.01 and 0.10, and the coefficient of friction for glass sliding on glass is 0.90. Diarthrodial joints have been called natural bearings that are nearly frictionless and nearly wear-resistant throughout our lives (Mow et al. 1992).

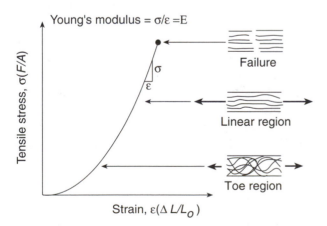

Figure 4.5 Recruitment of collagen fibers within the articular cartilage as a tensile load is applied to the tissue.

Reprinted, by permission, from V.C. Mow, A. Ratcliffe and A.R. Poole, 1992, "Cartilage and diarthrodial joints as paradigms for hierarchial materials and structures," *Biomaterials* 13(2): 84.

How is this elegant system of lubrication achieved? Several hypotheses have been proposed, but the answer is elusive. Two principal types of lubrication mechanisms have been proposed: boundary and fluid film. **Boundary lubrication** happens when a layer of molecules adheres to each of the two surfaces that are gliding past each other. **Fluid-film lubrication** exists when two nonparallel surfaces move past each other on a thin layer of fluid. The fluid wedge is trapped between the two moving surfaces and maintains a distance between the two surfaces. If the two moving surfaces are nondeformable, this produces a subtype of fluid-film lubrication called **hydrodynamic lubrication.** If one or both of the surfaces are deformable (usually considered as isotropic and linear elastic), this second subtype is called **elastohydrodynamic lubrication.** This latter type of lubrication commonly is used to model the lubrication in diarthrodial joints, because the articular cartilage is a deformable substance (Lo et al. 2003; Mow et al. 1992).

Additional theories of cartilage fluid-film lubrication have been proposed (Lo et al. 2003). One mechanism has been called **squeeze-film lubrication.** In this model, the two surfaces move at right angles to each other, as might happen in your knee joint at the instant of heel strike in walking. The weight of your body tends to bring the distal femur and proximal tibia closer together, and fluid is forced out of the cartilage to produce a fluid interface between the two surfaces. This type of lubrication would be effective only for a short duration and would work better with heavier loads.

Boosted lubrication is a potential mechanism that incorporates elastohydrodynamic and squeeze-film types of lubrication. Boosted lubrication may occur, for example, in the knee joint during the stance phase of walking or running. As the femoral and tibial articular surfaces assume load and slide past each other, the articular cartilage of both surfaces is deformed. As the deformation happens, matrix fluid is forced into the space between the surfaces, and this dynamic fluid flow increases the fluid's viscosity, which in turn boosts the effectiveness of the lubricating fluid film.

Articular Cartilage Adaptation

Articular cartilage is extremely well adapted for its purpose in synovial joints. As with many load-bearing connective tissues, articular cartilage has an amount of use that provides optimal function. If the cartilage is used too little (e.g., immobilization) or too much (e.g., excessive loading), the quality of cartilage breaks down. The active loading and unloading of articular cartilage may facilitate the diffusion of nutrients through the cartilage matrix, which is avascular in the adult. As described later, cartilage does adapt, but in many cases the adaptation is degenerative, leading to osteoarthritic changes to the joint.

Development and Maturation

Immature articular cartilage looks quite different from adult tissue. Immature cartilage appears blue–white and is comparatively thicker than adult tissue. The thickness appears to be a function of the substantially greater number of cells in young cartilage, which are found not only on the articular surface but also in the epiphyseal plate.

Besides the morphological differences in young versus older articular cartilage, a substantial difference is found in the biochemistry of articular cartilage as a function of age. The relative water content in immature articular cartilage is substantially greater than in the adult. Conversely, the collagen concentration increases with maturity, and proteoglycan content in articular cartilage is highest at birth and diminishes slowly throughout growth. The protein core and the glycosaminoglycan chains are longer in immature articular cartilage, and with advancing age, synthesis of proteoglycan decreases. The average length of the proteoglycan protein cores decreases with age (Mankin et al. 1994). Recall the analogy of the bottle brush (chapter 2). As articular cartilage ages, there are fewer and shorter bristles on the brush. This change in proteoglycan may account for some of the changes in the mechanical properties and resilience of articular cartilage as an individual ages.

Use Versus Disuse

In both animals and humans, exercise causes articular cartilage to swell (Walker 1996). Prolonged exercise in animals may produce chondrocyte (cartilage cell) **hypertrophy,** an increase in the pericellular matrix, and an increase in the number of cells per unit of cartilage (Englemark 1961). These effects are already evident after brief bursts of exercise, but long-term exercise produces a lasting change in the cartilage. In some instances, mechanically loaded chondrocytes responded positively by displaying increased proteoglycan and protein biosynthesis in cartilage explants (Jin et al. 2001), but more typical findings are that changes caused by wear and tear can accompany increased loading (Chen et al. 2003). Particularly with excessive loading, synthesis can decrease and degradation increase within articular cartilage. These changes may lead to **osteoarthritis, the most common joint disease in humans** and the leading cause of chronic disability in elderly persons. In describing the pathogenesis of osteoarthritis, Brandt (1992) stated,

> [Osteoarthritis] develops in either of two settings: (1) the biomaterial properties of the articular bone and cartilage are normal, but excessive loads applied to the joint cause the tissues to fail, or (2) the applied load is physiologically reasonable, but the biomaterial properties of the cartilage or bone are inferior. . . . In general, the earliest progressive degenerative changes in [osteoarthritis] happen at those sites within the joint that are subject to the greatest compressive loads. (pp. 75-76)

The exact cause of osteoarthritis unknown. It is likely that there is no single cause or common final pathway but rather a variety of factors that contribute to the end stage of osteoarthritis. Among the factors that may contribute are heredity; alterations in chondrocyte activity; changes in humeral-, synovial-, and cartilage-derived chemical mediators (e.g., interleukin-1); and altered joint mechanics—especially excessive joint laxity attributable to previous ligament injury (Mankin et al. 1994).

At the other end of the spectrum, if the loading of articular cartilage is substantially reduced, the cartilage can also significantly **atrophy,** or degenerate. When cartilage is left unloaded for substantial lengths of time, there can be a marked reduction in the synthesis and amount of proteoglycan in the cartilage, an increased fibrillation of the surface of the cartilage, and a decrease in the size and amount of the aggregated proteo-

glycan. Similarly, the mechanical properties of articular cartilage are degraded with prolonged immobilization. Cartilage that has been immobilized deforms much more rapidly when it is compressed, as the fluid rapidly exudes from the matrix. All of the biochemical and biomechanical changes that occur in articular cartilage as a consequence of immobilization are reversible, in part, after remobilization of the joint (Mankin et al. 1994).

Thus, articular cartilage can be compromised by too little or too much loading. Either a lack of stress or excessive stress causes degeneration in the articular cartilage, and there is a normal physiological zone of cyclic loading that promotes optimal cartilage health.

Tendon and Ligament Biomechanics

The average ultimate tensile stress of tendons and ligaments ranges between 50 and 100 MPa. The *ultimate load* is related to the cross-sectional area of the specific tendon or ligament. Ultimate loads in tendons, for example, can be extremely large, particularly in tendons such as the Achilles or patellar tendon. As explained in chapter 6, the patellar tendon of a competitive weightlifter was able to withstand an estimated 14.5 kN (more than 17.5 times the body weight of the weightlifter) before rupturing (Zernicke et al. 1977).

Like bone and articular cartilage, both tendon and ligament exhibit viscoelastic behaviors such as cyclic and static force relaxation (figure 3.43), hysteresis (figure 3.41b), and creep (figure 3.42). By comparison, bone is much more sensitive to changes in strain rate; nonetheless, the mechanical responses of both tendons and ligaments exhibit moderate strain-rate sensitivity. For example, if a tendon is stretched at a fast strain rate, its stiffness will be greater than if it is stretched at a slower strain rate. This differential effect of strain rate on bone and ligament can influence which structure is injured as a load is applied. If the tensile load is applied very quickly, it is more likely that the ligament will fail, whereas if the load is applied very slowly, it is more likely that a piece of bone at the ligament attachment site will fail (avulsion fracture).

A classical stress–strain curve for a ligament loaded in tension is shown in figure 4.6. At low stresses (toe region), the crimp, or waviness, of the collagen fibers begins to disappear. As the collagen fibers straighten, linearity develops, and in this relatively linear portion of the curve the material's elastic (Young's) modulus can be measured. As noted earlier, the elastic modulus *(E)* is the slope of the linear region of the stress–strain curve. Note that the strains in the toe and linear regions are relatively small (0-4% strain).

Near the latter part of the linear loading region, some of the collagen fibers may exceed their load-bearing capacity and rupture. If the load or stress was removed at that point, there would be a partial failure of the tendon or ligament, but the remaining intact fibers of the structure might still be able to carry out the load-transmission function. The partial tear may induce an inflammation response, and subsequent healing eventually will form scar tissue (chapter 5). If the tensile stress was increased even further, the remaining fibers of the tendon or ligament would fail (figure 4.6).

Like bone, a ligament or tendon has a specific geometry that affects its structural mechanical behavior. Figure 4.7, *a* and *b*, depicts two situa-

tions to highlight how ligament (or tendon) size could affect its mechanical function in the body. In figure 4.7*a*, we show how two ligaments of equal fiber length but different cross-sectional areas can have different mechanical behaviors. The 2*A* ligament has twice the cross-sectional area (six collagen fibers in parallel) of the *A* ligament (three collagen fibers in parallel). As a tensile load is applied to each of the ligaments, the 2*A* ligament with twice the cross-sectional area will have double the tensile strength and stiffness of the *A* ligament. The elongation to failure, however, will be the same in both ligaments.

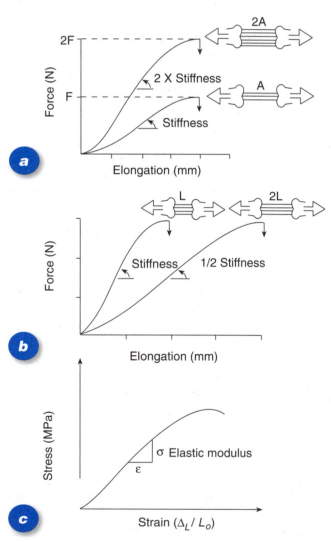

a

b

c

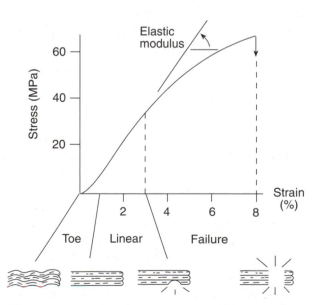

Figure 4.6 Example of a stress–strain curve for collagen fibers in a ligament. The stress–strain curve provides estimates of the material properties of the ligament independent of the size of the structure.

Reprinted, by permission, from D.L. Butler et al., 1978, "Biomechanics of ligaments and tendons" *Exercise and Sport Sciences Reviews* 6: 145.

Figure 4.7 Effects of a ligament s *(a)* cross-sectional area and *(b)* original length on its structural properties. *(c)* The normalizing effect of stress and strain calculations to estimate the material properties of ligaments of various sizes.

Reprinted, by permission, from D.L. Butler et al., 1978, "Biomechanics of ligaments and tendons" *Exercise and Sport Sciences Reviews* 6: 144.

Figure 4.7*b* shows the responses of two other ligaments—*L* and 2*L*. Here, we show the mechanical response of a long ligament versus a short ligament. Ligament 2*L* is twice the original (resting) length of ligament *L*, but each of the ligaments has the same number of collagen fibers in parallel (same cross-sectional area). If a similar load is applied to each ligament, the 2*L* ligament will have one-half the stiffness and twice the elongation at failure as the *L* ligament. But because each of the ligaments has the same cross-sectional area, the structural strength is the same for both ligaments.

Figure 4.7*c* shows that all four of these ligaments (even with their very different geometry) can have the same *material properties,* such as elastic modulus. Obviously, this is because the material properties of stress, strain, and elastic modulus are determined from values that are normalized to the differences in the geometry of the structures.

Tendon and Ligament Adaptation

Several decades ago, tendons and ligaments were considered to be passive and inert cords for transmitting loads. Since then, research has revealed the marked adaptive abilities of these fibrous connective tissues.

Development and Maturation

The mechanical properties and composition of tendons are greatly influenced by age. Prior to skeletal maturity, tendons and ligaments are slightly more viscous and are relatively more compliant (Frank 1996). With increasing age, the stiffness and modulus of elasticity increase within the linear range up until the point of skeletal maturity, and then these properties remain relatively constant (Woo et al. 1994). In middle age, the insertional attachments of ligaments or tendons into bone begin to weaken, viscosity begins to decline, and the collagen becomes more highly cross-linked and less compliant (Frank 1996; Woo et al. 1994). With age, bones also become more fragile, and the insertion between a ligament or tendon and the bone becomes a weak link. Avulsion fractures (in which a piece of bone pulls away from its attachment site) become more common as aging progresses.

Use Versus Disuse

Dense, fibrous connective tissues, such as ligaments and tendon, are sensitive to both training and disuse (Taylor et al. 2004; Woo et al. 2004). With exercise, normal tendons and ligaments adapt to the greater loads by becoming larger or by changing their material properties to become stronger per unit area (Kasashima et al. 2002; Kjaer 2004).

The primary effects of exercise on ligaments are increased structural strength and stiffness. Normal, everyday activity (without training) is apparently sufficient to maintain about 80% to 90% of ligament's mechanical potential (Frank 1996). Exercise appears to have the potential to increase ligament strength and stiffness by up to 20% (figure 4.8). For example, if ligament strength in the experimental immobilization group was only one-half of the strength of nonimmobilized normal control, then the experimental ligaments would be 50% compared with control. Exercise tends to moderately enhance the structural and mechanical properties of ligaments over time, whereas immobilization has a dramatic negative effect within 8 to 12 weeks (e.g., see second black

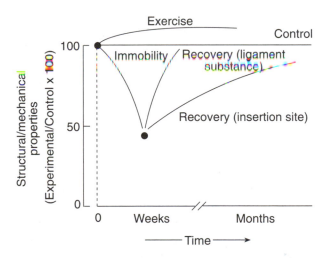

Figure 4.8 Ligament responses to exercise, immobilization, and remobilization. Starting at the first black dot (0 weeks), the *y* axis shows percentages, with 100% being a constant normal (control) level of structural and mechanical properties (e.g., stiffness or maximal tensile stress). The notation Experimental/Control ×100 on the *y* axis illustrates how the data for the curves were estimated. The *x* axis shows time, with early responses shown in weeks and later responses extending to months.

Adapted, by permission, from S.L. Woo et al., 1987, "The biomechanical and morphological changes in the medial collateral ligament of the rabbit after immobilization and remobilization," *J Bone Joint Surg Am* 69: 1200.

dot on figure 4.8). The two curves emanating from the second black dot show how moving (remobilizing) a joint after a period of immobilization is different for properties of the ligament substance (rapid recovery) versus ligament insertion site to bone (slower recovery).

Fewer quantitative data are available about the exercise-related adaptations of tendon than are available for ligaments. Data suggest that exercise can increase the number and size of collagen fibrils and increase the cross-sectional area of tendons compared with the tendons of sedentary controls (Kasashima et al. 2002). Exercise can lead to increased collagen synthesis in growing tendons (Kjaer 2004) and an increased number of fibroblasts in tendons (Benjamin and Hillen 2003).

Load deprivation or joint immobilization produces a rapid deterioration in ligament biochemical and mechanical properties, attributable in part to atrophy, which causes a net loss in ligament strength and stiffness (Frank 1996; Mullner et al. 2000). Immobilization or disuse decreases the glycosaminoglycans and water content of ligaments, increases the nonuniform orientation of the collagen fibrils, and increases reducible collagen cross links (Akeson et al. 1987). Collagen synthesis and degradation rates increase with immobilization, so the ratio of new collagen to old increases in immobilized ligaments (Harwood and Amiel 1992). Furthermore, decreases occur in the total collagen mass (Harwood and Amiel 1992) and stiffness (Kjaer 2004; Palmes et al. 2002) of ligaments. Tipton and colleagues (1975) concluded that the strength of the bone–ligament junction is related to the type of exercise regimen and not only the duration of the exercise—with endurance training being more effective. Exercise generally increases the strength of the bone–ligament junction (Doschak and Zernicke 2005). Similarly, Achilles tendon rupture without repair leads to calf muscle atrophy. These findings indicate that Achilles tendon ruptures should be surgically repaired soon after injury and early mobilization implemented to reduce range of motion loss, increase blood supply, and reduce the degree of muscle atrophy, especially in athletes (Sorrenti 2006).

The effects that exercise, immobilization, and remobilization have on ligaments and bone are illustrated in figure 4.8. The effect of immobilization or disuse is dramatic and rapid. With immobilization, major deterioration happens within a few weeks as ligament cells produce inferior ligament material, which in turn contributes to the structural weakening of the ligament complex (Frank 1996). At the same time, there is an osteoclast response at the bone–ligament junction that decreases the strength of the insertional bone.

Biomechanics of Skeletal Muscle

The fundamental properties of muscle are linked to force, length, and velocity. In chapter 2 we reviewed muscle contractile dynamics (figures 2.17 and 2.18) and length–tension and force–velocity relations (figure 2.21).

The fundamental mechanical events of a muscle contraction relate to twitch properties, unfused tetanic contraction, and tetanus. A single twitch develops in a muscle when the muscle is stimulated once. If another stimulus arrives to depolarize the muscle before it completely relaxes, the stimuli have an additive effect, and the force produced in the muscle is greater than with a single twitch. As the volleys of stimuli (impulses) arrive faster and faster, the tension developed in the muscle will continue to increase, until at maximal stimulation the muscle is in **tetany** (or **tetanus**) and a steady maximal tension is achieved. If the stimulation rate is slightly less than maximum, then fluctuations in the tension are seen; this is called **unfused tetanic contraction.**

The total force developed in a skeletal muscle is proportional to the number of cross-bridges in parallel, whereas the rate at which force can be developed in a muscle is proportional to the number of sarcomeres in series. As such, increasing muscle cross-sectional area through resistance training augments the recruitment of contractile proteins, such as *actin* and *myosin*, into each individual muscle cell, resulting in proportionally more cross-bridge connections, increasing force production.

To induce muscle cell hypertrophy, a threshold stimulus must be reached to induce adaptation in expression of contractile proteins, which will lead to subsequent muscle enlargement. This phenomenon is based on the training principle of overload, which states that muscles worked close to their force-generating capacity will increase in strength (McArdle et al. 2001). Thus resistance training programs are developed on the basis of exercise volume (number of repetitions and sets), frequency (number of times exercised per week), and intensity (usually measured as a percentage of a one repetition maximum) (Macaluso and De Vito 2004). The key is that the loads progressively increase to provide an adequate stimulus throughout the entire training regime. Rapid improvement in the ability to perform the strength exercises occurs in weeks 1 and 2 attributable to a learning effect, which involves improvements in motor skill coordination and level of motivation. During weeks 3 and 4, muscle strength increases without a corresponding increase in muscle size (Macaluso and De Vito 2004). These improvements are mainly attributable to neural adaptations, such as intermuscular coordination (efficient movements from agonist and antagonist muscle groups), intramuscular coordination (the number of motor units recruited and the synchronization of motor units), and an increased neural drive from the central nervous system. Beyond 6 weeks, muscle hypertrophy occurs both within the whole muscle (5-8% increase in size) and within the muscle fibers themselves (25-35% increase) (MacDougall 2003). This phenomenon occurs without **hyperplasia** (cell division) (McCall et al. 1999).

Muscle cell hypertrophy occurs through the activation of local support cells, called satellite cells, which are mitotically quiescent myoblasts located between the sarcolemma or the muscle fiber and its extracellular matrix (ECM). On physical stimulation, insulinlike growth factor-1 found circulating in the ECM is able to react with satellite cells, causing them to divide. The resultant daughter cells then fuse with the underlying muscle fiber, adding nuclei, cytoplasm, and proteins to the existing fiber. Because the number of nuclei within the muscle fiber increases, contractile protein synthesis can be up-regulated and hypertrophy of the muscle cell results (Chakravarthy et al. 2000; Ehrhardt and Morgan 2005; Kadi et al. 2005). As a result of hypertrophy, muscle cross-sectional area increases, and capillary and mitochondrial density decreases as a result of a dilution effect (Baldwin and Haddad 2002).

Adaptation of Skeletal Muscle

Skeletal muscle is capable of enormous adaptation, as can be readily appreciated by contrasting the muscular development of an Olympic gymnast or weightlifter with the wasted muscles of an individual who has been bedridden for years. The type of training influences the type of muscle adaptation. **Endurance training** enhances a muscle's oxidative potential, whereas **resistance training** increases a muscle's myofibrillar diameter (Platt 2005).

Development and Maturation

As explained in chapter 2, skeletal muscle cells are derived from the mesoderm (Brand-Saberi 2005; Tajbakhsh 2003). The typical muscle fiber type in the early fetus is a primitive fast-twitch fiber (Bandy and Dunleavy 1996). As the neurological and muscular systems mature, histochemically identifiable fiber types begin to emerge (Buckingham et al. 2003). Williams and Goldspink (1981) indicated that although muscle development is not fully complete at birth, the number of muscle fibers in the body is probably set at birth, and continued muscle growth is a result of the increase in fiber size (both in width and in length). After the first year of life, an individual-specific distribution of fast-twitch and slow-twitch fibers begins to emerge in the musculoskeletal system. The developmental addition of sarcomeres and the hypertrophy of muscle fibers continue until growth ceases and adult fiber size is reached (approximately 15 years of age).

Although a bone has growth plates that allow it to extend in a longitudinal direction, the length associated with skeletal muscle growth is usually derived from the addition of sarcomeres to the muscle fibers, primarily in the region of the myotendinous junction. If a muscle–tendon unit is stretched, additional sarcomeres are typically added at the region of the myotendinous junction (Garrett and Best 1994).

Compared with the amount of information on muscle adaptation in adults, the data related to muscle adaptation in children are sparse. Bandy and Dunleavy (1996) indicated that in prepubescent boys who underwent a program of progressive resistance training, for example, muscle strength increases happened without any appreciable change in cross-sectional area. The researchers suggested that the increased strength was a result of improved coordination of muscle groups responsible for the movements as opposed to changes in fiber size. Maximal strength in men and women is reached between the ages of 20 and 30 years, about the same time that the cross-sectional area of muscle is the greatest. The strength level tends to plateau through the age of 50, followed by a decline in strength that accelerates by 65 years of age and beyond.

The loss of strength with aging may be related to the loss of muscle mass associated with a reduction in the number of muscle fibers. Typically, the fast-twitch muscle fibers are hit harder than slow-twitch muscle fibers as aging progresses. Skeletal muscle of elderly individuals responds to resistance training with increases not only in performance but also in muscle mass and muscle fiber size, particularly in fast-twitch fiber areas. Individuals in their 60s and older are able to respond to progressive resistance training by increasing strength through hypertrophy of skeletal muscle. Such increases in strength (or maintenance of strength) may have significant consequences for maintaining neuromotor coordination and reducing the likelihood of falls and injury (Bandy and Dunleavy 1996).

Gender-Related Effects

Before puberty, gender differences in athletic performance are not obvious. Girls do not experience a significant change in muscle mass during puberty, and before the onset of puberty the muscular strength of girls and boys is essentially equal. At ages 11 to 12 years, girls are approximately 90% as strong as boys, and at ages 13 to 14 years, the girls' percentage strength decreases to about 85% that of boys (Komi 2003). By the ages of 15 to 16, girls are typically 75% as strong as boys. These diverging strength values are consistent with the differences that exist between men and women in the adult range. For example, the average percentage muscle mass (per total body mass) for a

conditioned female athlete is about 23%, whereas in the conditioned male athlete it is 40%. The average percentage fat (per total body mass) for conditioned women is in the range of 10% to 15%, whereas in conditioned men it is less than 7%. For the unconditioned woman, the average percentage body fat is 25%, whereas for the unconditioned man it is closer to 15%. Similarly, there are differences in average heart size; in adult women the heart is 10.7 cm in diameter, whereas in adult men it is 12.1 cm in diameter. Average total lung capacity for adult women is 4,200 ml versus 6,000 ml for men; the average vital lung capacity for women is 3,200 ml, whereas in men the value is closer to 4,800 ml (Åstrand et al. 2003).

The relative proportions of muscle fiber types are similar in men and women, but the total cross-sectional area of women's muscles is only about 75% that of men, which accounts for the differences in overall strength (Cureton et al. 1990). Strength differences between men and women are greater in the upper extremities than in the lower extremities (Hakkinen 2002); however, women are able to gain relative strength similarly to men (Cureton et al. 1990).

Wilmore (1979) indicated that although men have greater upper-body muscle strength, elite female athletes are nearly equivalent to their male counterparts in strength per unit size and muscle fiber type. Ikai and Fukunaga (1968), who examined the strength of muscle elbow flexors in men and women in comparison to muscle size, found that there is little or no difference in the relative strength of men and women. The quality of a muscle fiber in exerting force is independent of gender (Bandy and Dunleavy 1996). There is some indication that men and women can increase strength to a similar degree following resistance training, but muscle hypertrophy seems to be less pronounced in women. One of the factors that may contribute to the greater hypertrophy in men is their testosterone level, which is 20 to 30 times higher than that of women and has potent anabolic function in building muscle tissue (Linnamo et al. 2005).

Use Versus Disuse

Skeletal muscle adaptation is specific to imposed demands. Two common training modes are *resis-*

tance training (or *strength training)* and *endurance training* (Garrett and Best 1994). Strength training was discussed in the section on skeletal muscle biomechanics (pages 118-119), so this section focuses on endurance training.

Training to increase muscle endurance involves a different challenge to the muscle than does strength training, and different adaptations will occur. Endurance training enhances the muscle's energy supply rather than its size. An important element is the increase in oxidative metabolism associated with the mitochondria, which increase in size, number, and density in endurance-trained skeletal muscles. With endurance training, the metabolic pathways adapt to a more effective use of fatty acids for fuel instead of glycogen. Different muscle fiber types respond differently to endurance training, although the oxidative capacity of all three fiber types can increase with endurance training.

Dietary intake can also affect the endurance capacity of muscle. Glycogen, which is stored in muscle fibers and the liver, is used extensively in exercises of moderate to high intensity. If high quantities of carbohydrates are eaten prior to competition, muscle and liver glycogen stores can be replete and provide the maximum energy availability for contractions.

The converse of hypertrophy is atrophy. Muscular atrophy is a decrease in the size of muscle tissue that can result from several causes, such as immobilization, bed rest, or a sedentary lifestyle after a period of high-intensity training. The clinical signs of atrophy include decreases in muscle circumference, strength, and endurance. The decreases in strength and endurance become the principal concerns during rehabilitation after injury or when resuming training. The changes in fiber size that occur with atrophy are most likely related to a decreased rate of protein synthesis and an increased rate of protein degradation (catabolism). With immobilization and disuse, the atrophy that occurs in skeletal muscle is more likely to affect the more **tonically recruited** slow-twitch (type I) fibers than the fast-twitch (type II) fibers.

Finally, in terms of functional adaptation, a region in the muscle–tendon complex that is particularly important is the myotendinous junction. As noted earlier, this is a region where sarcomeres are added to increase the length of muscles. It is an active and very dynamic (and injury-prone) region. Surprisingly, very little is known about how the myotendinous junction adapts to training. With overload training, there is no change in the relative junctional angle between the myofibril and the collagen at the myotendinous junction, which may mean that its strength capacity is near optimal (Tidball 1983). Thus, adaptation at the myotendinous junction may be related to changes in the quality of the collagen and sarcolemmal membrane rather than to changes in the morphology of the junction.

■ CHAPTER REVIEW

Key Points

- For load-bearing connective tissues such as bone, cartilage, tendon, ligament, and muscle, form and function are inextricably linked.

- All of these tissues are able to adapt to their environment to a greater or lesser extent.

- The mechanical, biochemical, hormonal, and molecular ability of these tissues to adapt favorably to environmental influences is a primary attribute of healthy tissues.

- The inability of these tissues to adapt to excesses (either high or low) is a leading factor associated with degradation and injury.

In chapters 6, 7, and 8, we explore the types of injuries and damage that can occur to these tissues in major joints and regions of the body.

Questions to Consider

1. The chapter begins with a quote by D'Arcy Thompson (1917) stating that "the form…of matter…and the changes of form which are apparent in…its growth…are due to the action of force(s)." Based on your understanding of the chapter, assess the correctness of Thompson's statement.

2. Compare and contrast the conventional concept of homeostasis (physiological steady

state) with what the text alternatively describes as "continually changing equilibrium."

3. Describe the difference between *structural properties* and *material properties*, and the functional consequences of this difference.

4. Explain the advantages and disadvantages of a long bone's hollow structure.

5. Compare and contrast the *modeling* and *remodeling* of bone.

6. An inverted-U model often applies to concepts in human performance (e.g., in physiology, biomechanics, or psychology). Such a model predicts that the optimal effect is achieved by some moderate level of stimulus. Lower or higher stimulus levels result in diminished effect. Select a biological tissue and explain how an inverted-U model helps describe its response to loading.

7. Explain why a multidisciplinary approach is warranted in treating an athlete diagnosed with *female athlete triad*.

8. What training recommendations would you give to someone wanting to maintain healthy musculoskeletal tissues?

Suggested Readings

BONE

Bostrom, M.P.G., A. Boskey, J.J. Kaufman, and T.A. Einhorn. 1994. Form and function of bone. In *Orthopaedic Basic Science*, edited by S.R. Simon. Park Ridge, IL: American Academy of Orthopaedic Surgeons.

Currey, J.D. 2002. *Bones: Structure and Mechanics*. Princeton, NJ: Princeton University Press.

Dull, P. 2006. Hormone replacement therapy. *Primary Care* 33: 953-963.

Loitz-Ramage, B.J., and R.F. Zernicke. 1996. Bone biology and mechanics. In *Athletic Injuries and Rehabilitation*, edited by J.E. Zachazewski, D.J. Magee, and W.S. Quillen. Philadelphia: Saunders.

Maddalozzo, G.G., J.J. Widrick, B.J. Cardinal, K.M. Winters-Stone, M.A. Hoffman, and C.M. Snow. 2007. The effects of hormone replacement therapy and resistance training on spine bone mineral density in early postmenopausal women. *Bone* 40: 1244-1251.

Martin, R.B., D.B. Burr, and N.A. Sharkey. 1998. *Skeletal Tissue Mechanics*. New York: Springer.

ARTICULAR CARTILAGE

Lo, I.K.Y., G. Thornton, A. Miniaci, C.B. Frank, J.B. Rattner, and R.C. Bray. 2003. Structure and function of diarthrodial joints. In *Operative Arthroscopy* (3rd ed.), edited by J.B. McGinty. Philadelphia: Lippincott Williams & Wilkins.

Mankin, H.J., V.C. Mow, J.A. Buckwalter, J.P. Iannotti, and A. Ratcliffe. 1994. Articular cartilage structure, composition and function. In *Orthopaedic Basic Science*, edited by S.R. Simon. Park Ridge, IL: American Academy of Orthopaedic Surgeons.

Mow, V.C., A. Ratcliffe, and A.R. Poole. 1992. Cartilage and diarthrodial joints as paradigms for hierarchical materials and structures. *Biomaterials* 1: 67-97.

Walker, J.M. 1996. Cartilage of human joints and related structures. In *Athletic Injuries and Rehabilitation*, edited by J.E. Zachazewski, D.J. Magee, and W.S. Quillen. Philadelphia: Saunders.

TENDON AND LIGAMENT

Curwin, S.L. 1996. Tendon injuries: Patho-physiology and treatment. In *Athletic Injuries and Rehabilitation*, edited by J.E. Zachazewski, D.J. Magee, and W.S. Quillen. Philadelphia: Saunders.

Frank, C.B. 1996. Ligament injuries: Pathophysiology and healing. In *Athletic Injuries and Rehabilitation*, edited by J.E. Zachazewski, D.J. Magee, and W.S. Quillen. Philadelphia: Saunders.

Lo, I.K.Y., G. Thornton, A. Miniaci, C.B. Frank, J.B. Rattner, and R.C. Bray. 2003 Structure and function of diarthrodial joints. In *Operative Arthroscopy* (3rd ed.), edited by J.B. McGinty. Philadelphia: Lippincott Williams & Wilkins.

Maffuli, N., P. Renstrom, and W.B. Leadbetter. 2005. *Tendon Injuries: Basic Science and Clinical Medicine*. London: Springer.

Sorrenti, S.J. 2006. Achilles tendon rupture: Effect of early mobilization in rehabilitation after surgical repair. *Foot Ankle International* 27: 407-410.

Woo, S.L.-Y., K.-N. An, C.B. Frank, G.A. Livesay, C.B. Ma, J. Zeminski, J.S. Wayne, and B.S. Myers. 1994. Anatomy, biology, and biomechanics of tendon and ligament. In *Orthopaedic Basic Science*, edited by S.R. Simon. Park Ridge, IL: American Academy of Orthopaedic Surgeons.

SKELETAL MUSCLE

Alexander, R.M. 2003. *Principles of Animal Locomotion*. Princeton, NJ: Princeton University Press.

Bandy, W.D., and K. Dunleavy. 1996. Adaptability of skeletal muscle: Response to increased and decreased use. In *Athletic Injuries and Rehabilitation*, edited by J.E. Zachazewski, D.J. Magee, and W.S. Quillen. Philadelphia: Saunders.

Biewener, A.A. 2003. *Animal Locomotion*. Oxford, UK: Oxford University Press.

Garrett, W.E., Jr., and T.M. Best. 1994. Anatomy, physiology, and mechanics of skeletal muscle. In *Orthopaedic Basic Science*, edited by S.R. Simon. Park Ridge, IL: American Academy of Orthopaedic Surgeons.

Herzog, W. 2000. *Skeletal Muscle Mechanics: From Mechanisms to Function*. Toronto: Wiley.

Lieber, R.L. 2002. *Skeletal Muscle Structure, Function, & Plasticity* (2nd ed.). Philadelphia: Lippincott Williams & Wilkins.

CHAPTER 5

CONCEPTS OF INJURY AND HEALING

Kindnesses are easily forgotten; but
injuries!—what worthy man does not keep
those in mind?

William Makepeace Thackeray (1811-1863)

■ **OBJECTIVES**

■ To establish an overview of injury mechanisms, principles of injury, and contributing factors to injury occurrence

■ To examine the pathological and healing pathways involved in injury to tissues of the musculoskeletal system

In assessing injury, most people first ask, "How did it happen?" Accurately answering this query involves establishing a cause-and-effect relation between the events surrounding the injury and the injury itself. In biomechanical terms, this is the *mechanism of injury. Mechanism,* in this context, can be defined as the fundamental physical process responsible for a given action, reaction, or result. Retrospective and accurate identification of injury mechanisms is essential for diagnosis, effective treatment, and prevention of future injuries.

Most people have experience in identifying injury mechanisms. Consider, for example, a basketball player who, in descending from a jump shot, lands on an opponent's foot and crumples to the floor in pain. Her coach rushes over and asks, "What happened?" The player grimaces and responds, "I twisted my ankle." In simple terms, the player specified the mechanism of her injury. Knowledge of the mechanism provides the coach, trainer, and physician with insights that can help determine the proper course of action.

Now imagine an elderly man found lying on a sidewalk one January night. Responding paramedics, after checking vital functions, determine that the man is in no immediate danger, although he is in considerable pain. The man indicates that he slipped on an icy patch of sidewalk and landed on his tailbone and hands. Again, by describing the mechanism of the fall, the man gives paramedics valuable information to help them determine how to proceed. With further questioning, more details of the fall may emerge.

Trained professionals can translate the simple description of an injury, as illustrated in the preceding examples, into more discipline-specific terms. A physician, for instance, may explain a twisted ankle as rapid loading of the lateral aspect of the ankle and foot, with possible injury to the anterior talofibular and calcaneofibular liga-

ments. A biomechanist might approach a slip and fall from a mechanistic perspective, focusing on the coefficient of friction at the foot–ground interface and the velocity of the body at the instant of ground contact. In both cases, the practitioner uses knowledge of the mechanism of injury to establish a cause-and-effect relation.

The description of an injury mechanism depends in part on the perspective of the person involved. Physicians, athletic trainers, coaches, supervisors, physical therapists, and injury victims undoubtedly will describe the mechanism of an injury differently, each being correct from his or her perspective.

Overview of Injury Mechanisms

Although sometimes a single mechanism is responsible for an injury, mechanisms often act in combination. Accurate identification of the mechanisms of injury is important for appropriate conditioning, treatment, and rehabilitation. Many of these mechanisms have been alluded to in our earlier discussions and are explored in the next section and in subsequent chapters.

Do not confuse mechanisms with the related but different concept of predisposing or *contributory factors.* Mechanisms establish a cause-and-effect relation. Contributory factors increase or decrease the likelihood of occurrence and the level of the effect; contributory factors are discussed in more detail later in this chapter.

The mechanisms responsible for injury are many and varied, and there is no single system for categorizing them. Categorization of injury mechanisms is based on mechanical concepts, tissue responses, or a combination of the two. From a sports medicine perspective, for example, one useful classification system identifies seven mechanisms of injury: contact, dynamic overload, overuse, structural vulnerability, inflexibility, impact, and rapid growth (Leadbetter 2001). Another source lists crushing deformation, impulsive impact, skeletal acceleration, energy absorption, and the extent and rate of tissue deformation as causal mechanisms (Committee on Trauma Research 1985).

From yet another perspective, one could view every injury mechanism as a variation of overload. In chapter 3, we defined *load* as the application of an external force to a body and identified seven factors that characterize load: magnitude, location, direction, duration, frequency, variability, and rate. Body tissues continuously experience loads during normal activity with no obvious injury. Typical loads are said to be within a **physiological range.** The probability of injury increases when loads exceed the physiological range. If the tissue being loaded is already damaged by previous injury or disease, its physiological range will be reduced.

Injury can result when a single overload exceeds a tissue's maximal tolerance. *Use* is normal functional loading, whereas repeated overload is *overuse*. Many injuries (e.g., tendinitis and carpal tunnel syndrome) are called **overuse injuries** because they result from repeated overloads with insufficient time for recovery. Specific examples of overuse injuries are described in chapters 6 through 8.

Overuse injuries exemplify a broad class of conditions that are caused by repeated application of force. Such injuries are **chronic injuries** and may also be referred to as **cumulative trauma disorders** or **repetitive stress syndromes.** In contrast, injuries resulting from a single or a few loading episodes are called **acute injuries.** Chronic and acute injuries are usually distinguishable, but sometimes they are related. For example, chronic loading (overuse) may weaken a tissue, lower its maximal strength, and increase the likelihood of an acute injury. Thus, a person with a chronic inflammation of the Achilles tendon has an increased likelihood of an acute rupture of that tendon.

Principles of Injury

The mechanical how and why of injury are the keystones of our approach, and this section pres-

ents principles of injury important to the later discussions of specific injury mechanisms.

Injury Terminology

In the context of musculoskeletal biomechanics, we defined *injury* as damage caused by physical trauma sustained by tissues of the body. As you may expect, injury biomechanics has its own vocabulary that draws heavily from medicine and mechanics. Although agreement exists on most definitions, some exceptions lead to confusion and lack of clarity.

Confusion also arises when nonspecific, catch-all terms are used to describe an injury or group of injury conditions. *Tennis elbow, shin splints, jumper's knee, Little League elbow,* and *whiplash* are nebulous terms and have minimal clinical or biomechanical utility. We sometimes use these vague but all-too-common descriptors; nevertheless, we discourage their use and encourage more specific and appropriate terminology.

Injury Severity

Every injury is unique; although one injury may be similar to other injuries, they are never exactly the same. This presents challenges in assessing injuries and classifying their severity. All categorization systems create discrete groupings and assign similar characteristics to all injuries in that group. Thus, although two different head injuries categorized as mild concussions may share similar characteristics, they are not identical injuries. Diagnosis and treatment must remain specific to each injury based on its own characteristics.

Clinical classification schemes are useful, though, in assigning general or common characteristics to similar injuries. Many such schemes exist (and differ), based on the tissues (e.g., bone vs. ligament) and body regions (e.g., head vs. leg) involved. A typical, three-level classification system for ligament injury, for example, specifies the structural involvement, physical signs, and **level of dysfunction** (mild, moderate, and severe) (table 5.1). Similarly, a five-level system for classification of concussions has been proposed (table 5.2).

Injury severity is linked to the amount of damage experienced by the tissue. In mild and moderate injuries, the tissue structure typically

TABLE 5.1

CLASSIFICATION OF LIGAMENT INJURY SEVERITY

Grade	Severity	Degree	Structural involvement	Exam	Performance deficit
1	Mild	First	Negligible	No visible injury, locally tender only, joint stable	Minimal to a few days
2	Moderate	Second	Partial	Visible swelling, marked tenderness, +/− stability	Up to 6 weeks (may be modified by protective bracing)
3	Severe	Third	Complete	Gross swelling, marked tenderness, antalgic posture, unstable	Indefinite, minimum of 6-8 weeks

Reprinted, by permission, from W.B. Leadbetter, 1994, Soft tissue athletic injury. In *Sports injuries: Mechanisms, prevention, treatment*, edited by F.H. Fu and D.A. Stone (Baltimore, MD: Lippincott, Williams & Wilkins), 761.

TABLE 5.2

CLASSIFICATION OF CONCUSSIONS

Grade	AMNESIA		Loss of consciousness
	Posttraumatic	Retrograde	
I	No	No	No
II	Yes	No	No
III	Yes	Yes	No, or seconds
IV	Yes	Yes	Yes, 5-10 min
V	Yes	Yes	Yes, prolonged

Reprinted, by permission, from D.W. Marion, 1994, Head injuries. In *Sports injuries: Mechanisms, prevention, treatment*, edited by F.H. Fu and D.A. Stone (Baltimore, MD: Lippincott, Williams & Wilkins), 827.

is partially disrupted. The damaged tissue is still able to accept load, although of smaller magnitudes than before the injury. In cases of complete failure, the tissue's continuity is totally disrupted, and load transmission is not possible. In some cases appearances can be deceptive; a tissue may appear to be intact and capable of load acceptance, but in reality the fibers are disrupted and possess little or no ability to transmit loads.

In addition to understanding various classification systems, you should also be familiar with the terms *level of dysfunction* and *progression of injury:*

■ **Level of dysfunction.** Some injuries are simply annoying and relatively trivial. These injuries do not limit function appreciably, and they heal quickly. Increasing injury severity, however, produces greater dysfunction. At the extreme are catastrophic injuries that result in permanent disability or death.

■ **Progression of injury.** Relatively minor injuries that are ignored may, with repeated loading or insult, progress to more severe injuries. Delayed, improper, or inadequate treatment may also contribute to progression to a more serious injury. Athletes or workers may try to do too much too soon after injury. A minor injury, given inadequate time to heal, may progress to a more debilitating level.

Injury Type

Here are several important distinctions that help define injury type:

- **Primary versus secondary injury. Primary injury** means that the injury is a direct, immediate consequence of trauma. A skull fracture from blunt trauma and a torn medial collateral knee ligament from a violent lateral impact are two examples of primary injuries. A secondary injury can happen in one of two ways. A secondary injury can surface some time after the initial trauma. In cases of traumatic head impact, primary brain injury can occur as a direct and immediate result of the impact. Delayed (or secondary) brain injuries, such as diffuse axonal injury and local or regional ischemia (localized decrease in blood flow to a tissue), may not appear until days after the initial trauma. Effects of these secondary injuries are often potentially reversible. Alternatively, a secondary injury can develop as an accommodation to or as compensation for a primary injury. When an injured person alters his or her movement patterns in response to the pain or dysfunction of a primary injury, the altered movements redistribute loads through other joints in the body. These changes in loading can generate injury remote from the primary injury in what is termed a compensatory injury. A woman with a sprained ankle, for example, may alter her gait and place unaccustomed loads on both ipsilateral and contralateral joints and tissues. These sites, not accustomed to these redistributed forces, may be injured.

- **Chronic versus acute injury.** Injuries can result from a single insult (*acute injury*) or from long-term, repeated loading (*chronic injury*). Continuing chronic insults to tissues may lead progressively to degenerative conditions that set the stage for an acute injury.

- **Microtrauma versus macrotrauma.** Chronic injury can begin as microscopic damage to a tissue's structure. For example, the damage could be microscopic tendon fiber tears or bone microcracks (microtrauma). Repeated loading can exacerbate the injury, and eventually the injury becomes macroscopic. If left untreated, tendon microtears presage eventual tendon rupture (macrotrauma). Similarly, X-ray findings of a patient complaining of a dull ache in the foot may be negative because of the microscopic nature of the bone cracks, but continued reloading may generate a stress fracture.

COMPENSATORY INJURY

One notable example of indirect, secondary injury happened to Dizzy Dean, a Hall of Fame baseball pitcher. In the 1937 All-Star Game, the right-handed Dean's left big toe was struck by a line-drive hit. The toe was broken. Rather than wait until the toe healed completely, Dean returned to action too soon. He altered his pitching mechanics to accommodate the pain caused by the toe injury. In doing so, Dean suffered a career-ending injury to his pitching shoulder. Dean was the unfortunate victim of a secondary injury, now sometimes called Dizzy Dean syndrome.

Tissue Structure

The mechanical response of biological tissue depends largely on its noncellular structural makeup, including its constituent material, orientation, density, and connecting substances. Bone, for example, with its mineral latticework and collagen fibers, is designed to transmit compressive, shearing, and tensile loads of high magnitude. Any change in the relation between structure and load, as might occur in an osteoporotic bone or in a bone experiencing abnormal and unaccustomed loading, increases the likelihood of bone injury.

Similarly, the collagen fibers of tendon and ligament afford them exceptional load-bearing capacity parallel to the fiber orientation. Because of the *anisotropic* nature of these tissues, off-angle forces may expose them to a greater risk of injury.

VARUS, VALGUS, VEXATION!

In 1980, Houston and Swischuk published the paper "Varus and Valgus—No Wonder They Are Confused" in the *New England Journal of Medicine,* taking issue with the use of the terms **varus** and **valgus.** The excerpts from their notes presented next highlight the confusion caused by the use of these terms in the medical literature.

"In lecturing on pediatric bone disease to medical students, we have become painfully aware that the terms varus and valgus cause great confusion. Every year, when shown a radiograph of coxa valga and asked for the appropriate diagnosis, about a third of the students vote for coxa vara, a third for coxa valga, and a third admit that they do not know.

"Since radiologists, orthopedic surgeons, and pediatricians use varus and valgus regularly in their conversation and in their reports, we have attributed the ignorance of medical students to their inexperience, coupled with the unfortunate omission of high-school Latin as a prerequisite for entrance to medical school. This year one of us criticized the incorrect usage of the term varus in reviewing a book written by the other. This led to consultation with current and early dictionaries and to a survey of major orthopedic textbooks.

"To our surprise, we learned that the original Latin meaning of varus was knock-kneed, and of valgus bowlegged, exactly opposite to current pediatric and radiological usage.

"To show that current usage is consistently opposite to the derivation and the definitions of most dictionaries through the years, one of us checked 24 current orthopedic textbooks. . . . To our surprise, we could find a definition of the terms in only two of these texts. W.A. Crabb's *Orthopedics for the Undergraduate* (1969) provided this definition of varus: 'Deviation of a limb towards the midline of the body.' Robert B. Salter's *Textbook of Disorders and Injuries of the Musculoskeletal System* (1970) gave detailed and helpful definitions and was the only one of the 24 to mention the historical discrepancy. Under the heading, 'Varus and Valgus,' Salter stated, 'This particular pair of terms has caused more confusion than any other pair, partly because the original Latin terms had the opposite meaning to that which is now universally accepted.' All 24 texts referred to bowlegs as genu varum, even though the knee in this condition is away from the midline of the body. Thus, varus is now used to indicate a tilt toward the midline of the bone beyond the joint, regardless of whether the prefix is the name of the joint or the bone beyond it. This obviously is confusing.

"Since confusion is universal, since current usage is directly contrary to derivation, and since use of these terms in the directional sense is at the least misleading and at the most dangerous, we suggest that the simple English words bowlegged and knock-kneed are far superior to genu valgum and genu varum.

"In summary, it would seem best to avoid the terms varus and valgus altogether. Anyone who persists in using them should follow the lead of Crabbe and Salter and define them clearly. Furthermore, dictionaries should point out, in unambiguous fashion, not only the derivation of these terms, but the opposite way in which they are used in modern orthopedic literature."

Houston and Swischuk's recommendation that use of *varus* and *valgus* be avoided has apparently not found favor in the scientific and medical communities. Both terms are still widely used and continue to create confusion. We are tempted not to use these terms but believe that given their continued prevalence in the literature, you will be best served by our using the terms *varus* and *valgus* in the following chapters. In so doing, we provide precise definitions for each term.

The interaction between tissue structure and loading behavior creates a complex synergistic relation that has important consequences for proper function and potential injury.

When different tissues form a functional unit, the weakest-link phenomenon typically occurs during an injury. This means that when a combined structure is mechanically loaded, it will likely fail first at the weakest link in the structural chain. In the human body, the factors that contribute to making the weakest link are many, interrelated, and often not easily identified or well understood.

Contributory Factors

Simply stated, injury happens when an imposed load exceeds the tolerance (load-carrying ability) of a tissue. Many contributory factors, however, make this anything but a simple relation between load and injury. The following are examples of contributory factors.

■ **Age.** During our formative years, perhaps into our 20s, our tissues are growing and developing. Later, tissues may begin to degenerate and lose strength, compliance, density, and energy-carrying capacity. Acute injuries are more common in younger people, but as we age, chronic injuries happen more often, as do unintentional injuries from slips and falls. However there is a difference between chronological age and **physiological age.** The former is based on calendar years, whereas the latter is based on the physiological quality of the tissues. A 60-year-old person may have tissues with better physiological and mechanical properties than those of a 45-year-old. Although generalities are routinely used to describe aging responses, each person's response is unique and may be quite different from another person's response.

■ **Gender.** Gender-specific differences in structure, hormones, sociology, activity patterns, and many other measures dictate that one gender may be at greater or lesser risk of injury than the other in some circumstances.

– *Males.* The number of fatalities from unintentional injury is higher for men in all categories except for strangulation, ignited clothing, falls from the same level,

and certain types of poisoning. The male-to-female fatality ratio for injuries from machinery is 20:1, motorcycling 10:1, firearms 7:1, and suffocation 2:1 (Anderson et al. 2004). Similar ratios exist for nonfatal injuries as well.

– *Females.* Women do not hold an advantage in all areas, however. For example, they are more likely than men to suffer the consequences of **osteoporosis.** Osteoporotic bone has decreased density and diminished strength and is more susceptible to fracture. Men can also suffer from osteoporosis, but women are more likely to experience its effects, especially in the postmenopausal epoch. Estrogen deficiency is not the only factor involved in osteoporosis, making this condition a useful example of the complexity of analyzing injury-contributing factors. Other contributory factors in osteoporosis include disuse attributable to sedentary lifestyle, paralysis, or immobilization; chronic liver disease; rheumatoid arthritis; alcoholism; diabetes; malignancy; stress; poor diet; and smoking (Akesson 2003; Christiansen 1995; Sanders and Albright 1987).

■ **Genetics.** Genetic factors influence tissue matrix composition and are implicated in the predisposition toward certain injuries, including intervertebral disc and rotator cuff degeneration, carpal tunnel syndrome, and tendon ruptures.

■ **Physiological status and physical condition.** A person's physical condition is a primary factor in his or her chances of sustaining an injury. The fitter the person is, the less likely he or she will be injured. If injury does occur, a better-conditioned person probably will have a less severe injury and will recover more quickly.

■ **Nutrition.** Diet provides the raw materials to build, sustain, and repair the body's tissues and therefore plays an indirect yet essential role in injury biomechanics. Tissue homeostasis depends on remediating nutritional deficiencies, excesses, or imbalances.

■ **Psychological status.** Psychological parameters can influence the incidence of injury. These factors include stress levels, inattention, distraction, fatigue, depression, excitation, human error, risk evaluation, personality factors, and coping resources.

▪ **Fatigue.** Physical and mental fatigue increases the likelihood of injury because it compromises muscle strength, coordination, mental attentiveness, and concentration. Fatigue-related injuries tend to happen later in an activity period; for example, truck drivers were found to have three times the risk of crash involvement after 6 hr of driving compared with the first 2 hr and to be asleep at the wheel or inattentive in 45% of commercial vehicle accidents (Bunn et al. 2005). Athletes also tend to show greater risk of injury during the latter stages of a practice session or game. For example, ski injuries are more likely to happen after multiple runs down the mountain.

▪ **Environment.** Numerous environmental factors contribute to injury, including location (indoors vs. outdoors, urban vs. rural), weather conditions (temperature, humidity, visibility), time of day or night, terrain (flat vs. inclined, smooth vs. rough, slippery vs. sticky), altitude, and activity (work vs. recreation).

▪ **Equipment.** Equipment often plays a central role in injury, either in prevention, causation, or both. Equipment can include clothing, pads or protective devices, and implements such as tools, machinery, or computers. Apparel can be protective, especially in environments likely to produce injury. Implements such as bulletproof vests, helmets, and shields aid in injury prevention as well. Equipment can also be associated with acute injuries (e.g., finger severed in machinery) or chronic injuries (e.g., carpal tunnel syndrome caused by long-term, repetitive typing by a computer operator). The same piece of equipment can protect a person from or contribute to injury. A football player's helmet and pads, for example, protect the head and body from direct impact. However, the helmet and pads may also contribute to heat stress (by decreasing thermoregulatory capacity and increasing heat production and retention) and to cervical injury if the helmet fits improperly; helmets can inflict injury on another player if that player is hit with the top of the helmet.

▪ **Human interaction.** Interactions among people can be social, occupational, or competitive, as seen in sports. Whenever people interact, the potential for injury exists. The possibility may be remote at a dinner party, for example, but it becomes a prime factor in a rugby match.

▪ **Previous injury.** Following any serious injury, elements of the injury can persist, whether physically or psychologically. The repaired tissues are often not equal to their preinjury condition and for many reasons may be more susceptible to a subsequent injury. A person's psychological status following injury can be different, because the prior injury can stay in mind.

▪ **Disease.** Many diseases increase the risk of injury. For example, an osteosarcoma (malignant bone tumor) weakens bone, atherosclerosis damages arterial walls, and diabetes predisposes one to skin ulcers, particularly on the plantar surface of the foot.

▪ **Drugs.** The wide variety of drugs available for recreational or medical purposes, or as ergogenic aids, can produce positive or negative effects on the body's tissues and alter human performance so that risk of injury is changed. In addition, these chemical agents may also indirectly contribute to an injury because of their systemic effects.

– *Recreational.* Recreational drugs, either legal (alcohol) or illegal (marijuana, cocaine), may increase the risk of musculoskeletal injury. The injury risk associated with the use of tobacco or methadone is 1 in 100. Using heroin, morphine, barbiturates, or alcohol comes with an injury risk of 1 in 1,000 and can be associated with all types of violence and accidents, such as shaken baby syndrome and sudden infant death.

– *Medical.* Medical drugs can be obtained over-the-counter or by prescription and include everything from nonsteroidal anti-inflammatory drugs (NSAIDs)—for example, aspirin and ibuprofen—to prescription painkillers, anti-inflammatories, asthma medications, or any of thousands of other medications. The use of medicinal drugs may affect motor vehicle operation by decreasing response times or impairing vision, leading to accidents and the potential of musculoskeletal injury.

ERGONOMICS AND INJURY

The economic and personal costs of work-related injuries are staggering. For example, the U.S. National Safety Council (2004) estimated that the monetary cost of work-related injury was $156 billion in 2003 and that the total time lost because of injury was 115 million days. These figures indicate that each of the 139 million workers in the United States must produce goods and services valued at nearly $1,120 each year to offset the costs of injury (National Safety Council 2004).

Efforts to reduce these costs fall under the province of **ergonomics,** or *human factors,* which is the field of study that "discovers and applies information about human behavior, abilities, limitations and other characteristics to the design of tools, machines, systems, tasks, jobs, and environments for productive, safe, comfortable, and effective human use" (Sanders and McCormick 1993, p. 5). Ergonomics seeks to improve the things that people use and the environments in which they work and live. In many ways the problems associated with workplace injuries are more complex than the problems associated with sports-related injuries.

Sanders and McCormick (1993, p 5) identified several characteristics that uniquely define the human factors, or ergonomics, profession:

- Commitment to the idea that things and machines are built to serve humans and must be designed with the user in mind
- Recognition of individual differences in human capabilities and limitations and an appreciation for their design implications
- Conviction that the design of things and procedures influences human behavior and well-being
- Emphasis on empirical data and evaluation in the design process
- Reliance on the scientific method and the use of objective data to test hypotheses and generate data about human behavior
- Commitment to a systems orientation and a recognition that things, procedures, environments, and people do not exist in isolation

Despite the considerable progress that has been made in many areas of injury prevention, complex challenges remain for human factors professionals and others responsible for promoting society's health and well-being.

– *Ergogenic.* The use of drugs as ergogenic aids has become more and more insidious and widespread, perhaps most noticeably in competitive sports. *Ergogenic* refers to the work-generating or power-generating potential of these aids, which can include a host of substances or treatments that purportedly improve a person's physiological performance or remove the psychological barriers associated with more intense activity (Ellender and Linder 2005). Many of the pharmacological aids have been banned by official sports bodies because of the unfair advantage some substances give athletes during competition and because of the negative side effects that can occur, including a greater risk of injury in some instances. For example, since their infiltration into sport in the 1950s, anabolic steroids (e.g., oral or injectable forms of synthetic testosterone) have permeated many sports, from the high school level to the Olympic and professional levels. Anabolic steroids may have little or no effect on anaerobic performance. They may increase body size and muscular strength, although these increases are rapidly lost after discontinuing steroid use. Besides risking overuse or acute muscle–tendon injuries associated with the heightened training that may accompany anabolic steroid use, individuals who take anabolic steroids may experience increased acne on the face and upper body, testicular atrophy, changes in sex drive, irritability, aggression, and liver damage.

▪ **Inadequate rehabilitation.** After someone has sustained an injury, delaying even limited activity too long or returning to extremely high levels of activity too soon can result in further damage. Two areas in which this issue is important are inflammation and tissue repair.

– *Inflammation.* Inflammation is a primary result of injury to a joint or tissue of the musculoskeletal system. Although inflammation is a necessary step in the healing process (as described more fully later in this chapter), it can lead to joint pain, which carries the potential for tissue atrophy as a consequence of inactivity and lack of motion. On the other hand, if an inflamed tissue continues to be overused, there is a risk of developing a more serious injury.

– *Tissue repair.* Once inflammation is controlled, the body must rebuild tissue that has been disrupted. The optimum rate at which activity is resumed can be gauged by paying attention to signs of regression or robustness. If tasks that were once challenging become significantly easier, it is time to challenge the tissue more; if symptoms of pain or weakness increase, it is time to consider scaling back temporarily. Without such management, the tissue can be reinjured—and sometimes the secondary damage is even more serious than the original problem.

▪ **Anthropometric variability.** People come in many shapes and sizes, and these differences in body dimensions often play a critical role in injury. Analysis of our structural variability is anthropometry, the study of comparative measurements of the human body. Anthropometric measures such as height, weight, body composition, muscle mass, and shape (somatotype) can play a central role in assessing injury. Obese individuals, for example, are more likely to have knee problems, and this risk increases with age. Anthropometric measures are also involved in determining body posture and flexibility (joint range of motion), both of which—either alone or in concert—can affect the risk of injury.

▪ **Skill level.** The adeptness with which a person performs a task influences the risk of injury. Especially in high-risk activities (e.g., auto racing), the skill level of the performer may be the most important determinant of injury. Novice or less-skilled performers are more likely to be injured. The converse may prove true, however, when particularly skilled individuals believe that they are competent to attempt tasks with an unacceptably high risk of injury.

▪ **Experience.** Closely related to skill level is experience. Although related, skill and experience are not synonymous. An individual may be experienced, having performed a task many times, but be unskilled. Another person may be naturally gifted with a skill but have little experience at the task. Usually, however, the two factors are closely linked. Experienced performers typically exhibit efficiency of movement, along with sound judgment and decision-making abilities. These all combine to lower the risk of injury.

▪ **Pain.** The sensation of pain is fundamental to any discussion of injury. Pain, the body's message of distress, accompanies most injuries of consequence and often is the limiting factor in continued participation in an activity. Pain derives from various biomechanical and inflammatory sources. Pain is one factor used in determining an injury's severity and in prescribing and monitoring the rehabilitation during postinjury therapy. Pain also influences movement patterns (recall the Dizzy Dean syndrome). Pain may hinder further activity or preclude participation altogether.

Rehabilitation

Success in returning a person to preinjury status depends on the nature of the injury, the person's motivation, the expertise of the rehabilitation therapist, and the sophistication of the available rehabilitation methods.

One of the first priorities in treating musculoskeletal injuries is the control of excessive inflammation. Many effective modalities are used to counteract the inflammatory process, including cryotherapy (e.g., ice, cold compresses, and cooling sprays) and thermotherapy (e.g., moist hot packs, whirlpool baths, heating pads, and ultrasound). The injury may require other types of treatment such as reconstructive surgery or physical therapy.

Sometimes even the best rehabilitation methods are unable to return a person to her or his preinjury capacity. Severe injury may be permanently disabling and career threatening. On the other hand, rehabilitation has the potential for returning a person not only to preinjury condition but potentially to an even higher level of function.

Inflammation and Entrapment Conditions

Although the physical damage resulting from an injury is unique, the human body does have a generalized response to injury that occurs in all cases regardless of the specific body region or tissues affected. This immediate reaction to injury is termed the *inflammatory response*. Although the inflammatory response is essential to the healing process, it can lead to damage if uncontrolled. Because the inflammatory response is often the cause of compartment syndrome, that pathological condition is discussed later in this section.

Inflammatory Response

The inflammatory response, or **inflammation,** is a generalized pathological process affecting blood vessels and adjacent tissue. It happens in response to a variety of stimuli, especially injury. The cardinal signs of inflammation were identified long ago by Aulus Cornelius Celsus (30 BC to AD 38). Celsus described the inflammatory response as having *"rubor et tumor cum calore et dolore,"* or redness and swelling with heat and pain. A fifth sign, *functio laesa,* or functional loss,

was added by Galen (AD 129-199) and is often present in inflammation as well.

Inflammation (indicated by the suffix *-itis*) can develop in response to an acute injury or may develop from chronic irritation, as seen in arthritis, bursitis, or tendinitis. The redness and heat of inflammation are caused by blood vessel dilation and increased flow. Increases in intracapillary hydrostatic pressure and enhanced capillary permeability cause inflammatory swelling. Pain develops from the swelling-related increase in pressure on nerve endings and is most pronounced when a confined space (e.g., synovial joint) is inflamed. Swelling can restrict function and may persist as the damaged tissues heal (e.g., torn ligament or tendon, or fractured bone).

An injury produces a vasoconstrictive response known as the coagulation phase. This is followed within minutes by a vasodilatory phase, during which an increase in vascular permeability permits the flow of materials from the vessels into the surrounding tissues. These moving substances are termed **exudate** and consist of fluid and plasma proteins. Although the *edema* (swelling) caused by the exudate may contribute to pain, the exudate has a number of positive functions: It dilutes and inactivates toxins; provides nutrients for inflammatory cells; and contains antibodies, complement proteins, and fibrinogen (a precursor to fibrin, a protein instrumental in the coagulation process).

The inflammatory process is controlled by substances known as chemical mediators. This includes the immediately available histamine, as well as other mediators—such as serotonin, bradykinin, prostaglandins, leukotrienes, and plasmin—produced at the site of inflammation or by leukocytes (white blood cells) drawn to the injury site through **chemotaxis** (cellular attraction caused by chemical action).

The chemical mediators of the inflammatory response are joined by cells that perform specific functions. Among these are a class of cells known as **phagocytes** (which degrade bacteria and necrotic tissue), the most predominant of which are polymorphonuclear neutrophils, responsible for **phagocytosis** (the engulfing and destruction of particulate matter by phagocytes) and defense against fungal and bacterial infections. Several immune system cells—for example, accessory cells and lymphocytes (B cells, T cells, and NK cells)—also assist in defending against foreign substances (collectively known as antigens).

Inflammation is the body's first line of defense against insults such as those imposed by injury. The details of the process are complex and in some cases unknown. But for all the apparent complexity of the inflammatory process, "it seems . . . that the more we learn about inflammation, the simpler its message becomes: Our cells and humors defend the self against invisible armies of the other. We call our losses 'infection' and our victories 'immunity'" (Weissmann 1992, p. 5).

Compartment and Entrapment Conditions

The fundamental mechanical relation between mass and volume plays a central role in various injury conditions that are broadly termed *compartment*, *entrapment*, or *impingement syndromes*. The common element of all such conditions is the ratio of mass and volume, or density, and its mechanical consequences on biological tissues. Increasing the density of material within a confined space increases the pressure exerted on the boundaries of the space and on the material within the space. Increasing density, either by increasing the mass or decreasing the volume (or both), will increase pressure. This increase in pressure is transmitted to all structures within the enclosed space.

The pressure–density relation and its physiological effects can be seen in many instances of musculoskeletal injury, some of which are detailed in the following chapters. Included in the wide array of compartment–entrapment conditions are carpal tunnel syndrome, glenohumeral impingement syndrome, skeletal muscle compartment syndromes, synovial joint swelling, and cerebral edema. All are either caused or aggravated by the mechanical relations among mass, volume, and pressure.

The affected spaces may be compartments of the body completely enclosed in fascia (which does not stretch); joint capsules; or narrow apertures defined by bony tunnels through which vascular and neural tissues pass, such as the carpal tunnel or the thoracic outlet. Some of the causes of compartment syndrome are fractures,

casts, prolonged limb compression, car injuries, burns, hemorrhage, and intravenous drug use.

In biological systems, the structures affected are often muscles, nerves, and circulatory vessels. Pressure on nerves is felt as tingling, numbness, or pain. Pressure on circulatory vessels results in decreased arterial or capillary perfusion or restricted venous return. In the case of compartment syndrome, because the fascia that encloses the compartment does not stretch, even a small amount of bleeding inside the compartment can increase the pressure dramatically. The tissues that rely on proper neural and circulatory supply will be deleteriously affected if the syndrome is not treated promptly—even to the extent of nerve damage and muscle death. As noted previously, such situations are often either caused or worsened by inflammation—in particular the swelling that accompanies the inflammatory response. In many cases, the affected system becomes involved in a positive feedback loop, with increases in pressure causing restricted outflow, which in turn further increases pressure.

Our comments on injury thus far have applied to most biological tissues and structures. In addition, each tissue possesses unique characteristics determined by its own structure and function. The following sections examine the unique characteristics of the major tissues involved in musculoskeletal injury.

Bone Injuries

The role of bone in providing structural support and protection, facilitating movement, and serving as a site for **hematopoiesis** and mineral storage cannot be understated. Injury to bone can compromise any of these functions and interrupt our daily routines.

Pathology

The viability of bone as a tissue depends on the proper function of the bone's cellular component and the ability of these cells to produce extracellular matrix and perform other important physiological processes. Any disease or injury that compromises osteocyte performance jeopardizes the structural integrity of both the affected bone and the skeletal system. Three conditions that affect bone tissue are described briefly in the following sections. Specific injuries involving these conditions are examined in detail in chapters 6, 7, and 8.

Osteonecrosis

Osteonecrosis refers to the death of bone cells resulting from a cessation of the blood flow necessary for normal cellular function. When we describe the condition of bone cell death, the term *osteonecrosis* is preferred to the commonly used terms avascular necrosis (i.e., cell death caused by an absent or deficient blood supply) and aseptic necrosis (i.e., cell death in the absence of infection), because the term *osteonecrosis* best describes the histopathological processes involved and does not implicate any specific cause (Day et al. 1994).

The mechanisms of compromised circulation that may lead to osteonecrosis are mechanical disruption of vessels, occlusion of the arterial vessels, injury to or pressure on arterial walls, and occlusion of venous outflow. These conditions may result from bone fracture, joint dislocation, infection, arterial thrombosis, or a number of conditions that affect circulatory integrity. Although the bone's noncellular structures (e.g., organic and inorganic matrix) may not be immediately affected, over time they may suffer deleterious effects from the absence of cellular production. A decrease in extracellular matrix production, for example, may result in decreased bone strength and increased likelihood of fracture.

Osteopenia and Osteoporosis

Osteopenia (a loss of bone tissue) is a bone condition classified by having a *bone mineral density* (BMD) rating that is 1.0 to 2.5 standard deviations below the mean recorded for young, healthy adults. As such, osteopenia is not an injury in itself. However, if left untreated it can lead to the development of osteoporosis and subsequently predispose an individual to fracture.

Osteoporosis is a Latin-derived term that literally means "bone that is porous" (has more or larger holes than typical bone). Osteoporosis may be more accurately described as a condition "in which a progressive diminution of skeletal mass renders bone increasingly vulnerable to fracture" (Goltzman 2000). Clinically, osteo-

porosis is defined by a BMD rating more than 2.5 standard deviations below normative bone density values.

Osteoporosis was identified as a prominent public health issue more than 50 years ago and has been the subject of extensive research and scientific debate ever since. Histological, radiological, and clinical evidence has clearly demonstrated that progressive bone loss begins in the fourth decade of life, and the rate of loss increases with advancing age (figure 5.1).

Osteoporosis predominantly affects trabecular bone and the endosteal surface of cortical bone and is marked by reduced bone mineral mass and changes in bone geometry, leading to an increased probability of fractures, primarily of the hip, spine, and wrist. Progressive loss of bone mass can be a function of the normal aging process or it can be caused by other disease processes. Furthermore, the amount of bone mass at one site in the body is not necessarily correlated with bone mass at other sites. A longitudinal study on aging in the United States reported the prevalence of osteoporosis and hip fractures in people 70 years of age and older. These estimates are based on self-reports and are not confirmed through radiological detection; therefore, the true prevalence may be over- or underestimated. The World Health Organization has stated that the number of hip fractures, primarily from osteoporosis, will increase by more than 300% in the

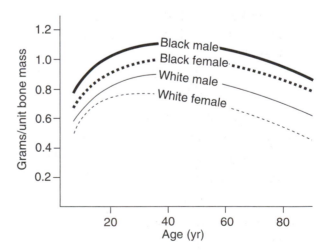

Figure 5.1 Bone mass as a function of age, gender, and race.

Reprinted from R. Pacifici and L.V. Avioli, 1993, *The osteoporotic syndrome: Detection, prevention, and treatment*, 3rd ed. (New York, NY: Wiley-Liss), 3.

next 3 to 4 decades. Unfortunately, many people are unaware of the existence of osteoporosis, particularly at the early stages.

Both men and women experience some loss of bone mass as part of normal aging, but osteoporosis progresses much more rapidly in postmenopausal women. After the age of 30, men typically lose bone mass at approximately the same rate for the remainder of their lives. In women, however, the loss of bone increases significantly for about 5 years after menopause and then slows to a more gradual loss. Just after menopause, the rate of bone mass loss in women is up to 10 times faster than in men of the same age (Reginster et al. 2006).

Osteoporosis is a multifactorial problem that continues to frustrate researchers in their attempts to identify the precise causes and pathogenic mechanisms involved in both the onset and the progression of the condition. Among the factors most often included in discussions of osteoporosis are dietary status, physical activity patterns, hormone dynamics, calcium absorption, and physical stress. Some of the clinical conditions associated with osteoporosis are listed here. With the average age of the population increasing, osteoporosis will undoubtedly continue to be a major public health concern and will provide fertile ground for research in the coming decades.

Clinical Conditions Associated With Osteoporosis

- Menopause of aging
- Disuse, immobilization, paralysis, weightlessness
- Corticosteroid treatment
- Cushing's syndrome
- Osteogenesis imperfecta
- Partial gastrectomy
- Malabsorption syndromes
- Chronic liver disease
- Heparin treatment
- Hyperparathyroidism
- Hyperthyroidism
- Rheumatoid arthritis
- Acromegaly
- Scurvy
- Alcoholism
- Diabetes
- Lactase deficiency
- Malignancy
- Mastocytosis
- Acid-rich diet
- Decreased physical stress
- Amenorrhea in runners
- Smoking

Reprinted from J.A. Albright and R.A. Brand, 1987, *The scientific basis of orthopaedics*, 2nd ed. (New York, NY: McGraw-Hill), 277, by permission of The McGraw-Hill Companies.

Fracture

The injury most commonly associated with bone is **fracture,** derived from the Latin *fractura,* meaning to break. Although the term *fracture* is also used to describe disruption to cartilage and the epiphyseal plate, it is most closely associated with breaks in the structural continuity of bony tissue. In a simple sense, fracture occurs when an applied load exceeds the bone's ability to withstand the force. The many factors involved in specifying the loading conditions and the response characteristics of the loaded bone, however, make the study of fracture mechanics anything but simple.

The fracture resistance of bone is determined by both the material properties of bone as a tissue and the structural properties of bone as an organ. Fracture resistance is influenced by the complex interaction of viscoelastic characteristics (e.g., strain rate), bone geometry (e.g., cross-sectional dimensions), anisotropic effects (e.g., microstructural orientation with respect to the loading direction), and bone porosity (Hipp and Hayes 2003).

The nature of bone loading determines in large part the potential for injury and the type of fracture produced. Fracture may occur in response to a single, large-magnitude loading, as occurs in a violent collision (acute loading). Alternatively, fractures may result from repeated application of lower-magnitude forces (chronic loading), as is characteristic of a metatarsal fatigue fracture (stress fracture) resulting from excessive running or jumping.

A fracture at the specific site of force application is termed a **direct injury.** When the fracture is remote from the location of force application, it is an **indirect injury.** Indirect injuries result from force transmission through other tissues. An example of indirect injury is when a force applied to a tendon or ligament is transferred to its bony attachment site and causes an **avulsion fracture** at that location (a piece of bone is pulled out at the insertion site).

The risk of fracture also depends on the type of bone being loaded. Cortical (compact) bone, because of its relatively low porosity (i.e., high density), is more fracture resistant than less dense trabecular (cancellous) bone. Factors that increase bone density also increase bone strength and thus decrease fracture risk. Conversely, factors that contribute to decreased bone density increase the risk of bone injury.

Fractures are often classified according to their mechanism of injury. Various fracture types and their commonly associated injury mechanisms are shown in figure 5.2.

A diagnosed fracture can be classified or described in various other ways, but the following factors are commonly considered (Salter 1999):

- **Injury site**—Fractures may be classified according to their location, such as diaphyseal, epiphyseal, or metaphyseal fractures.

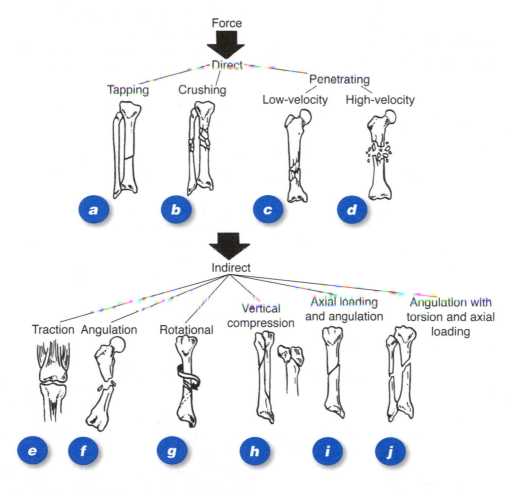

Figure 5.2 Classification of fractures according to the mechanism of injury. Direct force: *(a)* Tapping mechanism from relatively small forces over a small area causes transverse fracture. *(b)* Crushing from large forces over a large area results in extensive comminuted fracture. Large force over a small area at either *(c)* low velocity or *(d)* high velocity results in penetrating, comminuted fracture. Indirect force: *(e)* Traction mechanism from tensile loading results in either transverse fracture or avulsion fracture. *(f)* Angulation (bending) mechanism results in angulated, or butterfly, fracture. *(g)* Rotational mechanism from torsional loading results in spiral fracture. *(h)* Vertical, or axial compressive, loading causes oblique fracture. *(i)* Combination of axial compression and angulation produces a combination of transverse fracture (on the convex side) and oblique fracture. *(j)* Combination of angulation (bending) with torsion and axial loading results in complex fracture pattern.

Adapted, by permission, from J.W. Harkess, W.C. Ramsey and J.W. Harkess, 1996, Chapter name. In *Rockwood and Green's fractures in adults*, 4th ed., edited by C.A. Rockwood et al. (Philadelphia, PA: Lippicott Raven Publishers), 11.

- **Extent of injury**—Fractures may be either complete or incomplete, depending on whether the damage completely or only partially traverses the bone structure.

- **Configuration**—When there is only a single fracture line, the shape of the line may be either transverse, oblique, or spiral. When there is more than one fracture line, the fracture may be classified as a comminuted or butterfly fracture.

- **Fragment relations**—Bone fragments may be either undisplaced or displaced. In the latter case, the fragment displacement may occur in many ways, including angulation, rotation, distraction, overriding, impaction, or sideways shifting.

- **Environmental relations**—Fractures, even if displaced, that remain within the body's internal environment are termed closed fractures. Those penetrating the skin and resulting in exposure of bone to the external environment are open fractures. For obvious reasons, open fractures pose a much greater risk of infection than do closed fractures. Closed and open fractures sometimes have been called simple and compound fractures, respectively. This terminology somewhat misrepresents the nature of the fractures and can be misleading. Many clinicians and other professionals, therefore, discourage the use of *simple* and *compound* as fracture descriptors, preferring *closed* and *open*.

- **Complications**—Some fractures are accompanied by few, if any, complications. Many, however, have complications that may be immediate (e.g., skin, vascular, neurological, muscular, or visceral injury), early (e.g., tissue necrosis, infection, tetanus, pneumonia), or late (e.g., osteoarthritis, growth disturbances, posttraumatic osteoporosis, or refracture) (Salter 1999).

- **Etiological factors**—In some cases, fracture is preceded by conditions that predispose a bone to injury. Examples of predisposing etiological factors are repetitive-use microfractures that precede stress fractures and inflammatory disorders, bone disease, congenital abnormalities, and neoplasms.

- **Combination injuries**—Bone fracture may be associated with or caused by other injuries, such as multiple fractures and fracture-dislocation injuries. Another combination condition results from the "connectedness" of tissues. Forces applied to an osteotendinous junction, for example, can produce a tendon injury, bone fracture, or both. The weakest-link argument predicts that if the bone is stronger, the tendon will be strained. Conversely, if the tendon is relatively stronger, the bone will experience an avulsion fracture. Here, the viscoelastic properties of bone and tendon may come into play. If the injury load is applied slowly, the bone is more likely to be fractured (*avulsion fracture*), whereas the tendon is more likely to tear if the load is applied rapidly.

Fracture Healing

Fracture healing can be divided into three phases: inflammation, initial union of the bony ends, and remodeling of the callus. Immediately following injury, a hematoma (pool of blood) develops around the fracture site. Within 3 days, mesenchymal cells arrive in the area and produce a fibrous tissue that envelopes the fractured bone ends. The outer layer of the fibrous material begins to form the new periosteum. Until this point, stable fractures (which tend not to displace) and unstable fractures (which tend to slip out of place after reduction and immobilization) react similarly, but between 3 and 5 days after the fracture the degree of stability influences subsequent healing steps. Microscopic examination of the fibrous tissue reveals that in a stable fracture the tissue is well vascularized, but in an unstable fracture the fibrous tissue is poorly vascularized.

Where the fibrous tissue meets the original bony cortex—in both stable and unstable fractures—new trabeculae are formed by osteoblasts lying on the old bone surface. In a stable fracture, new bone forms along the periosteal surface of the fibrous layer and spans the fracture site. In an unstable fracture, new bone also forms along the periosteal surface of the fibrous material but does not span the fracture line. In humans, minimal periosteal bone formation occurs at this point in healing, and periosteal union is further delayed. As bony trabeculae continue to form, the bony collar becomes more compact, and the periosteum thickens.

In the gap between the bony ends (rather than along the periosteal surface), the first cells to invade after injury (approximately day 9) are macrophages, followed by fibroblasts and capillaries. Macrophages remove cell and matrix debris, whereas the fibroblasts generate the structural matrix for cells and vessels. Osteoblasts begin bone deposition by 2 weeks postfracture, and bony union across the fractured ends is established (optimally) by 3 weeks. If the bone adjacent to the fracture site dies secondary to disruption of its blood supply at the time of fracture, osteoclasts may be present to resorb the dead tissue. Otherwise, osteoclasts are not routinely present in all fractures.

In small gaps (<10 mm) or where fracture ends contact, healing is via direct Haversian remodeling: Osteoclasts resorb a cone of bone, osteoblasts deposit new Haversian bone, and osteocytes maintain the new bone after mineralization. In 10 to 30 mm gaps (too large for Haversian remodeling but too small for cells to move), osteoclasts may resorb the bone to increase the gap width. Osteoblasts later arrive to lay down disorganized lamellae across the gap. The disorganized bone is then remodeled.

In an unstable fracture, periosteal bone formation continues from the old bone ends toward the fracture line, but across the fracture line (where the fibrous material is avascular) chondrocytes proliferate and lay down a cartilage matrix. In a sequence identical to that which happens during endochondral ossification of long bones, the cartilage bridging the fracture ends is gradually replaced by bone. In humans, good stability is achieved by 6 weeks. With the improved stability, blood vessels and fibroblasts proliferate in the fracture gap.

Remodeling of the fracture callus begins as soon as the fracture site gains stability. The dynamics of this remodeling are similar to those of Haversian remodeling: Old bone is resorbed by osteoclasts, and new bone is deposited by osteoblasts. The process is vigorous in the area where the periosteal callus meets the surface of the old bone. Prior to remodeling, this line is clearly visible, but after remodeling, the junction is indistinguishable between the old bone and the callus.

Injuries in Other Connective Tissues

Articular cartilage, fibrocartilage, tendon, and ligament are specialized types of connective tissues that support and protect the body. They are characterized by a dense collagenous matrix and specialized cells, although each has a unique composition and function. For example, articular cartilage is composed principally of type II collagen and chondrocytes and is avascular. It functions principally by being compressed. Tendons and ligaments are primarily composed of fibroblasts surrounded by type I collagen, and although they are hypovascular, they have a significant neurovascular component. These two tissues are designed to bear tensile loads. Just as their compositions and functions vary, so too does their response to injury and healing.

Articular Cartilage

The articular surfaces of bones in synovial joints are, with few exceptions, covered with a thin (1 to 5 mm) layer of hyaline articular cartilage. This layer serves several important functions, including load distribution and minimization of friction and wear.

Injury

Injury to articular cartilage can severely compromise normal joint function and, in advanced cases, may necessitate joint replacement. Experimental data suggest that excessive joint loading leads to three types of articular damage: (1) loss of cartilage matrix macromolecules, alteration of the macromolecular matrix, or chondrocyte injury (any or all of these can occur with no detectable disruption of the tissue); (2) isolated damage to the articular cartilage itself in the form of chondral fracture or flap tears; and (3) injury to the cartilage and its underlying bone, a condition known as an osteochondral fracture.

Osteoarthritis

Degenerative joint disease (DJD), also called *osteoarthritis (OA),* is a noninflammatory disorder of synovial joints, particularly those with load-bearing involvement, that is characterized by deterioration of the hyaline articular cartilage and bone formation on joint surfaces and at the joint margins (figure 5.3). Strictly speaking, OA is not arthritis, because it does not initially manifest as an inflammatory condition. Inflammation may be secondary to the underlying tissue degeneration. Technically, DJD may be a more appropriate designation. However, because the condition is almost universally labeled OA, we use that term in our discussion. Osteoarthritis results in a degradation of articular cartilage and may be initiated by mechanical trauma and attendant chemical process alterations.

OA is a progressive condition, initially characterized by softening of articular cartilage attribut-

able to a decrease in matrix proteoglycan content. The degenerative process is characterized by cartilage fibrillation, cell loss, **chondromalacia** (cartilage softening), loss of elastic support, and disruption of the collagen framework (Fulkerson et al. 1987; Horton et al. 2005). These structural alterations increase the susceptibility of the articular cartilage to shearing loads and thereby predispose it to injury. Subsequently, the cartilage thins and its surface becomes rougher with characteristic pitting, fissuring, and ulceration. The cartilage damage results in enzyme release that causes further breakdown. Advanced cartilage degeneration is accompanied by subchondral bone necrosis and bony outgrowths of ossified cartilage **(osteophytes)** formed at the joint margin. The severity of OA is typically graded according to the degree of joint space narrowing, osteophyte formation, sclerosis, and joint deformity.

OA has been etiologically described as being either (1) primary, or **idiopathic** (i.e., of unknown origin); or (2) secondary, resulting from identifiable conditions such as trauma, metabolic disorders (e.g., calcium pyrophosphate dihydrate deposition disease and diffuse idiopathic skeletal hyperostosis), existing inflammatory conditions, or crystalline diseases. Among the factors implicated in the pathogenesis of osteoarthritis are obesity, genetics, endocrine and metabolic disorders, joint trauma, and activity patterns determined by a person's occupation or choice of recreation. Epidemiological evidence suggests that up to 90% of the population show some degree of osteoarthritic involvement by the age of 40, although in many cases with no clinical symptoms.

The causes of primary OA remain elusive. Postulated mechanisms include biomechanical, inflammatory, and immunological factors. Yet classifying OA as idiopathic may sometimes be a misstatement. In summarizing the evidence from a number of studies, Harris (1986) concluded that in the large majority of cases reported as primary OA of the hip, mild and unrecognized developmental abnormalities (e.g., acetabular dysplasia, pistol-grip deformity) were the likely causal factors. More recently, Tanzer and Noiseux (2004) reconfirmed that repetitive anterior femoroacetabular impingement often gives rise to anterior groin pain, labral tears, and chondral

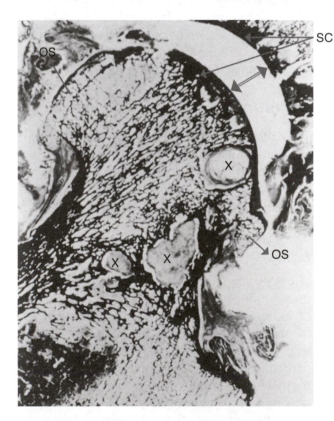

Figure 5.3 Histological section of the upper end of the femur showing principal features of osteoarthritis, including joint space narrowing (arrow), sclerosis of the bony end plates (SC), cysts (X), and osteophytes (OS) at the medial and lateral joint margins.

Reprinted from J.M. Fox and W. Del Pizzo, 1992, *The patellofemoral joint* (New York: NY: McGraw-Hill), 29, by permission of The McGraw-Hill Companies.

damage, which usually ends with the development of arthritis. Thus, a large proportion of OA is misclassified as primary OA.

In cases of OA with known cause, mechanical overuse plays a prominent role. This overuse can be acute (e.g., traumatic injury) or chronic (e.g., repeated heavy lifting). The mechanism of injury in chronic conditions often is occupationally related. For example, the incidence of OA among farmers has been linked to heavy lifting, walking on rough ground, and prolonged tractor driving. As expected, occupation-related OA has also been reported in the knees and spines of coal miners and the hands of cotton mill workers.

Somewhat surprisingly, obesity is not strongly associated with the onset of OA at the hip but rather may be more involved in the progression of already established OA (Croft et al. 1992). This contrasts with the knee joint, where obesity, along with repeated use and previous injury, is a strong risk factor for the occurrence of OA (Felson et al. 1988).

OA is strongly associated with advancing age. Radiological evidence of OA is rare in persons younger than 25, but by age 75 almost all persons exhibit evidence of OA in their hands and about half of these people show some degree of OA in their feet (Lawrence et al. 1989). The onset of OA at specific joints varies, occurring earliest at the metatarsophalangeal joints, next at the wrist and spine, later in the interphalangeals and first carpometacarpal, next in the tibiofemoral, and last in the hip. The reasons for this sequential appearance are unclear, but anatomical ultrastructural, biophysical, and biomechanical changes are likely involved. The development of OA in body joints is determined by a person's predisposition to OA, abnormalities in the joint, and patterns of mechanical loading and use. The precise relation between these biological and mechanical factors largely remains a mystery, and identification of the specific mechanisms is a challenge.

Healing

When significant damage occurs to articular cartilage, repair with new hyaline cartilage rarely takes place. The inability of articular cartilage to repair defects of any significant size is attributed to its lack of blood vessels and the relative lack of cells in cartilage. This inability of articular cartilage to effect substantial self-repair contributes to degenerative joint disease.

Fibrocartilage

Fibrocartilage serves as a transitional tissue at osteotendinous and osteoligamentous junctions, facilitating the distribution of forces at attachment sites and lowering the risk of injury. Fibrocartilage is also found in certain joints as a **meniscus,** an interposed fibrocartilage pad that acts as a shock absorber and a wedge at the joint periphery, thus improving the structural fit of the joint. Menisci are found in the joints of the tibiofemoral, acromioclavicular, sternoclavicular, and temporomandibular joints. Meniscal injuries at several of these joints are discussed in later chapters.

Fibrocartilage is also found in the outer covering **(annulus fibrosus)** of the intervertebral disks. Injury to the fibrocartilage of the annulus fibrosus plays a central role in the mechanisms of low back pain caused by so-called slipped discs.

Tendon

As the structure responsible for force transfer from skeletal muscle to bone, tendon is a critical link in the musculoskeletal system. Injury to the tendinous structures can restrict or even prevent normal movement and function. The connective structure of tendon creates three structural zones: (1) the body of the tendon itself (tendon substance), (2) the connections of tendon with bone *(osteotendinous junction)*, and (3) the connections with its accompanying muscle *(myotendinous junction;* figure 5.4).

Pathology

Tendon injury may result from a direct insult, as when tendons of the hand and fingers are lacerated by sharp implements such as knives, saws, and other bladed tools. Tendinous injury may also be indirect, resulting from excessive tensile loads applied to the tendon structure. Attempted transmission of loads exceeding the ultimate strength of fibers (or the whole tendon) leads to tendon injury.

ADVANCES IN JOINT REPLACEMENT

The debilitating pain of advanced osteoarthritis of the hip and knee severely limits a person's mobility. Joint replacement surgery **(arthroplasty),** in which the damaged structures are replaced by artificial materials, provides remarkable pain relief and restores function in most cases. Because of the load-bearing responsibilities of the lower extremities, it is not surprising that the hip and knee are the leading arthroplastic sites. In light of continued advances in biomaterials and surgical techniques, the advent of computer-assisted design and manufacture (Crowninshield 1990), and an aging population, the number of arthroplasties will continue to increase. The American Academy of Orthopaedic Surgeons estimates that total knee replacements in the United States will increase from 245,000 in 1996 to 454,000 by 2030 and total hip replacements from 138,000 to 248,000 in the same time period (AAOS 1999). The accuracy of this prediction is reinforced by the fact that in 2007, more than 193,000 total hip replacements were performed in the United States (AAOS 2007).

Total hip replacement (THR) involves excision of the femoral head and part of the neck and enlargement of the acetabulum. A metallic femoral prosthesis is inserted into the medullary canal of the femur. The prosthesis may be cemented into the canal using methyl methacrylate. An alternative, cementless technique uses a prosthesis with porous structure that encourages bony ingrowth. The success of cementless prostheses remains a subject of debate.

Although traditional THR is reasonably successful in elderly, relatively inactive patients, in younger, active patients it offers unacceptably poor long-term outcomes, often leading to multiple revisions and associated complications. A new British model of surface hip replacement called Birmingham hip resurfacing arthroplasty is a metal-on-metal prosthesis that minimizes bone loss, alleviates hip pain, and allows patients to return to work and leisure activities and to participate in sports. Revision surgery, when it becomes necessary, is much more successful. In resurfacing, the neck and head of the femur are not removed. The prosthesis is designed so that it fits like a cap on the head of the femur and a matching cup fits precisely into the groove of the acetabulum.

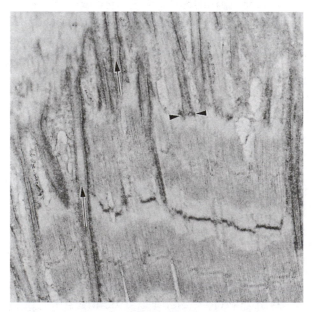

Figure 5.4 Transmission electron micrograph of the myotendinous junction. Collagen fibers (arrows) extend into the folded cell membrane. Densely bundled fibers meet at the terminal Z disc (arrowheads).

Reprinted, by permission, from J.G. Tidball, 1991, "Myotendinous junction injury in relation to junction structure and molecular composition," *Exercise and Sport Sciences Reviews* 19: 426.

Injuries to musculotendinous units are termed **strains.** Like skeletal muscle injuries, tendon injuries are categorized according to their severity. Mild strain is characterized by negligible structural disruption, local tenderness, and minimal functional deficit. Moderate strain exhibits partial structural defect, visible swelling, marked tenderness, and some loss of stability. Severe strains have complete structural disruption, marked tenderness, and functional deficits that typically necessitate corrective surgical intervention.

Severe tendon strain (complete rupture of the tendon's structure) is often preceded by undetected tissue damage that existed before the specific incident of rupture. Such cases have been termed *spontaneous tendon rupture* and are typically seen in middle-aged people engaged in strenuous activities. The injured often have no history of injury or any prior recognized distress. These spontaneous tendon ruptures are accompanied by a popping sensation and occur unexpectedly. Postinjury investigation during surgery often identifies preexisting pathology, suggesting that previous, undetected tendon degeneration may have facilitated the spontaneous failure (Woo et al. 1994). Examples of spontaneous ruptures in the Achilles and patellar tendons are described in chapter 6.

Repetitive overloading of a tendon may lead to an inflammatory response, or **tendinitis.** The reaction may be acute (in response to a limited session or event) or chronic (a result of repeated overuse). In addition to the tendon itself, related structures that facilitate tendon sliding may also become inflamed and subsequently injured (e.g., peritenon, tendon sheath, and accompanying bursa).

The terms used to describe tendon and tendon-related conditions vary. The terms *tendinitis* (also spelled *tendonitis*), *tenosynovitis*, *tendinosis*, and others may be used by clinicians and researchers in various contexts to describe various conditions; thus, caution is warranted in interchanging these terms. Specific structures under examination must be clearly identified to avoid confusion.

In addition to the tendon substance, the myotendinous and osteotendinous junctions can be injured. The myotendinous junction has been implicated as the site of many injuries. The structure of the junction shows an interdigitation of skeletal muscle fibers with tendinous collagen fibers in a characteristic membrane-folding pattern. This folding pattern reduces stress at the myotendinous junction during muscular contraction, lessening the likelihood of injury.

Injury also occurs at the *osteotendinous junction*, where the tendon attaches to the cortical bone surface. This attachment frequently features four zones of increasing mechanical stiffness: tendon, fibrocartilage, mineralized fibrocartilage, and bone, as shown in figure 5.5*a*. Cells within the tendon are tenocytes, whereas those in the fibrocartilage are fibrochondrocytes and those in the mineralized fibrocartilage are osteocytes. These transition zones distribute the tensile forces and reduce the chance of injury. Figure 5.5*b* is an electron micrograph showing an intact osteotendinous junction, and figure 5.5*c* shows a disrupted osteotendinous junction in which the applied force exceeded the maximum tolerance of the tissues.

Healing

A great deal is known about the healing of severed tendons that have been surgically repaired. After an initial inflammatory response, synthesis of glycosaminoglycans and collagen is triggered. These substances are used in restoring the matrix integrity of the tendon. Rest, ice, and immobilization are recommended to avoid additional tissue damage for the first week after a tendon is acutely

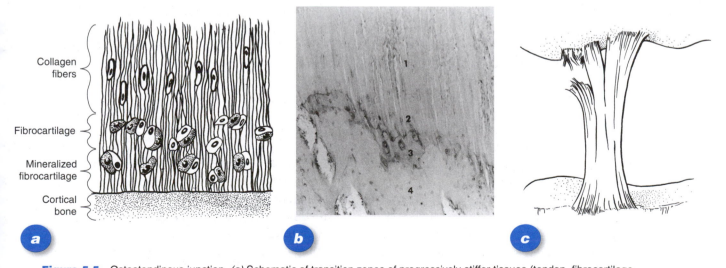

Figure 5.5 Osteotendinous junction. *(a)* Schematic of transition zones of progressively stiffer tissues (tendon, fibrocartilage, mineralized fibrocartilage, cortical bone). *(b)* Electron micrograph of patellar insertion from a dog with zones of (1) tendon, (2) fibrocartilage, (3) mineralized fibrocartilage, and (4) cortical bone. *(c)* Drawing of a disrupted osteotendinous junction.

(b) Reprinted, by permission, from R.R. Cooper and S. Misol, 1970, "Tendon and ligament insertion: A light and electron microscopic study," *J Bone Joint Surg Am.* 52(1): 3.

injured. In the second and third weeks, cyclic low loads applied to the healing tendons may help to align the new fibers and strengthen the repairing tendon. In addition, stretching and activation of the muscle–tendon unit may prevent excessive muscle atrophy and joint stiffness. After the third week, progressively increasing the stress on the tendon optimizes the tissue's healing. Much less is known about the processes and healing responses associated with chronic tendinitis.

Ligament

A ligament is a connective tissue that joins one bone to another. Ligaments protect the integrity of bone-to-bone connections by resisting excessive movements or dislocation of the bones. In that role, ligaments are sometimes called **passive joint stabilizers.**

Pathology

Injury to a ligament, termed a ligamentous **sprain,** may compromise a ligament's stabilizing ability and impair its ability to control joint movements. The severity of the sprain is clinically specified with a three-level scheme as shown in table 5.1. Mild and moderate sprains are most common. Complete ligament tearing (severe sprain) happens in a minority of cases.

Recall that ligaments can be **intracapsular** (located within the joint capsule), **capsular** (appearing as a thickening of the capsule structure), or **extracapsular** (extrinsic to the capsule). The location and attachments of these ligament types help determine their function, their response to mechanical loading, and their susceptibility to injury. Examples in chapters 6 and 7 illustrate these ligament-specific responses in lower- and upper-extremity joints.

The attachment of ligament to bone follows a structure similar to that at the osteotendinous junction. The osteoligamentous junction can also exhibit transitional zones of fibrocartilage and mineralized fibrocartilage intervening between the ligament and the cortical bone to which it attaches. This zonal structure facilitates load distribution at the ligament attachment sites. Additionally, fibers of the ligament can travel relatively parallel to the surface of the bone and gradually blend with the periosteum.

Healing

Frank (1996) outlined the three phases of ligament healing: bleeding and inflammation, proliferation of bridging material, and matrix remodeling. The first phase, the inflammatory response, parallels the description in the preceding section on inflammation. The platelets from the blood promote clotting, a fibrin clot is deposited, growth factors are released to promote the inflammatory cascade, local vessels dilate, acute inflammatory cells infiltrate, and the fibroblastic scar cells arrive on the scene.

The second phase of ligament healing is the generation of a scar matrix. The scar that is produced and eventually remodeled in phase 3, however, is not normal ligament. The collagen fibers in the scar are typically smaller in diameter than in normal ligament, and the alignment of the collagen fibers in the scar is more haphazard than in normal ligament.

Matrix remodeling constitutes the third phase of ligament healing: The scar matrix diminishes in size and becomes less viscous and more dense and organized. The scar, with time, may start to look and function more like an uninjured ligament, but it never becomes the same as a normal, uninjured ligament.

Skeletal Muscle Injuries

Skeletal muscles are the engines that provide the human body with the power needed to move. They have a unique ability among all the body's tissues to generate force and contract. Their specialized cells allow muscle to produce force (sometimes referred to as **tension,** or **tensile load)** and to change their shape by shortening, or contracting.

Pathology

Injury to skeletal muscle is common and can take several forms and involve various mechanisms. As seen in many occupations, sports, and physical activities, skeletal muscles have the capacity to generate high forces (and power) without sustaining injury. But if too much force is transmitted through a muscle–tendon unit, injury is likely to result. Which regions of the muscle–tendon unit are the most likely to be damaged?

Stimulated Versus Passive Muscle

Research has shown that when muscle–tendon–bone units are pulled to failure, the tears could happen at bone–tendon junctions, within the muscle belly, or at the myotendinous junctions (McMaster 1933). Later experimental studies confirmed tears in muscle–tendon units at the myotendinous junction (Garrett et al. 1988) or at sites within the muscle cell approximately 0.5 mm from the myotendinous junction (Tidball and Chan 1989).

Much of this work probed the failure characteristics of passive (unstimulated) muscle–tendon units, and certainly injuries can happen to non-contracting muscles. But two studies (Garrett et al. 1987; Tidball et al. 1993) provided interesting findings about the sites of muscle–tendon failure in stimulated versus quiescent muscle.

- Garrett and colleagues (1987) compared the biomechanical properties of passive versus stimulated muscle (rabbit extensor digitorum longus) that was rapidly lengthened to failure. The investigators found no significant difference in the amount of lengthening at which the failure happened, regardless of activation state. But when muscles were stimulated, they achieved 14% to 16% greater peak forces at failure than if the muscle was passive. Also, substantially more energy was absorbed to the point of failure in the stimulated than in the unstimulated muscle–tendon units. Garrett and colleagues reported that the site of failure was typically at the myotendinous junction.

- Tidball and colleagues (1993) used an electron microscope to locate the specific sites of the muscle–tendon interface that failed in tetanically stimulated versus unstimulated bone–tendon–muscle–tendon–bone units, using a frog semitendinous muscle. The specimens were strained at physiological strain rates to failure, and all failures happened at or near the proximal (tendon of origin) myotendinous junction in both stimulated and unstimulated muscle. Like Garrett and colleagues (1987), Tidball and colleagues (1993) found that the stimulated muscle–tendon units required

about 30% more force and about 110% more energy to reach failure. Interestingly, the electron-microscopic analysis revealed systematic differences in the sites of failure, which varied with the state of activation of the muscle cells at the time of injury. Also, the breaking strength of the Z disc when the muscle was stimulated was different than when it was unstimulated—suggesting that two load-bearing systems may be in parallel within the Z discs.

Figures 5.6, 5.7, and 5.8, taken from Tidball and colleagues (1993), show clearly the variation in failure site with activation state of the muscle. A transmission electron micrograph (figure 5.6a) illustrates a longitudinal section from an unstimulated muscle, and figure 5.6b illustrates a similar section from a stimulated muscle. In the unstimulated muscle, the site of failure was within the muscle, near the myotendinous junction. Failure happened in a single transverse plane of each cell within Z discs, and other Z discs in the area remained stretched—with residual strains of several hundred percent (figure 5.6a). In contrast, figure 5.6b was taken from a stimulated muscle at a site about 100 mm from the failure site. Here you can see the relatively normal-looking Z discs; they show no residual strains.

When the sites of failure were examined more closely in the unstimulated muscle, tears were found in the tendon near the myotendinous junction (outlined by the arrowheads in figure 5.7a). Identical-appearing tears were seen at the myotendinous junctions of both unstimulated and stimulated specimens. Large separations (denoted by S in figure 5.7b) were seen within the tendon near the myotendinous junction of an unstimulated muscle. In figure 5.8 (an electron micrograph taken through the myotendinous junction for a stimulated muscle), site of failure (complete separation) is immediately external to the membrane of the myotendinous junction, and no connective tissue appears connected to the junction membrane (see arrowheads).

Types of Injury

A *strain* injury (not to be confused with *mechanical strain*, discussed in chapters 3 and 4) happens when damage is inflicted on a musculotendinous unit. The specifics of tendon injury were

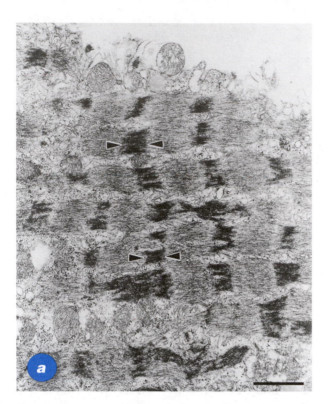

Figure 5.6 *(a)* Longitudinal section from an unstimulated muscle loaded to failure. This transmission electron micrograph shows an incomplete tear through a muscle fiber about 80 μm from the site of the complete tear. The Z discs (between the arrowheads) show extensive residual strain. *(b)* Longitudinal section taken through a stimulated muscle that was strained to failure. The site in the photograph is about 100 μm from the failure site. The Z discs are distinct and show no signs of persisting strain. The insert bar (1.5 μm) is the same in both photographs.

Reprinted, by permission, from J.G. Tidball, G. Salem and R.F. Zernicke, 1993, "Site and mechanical conditions for failure of skeletal muscle in experimental strain injuries," *Journal of Applied Physiology* 74: 1283.

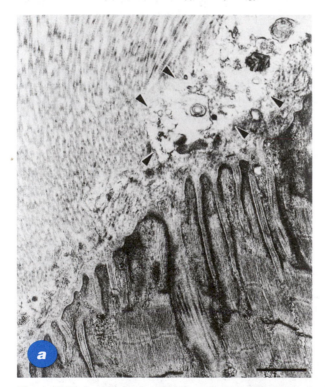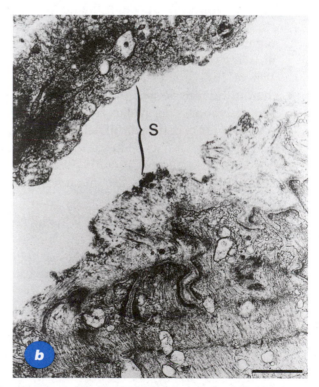

Figure 5.7 Longitudinal section through the myotendinous junction of origin of an unstimulated muscle that was strained to failure. *(a)* Tears existed in the tendon, near the myotendinous junctions (outlined by the arrowheads). Identical-looking tears were found at the myotendinous junctions of both unstimulated and stimulated fibers. The inset bar is 2.0 μm. *(b)* Large separations (S) within the tendon near the myotendinous junctions were observed in some preparations, although complete tears were located in the fibers for these preparations. The inset bar is 2.5 μm.

Reprinted, by permission, from J.G. Tidball, G. Salem and R.F. Zernicke, 1993, "Site and mechanical conditions for failure of skeletal muscle in experimental strain injuries," *Journal of Applied Physiology* 74: 1285.

Figure 5.8 Longitudinal section through the myotendinous junction of origin of a *stimulated* muscle that was loaded to failure. The site of complete separation is immediately external to the junctional membrane so that no connective tissue is found associated with the junctional processes (see arrowheads). Inset bar is 2.0 μm.

Reprinted, by permission, from J.G. Tidball, G. Salem and R.F. Zernicke, 1993, "Site and mechanical conditions for failure of skeletal muscle in experimental strain injuries," *Journal of Applied Physiology* 74: 1284.

discussed previously; here we concentrate on three forms of skeletal muscle injury: acute muscular strain, contusions, and exercise-induced muscle injury (Leadbetter 2001).

- **Acute muscular strain** typically results from overstretching a passive muscle or dynamically overloading an active muscle, either in **concentric** or **eccentric** action. The severity of tissue damage depends on the magnitude of the force, the rate of force application, and the strength of the musculotendinous structures. Mild strains are characterized by minimal structural disruption and rapid return to normal function. Moderate strains are accompanied by a partial tear in the muscle tissue (often at or near the myotendinous junction), pain, and some loss of function. Severe muscle strains are defined by complete or near-complete tissue disruption and functional loss, as well as marked **hemorrhage** and swelling.

- **Impact injuries** result from a direct compressive impact. Such contact may cause a muscle bruise (contusion), which is distinguished by intramuscular hemorrhage. Muscular contusion commonly occurs in contact sports (e.g., basketball, football, soccer), as when an athlete's thigh has a violent impact with another participant's knee. Such a thigh injury is often called a "Charley horse," but we again caution against the use of such catch-all designations because they are nonspecific and inconsistently applied. Repeated mechanical insult to a damaged muscle prior to healing may worsen the injury and lead to serious secondary conditions, such as **myositis ossificans** (deposition of an ossified mass within the muscle).

- **Exercise-induced muscle injury** results from connective and contractile tissue disruption following exercise. It is characterized by local tenderness, stiffness, and restricted range of motion. This type of injury, also referred to as **delayed-onset muscle soreness (DOMS)**, typically happens 24 to 72 hr after participation in vigorous exercise, especially following eccentric muscle action in contractile tissue unaccustomed to the activity's demands. Although the underlying mechanism of DOMS remains elusive, its symptoms and metabolic events (e.g., pain, swelling, presence of cellular infiltrates, increased lysosomal activity, and increased levels of some circulating acute-phase proteins) are similar to those of acute inflammation and suggest a relation between the two (Close et al. 2005).

Although not particularly injurious itself, the common muscle cramp may indicate conditions predisposing to injury. Excessive demands placed on such a sustained and often painful muscular spasm may result in muscle strain. The mechanisms of muscle cramps, despite the frequency of their occurrence, are not fully understood. Most cramps occur in a shortened muscle and are characterized by abnormal electrical activity. Many factors have been proposed to cause muscle cramps, including dehydration, electrolyte imbalances, direct impact, fatigue, and lowered levels of serum calcium and magnesium. Cramps happen in many muscles, especially the gastrocnemius, semimembranosus, semitendinosus, biceps femoris, and abdominals, and can be relieved by antagonistic muscle activity or manual stretching of the afflicted muscle. Stretching warrants care because excessive force applied to a muscle in spasm may result in muscle strain.

Joint Injuries

Skeletal *articulations (joints),* by virtue of their often intricate structure, are susceptible to injury and involve complex mechanical loading of multiple tissues. Although not well understood in many cases, the dynamics of joint mechanical loading determine the actual injuries that occur. Because numerous joint injuries are examined in detail in subsequent chapters, we limit our discussion here to basic concepts and terminology of joint injury.

When sufficient force is applied to a joint, the articulating bones may become displaced. This results in a complete dislocation **(luxation)** or a partial dislocation **(subluxation)** (figure 5.9). (Note: The term *subluxation* is defined in two ways. Here, subluxation refers to a partial, rather than a complete, dislocation. Other sources apply the term *subluxation* to cases where a dislocation [either partial or complete] is followed immediately by spontaneous joint **reduction** [i.e., the dislocated bone goes back into place].) Dislocations, whether partial or complete, often are accompanied by additional injuries, including ligamentous sprain and tears of the fibrous joint capsule.

The inner surface of the joint capsule is lined by the **synovial membrane,** a thin layer of tissue that has negligible biomechanical function but plays an important role in the physiology of both normal and injured joints. Irritation or trauma to the synovium may lead to **synovitis,** a condition with inflammatory symptoms that may in turn limit joint function.

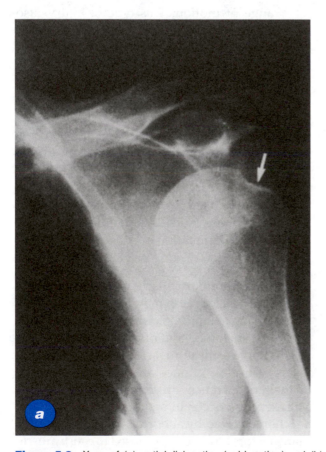
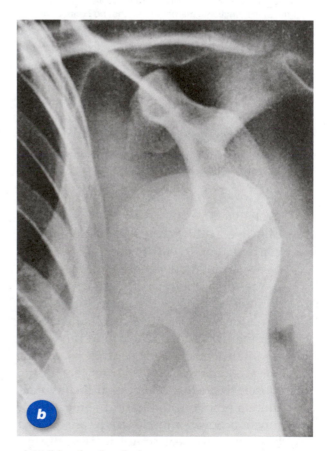

Figure 5.9 X ray of *(a)* partial dislocation (subluxation) and *(b)* complete dislocation (luxation).

(a) Reprinted, by permission, from M.J. Julian and M. Matthews, 1990, Shoulder Injuries. In *The team physicians handbook*, edited by M.B. Mellion (Philadelphia, PA: Hanley and Belfus, Inc.), 320. *(b)* Reprinted, by permission, from P.G. Gusmer and H.G. Potter, "Imaging of shoulder instability," *Clinics in Sports Medicine* 14(4): 780.

Arthritis refers to inflammation of a joint or a state characterized by joint inflammation. It includes many conditions that have either primary or secondary inflammatory involvement; more than 100 types of arthritis are known. Among the major types are those resulting from chronic and excessive mechanical loading (e.g., *osteoarthritis)*, systemic disease (e.g., **rheumatoid arthritis),** or biochemical imbalances (e.g., **gouty arthritis).** Arthritis may be a primary condition or may develop secondarily in response to a non-inflammatory insult, as is the case in osteoarthritis. In any case, arthritis and its sequelae have the potential to inflict debilitating pain and loss of joint function.

Nonmusculoskeletal Injuries

A wide range of nonmusculoskeletal injuries involve the skin and nervous tissue, which are often affected secondarily to musculoskeletal injury.

Skin

Skin forms the outer protective covering for most of the body's surface. As the body's outermost defense, skin is susceptible to a wide variety of injuries.

Pathology

Injury to the skin can take various forms and may involve many causal mechanisms, some of which are mechanical in nature. Sufficient friction between the skin and an opposing surface may result in superficial injury such as an **abrasion** (scraping away of the superficial skin layer, usually by mechanical action) or deeper injury such as a **blister** (a fluid-filled structure under or within the epidermis, caused by heat, chemical, or mechanical means).

Nonpenetrating skin injury is termed a **contusion,** or bruise, which usually results from a direct, violent impact. Internal hemorrhage can accompany contusion and, in severe cases, be quite debilitating.

Sharp implements that penetrate the skin create **puncture** wounds, which not only damage the dermal and epidermal layers but may, with sufficient penetration depth, injure interven-ing internal structures as well. The mechanics of puncture injury often result in a wound that appears on the skin surface to be much less severe than it may actually be, because the skin often closes on implement removal to obscure deeper damage. The wound often appears to be clean, leaving little visible external evidence of the damage created internally. Jagged tearing of the skin **(laceration)** is characteristic of cut wounds by a knife or other sharp implement, because the mechanism of injury is tearing rather than puncture. The external appearance of a laceration, compared with that of a puncture wound, is usually more indicative of the extent of injury.

Healing

Any injury that penetrates the skin, to any depth, carries the risk of infection. Caution is warranted in treating any penetrating injury, however slight, because the damaging effects of infection resulting from an improperly treated injury may far exceed the deleterious effects of the original injury itself.

Given the proximity of venous vasculature to the dermis, skin injuries are often characterized by considerable bleeding *(hemorrhage)*. Facial lacerations, for example, commonly produce effusive bleeding, especially in the areas immediately above (supraorbital) and below (infraorbital) the eye and on the chin.

Nervous Tissue

Nervous tissue is not classified as a musculoskeletal tissue, but because injury to nervous tissue can affect, either directly or indirectly, the function of musculoskeletal tissues, we present a brief outline of nervous tissue injury.

Pathology

Injury to nervous tissue has the potential to produce the most debilitating types of dysfunction. Damage to the brain and other supraspinal structures, spinal cord, spinal nerves, or peripheral nerves can impair the most essential of the body's communication systems, reducing or even eliminating sensory and motor processes. Injury to specific neural structures and the causal mechanisms of these injuries are considered in the three succeeding chapters. Injuries to structures of the peripheral nervous system are considered in chapters 6 and 7. Examples of injuries to the central nervous system (e.g., cerebral concussion) are presented in chapter 8. As a prelude, a brief review of nervous tissue injury is presented here.

Nerves are cordlike organs that serve as the communication conduits in the peripheral nervous system. Each nerve consists of many nerve fibers (axons) that are arranged in parallel bundles. These bundles are enclosed by layers of connective tissue. Each axon is covered by a delicate endoneurial sheath, or **endoneurium.** Axons are grouped into nerve bundles (fascicles) that are surrounded by **perineurium.** Finally, the fascicles collectively form the whole nerve (nerve trunk), which is surrounded by a tough covering of epineurial tissue **(epineurium).**

Nervous tissue can be injured through chemical, thermal, ischemic, or mechanical means; here we focus on only the mechanical means. Mechanical influences on nervous structures can take one of two forms, namely, entrapment or trauma. In the former, nervous tissue becomes entrapped in a confined anatomical space or between other anatomical structures. The resulting forces impinge on the nervous tissue and can produce damage and compromise function.

The second form of nerve injury, trauma, results from a direct mechanical insult to the tissue or from forces indirectly applied to the nervous tissue via surrounding structures. Each of the three principal loading types may be present, either alone or in combination. Compressive loads result in pressure on nervous tissue. Tensile loading creates tissue elongation, which may result in stretch injury. Shear loading may lead to friction-related injury. In entrapment and trauma, the nature of any resulting dysfunction depends on the characteristics of the mechanical environment at the time of injury.

Traumatic peripheral nerve injury may temporarily block nerve signal conduction either without axonal discontinuity or in actual severance (partial or full) of the axon. In the latter case, injury to the axon typically leads to axonal degeneration, in which the axon and its myelin sheath disintegrate **(Wallerian degeneration).** This degenerative process results in a separation (denervation) between the axon's neuron (nerve cell) and its target organ. The cell's response depends on the location and severity of injury and may enable a regenerative response or may result in cell death.

The severity of injury, based on nerve histopathology, is commonly assessed with a qualitative five-level classification system (Sunderland 1990). Only the basic elements of these levels are presented here:

▪ The least severe level is termed a first-degree injury and is characterized by the presence of a conduction block. This interruption of nerve signal transmission may be brief, mild, or severe. First-degree nerve injuries do not involve any denervation effects and result in full recovery, although this recovery may take several months. First-degree injury can happen after a prolonged, low-pressure compression (e.g., **carpal tunnel syndrome)** or an acute, high-compression event that results in a conduction block without axonal discontinuity, or **neurapraxia.**

▪ Second-degree injury involves **axonotmesis,** an interruption of axonal structure with accompanying Wallerian degeneration but without severance of the nerve's supporting structure. These lesions may result from pinching or crushing mechanisms or from prolonged pressure. During recovery from second-degree nerve injury, nerve regeneration normally occurs.

▪ Third-degree injury involves a loss of fiber continuity **(neurotmesis),** with damage to both the axon and the endoneurial sheath. Fascicular disorganization and intrafascicular hemorrhage, edema, and ischemia result. A third-degree injury results in complete loss of sensory and motor function. Recovery may be complete but typically is quite protracted.

▪ Fourth-degree nerve injury involves a loss of axonal, endoneurial, and fascicular continuity, leaving only epineurial tissue to provide structural continuity. Successful recovery seldom occurs spontaneously, and surgical repair is indicated in most cases.

▪ Fifth-degree injury, the most severe nerve injury, results in complete severance of the nerve trunk. Regeneration, if it occurs at all, is usually incomplete and disrupted. This category of nerve injury usually requires surgical repair.

Any level of compromised sensory or motor function can hasten or exacerbate musculoskeletal injury. Impaired sensory function, for example, may alter pain sensation, thus retarding the body's warning system and permitting a more severe injury than might occur with normal sensation. This can be seen in the peripheral neuropathies commonly found in the feet of persons with diabetes. These neuropathies can lead to foot ulcerations and eventual amputation.

Axonal damage that results in motor impairment can alter muscle recruitment patterns and produce uncoordinated and potentially dangerous movements. A runner, for example, with impaired motor control attributable to nervous tissue injury may experience selective muscle weakness that results in altered gait mechanics and eventual musculoskeletal injury.

■ CHAPTER REVIEW

Key Points

- We have described concepts of injury mechanics, terminology, and mechanisms of musculoskeletal injury, setting the stage for discussions in the next three chapters of specific injuries and their causes.

- The tenets and principles of mechanical load and overload, use and overuse, level and progression of injury, and the many contributory factors involved in injury form a complex and fascinating backdrop for viewing specific musculoskeletal injuries.

- Because it would be impossible to examine all injuries, the injuries discussed in succeeding chapters were selected because of their epidemiological pervasiveness and their value in illustrating the principles of injury biomechanics.

Questions to Consider

1. Select a specific injury that you or someone you know has experienced. Describe the injury, its mechanism(s), and any contributory factors that may apply to the injury you've described.

2. Explain the "weakest link phenomenon" as it applies to musculoskeletal injury and what factors may be involved in determining the weakest link for a particular anatomical structure or system.

3. Describe an injury situation in which two or more contributory factors (from those listed in the text or additional ones you identify) *interact* to increase or decrease the chance of injury.

4. Describe an example of how a single contributory factor might both increase *and* decrease injury risk.

5. Explain how inflammation provides a "first line of defense" against insult following an injury.

6. Osteoporosis and related conditions (e.g., fall-related fractures) are a growing public health concern. Describe the multifactorial nature of osteoporosis as a public health issue.

7. Select a tissue (e.g., bone, articular cartilage, tendon, ligament, skeletal muscle) and explain its pathology and healing.

Suggested Readings

Adams, J.C., and D.L. Hamblen. 1999. *Outline of Fractures* (11th ed.). Edinburgh, UK: Churchill Livingstone.

Albright, J.A., and R.A. Brand. 1987. *The Scientific Basis of Orthopaedics* (2nd ed.). Norwalk, CT: Appleton & Lange.

Finerman, G.A.M., and F.R. Noyes, eds. 1992. *Biology and Biomechanics of the Traumatized Synovial Joint: The Knee as a Model*. Rosemont, IL: American Academy of Orthopaedic Surgeons.

Fu, F.H., and D.A. Stone. 2001. *Sports Injuries: Mechanisms, Prevention, Treatment*. Baltimore: Williams & Wilkins.

Garrett, W.E., Jr., and P.W. Duncan. 1988. *Muscle Injury and Rehabilitation*. Baltimore: Williams & Wilkins.

Rockwood, C.A., D.P. Green, R.W. Bucholz, and J.D. Heckman. 2001. *Rockwood and Green's Fractures in Adults* (5th ed.). Philadelphia: Lippincott Williams & Wilkins.

Stanish, W.D., S. Curwin, and S. Mandel. 2000. *Tendinitis: Its Etiology and Treatment*. Oxford, UK: Oxford University Press.

Tencer, A.F., and K.D. Johnson. 1994. *Biomechanics in Orthopedic Trauma: Bone Fracture and Fixation*. Philadelphia: Lippincott.

Tidball, J.G. 1991. Myotendinous junction injury in relation to junction structure and molecular composition. In *Exercise and Sport Sciences Reviews*, edited by J.O. Holloszy. Baltimore: Williams & Wilkins.

Woo, S.L.-Y., and J.A. Buckwalter. 1988. *Injury and Repair of the Musculoskeletal Soft Tissues*. Park Ridge, IL: American Academy of Orthopaedic Surgeons.

Zachazewski, J.E., D.J. Magee, and W.S. Quillen, eds. 1996. *Athletic Injuries and Rehabilitation*. Philadelphia: Saunders.

CHAPTER 6

LOWER-EXTREMITY INJURIES

Fractured, hell! The damn thing's broken.

Dizzy Dean, Hall of Fame baseball pitcher,
commenting on his injured toe

■ OBJECTIVES

- ■ To describe the relevant lower-extremity anatomy involved in musculoskeletal injury
- ■ To identify and explain the mechanisms involved in musculoskeletal injuries to the major joints (hip, knee, ankle) and segments (thigh, lower leg, foot) of the lower extremity

Injuries to lower-extremity joints, in particular the knee and ankle, are among the most common of all musculoskeletal disorders. Given the importance of the lower extremities in everyday activities such as walking, running, and maintaining posture, injury to these joints has a huge effect on daily living. The circumstances of lower-extremity injury can vary, ranging from the acute, high-energy trauma of a sprained ankle to the more gradual onset of a metatarsal stress fracture. The lower-extremity injuries presented in this chapter were selected based on their prevalence and their value in illustrating specific injury mechanisms. Representative injuries are presented for the major lower-extremity joints (hip, knee, and ankle) and the regions spanning those joints (thigh, lower leg, and foot). In addition, we explore anterior cruciate ligament (ACL) injuries, expanding our discussion beyond the mechanisms of injury to describe the ACL's structure and tissue mechanics and explain clinical evaluation, treatment, and rehabilitation. Although space restrictions preclude this level of detail for all injuries, we believe it instructive to show, using the ACL, how any injury can be presented in a broader context and in greater detail.

Hip Injuries

The hip joint is formed by articulation of the femur with the pelvic girdle (coxal bone), specifically by the articulating surfaces of the femoral head and the acetabulum (figure 6.1). The bony fit is improved by the acetabular labrum, a fibrocartilage pad attached to the bony rim of the acetabulum. The hip joint is reinforced anteriorly by the iliofemoral ligament, posteriorly by the ischiofemoral ligament, and anteroinferiorly by the pubofemoral ligament. The ligament of the head of the femur provides limited structural support, serving primarily to contain the vasculature

supplying the femoral head. Additional support is provided by the articular joint capsule, whose fibers form a fibrous collar around the femoral neck and help secure the femoral head in the acetabulum.

The hip joint's ball-and-socket configuration permits movements in the three primary planes, referred to as flexion–extension (sagittal plane), abduction–adduction (frontal plane), and internal–external rotation (transverse plane). The muscles responsible for controlling movements about the hip joint are shown in figure 6.2; their actions are summarized in table 6.1. This considerable musculature further stabilizes the hip joint.

In the next two sections, we discuss hip fracture and dislocation. Although these injuries are presented in separate sections, conjoint fracture–dislocation injuries are not uncommon given that most severe hip injuries involve high-energy trauma.

Hip Fracture

Bone fractures in the hip typically result from high-energy trauma such as that associated with falls from heights and automobile crashes. The Canadian Institute for Health Information estimated the incidence of proximal femoral fractures in that country for fiscal year 1993-1994 to be more than 23,000, with an increase to more than 88,000 projected by 2041. The incidence in women was almost three times that in men and increased significantly with advancing age (Papadimitropoulos et al. 1997). Pelvic fractures, although not as prevalent as femoral fractures, nonetheless present a significant problem. For example, more than 60,000 pelvic fractures happen in the United States annually (www.wrongdiagnosis.com/p/pelvic_fracture/prevalence.htm).

In motor vehicle accidents, the direction of force largely determines the pattern of injury.

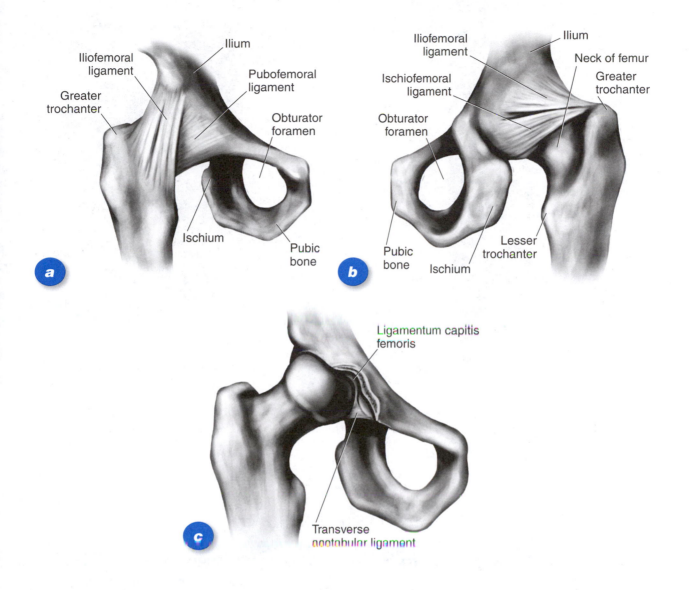

Figure 6.1 Ligaments of the hip joint: *(a)* the iliofemoral and pubofemoral ligaments, anterior view; *(b)* the ischiofemoral ligament, posterior view; and *(c)* the transverse acetabular ligament and ligamentum capitis femoris, anterior view.

Reprinted, by permission, from R.S. Behnke, 2005, *Kinetic anatomy*, 2nd ed. (Champaign, IL: Human Kinetics), 177.

Pelvic fractures resulting from severe motor vehicle accidents are significantly more frequent in side-impact collisions, occur at much lower speeds in side impacts than in frontal collisions, and have high mortality rates from associated injuries (Gokcen et al. 1994). Fracture-related mortality and morbidity are especially problematic for women. Trochanteric soft tissue thickness and total hip bone mineral density are significant determinants of fracture outcome in women with pelvic fractures resulting from side-impact collisions (Etheridge et al. 2005).

Proximal femoral (hip) fractures are a major health concern, with more than 352,000 fractures occurring annually in the United States (AAOS 2005a). Globally, hip fracture affects millions and is a significant cause of mortality and morbidity worldwide (Johnell and Kanis 2004). The statistics are all the more sobering when we consider that many hip fracture victims die within a year of injury, not usually because of the fracture per se but rather because of chronic conditions that worsen after the injury. Davidson and colleagues (2001) reported a 12-month mortality rate of 26%.

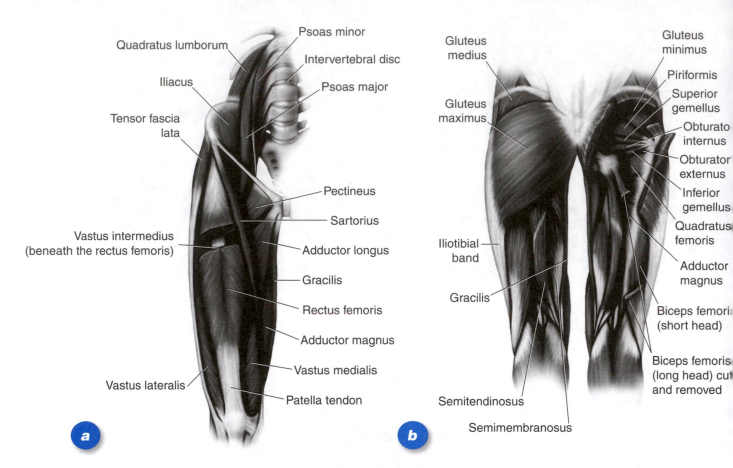

Figure 6.2 Musculature of the hip and knee joints: *(a)* anterior view and *(b)* posterior view.

(a) Adapted, by permission, from R.S. Behnke, 2005, *Kinetic anatomy*, 2nd ed. (Champaign, IL: Human Kinetics), 178. (b) Reprinted, by permission, from R.S. Behnke, 2005, *Kinetic anatomy*, 2nd ed. (Champaign, IL: Human Kinetics), 180.

TABLE 6.1

MUSCLES OF THE HIP

Muscle	Action
Adductor group	
Adductor brevis	Adducts and laterally rotates the thigh
Adductor longus	
Adductor magnus	
Biceps femoris (long head)	Extends the thigh
Gluteus maximus	Extends and laterally rotates the thigh
Gluteus medius	Abducts and medially rotates the thigh
Gluteus minimus	Abducts and medially rotates the thigh
Gracilis	Adducts the thigh
Iliopsoas (psoas major and iliacus)	Flexes the thigh; flexes the trunk when femur is fixed
Pectineus	Adducts, flexes, and laterally rotates the thigh
Piriformis	Laterally rotates the thigh; assists in extending and abducting the thigh
Rectus femoris	Flexes the thigh
Sartorius	Extends the thigh
Semimembranosus	Extends the thigh
Semitendinosus	Extends the thigh
Tensor fascia lata	Assists in flexion, abduction, and medial rotation of the thigh

Although there are encouraging projections of a leveling or even decrease in hip fracture in certain populations (Lofman et al. 2002), world-wide trends point to an increasing number of hip fractures in the foreseeable future, with an estimated overall increase in hip fracture incidence of 1% to 3% per year in most areas of the world for both women and men (Cummings and Melton 2002).

Although hip fractures are relatively rare in young individuals, the likelihood of hip fracture increases markedly with advancing age. Hip fractures are common in older populations, with a prevalence of 4.5 per 100 people 70 years and older (Cummings et al. 1985). Women are two to three times more likely to suffer a fractured hip than men. Among the many risk factors for hip fracture are falls, decreased bone mineral density and bone mass, small body size, decreased muscular strength, physical inactivity, environmental circumstances, drug use, chronic illnesses, and impairment of cognition, vision, and perception (Marks et al. 2003).

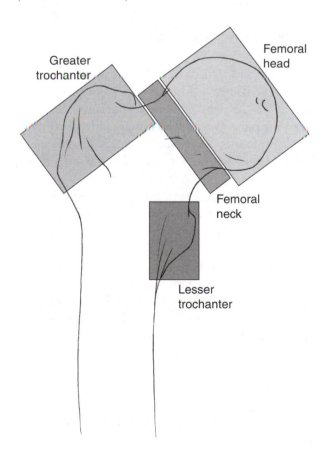

Figure 6.3 Classification of hip fractures based on location.

Classification and Causes

Proximal femoral fractures, or hip fractures, are classified according to their location (figure 6.3). Fractures to the femoral neck are considered intracapsular because that region is proximal to the distal joint capsule boundary. The extracapsular region is subdivided into trochanteric (also intertrochanteric) and subtrochanteric areas. Fracture incidence depends on region, with 49% of hip fractures occurring in the intertrochanteric region, 37% in the femoral neck (intracapsular), and 14% in the subtrochanteric region (Michelson et al. 1995).

Hip fractures in the young usually result from high-energy impacts, most commonly a result of motor vehicle crashes. These injuries are often associated with hip luxation. The most common mechanism for femoral neck fracture is direct trauma to the hip, as might be seen in a fall (Lauritzen 1997). A less common mechanism involves a lateral rotation of the leg while the body falls backward. In this case, the femoral neck fractures when the person falls backward with the foot planted on the ground; the iliofemoral ligament secures the femoral neck while the stiffened iliopsoas tendon provides a solid base against which the femoral neck fractures. On rare occasions, young people experience a stress fracture of the proximal femur as a result of repeated loading during strenuous activity.

Hip fractures in elderly people are associated with falls (see page 158), often caused by tripping or unsteady gait. This association raises an intriguing question: Does hip fracture cause the fall, or does the impact of landing from a fall cause the bone to break? In most cases, the force of impact precipitates the fracture, with only rare instances of a spontaneous fracture causing a fall. Hayes and colleagues (1993) showed that impact on the hip or side of the leg was the strongest determinant of fracture risk in elderly nursing home residents.

The energy created by a fall is much greater than that necessary to fracture a bone. Because hip fractures occur in fewer than 5% of falls, other tissues obviously absorb considerable energy. This observation is substantiated by the fact that risk of hip fracture is lower in people with a higher body mass index (weight/height^2). Robinovitch and colleagues (1995) concluded that although the thicker trochanteric soft tissues in obese people do cushion the fall, this force attenuation is insufficient by itself to prevent hip fracture. Additional absorptive mechanisms (e.g., breaking the fall with outstretched arm or eccentric action of the quadriceps during descent) are probably involved in preventing fracture.

Research suggests that even though osteoporotic bone exhibits diminished strength and increases the likelihood of fracture, the dynamics of the fall may be the dominant component in the incidence of fracture. Dynamic models of sideways falls predict peak trochanteric impact forces ranging from 2.90 to 9.99 kN. These forces are more than sufficient to cause fracture (van den Kroonenberg et al. 1995). Further research is needed to define the exact location, direction, and magnitude of forces experienced during falls. Osteoporosis appears to be but one piece in the puzzle and must be considered in conjunction with bone quality, muscular strength, soft tissue characteristics, and neuromuscular coordination. (Osteoporosis is usually associated more with *bone quantity. Bone quality* refers to the structural integrity of a given quantity of bone.)

Risks and Prevention in Older Persons

Impact forces from falling are the predominant mechanism of hip fracture in elderly persons. Reducing or eliminating the risk factors associated with falling is the best way to address the continuing health crisis of hip fracture in elderly people. Fall risk is a multifactorial problem. Among the many risk factors are the following:

- Neurological disorders (e.g., seizures)
- Cerebrovascular disorders (e.g., strokes)
- Cardiovascular disorders (e.g., cardiac arrhythmias)
- Cognitive disorders (e.g., Alzheimer's disease)
- Chronic illnesses
- Compromised balance, sensory perception, muscle strength, coordination, reflexes
- Drug and alcohol use
- Dizziness
- Postural hypotension

- Syncope (fainting)
- Previous history of falls
- Environmental factors (e.g., loose rugs and poor lighting in the home, slippery or uneven walking surfaces, stairs, obstacles)

Adapted from S.R. Cummings et al., 1985, "Epidemiology of osteoporosis and osteoporotic fractures," *Epidemiologic Review* 7: 178-208, R. Marks et al., 2003, "Hip fractures among the elderly: Causes, consequences and control," *Ageing Research Reviews* 2(1): 57-93 and L.Z. Rubenstein and K.R. Josephson, 2002, "The epidemiolgy of falls and syncope," *Clinics in Geriatric Medicine* 18(2): 141-158.

Researchers and clinicians are focusing efforts on reducing the incidence of falls. Several strategies are being used:

- Exercise to improve lower-extremity muscle strength
- Balance training
- Instruction in safe landing responses
- Reduction of impact forces during falls
- Hip pads and energy-absorbing floors
- Instruction in safe falling techniques

Adapted from S.N. Robinovitch et al., 2000, "Prevention of falls and fall-related fractures through biomechanics," *Exercise and Sport Sciences Reviews* 28(2): 74-79, and R. Sandler and S. Robinovitch, 2001, "An analysis of the effect of lower extremity strength on impact severity during a backward fall," *Journal of Biomechanical Engineering* 123(6): 590-598.

More than 50% of falls are caused by tripping (Pavol et al. 1999). Among the factors contributing to falls, especially in older persons, are quick walking pace, forward-leaning posture, and weak back and knee extensor muscles (Pavol et al. 2001). Prevention strategies include improving response time, practicing recovery responses, improving muscular strength, and walking more slowly (Grabiner et al. 2002; Owings et al. 2001; Pavol et al. 1999; Pavol et al. 2001; van den Bogert et al. 2002).

In 1985, Baker wrote that there was no major health problem with more potential for improvement than hip fracture. He hoped that "our generation and those that follow will not be subject to the same likelihood of morbidity, disability, and tragic changes in lifestyle that today characterize falls in the elderly that lead to hip fracture" (Baker 1985). Although progress has been made in the 2 decades since Baker's hopeful statement, much remains to be done to further reduce falls and consequent hip fractures.

Hip Dislocation

Hip dislocation *(luxation)* happens rarely, attributable in large part to the joint's strong ligamentous support and its substantial surrounding muscle mass. In most cases, hip luxation requires tremendous forces, so it should come as no surprise that motor vehicle accidents, falls from heights, and skiing accidents are among the most common causes. Although isolated luxations have been documented, the large forces involved often produce accompanying fracture of the acetabulum, proximal femur, or both.

Figure 6.4 Mechanisms of hip luxation in a motor vehicle collision. *(a)* Hip and knee joints at 90° of flexion at impact with dashboard. *(b)* Knee fully extended to brace for axially directed impact force that drives femoral head out of the acetabulum posteriorly.

Force application causing hip dislocation can arise in several ways. Force may be applied to the greater trochanter, flexed knee, foot with ipsilateral knee extended, and, rarely, posterior pelvis (Levin and Browner 1991). Depending on their location and direction, the applied forces tend to translate and rotate the femur. In most cases these forces cause posterior dislocation of the femur relative to the acetabulum.

Automobile crashes have long been the leading cause of hip luxation. Seven decades ago, Funsten and colleagues coined the term "dashboard dislocation" in describing 20 cases of traumatic dislocation (Funsten et al. 1938). The mechanism of injury is a violent collision of the occupant's knee against the dashboard (figure 6.4), resulting in posterior hip dislocation and often accompanied by acetabular or, less commonly, femoral fracture. Not surprisingly, the victims of vehicle-related hip luxation invariably are not wearing seat belts.

Although the dashboard impact mechanism has been accepted for decades, Monma and Sugita (2001) recently suggested an alternative mechanism. They proposed that the mechanism of traumatic posterior dislocation of the hip (TPDH) involves the brake pedal rather than the dashboard. Monma and Sugita hypothesized that in an impending head-on collision, the driver vigorously pushes on the brake pedal with the right hip slightly flexed, adducted, and internally rotated and that force is transmitted from the pedal through the lower extremity to the hip. In support of their hypothesis, Monma and Sugita presented evidence of 48 drivers involved in head-on collisions and noted that the right hip was involved in 45 of the cases; 31 cases did not involve knee injuries. Monma and Sugita posited that if the injury mechanism involved the dashboard, more left hip involvement would be found. Their data support a reasonable alternative explanation for TPDH in head-on collisions. The true mechanism is impossible to discern in actual accidents and may remain elusive, because in an experimental design, crash test dummies cannot actively push the brake pedal (Monma and Sugita 2001).

Although automobile collision is by far the most common cause of posterior hip dislocation, these injuries have been reported in sporting venues as well. Moorman and colleagues (2003) reported eight subluxations (i.e., partial dislocation) in American football and identified the most common mechanism as a fall on a flexed, adducted hip.

Matsumoto and colleagues (2003) found that snowboarders were five times more likely to suffer hip dislocation than skiers; there was a significantly higher incidence of anterior dislocation in skiers and a higher incidence of posterior dislocation in snowboarders.

Anterior hip dislocations occur infrequently (about 10% of all hip dislocations) and usually result in anteroinferior dislocation. Forcible abduction, the primary factor in anterior dislocation, presses the femoral neck or trochanter against the rim of the acetabulum and leverages the head of the femur out of its socket. Abduction combined with simultaneous hip flexion and external rotation result in obturator-type luxation. When combined with extension, abduction causes pubic- or iliac-type luxation.

As with most injuries, there are reports of unusual mechanisms of hip injury, usually in the form of case studies. One such study reported traumatic asymmetrical bilateral hip dislocation in an 18-year-old man who was the driver of a vehicle involved in a head-on collision (Lam et al. 2001). After the initial impact, his car spun 90° and was then hit by a second vehicle going in the opposite direction. The authors reported anteroinferior dislocation of the left hip (involving abduction and external rotation) and posterior dislocation of the right hip (with adduction and internal rotation).

Interestingly, there appears to be a relation between femoral structure and the likelihood of hip dislocation. Patients suffering dislocations have significantly less anteversion than a control group. Thus, it seems that patients exhibiting relative retroversion may be predisposed to hip dislocations (Upadhyay et al. 1985). Given the anatomical stability of the normal hip and the rare occurrence of hip dislocation, these observations suggest that people who experience recurrent hip dislocations have structural abnormalities (Levin and Browner 1991).

Although most hip dislocations result from trauma, there is a class of injury in young infants, known as congenital dislocation, in which the hip spontaneously dislocates. These occurrences depend on joint position: Hamstrings acting on a flexed hip and extended knee are associated with posterior dislocation, and iliopsoas acting on an extended hip is associated with anterior dislocation.

Hip Osteoarthritis

As described in the previous chapter, osteoarthritis (OA) is the most common form of arthritis and is a disabling condition worldwide. OA of the hip, in particular, is a major cause of disability, especially in older persons. The high mechanical loads placed on this major load-bearing joint put it at risk for OA.

Hip OA affects an estimated 10 million Americans (AAOS 2005b). Age is the greatest risk factor for hip OA. The exact mechanism responsible for the strong relation between OA and age remains speculative. One possible explanation implicates a molecular mechanism involving the accumulation of advanced glycation end products (AGEs) in the cartilage collagen. AGE cross-linking increases stiffness of the collagen network and may play a role in the network's decreased ability to resist damage (Verzijl et al. 2003; Verzijl et al. 2002). Other risk factors include these:

- **Bone mineral density**—Persons with hip OA have higher bone mineral density (BMD) than age-matched controls.
- **Developmental deformities**—Childhood hip disorders can directly cause premature hip OA.
- **Ethnicity**—There is conflicting evidence, but some evidence suggests a predisposition for hip OA in Caucasians of European ancestry.
- **Gender**—There is conflicting evidence; some studies show equal prevalence in men and women, whereas other studies report higher rates in women (possibly attributable to estrogen-related effects).
- **Genetic predisposition**—There is strong evidence of a genetic predisposition for OA, believed to be attributable to collagen structural defects or altered bone or cartilage metabolism linked to gene abnormalities.

- **Joint loading–physical activity**—Repetitive joint overload may be a predictive risk factor for hip OA. Participation in heavy physical activity and in high joint loading sports at an elite level is related to consequent hip OA. Farming, in particular, and occupations involving heavy lifting also are implicated. Some have hypothesized that drivers of vehicles with high levels of whole-body vibration are at higher risk of hip OA. One recent study, however, did not find evidence to support that hypothesis (Jarvholm et al. 2004).

- **Nutrition**—Limited literature is available relating hip OA to nutritional status. High intake of dietary antioxidants may provide some protection.

- **Obesity or body mass index (BMI)**—Obesity and BMI are often cited as risk factors for osteoarthritis. Evidence of a relation between obesity and knee OA is clear. The evidence associating obesity with hip OA is less clear and remains controversial. A recent evaluation of nine research studies concluded that there was moderate evidence for a positive association between obesity and hip OA (Lievense et al. 2002).

- **Smoking**—Evidence is mixed, but some research (e.g., Cooper et al. 1998) reports a negative association between smoking and hip OA (i.e., smoking associated with lower incidence of hip OA).

- **Trauma and injury**—Hip injury is associated with unilateral OA, likely attributable to injury-induced alterations in mechanical function and resultant abnormal joint loading.

Thigh Injuries

The thigh region spans the hip and knee joints and consists of the longitudinally aligned femur surrounded by three muscular compartments (anterior, medial, posterior), which are defined by their location and muscular actions (figure 6.2). The anterior compartment includes the iliopsoas (iliacus and psoas major), tensor fascia lata, pectineus, sartorius, and quadriceps group (vastus lateralis, vastus medialis, vastus intermedius, and rectus femoris). Included in the medial compartment are the three adductor muscles (adductor longus, adductor brevis, and adductor magnus) and the gracilis. The posterior compartment contains three muscles (semitendinosus, semimembranosus, biceps femoris), collectively termed the *hamstrings*.

Quadriceps Contusion

Contusions are among the most common injuries, especially in contact sports such as soccer, football, and rugby. The anterolateral aspect of the thigh is frequently involved, with resultant injury to the quadriceps muscle group. The predominant injury mechanism is compression from a nonpenetrating blunt force, most commonly in the form of an impactor's knee, helmet, or shoulder. The resulting capillary rupture, edema, inflammation, infiltrative bleeding, and muscle crush lead to pain, swelling, and decreased knee range of motion. Injury severity depends on impact site, level of muscle activation, age, and degree of fatigue (Beiner and Jokl 2001). Most quadriceps contusions are mild or moderate but in rare instances may be severe and may be accompanied by *compartment syndrome* (Diaz et al. 2003; Rooser et al. 1991).

Despite extensive description of the symptoms, treatment, and sequelae of quadriceps contusion, little is known about the underlying pathophysiological mechanisms of tissue injury. Crisco and colleagues (1994) examined selected biomechanical, physiological, and histological aspects of contusion injuries using a reproducible, single-impact model on the gastrocnemius muscle complex of anesthetized rats. Despite some limitations and the speculative nature of extrapolating the results of animal studies to human clinical observations, the results proved enlightening. Gross observation of the muscle surface within 2 hr after injury showed muscle disruption at the center of the impact site with extensive surrounding intramuscular–interstitial hematoma but no damage at either the proximal or distal myotendinous junction. An observed increase of 11% in muscle weight was attributed to hemorrhage and edema. Acute injury also resulted in a 38% decrease in maximal tetanic tension compared with uninjured contralateral controls.

The course of acute injury, degeneration, regeneration, and normalization also was examined microscopically. At day 0, injury to the gastrocnemius was localized near the site of impact and extended deep into the muscle complex. Intracellular **vacuolation** of intact myofibers was present, along with gross disruption of myofibers. Two days after injury, the muscle exhibited a marked inflammatory response, with evidence of macrophages, polymorphonucleocytes, and degenerating contractile proteins. By day 7, extensive cellular proliferation of myoblasts and fibroblasts was evident. Within 24 days the injured specimens were essentially indistinguishable from control muscles.

The study by Crisco and colleagues (1994) also provided the following instructive observations with respect to impact mechanics and tissue responses: (1) Impact pressures on the skin produce stress to the underlying muscle tissue. When these stresses exceed some critical value, damage results. The authors hypothesized that this is the mechanism of contusion injury. (2) Mass and velocity (see equation 3.24) of the impacting object alone are not sufficient to describe the event; the size and shape of the impacting object need to be considered as well. (3) Assuming that passive failure (i.e., inactive muscle loaded to failure) is indicative of tissue strength, contusion injuries may be more susceptible to subsequent strain injuries at the site of injury until fully healed.

Myositis Ossificans

Severe acute or repeated blunt force trauma to an injured muscle may lead to **myositis ossificans** (also called *myositis ossificans traumatica*), a condition characterized by calcified mass formation within the muscle, typically several weeks after the injury. The most common sites for myositis ossificans are the anterior upper arm (biceps brachii) and the anterior thigh (quadriceps group) (Leadbetter 2001).

Although data are limited, studies have reported the presence of calcification in 9% to 17% of patients with quadriceps contusion (Hierton 1983; Norman and Dorfman 1970; Rothwell 1982). Myositis ossificans is most commonly seen in young adults and is seen only rarely in children (Gindele et al. 2000). The advent of improved padding has limited the incidence of myositis ossificans in some sports (e.g., American football) but not in other sports (e.g., rugby) whose players wear limited protective gear (Beiner and Jokl 2002).

This pathogenic bone may form contiguously with normal bone (periosteal) or free of any connection with bone (heterotopic) within the muscle belly. The periosteal type may appear as flat new bone formation immediately adjacent to the diaphysis or as a mushroom-shaped formation attached to the bone (King 1998).

The precise mechanism of myositis ossificans remains unknown, but it has been theorized that the condition results from ossification of proliferating fascial connective tissue (Hait et al. 1970). Other theories (as reviewed by Booth and Westers 1989) include the following:

- Calcification of hematoma
- Intramuscular ossification following detachment of periosteal flaps
- Rupture of the periosteum with escape and proliferation of osteoblasts
- Transformation of intramuscular connective tissue into cartilage and bone
- Progressive transformation of fibrous tissue to cartilage and subsequently to bone
- Individual predisposition to myositis ossificans

Femoral Fracture

Although femoral neck fractures (discussed earlier) present one of the most urgent health concerns of older people, fracture to other regions of the femur is also of consequence. The incidence of femoral shaft fractures, for example, in the United States is 1:10,000 (DeCoster and Swenson 2002), which translates into nearly 30,000 cases annually. Most fractures of the femoral diaphysis (shaft) result from high-energy trauma and can thus be both life-threatening and a source of severe disability. One study of 520 femoral fractures reported that nearly 78% resulted from automobile, motorcycle, or automobile–pedestrian accidents (Winquist et al. 1984).

Fracture patterns typically are classified by the fracture's location, configuration (e.g., spiral, oblique, or transverse), and level of comminution or fragmentation. One system for classifying femoral fractures uses the categories of segmental and comminuted (grades I-IV; figure 6.5). In grade I comminuted fractures, a very small piece of bone is broken off. Grade II fractures are characterized by a larger fragment than grade I but with at least 50% contact between adjacent bone surfaces. Grade III comminuted fractures exhibit less than 50% cortical contact, which may allow further rotation, translation, and shortening of the fracture. In the most severe comminuted fractures (grade IV), the proximal and distal fragments are not in contact (Winquist and Hansen 1980).

A more current classification system, adopted by the Orthopaedic Trauma Association (OTA), classifies fractures according to the bone (e.g., tibia), location (e.g., proximal, diaphyseal, distal), fracture type (e.g., simple or multifragmented), fracture group (e.g., spiral, oblique, transverse, wedge), and subgroup (as defined for each specific bone) (Orthopaedic Trauma Association 1996). With this system, the OTA guidelines identify more than 100 femoral fracture classifications.

Femoral fractures in adolescents require special consideration, especially when the fracture occurs near joints or through the epiphyses. There is a potential for long-term problems associated with abnormal growth and development following injury and the possibility of subsequent osteoarthritis.

Gunshot, or ballistic, wounds provide a unique example of injury mechanisms. Obviously bullets can strike anywhere in the body, but we restrict our attention here to those that strike the femur. Gunshot fracture patterns depend on numerous factors, including bullet diameter (caliber), velocity, weight, shape, and tumbling characteristics (Brien et al. 1995). Low-velocity bullets (<600 m/s), typical of small-caliber handguns, tend to cause splintering of the femoral diaphysis. In one study, the vast majority (93%) of low-velocity gunshot-related fractures were classified as grade III or IV (Wiss et al. 1991). In addition to causing severe bone fracture, high-velocity bullets (>600 m/s) from rifles and close-range shotgun blasts cause more extensive soft tissue damage and considerable **cavitation.**

Long and colleagues (2003) classified gunshot injuries (grades 1-3) based on evidence of deep soft-tissue necrosis: These authors described grade 1 injuries as having small entry and exit wounds (<2 cm) and an absence of high-energy injury characteristics, grade 2 as having small

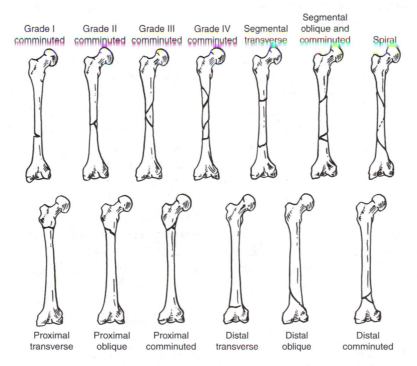

Figure 6.5 Types of femoral fractures. The upper left illustrates grades I to IV comminuted fractures (see text for description). The others show fractures based on location (e.g., proximal, distal) and configuration (e.g., transverse, oblique, spiral; see chapter 5).

wounds (<5 cm) with evidence of high-energy injury, and grade 3 as causing muscle tissue necrosis at the fracture site.

A final example of femoral fracture mechanisms is provided by skiing. Femoral fractures, like other skiing injuries, depend on skier ability, snow conditions, level of physical conditioning, age, and mechanism of injury. Sterett and Krissoff (1994) examined 85 cases of femoral fractures in alpine skiing, focusing on the mechanisms of injury as a function of skier age. These authors reported that in the youngest age group (3-18 years), femoral fracture tended to result from torsional loading of the femoral shaft, usually while skiers were moving at high speed and catching the ski in wet or heavy snow. In older skiers, such torsional loading would more likely result in soft tissue injury at the knee. Fractures in young adults (18-45 years) occurred mostly from high-energy, direct-impact collisions with an object (e.g., rock or tree) and not surprisingly resulted in high-grade comminuted fractures. In older skiers (>45 years), the majority of fractures were localized in the hip area (femoral neck and peritrochanteric) and were caused by low-energy impact falls on firm snow.

Expert skiers are much less likely to sustain injuries than novices. Femoral fracture, however, is one of the few injuries more prevalent in advanced skiers than in beginners. This is attributable largely to the high energy levels required for femoral fracture. Advanced skiers typically ski at higher speeds and over more difficult terrain than novices. The greater speeds result in higher kinetic energy (see equation 3.20), which on impact (e.g., with a rock or tree) is transferred to the musculoskeletal tissues, including the femur.

Hamstring Strain

Muscle *strain* (compare with *mechanical strain* in chapter 3) involves injury to a musculotendinous complex. Strain is classified as an indirect injury because it results from excessive tension loads and not from direct trauma.

Injury typically occurs during forced lengthening or eccentric muscle action used to control or decelerate high-velocity movements (e.g., sprinting or throwing). Hamstring strains are common.

In a study of strain injuries in Australian football from 1992 to 1999, 69% involved the hamstring group, followed by the quadriceps (17%) and calf (14%) muscle groups (Orchard 2001). Hamstring strains range from mild (slight pain, no muscle tearing) to severe (complete muscle rupture, usually resulting from an explosive movement) (Peterson and Renström 2001).

At the muscle level, animal studies implicate excessive sarcomere mechanical strain as the primary cause of injury (Lieber and Friden 2002). These authors hypothesize that excessive mechanical strain allows extracellular or intracellular membrane disruption that may permit hydrolysis of structural proteins. This leads to myofibrillar disruption and local inflammation that further degrade and weaken the muscle tissue.

The significance of active versus passive muscle regarding muscle strain has been well documented. It has been observed that force generated at failure was only 15% higher in stimulated rabbit extensor digitorum longus muscles compared with unstimulated ones, whereas the energy absorbed was approximately 100% higher at failure in the activated muscles (Garrett et al. 1987). This suggests that any compromise in a muscle's contractile capacity (e.g., fatigue) may reduce its ability to absorb energy and increase the risk of injury.

In addition to fatigue, many other risk factors for muscle strain have been identified. These include muscle imbalance, lack of flexibility, insufficient warm-up, age, history of injury, muscle weakness, poor training, use of inappropriate drugs, presence of scar tissue, and incomplete or overly aggressive rehabilitation (Croisier et al., 2002; Verrall et al. 2003; Worrell 1994).

Certain muscles seem more prone to strain injury than others. The hamstrings, in particular, are susceptible to muscle strain. Why is that? The muscles of the hamstring group, with the exception of the short head of the biceps femoris, have biarticular function. This structural arrangement dictates that muscle length is determined by the conjoint action of the hip and knee joints. Hip flexion and knee extension each lengthen the semitendinosus, semimembranosus, and biceps femoris (long head). Simultaneous hip flexion and knee extension place the hamstrings in a lengthened state that contributes to the muscles' susceptibility to injury (figure 6.6).

Figure 6.6 Simultaneous hip flexion and knee extension place the hamstring muscle group in an elongated position and increase the risk of strain injury.

The circumstances of hamstring strain in sprinters illustrate a common injury mechanism. Strain injury usually occurs late in the swing phase or early in the stance phase. During late swing, the hamstrings work eccentrically to decelerate both the thigh and lower leg in preparation for ground contact. Early in stance, the hamstrings act concentrically to extend the hip. Kinetic analyses have shown that peak torques at the hip and knee occur during these phases. An additional contributing factor may be the relatively high proportion of fast-twitch muscle fibers found in hamstring muscles, which allows for higher levels of intrinsic force production. This combination of factors places the hamstrings at elevated risk of injury in high-velocity movements.

Locating the precise injury site often proves difficult. Which of the hamstring muscles, for example, is most likely to sustain injury? Computed tomography (CT) used to localize hamstring muscle strain showed that injuries tended to be proximal and lateral within the hamstring group, most often in the long head of the biceps femoris (Garrett et al. 1989). More recent studies have confirmed that the biceps femoris is the most likely hamstring muscle to be injured. Koulouris and Connell (2003), for example, reported that 80% of 154 hamstring injuries involved the biceps femoris. The semimembranosus (14%) and semitendinosus (6%) were rarely implicated. Thelen and colleagues (2005) hypothesized that the high propensity for biceps femoris injury may be attributable, at least in part, to intermuscle differences in hamstring moment arms about the hip and knee joints.

Strain injury in the muscle–tendon unit occurs most often at the myotendinous junction (MTJ). The MTJ's microscopic structure makes it a likely site for focused mechanical loading *(stress riser).*

The MTJ has structural features that aim to reduce the stress riser effect. First, the structural folding of the junctional membrane (figure 6.7) increases the surface area (by at least 10 times and thus reduces stress. Second, the folding configuration aligns the membrane so that it experiences primarily shear forces rather than tensile forces. Third, the folding may increase the adhesive strength of the muscle cell to the tendon. Fourth, the sarcomeres near the junction are stiffer than those distant from the junction and thus have

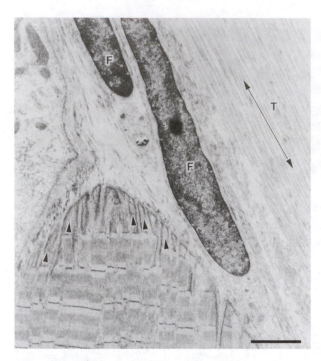

Figure 6.7 Electron micrograph of myotendinous junction. The muscle cell appears to interdigitate with the tendon (arrowheads). The tendon contains fibroblasts (F) and dense collagen fibers (T). (Inset bar = 3.0 μm.)

Reprinted, by permission, from J.G. Tidball, 1991, "Myotendinous junction injury in relation to junction structure and molecular composition," *Exercise and Sport Sciences Review* 19: 425.

limited extensibility (Noonan and Garrett 1992). Despite these structural characteristics that mediate against injury, the MTJ remains a likely site of musculotendinous injury.

Recurrent hamstring strains are quite common. Among the reasons for injury recurrence are premature return to action, weakened tissues, strength deficits, and altered mechanical characteristics. Recent evidence suggests that previously injured hamstring muscles reach their peak torque at significantly shorter lengths than do uninjured muscles (Brockett et al. 2004; Proske et al. 2004). This shift in torque-length profile may predispose the hamstring muscles to further injury and help account for the high rate of reinjury.

Although recurrent hamstring strains happen frequently, there is clear evidence that appropriate prevention programs can dramatically reduce injury risk. An effective injury prevention program may include eccentric overload (Askling et al. 2003), strength training (Croisier et al. 2002), mild eccentric exercise (Proske et al. 2004), stretching while fatigued, sport-specific training drills, and high-intensity anaerobic interval training (Verrall et al. 2005).

In summary, the hamstrings' gross and microscopic anatomical structure, biarticular arrangement, and involvement in controlling high-velocity movement all contribute to this muscle group's particular risk for muscle strain injury. This risk can be reduced, but not eliminated, by appropriate prevention, treatment, and rehabilitation protocols.

Knee Injuries

The knee joint contains three articulating surfaces: those between the medial and lateral condyles of the femur and tibia (tibiofemoral joint) (figure 6.8c) and between the patella and the femur (patellofemoral joint) (figure 6.8a). Although often classified as a hinge joint (which implies uniplanar movement), the knee is more correctly classified as *double condyloid*, because it has movement potential in both flexion–extension and rotation (when the knee is flexed). The muscles acting about the knee joint are depicted in figure 6.2 and summarized in table 6.2.

Knee joint movement is among the most complex of all joint movements in the human body. Although its primary movement is flexion–extension, the joint's structure dictates a nonhingelike motion characterized by a variable instantaneous axis of rotation and combined rotational, rolling, and gliding movements (figures 3.19b and 3.20). In addition to serving predominant flexion and extension functions, the tibiofemoral joint also has rotational capability when the joint is flexed as well as limited varus–valgus movement.

One unique motion feature of the tibiofemoral joint is the so-called **screw-home mechanism.** During the final few degrees of knee extension, a relative rotation between the tibia and femur locks the joint in place at full extension by rotationally screwing the bones together. If the tibia is fixed, the femur rotates medially at extension. Conversely, if the femur is fixed, the tibia experiences a relative lateral rotation into place. At the initiation of flexion, the tibiofemoral joint unscrews, reversing the relative tibial and femoral rotations.

As a synovial joint, the knee has a strong fibrous capsule that attaches superiorly to the femur and inferiorly to the articular margin of the tibia. Given its relatively poor bony fit, the

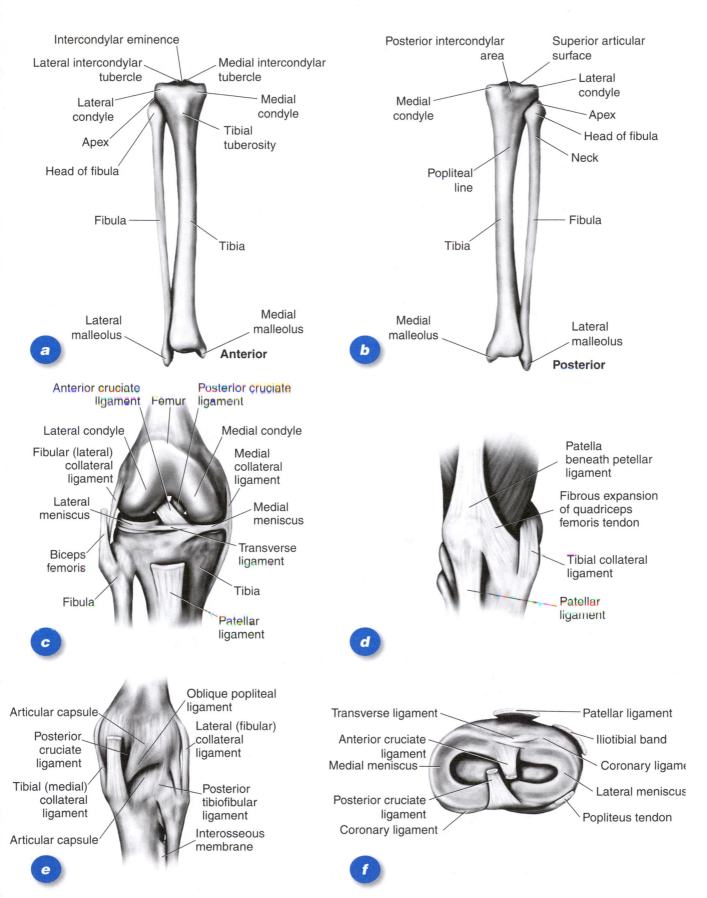

Figure 6.8 Anatomy of the knee: *(a and b)* skeletal structures, anterior and posterior views; *(c and d)* knee ligaments, anterior views; *(e)* knee ligaments, posterior view; *(f)* ligaments and menisci, superior view.

(a, b) Reprinted, by permission, from R.S. Behnke, 2005, *Kinetic anatomy*, 2nd ed. (Champaign, IL: Human Kinetics), 192. *(c)* Adapted, by permission, from R.S. Behnke, 2005, *Kinetic anatomy*, 2nd ed. (Champaign, IL: Human Kinetics), 197. *(d, f)* Adapted, by permission, from R.S. Behnke, 2005, *Kinetic anatomy*, 2nd ed. (Champaign, IL: Human Kinetics), 194. *(e)* Adapted, by permission, from R.S. Behnke, 2005, *Kinetic anatomy*, 2nd ed. (Champaign, IL: Human Kinetics), 195.

TABLE 6.2

MUSCLES OF THE KNEE

Muscle	Action
Gracilis	Flexes the leg
Sartorius	Flexes the leg
Quadriceps femoris group	
Rectus femoris	Extends the leg
Vastus intermedius	
Vastus lateralis	
Vastus medialis	
Hamstring group	
Biceps femoris	Flexes the leg
Semimembranosus	
Semitendinosus	

knee relies on ligaments (along with muscles and tendons) for much of its structural strength and integrity. Among the most important, and most often injured, are the collateral ligaments and the cruciate ligaments (figure 6.8, *c* and *d).* The collateral ligaments span the medial and lateral aspects of the knee and resist valgus and varus loading. *Valgus* loading at the knee results in an inward curvature of the leg at the knee (i.e., knock-kneed). *Varus* loading results in an outward curvature of the leg at the knee (i.e., bowlegged).

The lateral (fibular) collateral ligament (LCL) is extracapsular and extends from the lateral epicondyle of the femur to the lateral surface of the fibular head. The medial (tibial) collateral ligament (MCL) spans from the medial femoral epicondyle to the superomedial surface of the tibia. Unlike the LCL, which is extracapsular, the MCL is a capsular ligament that connects directly to the capsule and the medial meniscus. This structural arrangement, as discussed later, has important implications for the MCL's susceptibility to injury. Reference to the MCL as a singular ligament understates its anatomical complexity. It is usually described as having superficial and deep components that act synergistically to restrict knee joint movements. Some authors also include the posterior oblique ligament as part of the MCL complex. The posterior oblique ligament is a capsular thickening just posterior to the superficial MCL (Peterson and Renström 2001).

The two cruciate ligaments, named for their oblique or X-shaped orientation to one another, extend between the femur and tibia. The weaker of the two, the anterior cruciate ligament (ACL), attaches proximally on the posteromedial aspect of the lateral condyle of the femur and distally on the anterior portion of the intercondylar surface of the tibia. The ACL is composed of two major bundles: the anteromedial bundle, which is tight in flexion and relatively lax in extension, and the posterolateral bundle, which is tight in extension and lax in flexion. The likelihood of injury to a bundle obviously depends on the degree of knee flexion at the time of injury. The ACL's function is primarily to restrict anterior movement of the tibia relative to the femur (or conversely to limit posterior movement of the femur relative to the tibia) and secondarily to provide resistance to valgus, varus, and tibial rotation.

The stronger posterior cruciate ligament (PCL) attaches proximally on the anteromedial aspect of the medial condyle of the femur, passes medial to the ACL, and secures distally to the posterior portion of the intercondylar area of the tibia. The PCL also consists of two bundles: The larger anterolateral bundle tightens in flexion and is relatively lax in extension, and the smaller posteromedial bundle is tightest in extension and relatively lax in flexion. The PCL limits posterior movement of the tibia relative to the femur (or conversely restricts anterior movement of the femur relative to the tibia). In addition, the PCL limits hyperflexion of the knee and assists in stabilizing the femur when a flexed knee is bearing weight.

In addition to receiving capsular and ligamentous support, the knee contains two menisci. Each meniscus is a wedge-shaped fibrocartilage pad with peripheral attachment to the joint capsule. The menisci project centrally and are freely moving in non-weight-bearing flexion. Both menisci have a characteristic crescent shape (figure 6.8*e).*

The medial meniscus has a larger radius of curvature than the lateral meniscus and a semilunar shape. It averages about 10 mm in width in the posterior horn and is narrower in the middle and anterior zones. Several structural characteristics of the medial meniscus increase its risk of injury. These include its tight connection with the joint capsule and MCL and its frequent connection with the ACL.

The lateral meniscus exhibits a tighter curvature and forms a nearly closed curve. Its posterior horn is wider than the corresponding region of the medial meniscus. The lateral meniscus attaches posteriorly to the femoral intercondylar fossa and has only a loose attachment to the joint capsule and no direct connection with the LCL. The nature of these attachments allows more mobility for the lateral meniscus than for the medial meniscus in both unloaded (figure 6.9) and loaded conditions.

The patella, located between the **quadriceps tendon** and its attachment on the tibial tuberosity (via the **patellar tendon,** also known as the **patellar ligament),** increases the mechanical advantage of the **knee extensor mechanism.** The articulation of the patella with the femur forms the patellofemoral joint (PFJ). The PFJ experiences large loads when the knee is in a flexed position (figure 6.10) and thus is predisposed to certain injuries.

Posterior Cruciate Ligament Injury

Considerable literature on cruciate ligament injury has accumulated since the first description of ACL rupture in the mid-19th century and initial attempts at surgical reconstruction early in the 20th century. Growing participation in exercise and sports in recent years has presaged an increased incidence of cruciate injuries, especially in girls and women.

The PCL was once thought to be rarely injured. Advances in diagnostic techniques, along with knowledge gained from recent anatomical and biomechanical studies, have improved our understanding of PCL injury (Wind et al. 2004). Despite these diagnostic advances, the prevalence of PCL injury still is much lower than that of ACL injury. Of PCL injuries, fewer than half are isolated (PCL only), with the remainder (60%) combined injuries involving the PCL and other ligaments and knee structures (e.g., menisci) (Fanelli and Edson 1995).

About half of the cases of PCL injury are attributable to trauma resulting from motor vehicle accidents. Most of the remaining cases happen during sporting activities (40%) and industrial accidents (10%). Motor vehicle accidents and industrial accidents typically involve high-energy dynamics, whereas sport-related injuries are considered to involve low-energy dynamics.

There are many mechanisms of PCL failure. Some are quite common; others are relatively rare. Five of these mechanisms are illustrated in figure 6.11. Most commonly, PCL injury occurs in vehicular crashes when an unrestrained occupant is thrown into the dashboard. With the knee flexed 90°, the PCL is taut, and the posterior capsule is lax. The impact force drives the tibia posteriorly and causes PCL rupture (figure 6.11a). In a fall onto a flexed knee with a plantar flexed foot, the impact occurs on the tibial tuberosity. The proximal tibia is again driven posteriorly (figure 6.11b). Forced

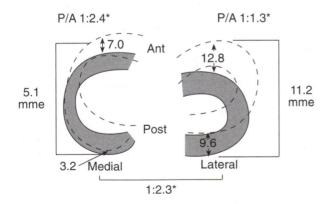

Figure 6.9 Diagram of mean meniscal excursion (MME) along the tibial plateau. Ant, anterior; Post, posterior; P/A, ratio of posterior to anterior meniscal translation during flexion. Values other than ratios are given in millimeters. *$p < .05$ by student's *t*-test analysis.

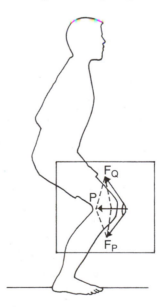

Figure 6.10 Patellofemoral joint reaction force (*P*). Vector *P* is formed by the vector sum of the force vector of the quadriceps tendon (F_Q) and the force vector of the patellar tendon (F_P).

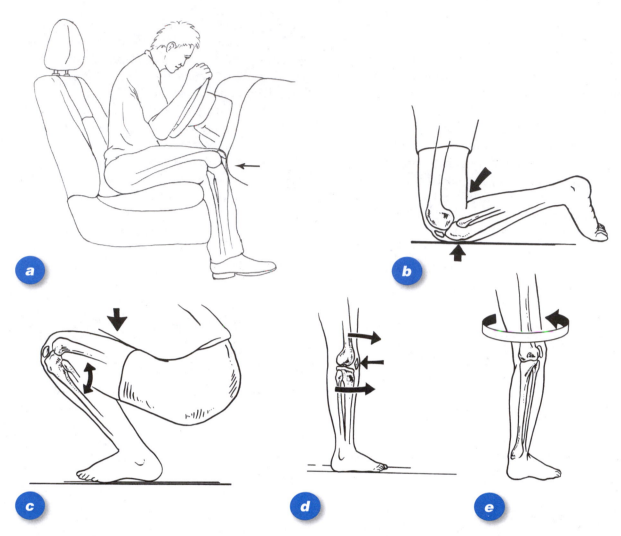

Figure 6.11 Mechanisms of posterior cruciate ligament injury. *(a)* Motor vehicle collision in which impact with the dashboard forces the tibia posteriorly relative to the femur. *(b)* Fall on a flexed knee pushing the tibia posteriorly. *(c)* Forced knee flexion. *(d)* Forced knee hyperextension. *(e)* Cutting (changing direction) on a minimally flexed knee.

knee flexion with the foot either plantar flexed or dorsiflexed can result in PCL injury (figure 6.11*c*). Sudden and violent hyperextension of the knee can cause PCL rupture, and it is not unusual in this situation to have accompanying ACL damage (figure 6.11*d*). Rapidly shifting weight from one foot to another while rotating the body quickly on a minimally flexed knee causes internal rotation and anterior translation of the femur and resulting PCL damage (figure 6.11*e*) (Andrews et al. 1994).

Most PCL injuries involve some level of posteriorly directed tibial force. The amount of this posterior tibial force transmitted to the PCL is highly sensitive to knee flexion angle: Greater knee flexion angles increase the proportion of posterior tibial force seen in the PCL (Markolf et al. 1997). Addition of a valgus moment to the posterior tibial force greatly increases PCL forces and increases the likelihood of injury.

PCL injury can occur in the absence of posterior tibial forces, although this is relatively rare. Markolf and colleagues (1996) reported highest PCL forces with combined internal tibial torque and a varus moment at 90° of knee flexion (in the absence of posterior tibial force). As in this case, combined loading frequently increases injury risk.

Anterior Cruciate Ligament Injury: A Closer Look

Few injuries, if any, have received more clinical and research attention than those of the anterior cruciate ligament (ACL) in the past few decades. In the last 20 years, more than 5,000 articles about the ACL have been published. This is not at all surprising given the important role the ACL plays in knee joint function and the high economic toll and morbidity associated with ACL injury. Although there is no doubt about the functional importance of the ACL, controversy persists with respect to its injury diagnosis, treatment, and rehabilitation. Because of the frequency of injury and the extensive research available, we present a more detailed review of ACL injuries than of other injuries in this book.

Epidemiology

Estimates vary on the incidence of ACL injuries, with reported annual rates of 1:3,000 in the United States (Miyasaka et al. 1991) and 1:5,000 in the United Kingdom (Dandy 2002). Estimates of ACL pathology in the United States vary considerably, ranging from 80,000 tears (Griffin et al. 2000) to 200,000 injuries (Hubbell and Schwartz 2005). Many ACL injuries are sports related, with the highest incidence seen in 15- to 25-year-olds participating in sports that involve pivoting (Griffin et al. 2000).

In light of the multifactorial nature of ACL pathology, ACL injury rates are, not surprisingly, sport and gender specific (e.g., Agel et al. 2005; Bradley et al. 2002; Bjordal et al. 1997).

Tissue Structure and Function

The ACL is a complex ligament that connects the femur and tibia (figure 6.8c). Proximally, the ACL attaches to the medial surface of the lateral femoral condyle. The ACL attaches distally to the anterior surface of the midtibial plateau. The ACL presents through its midsubstance as a band of regularly oriented connective tissue; at its attachment sites the ACL fans out to create a broader attachment area.

The ACL consists of two bands or bundles: anteromedial (AM) and posterolateral (PL). (Note: some authors include a third intermediate band.) Each band plays a unique role in stabilizing the tibiofemoral articulation. In knee flexion, the AM band is taut whereas the PL band is relatively lax. With knee extension, the PL band becomes taut and the AM band remains taut, but less so than the PL (Dienst et al. 2002). The role of each band was confirmed in an in situ cadaver study that reported higher PL band forces, compared with AM band forces, at knee flexion angles less than 15° in response to a 110 N anterior tibial load (Allen et al. 1999).

The ACL acts as the primary restraint to anterior tibial translation (ATT). In this role, the ACL limits ATT relative to a fixed femur or, conversely, restricts posterior movement of the femur on a fixed tibia. As the primary restraint to ATT, the ACL accepts 75% of anterior forces at full knee extension and 85% at 90° of flexion (Peterson and Renström 2001).

The ACL acts as a secondary restraint to internal tibial rotation. The ACL's role as a secondary restraint to varus–valgus angulation and external rotation is less clear, although generally accepted. The ACL also works in concert with the posterior cruciate ligament (PCL) to limit knee hyperextension and hyperflexion.

Tissue Mechanics

Woo and colleagues (1991) tested paired femur-ACL-tibia complexes to assess the effect of load orientation and age on ACL mechanical properties. These authors performed tensile tests under two conditions: (1) tensile loading along the preserved anatomical axis of the ACL (anatomical orientation) and (2) tensile loading along the long axis of the tibia (tibial orientation). The researchers reported that young specimens (22-35 years) tested in anatomical orientation had significantly higher ultimate loads (2160 N or 485 lb) than those tested in tibial orientation (1602 N or 360 lb). Specimens tested in anatomical orientation also were stiffer (242 N/mm) than those in tibial orientation (218 N/mm). The authors explained these results by noting that in anatomical orientation, more of the ACL's fibers are aligned to accept tensile loads. The authors also reported significant decreases in ultimate load and stiffness with age, regardless of load orientation.

The ACL's complex structure makes it extremely difficult to load all ligament fibers uniformly while using the entire complex. Investigators therefore have also tested each of the ACL's bands separately to assess their individual mechanical properties. For example, in tests of seven cadaveric ACL–bone units, the anterior bundles developed significantly higher moduli, maximal stresses, and strain energy densities compared with the posterior bundles (Butler et al. 1992).

Injury Mechanisms

ACL injury happens most often in response to valgus loading in combination with external tibial rotation or to hyperextension with internal tibial rotation. Combined loading conditions place the ACL at great risk of injury (Markolf et al. 1995).

The first mechanism (valgus rotation) typically happens in what is termed a *noncontact injury* in which the foot is planted on the ground, the tibia is externally rotated, the knee is near full extension, and the knee collapses into valgus (Myer et al. 2005). The collapse into valgus appears to be a critical element. Recent evidence using biomechanical models suggests that sagittal plane knee joint forces cannot rupture the ACL during side-step cutting maneuvers and that valgus loading is a more likely mechanism (McLean et al. 2004).

The situation is exacerbated if a force is applied to the knee while the foot is in contact with the ground *(contact injury)*. This is common in contact sports such as American football, rugby, and soccer when another player impacts the lateral aspect of the knee, accentuating the valgus loading and rotation. Loaded knee valgus, combined with external tibial rotation, severely stresses the ACL. Although contact injuries are much less common than noncontact injuries (Boden et al. 2000), the added force of impact in contact injuries greatly increases the likelihood of injury occurrence and increases injury severity.

The second mechanism involves knee hyperextension with internal tibial rotation. Although a less common mechanism overall, hyperextension may be the predominant mechanism in certain populations such as basketball players or gymnasts, whose injuries often occur as they land following a jump and violently hyperextend their knee (figure 6.12*a*). Identification of the injury mechanism is important, because jumping-related

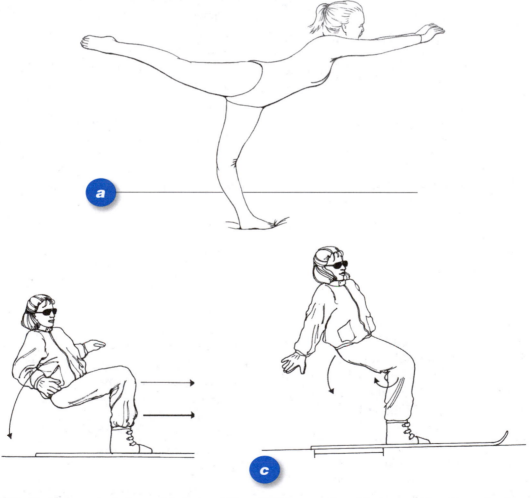

Figure 6.12 Mechanisms of anterior cruciate ligament injury. *(a)* Hyperextension of the knee. *(b)* Anterior cruciate ligament (ACL) injury caused by a backward fall. This mechanism forcibly pushes the tibia anteriorly relative to the femur and stresses the ACL. *(c)* Anterior cruciate ligament injury caused by the *phantom foot* provided by the section of ski posterior to the boot.

injuries are associated with a higher incidence of meniscus tears and possible predisposition to subsequent degenerative changes (Paul et al. 2003).

One approach to determine mechanisms of ACL injury uses magnetic resonance imaging (MRI) to assess the bone lesions associated with these injuries. Valgus loading, as previously described, results in compressive forces between the lateral femoral condyle and the lateral tibial plateau. Speer and colleagues (1992) identified an injury pattern they termed the "MRI triad" (ACL rupture, terminal sulcus osseous lesion, and bone or soft tissue injury [or both] at the posterolateral corner) and proposed several mechanisms consistent with this pattern of lesions. These mechanisms include valgus loading with subsequent lateral joint compression and hyperextension. The authors were prudent in their conclusions by making a statement that holds true for many injuries: "The great difficulty in resolving the injury into a single mechanism may be that the issue is not resolvable; that is, both mechanisms may come into play depending on the nature of applied extrinsic and intrinsic forces" (Speer et al. 1992, p. 387).

A similar approach has been taken to investigate the hypothesis that for downhill skiers, various mechanisms may be involved in ACL injury. Bone bruise patterns suggest that there is less valgus loading in skiers, as evidenced by greater variety of presentation of injury across the posterior tibial rim (Speer et al. 1995). These data are consistent with observations that ACL injury in skiers is often associated with a backward fall. In such a fall, the skis and boots accelerate forward relative to the body and, because of modern boot design, take the tibia along. This creates an **anterior drawer mechanism** (i.e., anterior tibial translation) that is consistent with ACL failure in what is termed a *boot-induced* ACL injury.

A second mechanism of ACL injury unique to skiing has been described as the *phantom foot,* referring to the lever formed by the rear section of the ski that effectively forms another "foot" in the posterior direction. In a backward fall, this phantom foot levers the flexed knee into internal rotation of the tibia relative to the femur and amplifies stress in the ACL. Research has indicated that a change in ski binding design may help reduce the incidence of ACL injury resulting from the phantom foot mechanism (St-Onge et al. 2004).

Injury to either cruciate ligament may be an isolated injury or may occur in concert with damage to other structures. An example of a combination injury is one known variously as O'Donoghue's triad, or the **unhappy triad,** in which the ACL, medial collateral ligament (MCL), and medial meniscus sustain damage. This injury typically involves the valgus–external rotation mechanism described earlier.

Conservative Treatment

Following ACL rupture, the decision whether to pursue conservative treatment or surgical repair is difficult. In a review of studies comparing conservative versus surgical outcomes, Linko and colleagues (2005) concluded that there is insufficient evidence to determine which approach is best overall. Many factors are involved in the decision-making process, including patient age, physical condition, related injuries, and anticipated postsurgery activities. Conservative treatment usually is best suited for very young and very old patients.

The decision for surgical repair may be easier for young who want to return to an active lifestyle involving sport activities. In older persons, however, the decision is more difficult, because ACL deficiency may not compromise performance of low-load activities such as walking. Pursuit of conservative (i.e., nonsurgical) treatment is not without risk, however, because ACL-deficient knees have been associated with knee instability, secondary injury to other structures (e.g., menisci, articular cartilage), and premature onset of osteoarthritis (Woo et al. 2001).

Conservative treatment initially involves control of swelling and pain management and may also include use of a brace for support, **cryotherapy,** and anti-inflammatory medications. Exercises to increase range of motion and strengthen muscles are added progressively, as are noncutting exercises such as swimming, cycling, and sagittal-plane walking, and light jogging.

Surgical Repair

Surgical ACL reconstruction is indicated in many cases, especially in patients who plan to resume an active lifestyle that includes cutting or pivoting movements (e.g., tennis or basketball). Once the decision in favor of surgery is made, the next critical question involves the source of the replacement tissue, or **graft.** The choices include tissue taken from the patient **(autologous graft, or autograft),** tissue from a cadaver **(allograft),** or artificial synthetic grafts. Although each choice has its advantages and limitations, the great majority of ACL replacement surgeries use autografts involving either a bone–patellar tendon–bone (BPTB) graft or a hamstring (e.g., semitendinosus and gracilis) tendon (HT).

A primary advantage of the BPTB approach is the presence of bone plugs at each end of the donor graft (figure 6.13, *a-c).* These plugs provide good fixation of the graft within the femoral and tibial attachment sites and provide initial stability. A limitation of the BPTB procedure is the morbidity seen at the graft excision site.

Proponents of the HT procedure cite the advantages of lower donor site morbidity and comparable graft strength and stiffness (using current techniques). The absence of bone plugs in the HT graft, however, reduces the initial integrity of attachment site fixation.

Debate continues as to which graft source is superior; each has its proponents. A recent meta-analysis of 11 reports comparing BPTB and HT found no significant difference between the two grafts in the incidence of instability and concluded that graft choice should be based on patient needs and the surgeon's preference and experience; before the decision is made, the patient and surgeon should consider the benefits and risks of each approach (Goldblatt et al. 2005).

Rehabilitation

Although protocols for rehabilitation after ACL reconstructive surgery must be individualized, rehabilitative guidelines apply. These guidelines include restoration of knee joint range of motion, reduction of muscle inhibition, control of pain and swelling, muscle strengthening, and progressive return to normal activity. The time course of rehabilitation varies. Older and less conditioned patients may require longer rehabilitation than younger athletes who want to return to their sport as quickly as possible. The appropriate level of rehabilitation aggressiveness, especially for athletes, is controversial (Cascio et al. 2004).

Following are the phases and goals of a standard rehabilitation protocol preceding and following ACL reconstruction (BPTB) as described by Peterson and Renström (2001, pp. 518-519):

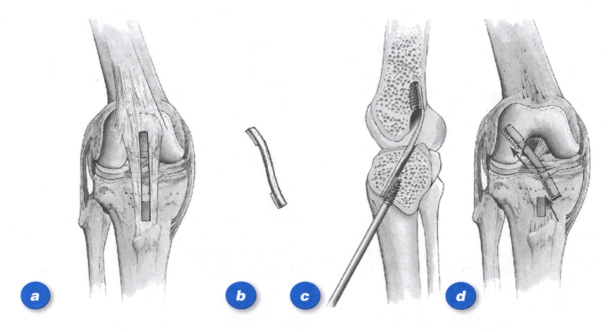

Figure 6.13 Surgical ACL repair with bone–patellar tendon–bone (BPTB) graft. *(a)* The graft is excised from the middle of the patellar tendon (also called the *patellar ligament). (b)* The graft after it is removed from the patellar tendon. The bone plugs are visible on the undersurface of each end. *(c)* The graft is threaded through an osseous tunnel created by drilling from the posterior aspect of the tibia into the inferior aspect of the femur. *(d)* The graft is fixed using screws.

From *Sports Injuries: Their prevention and treatment,* 3rd ed., L. Peterson and P. Renstrom, Copyright 2001, Human Kinetics. Reproduced by permission of Taylor & Francis Books UK.

Phase 1—*Preoperative*

- Reduce pain and swelling
- Improve range of motion
- Improve muscle strength
- Maintain overall strength and aerobic capacity

Phase 2—*Postoperative (first 7-10 days)*

- Reduce pain and swelling
- Improve range of motion (full knee extension and at least 90° knee flexion)
- Improve quadriceps–hamstring function

Phase 3—*Return to weight bearing–full motion (up to 11 weeks)*

- Normal gait without brace
- No swelling
- Full range of motion
- Improved muscular strength (hip, knee, ankle)
- Improve proprioception, balance, and postural stability
- Maintain overall strength and aerobic capacity

Phase 4—*Return to straight-line activities (3-4 months)*

- Active quadriceps training against resistance without pain
- Normal running gait
- Maintain overall strength and aerobic capacity

Phase 5—*Return to cutting activities (4-6 months)*

- >90% of normal muscle torque (measured isokinetically)
- All types of running without pain and in full control
- Jumping without pain and in full control

Phase 6—*Return to competition (6 months and beyond)*

- Return to full unrestricted activity
- Must be accompanied by

– full range of motion and muscular flexibility compared with contralateral leg,

– isokinetic strength >90% of contralateral leg,

– no pain or swelling as a result of training, and

– good results of all functional testing.

Prevention

Numerous studies have noted an alarming increase in the incidence of ACL injuries in female athletes, disproportionate to those seen in men. Female intercollegiate soccer players, for example, are three times more likely to sustain ACL lesions than their male counterparts in the same sport. In basketball, the difference jumps to 3.6 times (Agel et al. 2005). Similar gender differences have also been reported for competitive alpine ski racers. Stevenson and colleagues (1998) found that female alpine racers were 3.1 times more likely to suffer ACL injury than males.

The reasons for these differences in ACL injury rates are unclear, and consensus now points to the multifactorial nature of the problem (Ireland 2002). Among the suggested predisposing factors have been a woman's wider pelvis, greater flexibility, less-developed musculature, hypoplastic vastus medialis obliquus, narrow femoral notch, genu valgum, and external tibial torsion. Body movement in sport, muscular strength and coordination, shoe-surface characteristics, level of conditioning, joint laxity, limb alignment, and ligament size must also be considered (Arendt and Dick 1995; Ireland and Ott 2001). In addition, hormonal (Dragoo et al. 2003) and menstrual cycle (Arendt et al. 2002; Wojtys et al. 2002) effects may play a role. The exact combination of factors responsible for the gender discrepancies in ACL injury remains elusive and is a subject of intense clinical debate.

Video analysis of actual ACL injuries implicates a "position of no return" that is typically seen in rapid yet awkward stops or landings (Ireland 2002). This position of no return involves loss of hip and pelvis control, internal femoral rotation, knee valgus, and external tibial rotation on a pronated foot. Avoiding the position of no return may dramatically reduce ACL injury risk.

One element of the position of no return, knee valgus, has been studied extensively. McLean and colleagues (2005), for example, found that female intercollegiate athletes had significantly larger peak valgus moments, initial hip flexion, and internal rotation in a sidestepping task compared with male athletes. The authors concluded that a training program to reduce knee valgus loading may help prevent ACL injury, especially in females.

In summary, "Multiple factors are responsible for ACL tears. The key factor in the gender discrepancy appears to be dynamic, not static, and proximal, not distal. The factors involved in evaluating the female ACL are multiple. However, it is the dynamic movement patterns of hip and knee position with increased flexion and a coordinated proximal muscle firing pattern to keep the body in a safe landing position that are the most critical factors. An ACL injury at an early age is a life-changing event. We can very successfully reconstruct and rehabilitate an ACL, but we cannot stop there. We must now go into the prevention arena" (Ireland 2002, pp. 648-649).

Emerging evidence strongly suggests that the incidence of ACL injury can be dramatically reduced through training programs targeting specific movements and neuromuscular control strategies (Hewett et al. 2005). Positive results have been reported in female athletes in soccer (Mandelbaum et al. 2005) and team handball (Myklebust et al. 2003; Petersen et al. 2005), and across genders in skiing (Urabe et al. 2002).

One study (Mandelbaum et al. 2005), involving a 2-year follow-up of female soccer players (age 14-18 years) who received a sport-specific training intervention, reported a 74% to 88% decrease in ACL injuries. The intervention included education, stretching, strengthening, plyometric training, and sport-specific agility drills. These elements replaced the traditional warm-up. The authors concluded that a neuromuscular training program may directly reduce the number of ACL injuries in female soccer players.

Meniscus Injury

The menisci once were believed to be useless remnants of intra-articular attachments. We now know that the menisci serve an essential role in maintaining normal knee function. Compression of the menisci facilitates the distribution of nutri-

ents to adjacent structures. Of greater interest in our discussion of injury mechanisms are the mechanical functions of the menisci, specifically weight bearing, shock absorption, stabilization, and rotational facilitation.

The menisci transmit varying percentages of forces across the knee joint depending on knee position. In full extension the menisci accommodate 45% to 50% of the load, whereas in 90° of flexion they accept 85% of the load (Ahmed and Burke 1983; Ahmed et al. 1983). The load distribution between the medial and lateral menisci differs. Medially, the meniscus and articular cartilage share the load equally. Laterally, the meniscus assumes 70% of the load transmission (Seedhom and Wright 1974; Walker and Erkman 1975).

The tibiofemoral joint experiences a combination of compressive, tensile, and shearing forces that vary according to both the individual and the task. Most obvious is the compressive force created by the ground reaction forces of contact (e.g., at foot strike in walking or running). These compressive loads are typically accommodated through a circumferential hoop effect in which the forces are directed peripherally along the lines of greatest collagen fiber stiffness. Tensile forces are seen in the structures resisting **distraction** between the tibia and femur. Shear forces arise from the rotational loads in movements involving rapid change of direction.

Because meniscal injury often is caused by high rates of force application, the biphasic (i.e., solid and fluid) character of the meniscus plays a fundamental role in determining the mechanical response. Even though circumferential stresses dominate the tissue's response (attributable to the hoop effect), circumferential strains are relatively small, and the fluid phase carries a significant part of the applied load (Spilker et al. 1992).

In broad terms, meniscal injury is either traumatic or degenerative. Traumatic injuries arise from acute insult to the meniscus and are usually seen in young, active individuals. Degenerative meniscal tears are chronic injuries, usually found in older persons, which typically result from simple movements, such as deep knee bends, that load the weakened meniscal tissue.

The combination of complex joint movement and continuously varying loading patterns creates a formidable puzzle in terms of identifying specific meniscal injury mechanisms. Nonetheless, certain mechanisms are implicated. Damage usually occurs when the meniscus is subjected to a combination of flexion and rotation or extension and rotation during weight bearing and resultant shear between the tibial and femoral condyles (Silbey and Fu 2001; Siliski 2003). For example, when an athlete whose foot is planted on the ground attempts a rapid change of direction, internal femoral rotation on a fixed tibia causes posterior displacement of the medial meniscus. The meniscal attachment to the joint capsule and MCL resists this movement and places the meniscus under tensile loading. Because of structural considerations (e.g., medial meniscus attachment to other medial structures) and movement characteristics (e.g., valgus rotation loading during cutting movements), the medial meniscus is five times more likely to be injured than its lateral counterpart. External rotation of the foot and lower leg (relative to the femur) predispose to medial meniscus injury. In contrast, internal rotation of the foot and leg makes the lateral meniscus

more vulnerable (Peterson and Renström 2001).

Rapid extension of the knee generates forces sufficient to cause a longitudinal tear of the medial meniscus. On occasion the loading is large enough to cause a vertical longitudinal tear that extends into the anterior horn, an injury termed a **bucket-handle tear.** The bucket-handle pattern more typically arises from repeated insult to a partial tear that progresses to span a large portion of the meniscus (figure 6.14a). Bucket-handle tears happen predominantly in skeletally immature persons. Degenerative, transverse (radial), and horizontal tears (figure 6.14b) are more common in adults (Stanitski 2002).

Predisposition to meniscal injury is activity dependent; that is, certain sports are associated with high incidence of meniscal injury. Leading the list is soccer, in which players experience frequent collisions with opponents and often change direction and body position while their cleats are embedded in the turf. Meniscal injury is also common in track and field (e.g., knee torsion in the shot put or discus) and skiing (e.g., ski

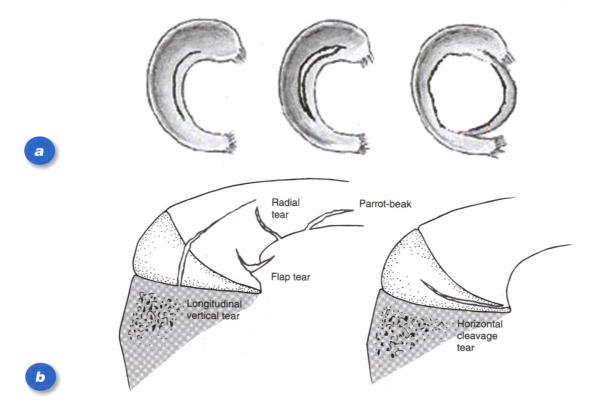

Figure 6.14 *(a)* Bucket-handle tear of the meniscus. *(b)* Various meniscal tears. A "parrot beak" tear is a radial tear that splits the meniscus in the shape of a parrot's beak.

(a) From *Sports injuries: Their prevention and treatment*, 3rd ed., L. Peterson and P. Renstrom, Copyright 2001, Human Kinetics. Reproduced by permission of Taylor & Francis books UK.

(b) This article was published in *Skeletal trauma, 3rd Edition*, B.D. Browner et al., p. 2064, Copyright Elsevier (2003).

slippage or catching that imparts a sudden twist to the knee). Occupations involving sustained or repeated squatting (e.g., mining, carpet laying, or gardening) are also implicated in meniscus injury, often attributable to the degenerative processes that accompany prolonged knee flexion and its attendant structural loads.

Collateral Ligament Sprain

Injury to the MCL complex is quite common, with involvement of the LCL much less frequent. Both injuries result from sudden and violent loading. Proper physical examination, in conjunction with MRI evaluation, is required for treatment and rehabilitation of both MCL and LCL injuries (Quarles and Hosey 2004).

Medial Collateral Ligament

Impact on the lateral side of the knee causes knee valgus and tensile loading of the medial aspect (figure 6.15a), a mechanism that can produce MCL injury. The MCL is most effective in resisting valgus loading when the knee is flexed 25° to 30° (Swenson and Harner 1995). Other structures play a relatively greater role when the knee is at full extension than when the knee is partially flexed. Each part of the MCL is differentially loaded at varying angles of knee flexion. At 5° of knee flexion, the superficial MCL accounts for 57% of medial stability, with the deep MCL at 8% and the posterior oblique ligament at 18%. When the knee is further flexed to 25°, the superficial MCL increases to 78% while the deep MCL and

posterior oblique each drop to 4% (Peterson and Renström 2001).

The role of the MCL in resisting valgus loading has been demonstrated experimentally (Grood et al. 1981; Piziali et al. 1980; Seering et al. 1980). Research suggests that the MCL is the primary valgus restraint, with only secondary involvement provided by the cruciate ligaments. In cases of isolated MCL failure, however, residual structures, particularly the ACL, are able to resist varus–valgus moments (Inoue et al. 1987). Although most MCL injuries are acute and traumatic, overuse syndromes also have been implicated, specifically associated with the whip-kick technique used by swimmers performing the breaststroke.

Lateral Collateral Ligament

LCL injury usually results from varus loading, often in combination with hyperextension. Varus loading happens when an impact is applied to the knee's medial aspect while the foot is planted on the ground (figure 6.15b). The varus loading creates tensile forces in the lateral knee structures. Given its extracapsular structure, the LCL is more likely than the MCL to sustain isolated injury. Nonetheless, combined injury to the LCL and one of the cruciate ligaments is not uncommon.

The degree of impact affects the progression of injury. Moderate impact results in isolated LCL rupture. Violent impact causes LCL rupture in concert with ACL failure. Extremely violent impacts rupture the LCL, ACL, and PCL (Peterson and Renström 2001).

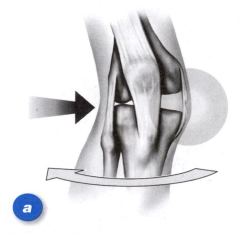
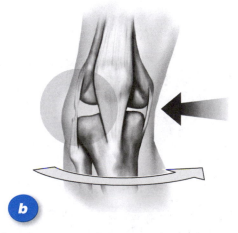

Figure 6.15 Collateral ligament injury at the knee. *(a)* Valgus loading of the medial collateral ligament. *(b)* Varus loading of the lateral collateral ligament.

Reprinted, by permission, from R.S. Behnke, 2005, *Kinetic anatomy*, 2nd ed. (Champaign, IL: Human Kinetics), 195.

Knee Extensor Disorders

The knee joint complex is the critical middle link in the kinetic chain of the lower extremity, and its loading and motion characteristics dictate effective limb function. Aberrations in any of the many functional components of the knee joint complex increase the risk of injury. Arguably the most important component is the so-called *knee extensor mechanism* (KEM), which consists of the quadriceps muscle group, the patellofemoral joint, and the tendon group connecting these elements.

The patella serves as the central structure in the knee extensor mechanism. In that role it acts as a fulcrum, or pivot, to enhance the mechanical advantage of the quadriceps during knee flexion and extension. The patella effectively displaces the tendon line of action away from the instantaneous joint center (axis) and thus increases the moment arm (figure 6.16). A given force then produces a greater *moment of force* or *torque*.

Force created by the quadriceps is transmitted through the *quadriceps tendon* and *patellar tendon* (also *patellar ligament)* to the tibial tuberosity. Some researchers erroneously have assumed that the force in the quadriceps tendon (F_Q) is the same as that in the patellar tendon (F_p). However,

this has been disproved by research showing that F_Q and F_p are not equal. The actual forces in each tendon depend on the knee joint angle (figure 6.17). Tendofemoral contact in positions of extreme flexion (e.g., in a deep squat) carries a significant portion of the contact force, thus reducing the load on the patella.

Patellofemoral Disorders

Patellofemoral pain (PFP) is one of the most common lower-extremity pathologies. Despite its prevalence, controversy persists regarding its cause, evaluation, and treatment, and much remains to be done in identifying the mechanisms involved. Powers (2003) suggested that comprehensive evaluation of PFP should include consideration of hip–pelvic motion and foot–ankle motion and should not focus solely on structures of the knee or events involving the knee.

As forces are transmitted through the knee extensor mechanism, a component of the force is directed through the patella toward the joint center and pushes the patella against the femur. Near full extension, the patella rides high on the femur. As the knee flexes, the patella slides into the intercondylar groove. This movement of the patella along the femur is referred to as **patellar tracking.**

Effective patellar tracking depends on congruence between the patella and femur. This congruence is typically measured by congruence angle, lateral patellofemoral angle, and patellar tilt angle (figure 6.18). Caution is warranted in

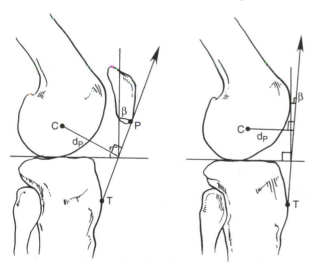

Figure 6.16 Effect of the patella in increasing the mechanical advantage of the knee extensor mechanism. *(a)* The patella effectively moves the tendon line of action away from the knee joint instantaneous center (axis of rotation, *C*), increasing the moment arm *(d_p)* of the quadriceps group and thus enhancing its mechanical advantage. *(b)*Without a patella, the moment arm *(d_p)* is shorter, the angle of pull *(β)* is smaller, and the mechanical advantage is reduced. *T* is the patellar tendon attachment site at the tibial tuberosity.

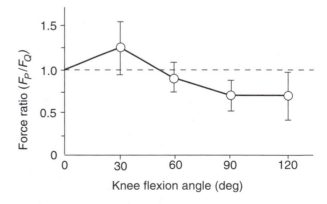

Figure 6.17 Ratio of patellar tendon force *(F_p)* to quadriceps tendon force *(F_Q)* as a function of knee flexion angle.

Adapted, by permission, from H.H. Huberti et al., 1984, "Force ratios in the quadriceps tendon and ligamentum patellae," *Journal of Orthopaedic Research* 2(1): 51.

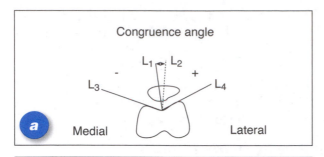

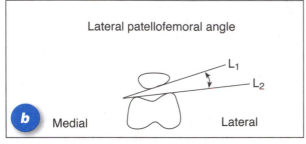

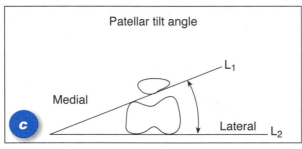

Figure 6.18 Patellofemoral angle measures. *(a)* Congruence angle (between lines L_1 and L_2). *(b)* Lateral patellofemoral angle. *(c)* Patellar tilt angle.

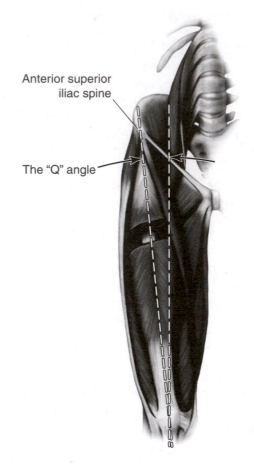

Figure 6.19 Quadriceps (Q) angle, measured as the angle formed by a line drawn from the anterior superior iliac spine to the midpatella and a line drawn from the midpatella to the tibial tuberosity.

Reprinted, by permission, from R.S. Behnke, 2005, *Kinetic anatomy*, 2nd ed. (Champaign, IL: Human Kinetics), 200.

interpreting these measures based on osseous landmarks, because the surface geometry of the cartilage may not match the osseous morphology (Staubli et al. 1999).

Proper tracking also depends on a complex interaction of muscle forces (i.e., vector sum of the individual force vectors of the vastus medialis, vastus lateralis, vastus intermedius, and rectus femoris), other forces (e.g., retinacula) and structural considerations (e.g., **Q angle;** figure 6.19), **patella alta** (patella positioned abnormally high on the femur), and geometry of the intercondylar groove. As the patella moves, contact pressures develop between the patella and femur. These pressures, as well as the contact area and contact force, vary with the amount of knee flexion (figure 6.20).

Patellar movement simultaneously changes the location of the patellofemoral joint reaction force and the moment arm about the instanta-

neous knee joint axis. As the patella slides down within the groove, the retropatellar contact area changes (figure 6.21). Brechter and colleagues (2003) established the efficacy of using MRI to quantify patellofemoral joint contact area. MRI has been used (Salsich et al. 2003) to confirm the results of earlier studies (e.g., Hungerford and Barry 1979) reporting that as the knee flexes, patellofemoral joint contact area increases and migrates superiorly on the retropatellar surface as the patella glides deeper into the intercondylar groove. Most of the increase in contact area happens in the first 45° to 60° of flexion (Salsich et al. 2003). The increase in contact area permits wider distribution of the escalating contact forces and thus helps moderate the patellofemoral joint pressure (and stress). At knee angles

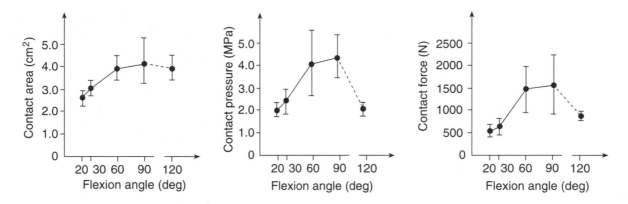

Figure 6.20 Patellofemoral contact area, pressure, and force vary with knee flexion angle.

Reprinted, by permission, from H.H. Huberti and W.C. Hayes, 1984, "Patellofemoral contact pressures. The influence of q-angle and tendofemoral contact," *J Bone Joint Surg Am.* 66(5): 717.

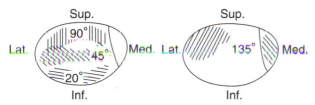

Figure 6.21 Patellofemoral contact areas as a function of change in degree of knee flexion. As the knee is flexed from full extension (0°) through 90°, the contact area migrates from the inferior retropatellar surface to the superior region. At 135° of flexion, the contact area is on both the superolateral surface and the medial odd facet.

Adapted, by permission, from D.S. Hungerford and M. Barry, 1979, "Biomechanics of the patellofemoral joint," *Clinical Orthopaedics and Related Research* 144: 11.

greater than 60°, contact area levels off and may actually decrease with extreme knee flexion at 120° under certain loading conditions (Huberti and Hayes 1984).

Patellofemoral joint stress and contact area have important clinical ramifications, as demonstrated in the following studies:

- Brechter and Powers (2002) reported significantly higher patellofemoral joint stress (attributable to smaller contact area) in subjects with PFP compared with subjects without pain.

- In a study of the influence of patella alta on patellofemoral joint stress, Ward and Powers (2004) showed that subjects with patella alta had significantly less contact area and greater patellofemoral stress compared with controls.

- Salem and Powers (2001) studied collegiate female athletes performing squats to three

depths (70°, 90°, 110° of knee flexion) using 85% of their one repetition maximum (1RM). The results suggested that peak patellofemoral joint reaction force and patellofemoral joint stress do not vary significantly between 70° and 110°. Thus, the authors concluded that deep squats are no more challenging to the patellofemoral joint than shallow squats.

Controversy still exists regarding the causes of patellofemoral pain. Powers (2003, p. 639) suggested that "abnormal motion(s) of the tibia and femur in the transverse and frontal planes are believed to have an effect on patellofemoral joint mechanics and therefore PFP." Thomee and colleagues (1995, p. 237) suggested that "chronic overloading and temporary overuse of the patellofemoral joint, rather than malalignment, contribute to patellofemoral pain."

Dye (2004, p. 5) asked, "If the presence of malalignment is crucial in the genesis of anterior knee pain, why does one find patients with bilateral radiographically determined patellofemoral malalignment (i.e., patellar tilts) with only unilateral symptoms? Why do more than 90% of patients with patellofemoral pain who have a diagnosis of malalignment as the cause have a successful response to conservative therapy, even though there has been no 'correction' or restoration of the supposed underlying indicators of malalignment (e.g., a high Q angle or a shallow trochlea)?" Dye took issue with the malalignment theory for PFP and concluded that "one can logically assume that the perception of patellofemoral pain, in most instances, is a function of nociceptive neurological output of any combination of innervated patellar and peripatellar tissues" (Dye 2004, p. 7). There clearly is disagreement on the sources of PFP, and given the multifactorial nature of PFP, controversy will persist.

In summary, the integrity of patellofemoral movement is dictated by the neuromechanical synergy between patellofemoral tracking, patellofemoral contact pressures, and neuromotor control of patellofemoral agonists. The precise relation between patellofemoral movement and patellofemoral pain remains unclear.

Disturbance of patellofemoral integrity often leads to injuries. Injuries to the knee extensor mechanism result from direct trauma, indirect trauma, or chronic overuse. Whatever the cause, there is little doubt that the injuries are both myriad and prevalent.

In chapter 5, we cautioned against the use of nonspecific injury descriptors. Two such terms are used in referring to patellofemoral pathologies: *jumper's knee* and *chondromalacia patella*. The former term refers to tendon pain of the knee extensor mechanism developed through repeated jumping. We suggest the use of more clinically useful terms that identify the location and condition of the involved tissue (e.g., *quadriceps tendinitis, patellar tendinitis, apophysitis of the tibial tuberosity,* or *Osgood-Schlatter disease).*

The second term, chondromalacia patella, has evolved into an all-too-common descriptor for generalized patellar pain. The term is best reserved to describe specifically the degeneration of retropatellar articular cartilage. Once believed to be a primary condition of unknown etiology, chondromalacia patella is now thought to occur most often secondary to other mechanisms. These include both traumatic (e.g., patellar fracture) and chronic (e.g., patellar malalignment, chronic subluxation, pathological patellar tracking) events.

Quadriceps Tendon and Patellar Tendon Ruptures

Extreme forces or continued mechanical insult to an already weakened knee extensor mechanism may lead to tendon rupture (see case study). Quadriceps tendon rupture typically occurs in people older than 40 years and is localized at the osteotendinous junction of the superior patellar pole. Calcification at the rupture site suggests that quadriceps tendon rupture tends to occur in areas of previous microtrauma. In contrast, patellar tendon ruptures tend to afflict those younger than 40 years, most often tearing at the inferior patellar pole.

The injury mechanism typically involves a violent quadriceps contraction against resistance at the knee and requires substantial loads to induce rupture. Although rare, several cases of *bilateral* patellar tendon ruptures have also been reported in the literature (e.g., Kellersman et al. 2005; Rose and Frassica 2001). Some of these injuries have identifiable mechanisms (e.g., jumping, tripping, or specific athletic tasks) and predisposing risk factors (e.g., systemic disease or steroid use), whereas others appear as *idiopathic* events in otherwise healthy individuals.

Osgood-Schlatter Disease

Osgood–Schlatter disease (OSD), named after two physicians (Robert Bayley Osgood and Carl Schlatter) who in 1903 simultaneously and independently identified the disorder (Nowinski and Mehlman 1998), is a traction apophysitis of the tibial tuberosity. Commonly found in adolescent athletes, OSD is characterized by inflammation of the bone where the patellar tendon attaches to the tibial tuberosity (figure 6.22). OSD is distinguished from patellar tendinitis, which manifests as inflammation of the patellar tendon near its attachment at the tibial tuberosity.

The injury mechanism most associated with OSD is repetitive high-load quadriceps action in adolescents during periods of rapid growth (Peterson and Renström 2001). Sports involving jumping and running (e.g., basketball and volleyball) are commonly implicated. In addition to activity, anatomical structure may play a role in the risk of OSD. Demirag and colleagues (2004) concluded that OSD may be caused, at least in part, by a patellar tendon that attaches more proximally and across a broader area of the tibia.

Magnetic resonance imaging has been used to describe the progression of OSD: swelling (edema) and inflammation at the tibial tuberosity, followed by tears at the secondary ossification center and ossicle formation from a partially avulsed portion of the tuberosity (Hirano et al. 2002).

OSD most commonly happens in boys between the ages of 10 and 16 and usually disappears

CASE STUDY: PATELLAR TENDON RUPTURE

Musculoskeletal injuries typically do not happen under conditions that permit quantitative assessment of the injury dynamics. In most cases, clinicians and researchers are limited to qualitative evaluation. On rare occasions, however, circumstances do allow for quantitative examination. One such case, in which a human patellar tendon ruptured during a sports competition, was reported by Zernicke and colleagues (1977). A world-class light heavyweight lifter was attempting to complete the second phase of a clean-and-jerk movement using 175 kg when his right patellar tendon ruptured. The tendon was torn completely with evidence of damage at the distal pole of the patella and at the patellar tendon insertion on the tibia.

The lift was attempted during a national weightlifting competition and was being filmed for biomechanical analysis. Using a multilink, rigid-body model, Zernicke and colleagues (1977) were able to estimate the tensile force in the tendon at the instant of rupture. The force was approximately 14.5 kN, or more than 17.5 times the lifter's body weight.

This exceptional example highlights the complex dynamics of injury and provides evidence of the high stresses applied to and tolerated by biological tissues and the importance of loading rate on tissue response to mechanical loading.

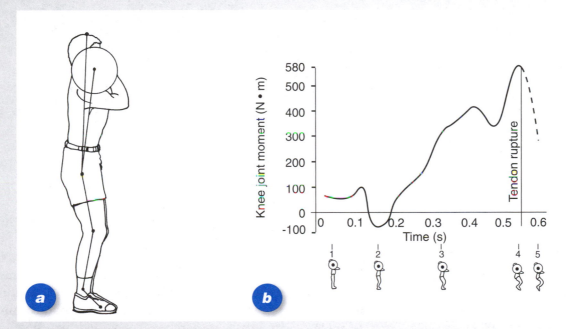

Patellar tendon rupture. *(a)* Five-segment, rigid-body model of weightlifter. *(b)* Mean resultant knee joint moments from the beginning of the jerk movement until after tendon failure.

(a) Adapted, by permission, from R.F. Zernicke, J. Garhammer and F.W. Jove, 1977, "Human patellar-tendon rupture," *J Bone Joint Surg Am.* 59(2): 180. *(b)* Adapted, by permission, from R.F. Zernicke, J. Garhammer and F.W. Jove, 1977, "Human patellar-tendon rupture," *J Bone Joint Surg Am.* 59(2): 181.

when the adolescent reaches full maturity. One study, however, suggested that persons with a history of OSD may experience a higher level of knee disability in later years than those with no history of OSD (Ross and Villard 2003).

Iliotibial Band Syndrome

The iliotibial band (ITB) is a thickened band of fascial tissue spanning from the iliac crest (with partial insertion from the tensor fascia lata) to

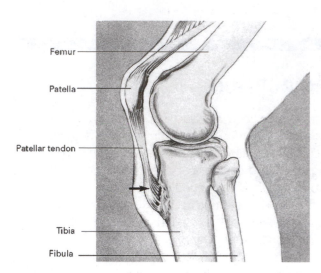

Figure 6.22 Knee with Osgood–Schlatter disease. A traction apophysitis at the tibial tuberosity (arrow).

From *Sports injuries: Their prevention and treatment*, 3rd ed., L. Peterson and P. Renstrom, Copyright 2001, Human Kinetics. Reproduced by permission of Taylor & Francis Books UK.

the lateral tibial tubercle (figure 6.23). In full extension (0° of flexion), the ITB lies anterior to the tibial tuberosity. As the knee flexes through about 30°, the ITB passes over the lateral tibial epicondyle to a position posterior to the tuberosity (figure 6.23). The area where the ITB rubs on the lateral epicondyle is sometimes referred to as the *impingement zone* (Farrell et al. 2003).

When the ITB passes over the epicondyle, friction is created between the ITB and the lateral epicondylar surface, and with repeated flexion–extension cycles, the ITB can become irritated, resulting in a painful inflammatory condition called **iliotibial band syndrome (ITBS),** also termed *iliotibial band friction syndrome* (ITBFS).

ITBS is commonly found in runners, cyclists, and military personnel (Kirk et al. 2000) whose knees experience repeated flexion–extension through limited range of motion. In a study of military recruits, ITBS was second only to ankle sprains as the most frequent injury diagnosis and was associated with running and abrupt increases in training volume (Almeida et al. 1999).

Excessive foot pronation has been associated with ITBS (Peterson and Renström 2001), but this relation is controversial, because some (e.g., Khaund and Flynn 2005) contend that research support for this theory is lacking. Other potential risk factors include ITB tightness; high running mileage; interval training; muscle weakness in the knee extensors, knee flexors, and hip abductors; genu varum (bowleggedness); tibial rotation; and

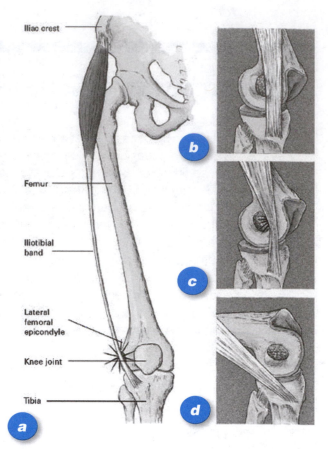

Figure 6.23 Iliotibial band syndrome. *(a)* Iliotibial band friction. *(b)* Iliotibial band anterior of epicondyle during knee extension. *(c)* Iliotibial band rubs over epicondyle with knee flexed 30°. *(d)* Iliotibial band behind epicondyle when knee is flexed more than 30°.

From *Sports injuries: Their prevention and treatment*, 3rd ed., L. Peterson and P. Renström, Copyright 2001, Human Kinetics. Reproduced by permission of Taylor & Francis Books UK.

leg-length discrepancy (Khaund and Flynn 2005; Peterson and Renström 2001).

Cyclists commonly develop ITBS. In cycling, the knee moves through the 30° flexion range with each pedaling cycle. Farrell and colleagues (2003) found that foot–pedal forces at the impingement zone during cycling were only 18% of those found at the foot–ground interface in running, and these researchers concluded that repetition appears to be more important than force levels in hastening the onset of ITBS. The authors also noted that ITBS may be aggravated by improper cycle seat height, anatomical differences, and training errors.

Lower-Leg Injuries

The lower leg (also called *leg* or *shank)* spans the knee and ankle joints and contains two longitudinally aligned bones, the tibia (medial) and fibula (lateral). Four muscle compartments (anterior, lateral, superficial posterior, deep posterior) surround these bones, with tight fascia enclosing each compartment. The anterior compartment contains the tibialis anterior, extensor hallucis longus, extensor digitorum longus, and peroneus tertius. The lateral compartment contains the peroneus longus and peroneus brevis. The largest compartment in terms of muscle mass is the superficial posterior compartment, which contains the gastrocnemius and soleus (together termed the *triceps surae)* and plantaris. The deep posterior compartment houses the flexor hallucis longus, flexor digitorum longus, and tibialis posterior (figure 6.24). The actions of the foot and ankle muscles are summarized in table 6.3.

Compartment Syndrome

Acute injury or chronic exertion often increases fluid accumulation within muscle compartments of the arms, feet, and legs. The excess fluid may be attributable to hemorrhage, edema, or both. Given the relative inextensibility of the surrounding fascia, the fluid increase results in greater compartmental pressure. This creates a **compartment syndrome (CS),** defined as "a pathologic condition of skeletal muscle characterized by increased interstitial pressure within an anatomically confined muscle compartment that interferes with the circulation and function of the muscle and neurovascular components of the compartment" (Garrett 1995, p. 48).

From a mechanical perspective, compartment syndromes are a consequence of the relations among mass, volume, and pressure (see chapter 5). Increasing the mass within a fixed volume increases the internal pressure. This is the essence of a compartment syndrome.

The mechanism of CS may be either chronic *(chronic compartment syndrome,* also *chronic exertional compartment syndrome)* or, less commonly, acute *(acute compartment syndrome).* Many conditions can lead to a compartment syndrome: soft tissue contusion, crush, bleeding disorders, venous obstruction, arterial occlusion, burn, prolonged compression after drug overdose, surgery, apparel (e.g., medical antishock trousers), and exercise.

In the lower leg, any of the four muscle compartments (anterior, lateral, superficial posterior, deep posterior) may be affected. Most commonly, the anterior compartment is involved (Schepsis et al. 2005), with all muscles in the compartment affected. In rare instances, an isolated muscle (e.g., tibialis anterior) is involved (Church and Radford 2001).

Increased compartmental pressure compromises vascular and neural function and sets the stage for ischemia and a self-perpetuating cycle of fluid accumulation and restricted flow. The situation is exacerbated by the mechanical properties of the fascia, which has been shown to increase in thickness and stiffness in response to chronic compartment syndrome (Hurschler et al. 1994). The situation is worsened by a decrease in compartment volume, as might be caused by compression wraps or tight clothing.

Sufficiently large compartment pressures result in vessel closure and potentially catastrophic physiological consequences. Venous collapse severely reduces blood return and leads to capillary congestion and decreased tissue perfusion. Local tissues then suffer the consequences of hypoperfusion (e.g., ischemia and eventual necrosis).

Transient increases in compartment pressures are normally seen in response to exertion. In people without chronic compartment syndrome (CCS), resting pressures vary, ranging from 0 to 20 mmHg (Dayton and Bouche 1994). During exertion, pressures may exceed 70 mmHg but quickly return to resting levels within minutes of exercise cessation. A person with CCS, in contrast, may exhibit resting pressures of 15 mmHg that climb to more than 100 mmHg during exercise, with prolonged postexercise decline (figure 6.25).

Relief from CCS is achieved surgically by fascial incision (fasciotomy) to release the compartment and effectively increase its volume and also to reduce internal pressure. Some controversy exists over the threshold pressure above which fasciotomy is indicated. Suggested values range from 30 to 45 mmHg. However, many other factors should be considered as well:

- Intracompartmental pressures do not measure neuromuscular ischemia.
- Ischemic development depends on both the magnitude and duration of the elevated pressure.
- Patient tolerance to ischemia may vary.
- Injured muscle may be less tolerant of ischemia and elevated pressure than uninjured muscle (Gulli and Templeman 1994).

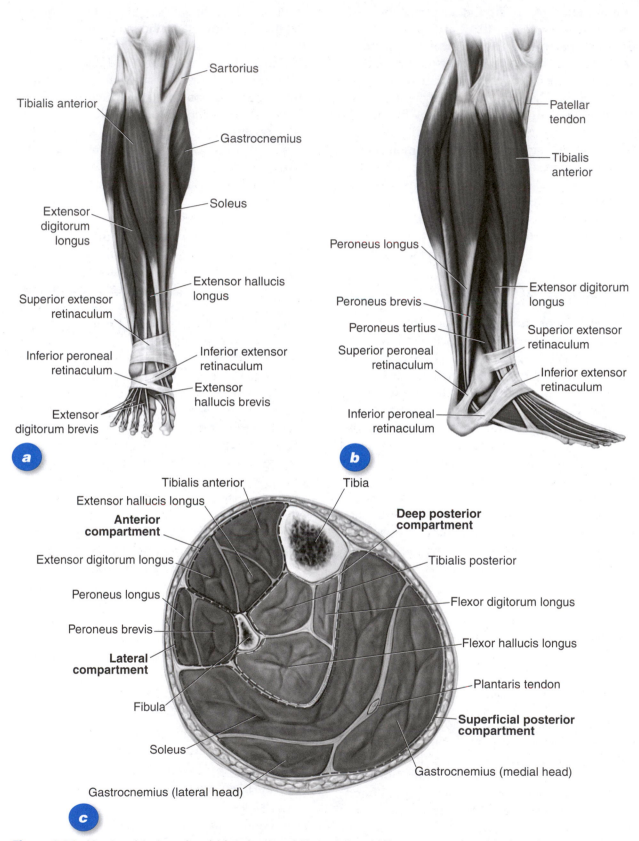

Figure 6.24 Muscles of the lower leg. *(a)* Anterior view. *(b)* Lateral view. *(c)* Four compartments of the lower leg.

(a, b) Reprinted, by permission, from R.S. Behnke, 2005, *Kinetic anatomy*, 2nd ed. (Champaign, IL: Human Kinetics), 216. *(c)* Reprinted, by permission, from R.S. Behnke, 2005, *Kinetic anatomy*, 2nd ed. (Champaign, IL: Human Kinetics), 219.

TABLE 6.3

MUSCLES OF THE LOWER LEG AND ANKLE

Muscle	Action
Anterior compartment	
Tibialis anterior	Dorsiflexes the ankle and inverts the foot
Extensor hallucis longus	Dorsiflexes the ankle and inverts the foot
Extensor digitorum longus	Dorsiflexes the ankle and everts the foot
Peroneus tertius	Dorsiflexes the ankle and everts the foot
Lateral compartment	
Peroneus longus	Plantar flexes the ankle and everts the foot
Peroneus brevis	Plantar flexes the ankle and everts the foot
Superficial posterior compartment	
Gastrocnemius	Plantar flexes the ankle and flexes the leg
Soleus	Plantar flexes the ankle
Plantaris	Plantar flexes the ankle and flexes the leg
Deep posterior compartment	
Popliteus	No action at ankle or foot; flexes and medially rotates the leg
Flexor hallucis longus	Plantar flexes the ankle and inverts the foot
Flexor digitorum longus	Plantar flexes the ankle and inverts the foot
Tibialis posterior	Plantar flexes the ankle and inverts the foot

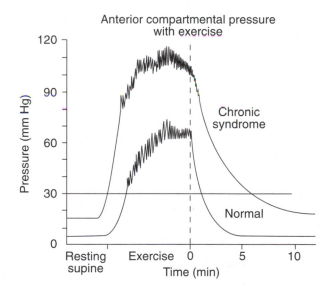

Figure 6.25 Anterior compartment pressures recorded in a patient with chronic anterior compartment syndrome and in a normal subject.

Reprinted, by permission, from S.J. Mubarak, 1981, *Compartment syndromes and Volkmann's contracture* (Philadelphia, PA: W.B. Saunders Company), 218.

Medial Tibial Stress Syndrome

Medial tibial stress syndrome (MTSS), often seen in runners and military personnel, is the most common source of exercise-induced lower-leg pain. Yates and White (2004) reported an MTSS incidence of 35% in a group of 124 naval recruits undergoing a 10-week basic training program. Women recruits were more likely than men (53% vs. 28%) to develop MTSS.

Historically, MTSS has been described as an inflammatory reaction of the deep fascial tibial attachments to chronic loads, with pain localized to the posteromedial crest of the tibia. Recent evidence, however, suggests that MTSS may be associated with lower regional bone mineral density (Magnusson et al. 2001) and related to tibial bending during chronic weight-bearing activity (Beck 1998).

MTSS results from excessive tensile forces applied to the fascia by the eccentric action of musculotendinous units, most often the soleus and flexor digitorum longus. Controversy exists over whether the tibialis posterior may be involved. The condition initially manifests as fasciitis and with continued loading may progress

SHIN SPLINTS

Of the many catch-all terms in the medical literature, perhaps none can match *shin splints* when it comes to nonspecificity, lack of consensus on meaning, and continuing misunderstanding and confusion. As evidence, we present a few of the many and varied descriptions of shin splints.

- "Diffuse areas of increased tenderness over the anterior or posterior bony attachments of the tibialis anterior muscles to the tibia. . . . This relatively mild condition must be distinguished from its two more disabling cousins, tibial stress fracture and chronic exertional compartment syndrome" (Kibler and Chandler 1994, p. 549).

- "Pain in the shin may be related to overuse or stress of the muscles within the extensor or flexor groups, stress fracture, or induced ischemia within muscular compartments leading to compartment syndrome" (Ciullo and Shapiro 1994, p. 661).

- "Painful injury to and inflammation of the tibial and toe extensor muscles or their fasciae that is caused by repeated minimal traumas (as by running on a hard surface)" (Merriam-Webster 2005, p. 758).

- "A nondescript pain in the anterior, posterior, or posterolateral compartment of the tibia. It usually follows strenuous or repetitive exercise and is often related to faulty foot mechanics such as pes planus or pes cavus. The cause may be ischemia of the muscles in the compartment, minute tears in the tissues, or partial avulsion from the periosteum of the tibial of peroneal muscles" (Venes 2005, p. 1987).

And among the most confusing of all, from the same source (*Stedman's* 2005, p. 1338):

- *Shin-splints* (hyphenated)—"Tenderness and pain with induration and swelling in the anterior tibial compartment, particularly following athletic overexertion by the untrained."

- *Shin splints* (two words, nonhyphenated)—"A collective term for various injuries to the leg including acute and chronic exertional compartment syndrome, medial tibial stress syndrome, and periostitis."

O'Donoghue (1984, p. 591) astutely noted, "As with many names in common use, there is considerable and often heated argument as to what is actually meant by the term. As is usual in these circumstances, the term 'shin splints' is a wastebasket one including many different conditions. The authors of various articles on the subject are inclined to state very definitely that it is caused by one particular thing to the exclusion of all others, which causes great confusion."

We recommend relegating the term *shin splints* to O'Donoghue's wastebasket and instead using terms that are clinically correct, specific, and useful. We echo the sentiment of Batt (1995, p. 53) that "the term shin splint be recognized as generic, rather than diagnostic, and that specific conditions that currently exist under this term be differentiated."

to periostitis and ultimately to changes in bone mineral density in the affected area (Magnusson et al. 2001)

Commonly seen in runners, MTSS is a multifactorial overuse syndrome related to the runner's anatomical structure, training program, flexibility, muscle strength, footwear, and running mechanics. Changes in any of these variables may lead to an MTSS-related injury. Despite the prevalence of MTSS, its diagnosis is problematic in light of differential diagnoses of stress reaction and stress fracture (see following section), tendinitis, musculotendinous strain, and chronic compartment syndromes.

MTSS treatment begins with conservative management (e.g., rest and ice) but may require surgical treatment if conservative measures fail. The surgery involves a deep posterior compartment **fasciotomy** to relieve pressure in the affected area. Surgical outcomes are mixed. In a study of surgical treatment of MTSS, Yates and colleagues (2003) reported excellent (35%), good (34%), fair (22%), and poor (9%) results in 78 patients. Although surgery may relieve pain in a majority of patients, return to full activity may be precluded for many. Yates and colleagues reported significant pain reduction in 72% of patients, but only 41% were able to fully return to their presymptom activity level.

Tibial Stress Reaction and Stress Fracture

Bone responds to repetitive loading by adapting its structure according to what is known as Wolff's law (cf. p. 107). This adaptation or remodeling process includes resorption of bone where the loading conditions render bone unnecessary and the deposit of bone in regions needed to sustain the new mechanical loads. If, however, the magnitude and frequency of loading exceed the bone's ability to adapt, injury occurs. The most recognizable form of injury is bone fracture. As discussed in the previous chapter, fracture may occur acutely (traumatic fracture) or in response to chronic loading **(stress fracture).** Chronic fractures are most often associated with a sudden increase in activity (e.g., athletes, military recruits). These stress fractures are termed **fatigue fractures.** Less recognized are chronic fractures found in persons with no increase in activity but with decreased bone density. These stress fractures are called **insufficiency fractures.**

The term *stress fracture* itself suffers from overuse, or perhaps misuse, because it is used frequently to describe bone with no clear evidence of discontinuity or line of fracture. The term **stress reaction** describes bone with evidence of remodeling but with an absence of radiological evidence of fracture. Such stress reactions are quite common and are detectable using a combination of radiographs, bone scans, and magnetic resonance imaging scans (figure 6.26).

Actual fractures occur much less frequently than pure mechanical loading (i.e., material fatigue failure) alone would predict, suggesting that the process leading to stress reaction and subsequent stress fracture involves physiological processes of bone adaptation to mechanical loading. This is not to discount completely the role of mechanical fatigue, however, because microfractures have been detected at remodeling sites.

Verifiable stress fractures most frequently happen in the tibia, accounting for up to 50% of all stress fractures. Most long-bone stress fractures are oriented transversely to the longitudinal axis of the bone; longitudinally directed stress fractures are uncommon and usually found on the anterior cortex of the distal tibia (Tearse et al. 2002).

Stress fracture location depends somewhat on activity. The mechanical demands of specific movements appear to play a prominent role in determining the fracture site. Runners, the most common victims of tibial stress fracture, exhibit fractures focused between the middle and distal thirds of the tibia. Athletes in jumping sports (e.g., basketball and volleyball) tend to experience proximal fractures. Dancers, in contrast, sustain more midshaft fractures.

Traumatic Fractures of the Tibia and Fibula

Mechanical insult to the lower leg may result in traumatic fracture of the tibia, fibula, or both. The sources of the applied force vary but most commonly involve vehicle–pedestrian accidents and sports-related movements. Other causes include slip-and-fall accidents, falls from a height, direct blows, crushing, gunshot, and overuse (Court-Brown and McBirnie 1995).

Causal mechanisms can be classified as either low energy (e.g., slip and fall, sports-related injury) or high energy (e.g., direct blows, motor vehicle incidents). Low-energy fractures often involve torsion or bending of the tibia with minimal soft tissue involvement. Torsional loading occurs when the lower leg is twisted about its long axis (see chapter 5), as in skiing when the ski provides an extended moment arm for applying torques to the tibial shaft. Details of a case of bilateral spiral fracture are presented in the

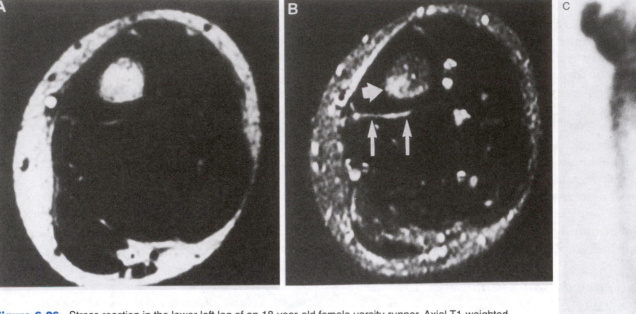

Figure 6.26 Stress reaction in the lower left leg of an 18-year-old female varsity runner. Axial T1-weighted magnetic resonance image *(a)* shows no detectable abnormality, but the T2-weighted image *(b)* shows moderate periosteal edema (long arrows) along the posterior and medial aspect of the tibia. There is also marrow edema (short arrow) in the adjacent part of the tibia. The bone scintigraphy *(c)* shows increased activity along the distal half of the tibial diaphysis (arrows).

Reprinted, by permission, from M. Fredericson et al., 1995, "Tibial stress reaction in runners: Correlation of clinical symptoms and scintigraphy with a new magnetic resonance imaging grading system," *The American Journal of Sports Medicine* 23(4): 472-481.

sidebar on page 191. Bending loads are created when parallel and oppositely directed forces are applied simultaneously to the bone. The classic *boot-top fracture* illustrates this bending mechanism (figure 3.35*c*).

High-energy injuries, in contrast, involve direct impact or high bending forces and result in transverse fractures and considerable fragmentation or comminution (Watson 2002). Motor vehicle crashes account for the majority of direct impact, or crushing, fractures.

One study identified baseball bats as a causal agent in tibial fractures. The fractures did not occur during athletic competition but rather resulted from the bats being used as weapons. Levy and colleagues (1994) reported 47 such bat-induced fractures during a 1-year period at an urban trauma center. Eleven of these fractures were to the tibia, and many involved extensive comminution and complications (e.g., delayed union or compartment syndrome).

Whatever the mechanism, tibiofibular fracture is a serious injury because of the injured bone's compromised ability to carry loads in its role as a critical link in the lower extremity's kinetic chain.

Ankle and Foot Injuries

Given their numerous bones, ligaments, and articulations, the foot and ankle are arguably the human body's most complex area. The ankle joint is formed by the articulation of the tibia, fibula, and talus. The tibia and fibula create a deep socket, or mortise, that contains the talus. In a dorsiflexed position, the talus fits snugly within the mortise and is quite stable. As the ankle plantar flexes, the narrower posterior section of the talus rotates into the area between the malleoli. This looser fit compromises joint stability, resulting in a relatively unstable ankle in the plantar flexed position.

Many ligaments reinforce the ankle. Medially, the strong deltoid ligament complex provides resistance to forceful eversion. On the lateral aspect, three ligaments are primarily responsible for restricting inversion. The weakest of the three, the anterior talofibular ligament (ATFL), extends anteromedially from the fibular malleolus to the neck of the talus. The calcaneofibular ligament (CFL) passes posteroinferiorly from the tip of the fibular malleolus to the lateral surface of the cal-

CASE STUDY: BILATERAL SPIRAL FRACTURE

Many thousands of skiing-related injuries happen every year, sometimes because of excessively tight bindings or binding release malfunctions (Hull and Mote 1980). Zernicke (1981) reported the case of an injury in which a skier, on his first run of the day on the beginner's slope, unintentionally began to snowplow uncontrollably. Both of his legs were forced into extreme internal rotation. Failure of the bindings to release resulted in spiral fractures in both tibias. Subsequent testing of the skis, boots, and bindings using a biomechanical model (see figure) provided quantitative estimates of the torques transmitted to the lower leg as lateral forces were applied to the skis. The data showed that the bindings placed the skier at high risk of injury, because "for nearly all applications of lateral loads to the ski, this skier's bindings would not have released prior to exceeding the torsional elastic threshold of the tibia" (Zernicke 1981, p. 243).

Fortunately, poor bindings such as the ones in this case are no longer in use. This example nonetheless demonstrates the potential for injury in skiing and points to the importance of proper equipment selection and maintenance in reducing the risk of musculoskeletal injury.

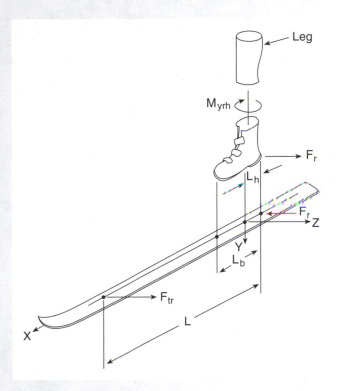

Schematic diagram of leg–boot–ski binding system. M_{yrh} = torque transmitted to skier's lower leg; F_r = lateral release force applied to heel; L_h = distance from vertical axis of the tibia to the heel point; L_b = distance from the heel to the toe pivot point (i.e., length of the boot); F_{tr} = lateral force applied to the ski at a distance (L) in front of the heel-release point; Z = z-coordinate axis; Y = y-coordinate axis; X = x-coordinate axis.

caneus. The posterior talofibular ligament (PTFL) connects the fibular malleolar fossa to the lateral tubercle of the talus (figure 6.27).

Each foot contains 26 bones (figure 6.28). The largest of these, the calcaneus, serves as the attachment for the calcaneal (Achilles) tendon, which transmits the force of the triceps surae muscles in plantar flexing the ankle. The articulation of the calcaneus with the talus forms the subtalar (talocalcaneal) joint, an articulation essential to proper function of the foot and ankle complex during load bearing. The subtalar joint axis runs obliquely, as shown in figure 6.29.

The bones of the foot form two primary arches: the longitudinal arch, running from the calcaneus to the distal ends of the metatarsals, and the transverse arch, which extends from side to side across the foot (figure 6.30). The longitudinal arch is divided into a medial portion that includes the calcaneus, the talus, the navicular, three cuneiforms, and the three most medial metatarsals. The lateral portion is much flatter and is in contact

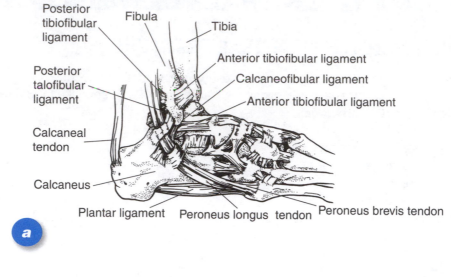

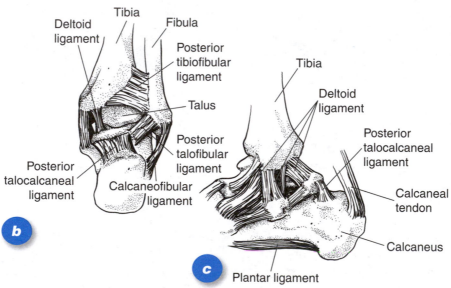

Figure 6.27 Tendons and ligaments of the ankle and foot: *(a)* Lateral view. *(b)* Posterior view. *(c)* Medial view.

with the ground during standing. The transverse arch is formed by the cuboid, cuneiforms, and bases of the metatarsals.

During weight bearing, the arches compress to absorb and distribute the load. Several ligaments assist in this force distribution. These include the plantar calcaneonavicular ligament (spring ligament), the short plantar ligament, and the long plantar ligament. The integrity of the arches and their ability to absorb loads are maintained by the tight-fitting articulations between foot bones, the action of intrinsic foot musculature, the strength of the plantar ligaments, and the plantar aponeurosis (plantar fascia).

Ankle Sprain

As a result of its relative anatomical instability and its supportive function, the ankle joint (figure 6.31) is frequently injured. In certain sports (e.g., basketball), ankle sprains are the most common injury. Despite their prevalence, ankle sprains continue to present clinicians with diagnostic and therapeutic difficulties (Safran et al. 1999).

To present a meaningful discussion of ankle injury mechanisms, we must review several anatomical structures and their functional characteristics. As briefly described earlier, the ankle joint is formed by the articulation of the tibia,

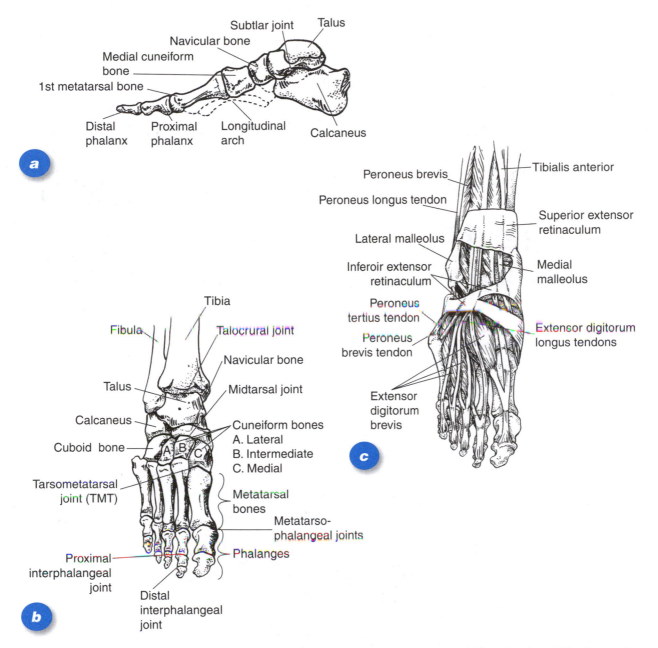

Figure 6.28 Bones, tendons, and ligaments of the foot. Bones of the foot: (a) Medial view. (b) Superior view. (c) Tendons and ligaments of the foot (superior view).

fibula, and talus (6.31a). The talar body is wedge shaped, with its anterior portion being wider than its posterior. This irregularity contributes directly to the joint's positional stability. In dorsiflexion, the wider part wedges between the malleoli, lending stability to the joint. The narrow portion of the talus, however, moves between the malleoli in plantar flexion, permits talar translation and tilt, and results in lateral instability. The juxtaposition of the tibia and fibula is maintained by a tibiofibular *syndesmosis*, which consists of an interosseous ligament (thickening of the interosseous membrane), an anterior inferior tibiofibular ligament, a posterior inferior tibiofibular ligament, and an inferior transverse tibiofibular ligament (Carr 2003).

Ankle sprain is a misnomer because the injury typically involves both the ankle and subtalar joints (6.31b). These two joints move in concert to execute what should be correctly viewed as combined ankle–foot movement. Our movement conventions are described and illustrated in figure 6.32.

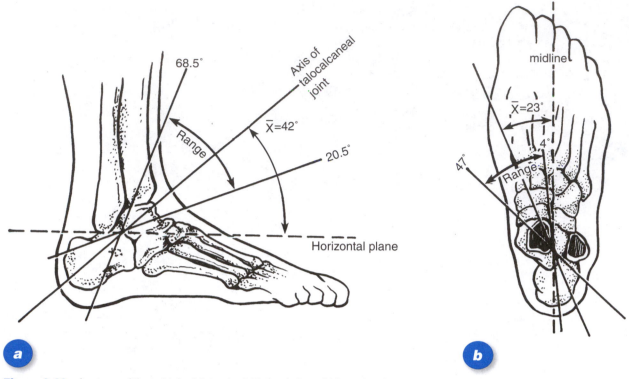

Figure 6.29 Anatomy of the subtalar joint axis. *(a)* Lateral view. *(b)* Superior view.

Reprinted, by permission, from B.J. Sangeorzan and J.B. Stiehl, 1991, Subtalar joint: Morphology and functional anatomy. In *Inman's joints of the ankle*, 2nd ed., edited by J.B. Stiehl (Baltimore, MD: Lippincott, Williams & Wilkins), 34-35.

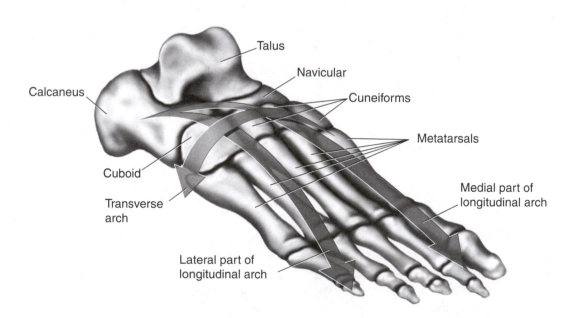

Figure 6.30 Arches of the foot.

Reprinted, by permission, from W.C. Whiting and S. Rugg, 2006, *Dynatomy: Dynamic human anatomy* (Champaign, IL: Human Kinetics), 31.

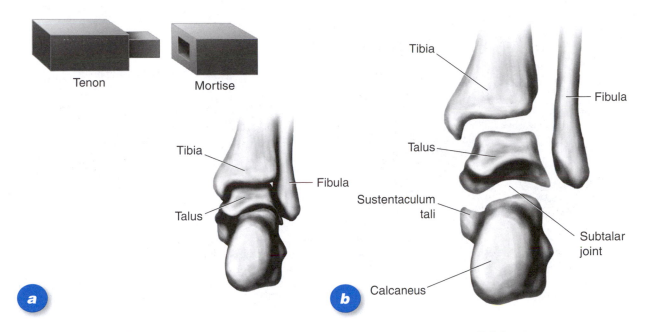

Figure 6.31 Skeletal anatomy of the ankle joint. *(a)* The talocrural joint as a mortise-(tibia and fibula) and-tenon (talus) joint. *(b)* Posterior view of the ankle, showing the subtalar joint.

(a) Reprinted, by permission, from R.S. Behnke, 2005, *Kinetic anatomy*, 2nd ed. (Champaign, IL: Human Kinetics), 213. *(b)* Reprinted, by permission, from R.S. Behnke, 2005, *Kinetic anatomy*, 2nd ed. (Champaign, IL: Human Kinetics), 211.

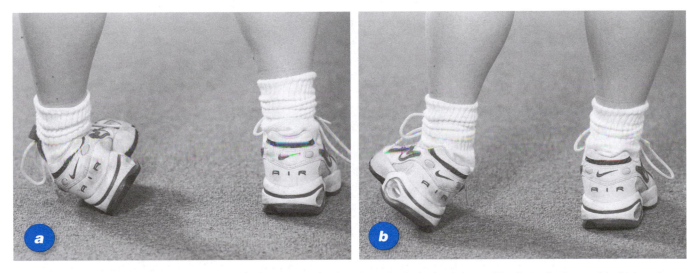

Figure 6.32 Foot and ankle movement. *(a)* Supination (combined motions of subtalar inversion, ankle plantar flexion, and foot internal rotation). *(b)* Pronation (combined motions of subtalar eversion, ankle dorsiflexion, and foot external rotation).

The determining factors in ankle injury, as in most injuries, are the joint position at the time of injury; the magnitude, direction, and rate of applied forces; and the resistance provided by joint structures. The joint motions commonly involved in ankle–foot injuries are precipitated by walking on uneven surfaces, stepping in holes, rolling the ankle during a cutting maneuver, or landing on another player's foot when descend-

ing from a jump in sporting events (figure 6.33). Resulting injuries range from fracture–dislocation to ligamentous damage (sprain).

The vast majority (85%) of ankle sprains result from what are termed *inversion injuries*. According to our nomenclature, the mechanism is actually **supination** (i.e., a combination of ankle plantar flexion, subtalar inversion, and internal rotation of the foot in which the longitudinal midline of

Figure 6.33 Common mechanism of ankle injury.

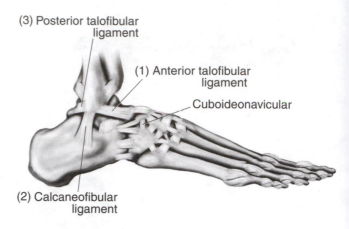

Figure 6.34 Sequential failure pattern of the lateral ankle ligaments. Typically the ATFL (1) fails first, followed by the CFL (2) and the PTFL (3).

Adapted, by permission, from R.S. Behnke, 2005, *Kinetic anatomy*, 2nd ed. (Champaign, IL: Human Kinetics), 213.

the foot deviates, or rotates, medially). The term **inversion sprain** is so entrenched in the literature, however, that its extinction seems unlikely. An alternative term, **lateral ankle sprain,** may be more appropriate.

In most cases there is an orderly sequence of ligament failure (figure 6.34). The anterior talofibular ligament (ATFL) fails first because of its orientation at the instant of loading and its inherent weakness (Siegler et al. 1988). When the ankle assumes a plantar flexed position (as it does in ankle–foot supination), the ATFL aligns with the fibula and functions as a collateral ligament (Carr 2003). This alignment, taken together with the ATFL's relative weakness, predisposes the ATFL to injury.

The calcaneofibular ligament (CFL) is next injured, followed by rare failure of the posterior talofibular ligament (PTFL). When the ankle is dorsiflexed, the CFL aligns with the fibula and provides collateral reinforcement. About 65% to 70% of ligament injuries at the ankle involve isolated ATFL damage. Isolated CFL injuries are rare; the CFL is most often injured in combination with the ATFL (20%) (Peterson and Renström 2001).

An interesting structural relation between the ATFL and CFL was described more than three decades ago by Inman (1976). He described considerable variation (70-140°) in the angle between the ATFL and CFL (figure 6.35). Inman hypothesized that a larger angle may be associated with lateral ankle joint laxity and possibly a greater injury risk.

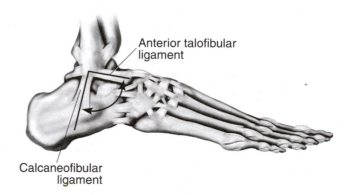

Figure 6.35 Angle between the anterior talofibular ligament (ATFL) and calcaneofibular ligament (CFL).

Adapted, by permission, from R.S. Behnke, 2005, *Kinetic anatomy*, 2nd ed. (Champaign, IL: Human Kinetics), 213.

Occasionally, the anterior portion of the deltoid ligament (DL) is injured during an inversion injury. At first glance, this may appear incongruous. Why would a medial structure incur damage from forcible inversion? The answer lies in the complexity of joint action, specifically that the anterior portion of the DL is taut in ankle plantar flexion. Because the ankle is plantar flexed at the time of injury, the anterior portion of the deltoid ligament becomes a candidate for injury. As an inherently strong ligament, however, the DL is rarely injured in so-called inversion sprains. The opposite movement pattern creates **eversion sprain (pronation** by our definition); the injury mechanism involves ankle dorsiflexion, subtalar eversion, and lateral rotation of the foot. An alternative term, medial ankle sprain, may be more appropriate.

Given the inherent strength of the medial collateral (deltoid) ligament group, injuries resulting from this mechanism are both less frequent (about 5%) and less severe. In this mechanism the talus is forced against the lateral malleolus. Because the lateral malleolus is longer and thinner than the medial malleolus, the talus cannot rotate over the lateral malleolus. This may result in malleolar fracture. Rupture of the deltoid ligament may occur, although this is rare and is always seen in conjunction with other ligament tears.

In some cases applied loads drive the fibula away from the tibia with sufficient force to tear the syndesmosis (interosseous membrane and tibiofibular ligaments) in what is termed a high ankle sprain. Likely mechanisms for a high ankle sprain include talar torsion and forced ankle dorsiflexion. Unrecognized and untreated syndesmotic injuries may lead to chronic instability and ankle arthrosis.

Calcaneal Tendon Pathologies

Ever since the Greek warrior Achilles was felled by an arrow judiciously aimed at his unprotected heel, the calcaneal region has been associated with susceptibility to injury.

The calcaneal (Achilles) tendon, the largest and strongest tendon in the body, is formed by merging of the distal tendons of the gastrocnemius and soleus about 5 to 6 cm proximal to its insertion site on the posterior surface of calcaneus. At the insertion site, tendon width varies from 1.2 to 2.5 cm. Approximately 12 to 15 cm proximal to the insertion, the calcaneal tendon begins to spiral, twisting about 90° as it approaches its calcaneal insertion (Schepsis et al. 2002).

Frequent and repeated loading of the calcaneal tendon predisposes it to overuse pathologies, most commonly peritenonitis (inflammation of the peritenon), insertional disturbances (e.g., bursitis or insertion tendinitis), myotendinous junction injury, or tendonopathies (Kvist 1994).

The calcaneal tendon transmits substantial loads from the triceps surae muscle group (gastrocnemius and soleus) to its attachment on the posterior calcaneus. A sample of studies confirms the high loads transmitted by the calcaneal tendon:

- Burdett (1982), using a biomechanical model, estimated peak Achilles tendon forces ranging from 5.3 to 10.0 times body weight during the stance phase of running.

- Fukashiro and colleagues (1995), using an implanted tendon force transducer, reported peak Achilles tendon force of 2233 N (502 lb) in the squat jump, 1895 N (426 lb) in the countermovement jump, and 3786 N (851 lb) in hopping.

- Giddings and colleagues (2000), using experimental data and a quantitative model, predicted maximal Achilles tendon force 3.9 times body weight for walking and 7.7 times body weight for running, with the peak loads at 70% of the stance phase for walking and 60% of stance for running.

- Bogey and colleagues (2005), using an electromyograph-to-force processing technique, estimated peak Achilles tendon force of 2.9 kN (652 lb) during gait.

- Pourcelot and colleagues (2005), using a noninvasive ultrasonic technique, found peak tendon forces of about 850 N (191 lb) during the stance phase of walking.

Although the calcaneal tendon clearly is subjected to high magnitude loads across a spectrum of activities, Komi and colleagues (1992) suggested that the loading *rate* may be more clinically relevant than the loading *magnitude*.

Four primary mechanisms have been implicated in calcaneal tendon rupture (Mahan and Carter 1992; figure 6.36): (1) sudden dorsiflexion of a plantar flexed foot (e.g., a football quarterback dropping back and planting his rear foot as he throws), (2) pushing off the weight-bearing foot while extending the ipsilateral knee joint (e.g., a basketball player executing a rapid change of direction), (3) sudden excess tension on an already taut tendon (e.g., catching a heavy weight), and (4) a taut tendon struck by a blunt object (e.g., baseball bat). These mechanisms suggest, and epidemiological evidence confirms, that most calcaneal tendon ruptures are unilateral. Although rare, bilateral Achilles tendon ruptures have been reported (Garneti et al. 2005).

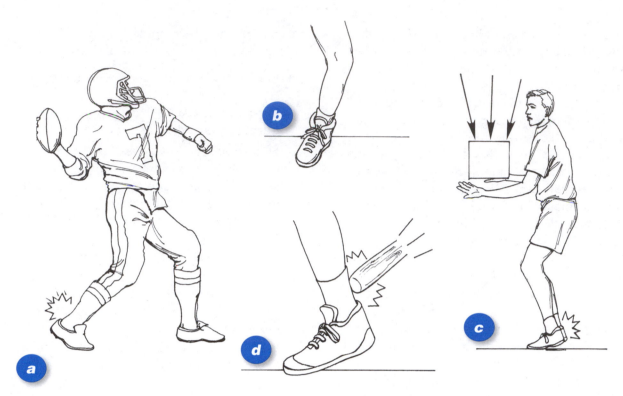

Figure 6.36 Mechanisms of calcaneal (Achilles) tendon rupture. *(a)* Rapid dorsiflexion of the ankle by a football quarterback. *(b)* Cutting maneuver with rapid change of direction. *(c)* Catching a falling weight. *(d)* Blunt trauma to a taut tendon.

The cause of calcaneal tendonopathies is multi-factorial; contributing factors include training errors, running terrain, malalignments (e.g., combined flat foot and excessive pronation) and biomechanical faults, improper footwear, trauma, age, gender (men are at five to six times greater risk of rupture than women), anthropometrics, environment, and psychomotor factors. The tendon also may be weakened, and thus put at even greater risk, by systemic diseases, steroids, and fluoroquinone antibiotics (Casparian et al. 2000; Maffulli and Wong 2003; Vanek et al. 2003). Nonuniform tendon stresses attributable to individual muscle contributions may also contribute to injury risk (Arndt et al. 1998).

Tendon degeneration may eventually lead to complete tendon rupture. Calcaneal tendon ruptures typically happen in sedentary, 30- to 40-year-old men who suddenly exert themselves in a sporting task that involves running, jumping, or rapid change of direction (Jarvinen et al. 2001; Schepsis et al. 2002; Yinger et al. 2002). In many instances these spontaneous tendon ruptures seem to "just happen." Postinjury assessment, however, shows evidence of degeneration in the substance of the ruptured tendon. Tendon rupture thus often seems to be secondary to degenerative

processes rather than a primary injury that happens spontaneously.

Calcaneal tendon rupture usually occurs about 2 to 6 cm proximal to the calcaneal insertion in a region known to be **hypovascular.** This fact, combined with decreased blood flow associated with age, helps explain the frequency of rupture in middle-aged people. The calcaneal tendon has a poor blood supply that previously was thought not to vary along its length (Ahmed et al. 1998). A recent study using a new method, however, reported an avascular region close to the calcaneal insertion site and regional differences in vascular density (Zantop et al. 2003) along the tendon length. The middle part of the calcaneal tendon had a much lower vascular density (28.2 vessels/cm²) than either the proximal part (73.4 vessels/cm²) or the distal part (56.6 vessels/cm²). The authors identified this reduced vascularization (and resulting **hypoxia)** as a predisposing factor for calcaneal tendon degeneration and eventual rupture.

As a side note of interest, there may be a relation between blood type and increased incidence of tendon rupture. Persons with type O blood seem to be more likely to suffer from tendon rupture in general (Jozsa et al. 1989) and calcaneal tendon rupture in particular (Kujala et al. 1992) compared with people who have other blood types, suggesting a genetic link between one's ABO blood group and the molecular structure of tendon tissue. Another study, however, found no significant relation between the proportions of ABO blood groups and Achilles tendon rupture (Maffulli et al. 2000). Maffulli and colleagues concluded that the association between blood group and tendon rupture may be attributable to differences in blood group distribution in genetically segregated populations.

Predisposition to rupture may also be affected by collagen type. Eriksen and colleagues (2002) found type III collagen accumulation at the rupture site, likely attributable to microtrauma and healing events. Increased type III collagen content may contribute to lower tendon tensile strength and enhanced rupture risk.

> "In summary, the theoretical explanations for Achilles tendon ruptures and other chronic pathology suggest a sequence of events that is initiated with an intrinsic tendon pathology associated with disuse, age-related tendon change, and hypovascularity, resulting in localized degeneration and tendon weakening. This decreases the tendon's threshold to rupture. The precise proprioceptive and pathomechanical position and load that causes the injury remains obscure. It is probably a complex equation of neuromuscular control and endocrine factors that result in the Achilles tendon rupture."
>
> (Yinger et al. 2002, p. 234)

Plantar Fasciitis

Plantar fasciitis (PF) has been described as an inflammatory condition of the plantar fascia in the midfoot or at its insertion on the medial tuberosity of the calcaneus that involves microtears or partial rupture of fascial fibers. Once again we encounter a catch-all term, *plantar fasciitis*, which has become entrenched in the literature as a descriptor of pain in the plantar area of the posterior foot. The more appropriate nonspecific designation is *heel pain syndrome*, with *plantar fasciitis* reserved for inflammation to the plantar fascia alone.

In most cases, PF develops in response to repeated loading (e.g., running) in which compressive forces flatten the longitudinal arch of the foot. Forces in the plantar fascia during running have been estimated to be 1.3 to 2.9 times body weight (Scott and Winter 1990). This flattening of the arches stretches the fascia and absorbs the load in much the same way as a leaf spring bends to accommodate heavy weights, in what is termed a **truss mechanism** (figure 6.37*a*). Extension of the toes puts added stress on the structures by way of a **windlass mechanism,** as depicted in figure 6.37*b*.

Plantar fasciitis is hastened or worsened by lack of flexibility. Tightness of the calcaneal tendon, for example, limits ankle dorsiflexion and results in greater plantar fascial stress. Ankle strength

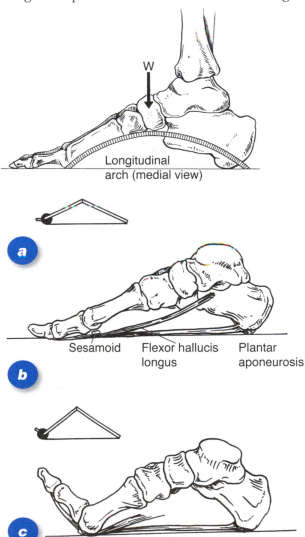

Figure 6.37 *(a and b)* Truss mechanism in which body weight compresses the longitudinal arch. *(c)* Windlass mechanism in which hyperextension of the toes increases tension on the plantar structures.

and flexibility deficits have been observed in the symptomatic limbs compared with the unaffected limbs and with an asymptomatic control group (Kibler et al. 1991).

In addition to strength and flexibility, other factors are associated with PF, including overtraining, leg length discrepancies, fatigue, fascial inextensibility, and poor movement mechanics. Excessive pronation during running provides a good example of how a pathological movement pattern contributes to PF. During pronation the subtalar joint everts, causing plantar fascial elongation and increased tissue stress. Repetition of this pathological loading leads to microdamage and attendant inflammation. Although overpronation has been associated with PF, there is not a clear link between PF and **pes cavus** (high foot arch) or **pes planus** (flat foot) (Peterson and Renström 2001). The cause of plantar fasciitis remains unclear.

Toe Injuries

Jacques Lisfranc, a field surgeon in Napoleon's army, described amputation through the tarsometatarsal joint of a gangrenous foot (Vuori and Aro 1993). Although his description did not include reference to fracture–dislocation of the joint, his name is now given to these injuries to the tarsometatarsal region. The circumstances of Lisfranc joint injury vary and include both low-energy injuries (e.g., tripping or stumbling) and high-energy trauma (e.g., fall from a height, direct crush, vehicular crash). Several mechanisms have been suggested to explain Lisfranc fracture–dislocations. One relatively uncommon mechanism is direct force, as when a heavy object is dropped on the foot. Direct force applied to the metatarsal pushes the bone down and causes plantar dislocation and possible accompanying fracture. Force applied proximal to the tarsometatarsal joint results in dorsal dislocation.

A second mechanism involves axial loading of the region when indirect forces (e.g., ground reaction force) are applied to a foot in extreme plantar flexion. This occurs when a person is in a tiptoe position at the instant of load application. A similar loading may occur in dorsiflexion as well. In both cases the metatarsal is forcibly pushed out of joint. Such injury is typically accompanied by capsular rupture and metatarsal fracture.

Violent abduction, induced by a twisting mechanism, is another cause of Lisfranc injury. This is classically illustrated by an equestrian injury in which the rider's foot is fixed in the stirrup while the rider falls. The force of the fall pushes the metatarsals into extreme abduction.

Although Lisfranc fracture–dislocations have instructive value in demonstrating mechanisms of injury, their incidence is quite low. A review of nearly 700 cases of metatarsal fracture found that less than 10% involved Lisfranc joint injuries (Vuori and Aro 1993).

Other foot and toe injuries are more prevalent. Among these is **turf toe,** an injury involving damage to the capsuloligamentous structures of the first metatarsophalangeal (MP) joint. Multiple mechanisms have been implicated in turf toe injuries, with hyperextension the most common (Allen et al. 2004). This injury typically occurs when the foot is planted on the ground with the first MP joint in extension. A load, such as another player falling on the foot, forces the joint into hyperextension and damages joint structures (figure 6.38). Much less frequently, turf toe results from a hyperflexion mechanism. Turf toe also may happen secondarily in response to excessive valgus and varus loading of the first MP joint. Although once thought to be a relatively minor injury, turf toe is now recognized as a condition with significant short-term effects and potentially serious long-term consequences.

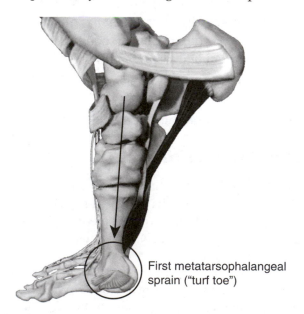

First metatarsophalangeal sprain ("turf toe")

Figure 6.38 Turf toe injury caused by hyperextension of the hallux (big toe) with simultaneous compressive loading.
Copyright Primal Pictures Ltd.

■ CHAPTER REVIEW

Key Points

- Lower-extremity injuries are common and affect posture, walking, running, and other load-bearing tasks.

- Increased participation in competitive and recreational sports, especially by girls and women, in recent decades has contributed to the increased incidence of lower-extremity injuries.

- Injury risk is multifactorial.

- Demographic and sociological changes logically predict future increases in certain lower-extremity injuries (e.g., hip fractures in older persons).

- Injury prevention programs have proven effective in reducing the incidence of certain injuries (e.g., ACL ruptures).

Questions to Consider

1. This chapter's "A Closer Look" examined anterior cruciate ligament injury in detail. Select another injury presented in the text and write your own "A Closer Look" for that injury.

2. Explain, using specific examples, how lower-extremity injuries may affect posture, walking, running, and other load-bearing tasks.

3. Select a lower-extremity injury described in the text and explain how and why it might be considered a multifactorial problem.

4. The text describes how predicted demographic changes will likely increase the incidence of hip fracture. Select another injury and explain how demographic changes may predict future increases (or decreases) in risk for the injury you have selected.

Suggested Readings

Baker, S.P. 1995. Fall injuries in the elderly. *Clinics in Geriatric Medicine* 1: 501-511.

Baxter, D.E. 1995. *The Foot and Ankle in Sport*. St. Louis: Mosby-Year Book.

Biedert, R.M., ed. 2004. *Patellofemoral Disorders: Diagnosis and Treatment*. West Sussex, UK: Wiley.

Browner, B.D., J.B. Jupiter, A.M. Levine, and P.G. Trafton. 2003. *Skeletal Trauma* (3rd ed.). Philadelphia: Saunders.

Bucholz, R.W., ed. 2005. *Rockwood, Green, and Wilkins' Fractures* (6th ed). Philadelphia: Lippincott Williams & Wilkins.

Bulstrode, C., J. Buckwalter, A. Carr, L. Marsh, J. Fairbank, J. Wilson-MacDonald, and G. Bowden, eds. 2002. *Oxford Textbook of Orthopedics and Trauma*. Oxford, UK: Oxford University Press.

Daniel, D.M., J.J. O'Connor, W.H. Akeson, and R.A. Pedowitz, eds. 2003. *Daniel's Knee Injuries: Ligament and Cartilage Structure, Function, Injury, and Repair*. Philadelphia: Lippincott Williams & Wilkins.

Fanelli, G.C., ed. 2001. *Posterior Cruciate Ligament Injuries: A Practical Guide to Management*. New York: Springer.

Finerman, G.A.M., and F.R. Noyes. 1992. *Biology and Biomechanics of the Traumatized Synovial Joint: The Knee as a Model*. Rosemont, IL: American Academy of Orthopaedic Surgeons.

Fu, F.H., and D.A. Stone. 2001. *Sports Injuries: Mechanisms, Prevention, Treatment* (2nd ed.). Philadelphia: Lippincott Williams & Wilkins.

Marder, R.A., and G.J. Lian. 1997. *Sports Injuries of the Ankle and Foot*. New York: Springer.

Moore, E.E., D.V. Feliciano, and K.L. Mattox, eds. 2003. *Trauma* (5th ed.). New York: McGraw-Hill.

Nicholas, J.A., and E.B. Hershman, eds. 1995. *The Lower Extremity and Spine in Sports Medicine*. St. Louis: Mosby-Year Book.

Scott, W.N. 1991. *Ligament and Extensor Mechanism Injuries of the Knee: Diagnosis and Treatment*. St. Louis: Mosby-Year Book.

Steinberg, M.E., ed. 1991. *The Hip and Its Disorders*. Philadelphia: Saunders.

Wolman, R., D. Singh, J. Brodsky, S. Costain, A. Betts, A. Saifuddin, and V. Mahadevan, eds. 2002. *Sports Injuries: Foot, Ankle & Lower Leg*. London: Primal Pictures.

Woo, S.L.-Y., and J.A. Buckwalter. eds. 1988. *Injury and Repair of the Musculoskeletal Soft Tissues*. Park Ridge, IL: American Academy of Orthopaedic Surgeons.

CHAPTER 7

UPPER-EXTREMITY INJURIES

There's nothing wrong with his shoulder
except some pain—and pain don't hurt you.

Sparky Anderson, former Detroit Tigers manager

■ OBJECTIVES

■ To describe the relevant upper-extremity anatomy involved in musculoskeletal injury

■ To identify and explain the mechanisms involved in musculoskeletal injuries to the major joints (shoulder, elbow, wrist, and fingers) and segments (upper arm, forearm, and hand) of the upper extremities

A day rarely passes without media reports of some notable musculoskeletal injury. Headlines such as "Child Injures Hand Severely in Fireworks Accident" or "High Incidence of Carpal Tunnel Syndrome Found in Keyboard Operators" or "World Series Pitcher Suffers Rotator Cuff Tear" are all too common. Upper-extremity injuries are of special concern because they impair one's ability to manipulate the environment. Even simple tasks such as opening a jar or putting a key in a lock become difficult for a person with impaired dexterity. Significant injury to the shoulder, elbow, wrist, or fingers can end a career or mandate a change of occupation or recreational involvement.

Effective diagnosis, treatment, and rehabilitation of upper-extremity injuries depend on a sound understanding of injury mechanisms. Only when the causal relations between applied forces and resultant injury are established and understood can appropriate programs of intervention and prevention be designed and implemented.

As was done for the ACL in chapter 6 on lower-extremity injuries, we present in this chapter a detailed account of upper-extremity injuries. Glenohumeral impingement and rotator cuff injuries are presented in the section titled Rotator Cuff Pathologies: A Closer Look. In this section (p. 212) we expand our discussion beyond the mechanisms of injury to include detailed description of shoulder complex structure and tissue mechanics and explanation of clinical evaluation, treatment, and rehabilitation of rotator cuff pathologies.

Shoulder Injuries

The shoulder (or pectoral) girdle contains two bones: the scapula and clavicle. The clavicle attaches medially to the sternal manubrium (ster-noclavicular joint) and laterally to the acromion process of the scapula (acromioclavicular joint) (figure 7.1). The acromioclavicular (AC) joint is a plane (gliding) synovial joint with articular surfaces separated by an articular disc. The acromioclavicular ligament supports the AC joint superiorly, with inferior support provided by the coracoclavicular ligament.

The humerus of the upper arm articulates with the scapula at the glenohumeral (GH) joint (also called shoulder joint), the body's most mobile joint, where the humeral head fits loosely into the shallow glenoid fossa of the scapula. The glenoid labrum attaches to the rim of the glenoid fossa and improves the joint's bony fit. Two ligaments strengthen the GH joint: the glenohumeral ligament (a thickening of the anterior joint capsule) and the coracohumeral ligament, which anchors the humerus to the coracoid process of the scapula.

The GH joint's ball-and-socket structure permits triplanar motions referred to as flexion–extension (sagittal plane), abduction–adduction (frontal plane), and internal–external rotation (transverse plane). The muscles responsible for producing and controlling movements at the GH joint are shown in figure 7.2 and summarized in table 7.1. Among the most important (and often injured) GH muscles are the four muscles of the rotator cuff group (subscapularis, supraspinatus, infraspinatus, and teres minor). These muscles assist in stabilizing the GH joint by forming a cuff around the humeral head and pulling the humerus into the glenoid fossa.

Significant injuries to the shoulder include acromioclavicular sprain, glenohumeral instability and dislocation, biceps tendinitis, impingement syndrome, rotator cuff rupture, and labral pathologies. These injuries often are associated with specific sports (table 7.2) and often are position dependent (Perry and Higgins 2001).

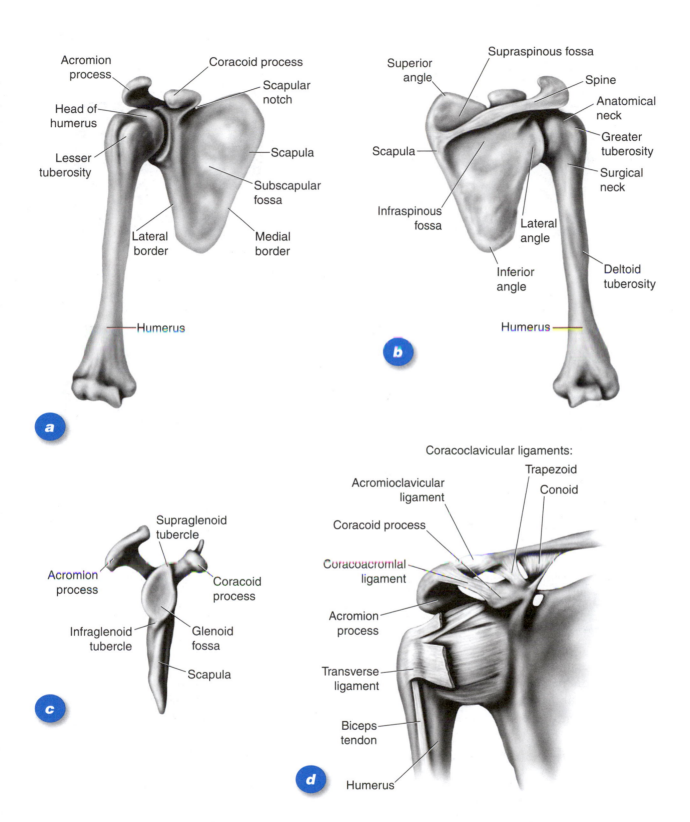

Figure 7.1 Bones and ligaments of the shoulder girdle. *(a)* Anterior view. *(b)* Posterior view. *(c)* Lateral view. *(d)* Shoulder ligaments.

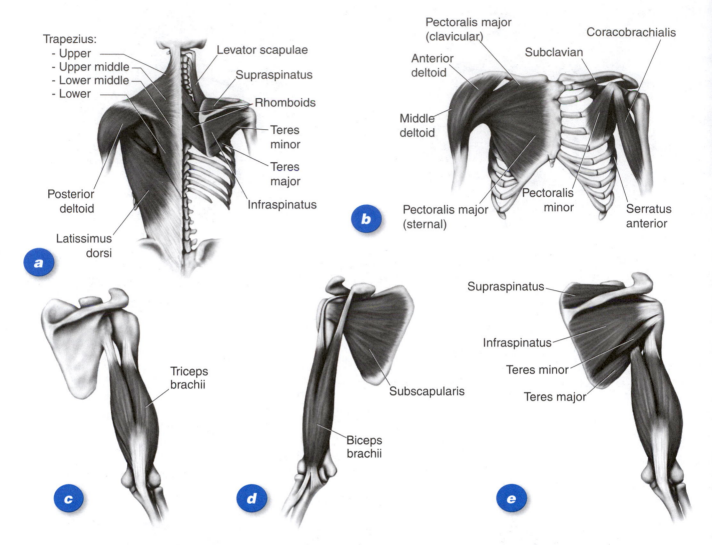

Figure 7.2 Musculature of the shoulder: *(a)* Posterior muscles. *(b)* Anterior muscles. *(c)* Triceps brachii. *(d)* Biceps brachii and subscapularis. *(e)* Posterior view of supraspinatus, infraspinatus, teres minor, and teres major.

(a, b) Reprinted, by permission, from R.S. Behnke, 2005, *Kinetic anatomy*, 2nd ed. (Champaign, IL: Human Kinetics), 47. *(c, d)* Reprinted, by permission, from R.S. Behnke, 2005, *Kinetic anatomy*, 2nd ed. (Champaign, IL: Human Kinetics), 50. *(e)* Reprinted, by permission, from R.S. Behnke, 2005, *Kinetic anatomy*, 2nd ed. (Champaign, IL: Human Kinetics), 51.

For example, shoulder injuries are more common in baseball pitchers and swimmers competing in the butterfly and freestyle events than in other baseball players and swimmers.

Acromioclavicular Sprain

Acromioclavicular (AC) sprain results from applied forces that tend to displace the scapular acromion process from the distal end of the clavicle. This injury is commonly referred to as a *separated shoulder* or **shoulder separation** and should not be confused with a shoulder dislocation (see next section). The synovial AC joint is classified as a plane-type joint and contains an intra-articular disc that normally degenerates with age (Horvath and Kery 1984). Superior and inferior AC ligaments provide horizontal stability, with vertical stability provided by the coracoclavicular ligaments.

Acromioclavicular injury results from either direct or indirect forces. Direct force applied to the point of the shoulder with the arm in an adducted position is the most common cause of AC injury (Buss and Watts 2003). This mechanism is seen when a person collides with a solid object or surface (figure 7.3*a*). The impact force drives the acromion inferiorly relative to the clavicle. In the absence of fracture, increasing force levels cause AC injury progression as follows: (1) mild sprain

TABLE 7.1

MUSCLES OF THE SHOULDER

Muscle	Action
Biceps brachii	Flexes the arm
Coracobrachialis	Flexes and adducts the arm
Deltoid	Abducts the arm; posterior fibers extend and laterally (externally) rotate the arm; anterior fibers flex and medially (internally) rotate the arm
Infraspinatus[a]	Laterally rotates and slightly adducts the arm
Latissimus dorsi	Adducts, extends, and medially rotates the arm
Pectoralis major	Adducts, flexes, and medially rotates the arm
Subscapularis[a]	Medially rotates the arm
Supraspinatus[a]	Abducts the arm
Teres major	Adducts, extends, and medially rotates the arm
Teres minor[a]	Laterally rotates, slightly adducts, and extends the arm
Triceps brachii (long head)	Extends the arm

[a]Muscles included in the rotator cuff group.

TABLE 7.2

INCIDENCE OF SPORT-SPECIFIC SHOULDER INJURIES

Sport	Percent injury	Typical injuries seen involving the shoulder
Baseball	11-57	Acromioclavicular joint injuries, impingement syndrome, rotator cuff tendinitis
Wrestling	17	Glenohumeral subluxation, acromioclavicular joint injuries, glenohumeral dislocation
Tennis	56	Rotator cuff tendinitis, impingement syndrome
Football	8-14	Acromioclavicular joint injuries, glenohumeral instability
Gymnastics	1-18	Impingement syndrome, biceps tendinitis, glenohumeral instability
Swimming	3-50	Impingement syndrome, glenohumeral instability
Golf	3-13	Impingement syndrome
Basketball	3	Glenohumeral subluxation
Skiing	6-9	Glenohumeral dislocation, glenohumeral subluxation, acromioclavicular joint injuries
Volleyball	44	Biceps tendinitis, impingement syndrome
Javelin	29	Biceps tendinitis, impingement syndrome

Reprinted, by permission, from J.J. Perry and L.D. Higgins, 2001, Shoulder injuries. In *Sports injuries: Mechanisms, prevention, treatment*, edited by F.H. Fu & D.A. Stone (Baltimore, MD: Lippincott, Williams & Wilkins), 1015.

Figure 7.3 Mechanisms of acromioclavicular sprain (separated shoulder). *(a)* Collision with an object or surface. *(b)* Falling on an outstretched arm.

Adapted by permission, from C.A. Rockwood Jr., G.A. Williams and D. Christopher Young, 1996, Injuries to the acromioclavicular joint. In *Rockwood and Green s fractures in adults*, 4th ed., edited by C.A. Rockwood et al. (Philadelphia, PA: Lippincott, Williams & Wilkins), 1351.

of the AC ligament, (2) moderate AC ligament sprain with coracoclavicular ligament involvement, and (3) complete AC dislocation with tearing of clavicular attachments of the deltoid and trapezius muscles and complete rupture of the coracoclavicular ligament.

Less frequently, AC injuries result from indirect forces, as when a person falls on an outstretched arm (figure 7.3*b*). In this mechanism, contact forces are transmitted up the arm, through the humerus, and to the acromion. These superiorly directed loads force separation of the acromion and clavicle. On rare occasions, extreme traction forces applied to the arm may separate the acromion from its clavicular attachment (Rockwood et al. 1996).

AC injuries are classified as six types of AC sprain and dislocation (Williams et al. 1989). The mechanism and resulting injuries for each of the six types are illustrated in figure 7.4. The first three types (types I-III), originally described by Tossey and colleagues (1963) and Allman (1967), are most common, accounting for nearly 98% of all AC injuries (Lambert and Hertel 2002). Types IV through VI are very rare.

The six types of AC injury are characterized as follows (Peterson and Renström 2001):

■ Type I: Isolated sprain of the acromioclavicular ligament, with pain over the AC joint; minimal pain with shoulder motion; mild tenderness

■ Type II: Widening of the AC joint with elevation of the distal end of the clavicle with disruption of acromioclavicular ligament; moderate to severe pain; limited shoulder motion

■ Type III: Dislocation of AC joint with superior displacement of clavicle; disruption of coracoclavicular ligaments; widened coracoclavicular space; moderate to severe pain; upper extremity depressed with possible free-floating clavicle

■ Type IV: Dislocated AC joint, with posterior displacement of clavicle into or through trapezius muscle; complete disruption of coracoclavicular ligament; clinically similar to type III but with greater pain and posterior clavicular displacement

■ Type V: Disruption of acromioclavicular and coracoclavicular ligaments; gross displacement of AC joint; clinically similar to type III but with more pain and displacement

■ Type VI: Disruption of acromioclavicular and coracoclavicular ligaments; inferior displacement of clavicle into subacromial or subcoracoid position

FROZEN SHOULDER

In 1872, French physician E.S. Duplay described a "peri-arthritis" characterized by shoulder stiffness and limited joint movement (Duplay 1872). Sixty-two years later, E.A. Codman (1934) coined the term *frozen shoulder* to describe the same condition and its attendant pain and reduced external rotation and abduction.

Neviaser (1945) identified the condition as ***adhesive capsulitis,*** a term still sometimes used to describe the condition. Frozen shoulder is an idiopathic condition involving shoulder stiffness and pain affecting the anterior joint capsule and the rotator cuff interval (between the subscapularis and supraspinatus). Recent magnetic resonance arthrographic assessment points to a thickening of the coracohumeral ligament in the rotator cuff interval (Mengiardi et al. 2004).

The three primary symptoms of frozen shoulder are insidious shoulder stiffness, severe pain, and loss of both passive and active external rotation (Dias et al. 2005). Frozen shoulder most commonly afflicts persons 40 to 70 years of age and is more common in women than men.

Although the cause of frozen shoulder is unknown, several factors have been associated with the condition. These include rotator cuff pathology, diabetes, thyroid and autoimmune disease, cervical spine disease, trauma, chest disease, or hyperlipidemia (Clasper 2002).

Frozen shoulder typically resolves in 1 to 3 years but not always completely. Three clinical stages have been described (Clasper 2002; Dias et al. 2005):

- Stage 1: Painful (freezing) stage—gradual pain onset, especially at night. Stiffness results in less arm use.
- Stage 2: Stiffening (adhesive) stage—pain reduction with residual stiffness. Reduced joint movements, especially shoulder external rotation.
- Stage 3: Thawing (resolution) stage—improved joint range of motion.

Despite progress in the diagnosis and treatment of frozen shoulder, much of what Codman said in 1934 still holds true, "This is a class of cases that I find difficult to define, difficult to treat, and difficult to explain from the point of view of pathology" (Codman 1934).

Glenohumeral (Shoulder) Instability and Dislocation

The ability of any joint to resist dislocation is directly related to its inherent stability. What the shoulder gains in mobility, it sacrifices in stability. As discussed in chapter 6, joints such as the hip, with good bony fit and extensive surrounding musculature, rarely dislocate. The glenohumeral joint, in contrast, is prone to dislocation (luxation) because of its poor bony fit and limited supporting musculature. The shallowness of the glenoid fossa and limited contact area between the fossa and the humeral head contribute to the joint's instability.

The glenoid labrum improves the joint fit to a limited extent by increasing surface area and deepening the fossa, but nonetheless the glenohumeral joint is arguably the least stable articulation in the body—a dubious distinction substantiated by its frequent dislocation. The factors contributing to its stability, therefore, must be understood to discuss adequately the mechanisms of glenohumeral luxation.

Stabilizing factors are classified as active (dynamic) or passive (static). **Active stabilization** is provided by muscles surrounding and acting at the glenohumeral joint. These include the deltoid, trapezius, latissimus dorsi, pectoralis major, and muscles of the rotator cuff group (subscapularis,

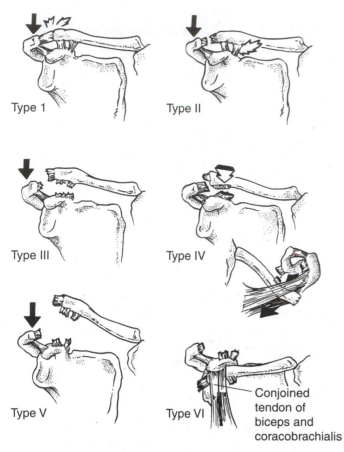

Figure 7.4 Types (I-VI) of acromioclavicular joint injury.
Reprinted, by permission, from C.A. Rockwood, G.R. Williams and D. Christopher Young, 1996, Injuries to the acromioclavicular joint. In *Rockwood and Green s fractures in adults*, 4th ed., edited by C.A. Rockwood et al. (Philadelphia, PA: Lippincott, Williams & Wilkins), 1354.

supraspinatus, infraspinatus, and teres minor). In midrange positions, shoulder muscles act as powerful glenohumeral stabilizers. At end-range positions, however, certain muscles (e.g., deltoid and pectoralis major) may contribute to glenohumeral instability (Labriola et al. 2005).

Passive stabilization is provided by the joint capsule and supporting ligaments. At the extremes of joint movement, tension in the capsuloligamentous structures provides resistance to dislocation. The laxity in these structures necessary for the exceptional movements at the glenohumeral joint precludes their involvement as stabilizers throughout normal ranges of motion. During normal ranges, other stabilizing mechanisms are necessary. These include the combined effects of negative intracapsular pressure and the mechanisms of concavity compression and scapulohumeral balance.

In a normal (i.e., undamaged capsule) glenohumeral joint, a small negative intracapsular

pressure helps stabilize the joint (Speer 1995). Although not especially large, this force (90-140 N) nonetheless contributes to maintaining glenohumeral stability throughout its range of motion.

The mechanisms of concavity compression and scapulohumeral balance contribute significantly to joint stabilization. **Concavity compression** refers to the stability created when a convex object is pressed into a concave surface (Lippitt and Matsen 1993). When the surfaces are pressed together, there is greater resistance to translational movement between the surfaces. At the surface between the humeral head and the glenoid fossa, numerous muscle forces increase the articular pressure and stabilize the joint. Translational resistance at the glenohumeral articulation is greater in the superior–inferior direction than in the anterior–posterior direction and increases with greater compressive loads (table 7.3). Despite the resistance, translation occurs with combined abduction, extension, and external rotation of the glenohumeral joint.

Harryman and colleagues (1990), using a three-dimensional position sensor and force and torque transducers, reported translation of the humeral head with passive glenohumeral motion. They found significant anterior translation during glenohumeral flexion and cross-body movement and posterior translation with extension and external rotation. Their results have clinical relevance because they indicate that the glenohumeral joint does not function purely as a ball-and-socket mechanism and that even passive manipulation of the joint causes significant translation of the humeral head across the glenoid fossa.

Scapulohumeral balance refers to the coordinated muscle action that maintains the net joint reaction force within the fossa. As seen in figure 7.5, when the reaction–force line of action is directed into the glenoid, the joint is stable. The joint becomes unstable as the line of action moves away from the geometric center of the glenoid and beyond the surface boundary. Responsibility for maintaining appropriate glenohumeral congruity lies most immediately with the rotator cuff muscles and secondarily with the deltoid, trapezius, serratus anterior, rhomboids, latissimus dorsi, and levator scapulae. Fatigue in any of these muscles (e.g., from repeated throwing)

TABLE 7.3

EFFECT OF INCREASING COMPRESSIVE LOAD
ON GLENOHUMERAL JOINT STABILITY

Direction	Compressive load, N	Maximum translating force, N
Superior: 0°	50	29 ± 7
	100	51 ± 9
Anterior: 90°	50	17 ± 6
	100	29 ± 5
Inferior: 180°	50	32 ± 4
	100	56 ± 12
Posterior: 270°	50	17 ± 6
	100	30 ± 12

Adapted, by permission, from S. Lippitt and F. Matset, 1993,"Mechanisms of glenohumeral joint stability,"*Clinical Orthopaedics and Related Research* 291: 24.

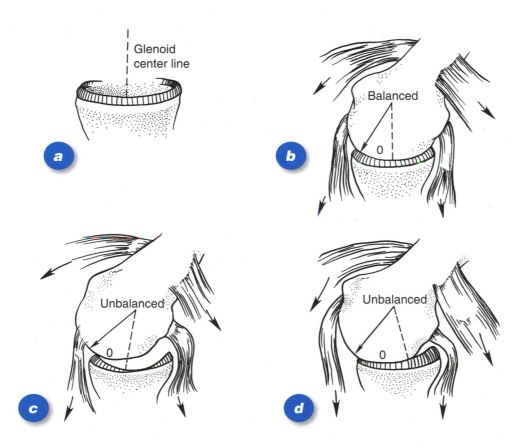

Figure 7.5 *(a)* The glenoid center line is defined as a perpendicular to the midpoint of the glenoid fossa. *(b)* The glenohumeral joint is stable in positions where the net joint reaction force is balanced within the glenoid fossa. *(c)* The glenohumeral joint is unstable in positions where the net joint reaction force is not balanced within the glenoid fossa. *(d)* Abnormal orientation of the glenoid fossa can contribute to an unbalanced and therefore unstable joint. *O* represents the angle between the glenoid center line and the net joint reaction force.

Reprinted, by permission, from S. Lippitt and F. Matset, 1993, "Mechanisms of glenohumeral joint stability," *Clinical Orthopaedics and Related Research* 291: 26.

compromises the compensatory capability of the musculoskeletal complex at the shoulder and predictably increases the potential for injuries such as tendinitis, impingement, rotator cuff pathology, joint instability, and glenohumeral luxation.

Individuals with congenitally lax shoulders may experience atraumatic luxations in which minimal forces cause glenohumeral dislocation. Most cases of shoulder luxation, however, arise from traumatic insult to the glenohumeral complex. In the vast majority of these cases (>90%), dislocation occurs anteriorly. Anterior luxation occurs most often from indirect forces when axial loads are applied to an abducted, extended, and externally rotated arm (figure 7.3b). Less frequently, anterior dislocation results from direct forces applied to the posterior aspect of the humerus (figure 7.6a).

In a unique case, a young man experienced *bilateral* anterior glenohumeral dislocation while he was performing a bench press. The lifter became fatigued, causing the weight of the bar to force his arms into hyperextension while in a midabducted position. His humeral shafts pivoted on the bench and forced the humeral heads to dislocate anteriorly (Cresswell and Smith 1998).

Approximately 25% of anterior shoulder dislocations are complicated by concomitant fracture. Risk factors for fracture–dislocations are age (≥40 years), first episode of dislocation, and mecha-nism (e.g., fall from heights, fight or assault, or motor vehicle crash) (Emond et al. 2004).

Mechanisms of posterior dislocation essentially reverse those just described for anterior luxation. Indirect forces transmitted through a flexed, adducted, and internally rotated shoulder drive the humerus posteriorly (figure 7.6b). Posterior dislocation also results from direct trauma to the anterior aspect of the humerus (figure 7.6c) or to an arm subjected to forceful internal rotation and adduction (Samilson and Prieto 1983). Cases have been reported in which violent muscle contractions during electrical shock or seizures have caused posterior dislocations. In such cases, the substantial forces of fully activated internal rotators (subscapularis, latissimus dorsi, and pectoralis major) overwhelm the external rotators (infraspinatus and teres minor) and leverage the humeral head from the glenoid fossa. Occasional inferior dislocations occur from a hyperabduction mechanism that creates a fulcrum between the humeral neck and the acromion process and levers the head out inferiorly.

Rotator Cuff Pathologies: A Closer Look

Rotator cuff problems are a common source of pain and dysfunction in persons who use overhead movements in their work or play. Because of the prevalence of rotator cuff pathologies and

Figure 7.6 Mechanisms of glenohumeral dislocation (luxation). *(a)* Anterior glenohumeral dislocation from direct force applied to the posterior aspect of the shoulder. *(b)* Posterior glenohumeral dislocation from indirect force applied through the arm in a flexed, adducted, and internally rotated position. *(c)* Posterior glenohumeral dislocation from direct force applied to the anterior aspect of the shoulder.

(b) Adapted, by permission, from K.P. Butters, 1996, Fractures and dislocations of the scapula. In *Rockwood and Green s fractures in adults*, 4th ed., edited by C.A. Rockwood et al. (Philadelphia, PA: Lippincott, Williams & Wilkins), 1280.

(c) Adapted, by permission, from K.P. Butters, 1996, Fractures and dislocations of the scapula. In *Rockwood and Green s fractures in adults*, 4th ed., edited by C.A. Rockwood et al. (Philadelphia, PA: Lippincott, Williams & Wilkins), 1279.

the resultant potential for disability, we present in this section a detailed review of glenohumeral impingement and rotator cuff injury.

Shoulder symptoms are third after back and knee problems regarding the number of physician visits prompted in the United States (Praemer et al. 1999). These shoulder problems include both glenohumeral impingement and rotator cuff lesions. Glenohumeral impingement and attendant rotator cuff lesions are the most common shoulder conditions seen in patients visiting a physician's office (van der Windt et al. 1995; Vecchio et al. 1995). In 2003, there were more than 4.4 million physician visits in the United States for rotator cuff problems (AAOS 2007).

Tissue Structure and Function

The morphology of the rotator cuff musculotendinous unit and surrounding structures dictates glenohumeral motion and the joint's susceptibility to impingement syndrome and rotator cuff lesions. The rotator cuff muscle group consists of four muscles: subscapularis, supraspinatus, infraspinatus, and teres minor. Table 7.4 summarizes some of the structural and functional characteristics of the rotator cuff muscles.

The rotator cuff muscles receive their vascular supply from a number of arteries. A critical zone of hypovascular tissue has been demonstrated in the distal 1.0 to 1.5 cm of the supraspinatus and infraspinatus tendons. The hypovascular-

ity in this area has been suggested by some as a predisposing factor in rotator cuff pathology, but "the existence and extent of a true critical zone, and its significance relative to the pathological changes occurring with the rotator cuff, remains in question" (Malcarney and Murrell 2003, p. 995).

Healthy rotator cuff tendons contain predominantly water and type I collagen fibers (with trace quantities of type III collagen). Type I collagen fibers present in a parallel orientation. Type III fibers, in contrast, tend to orient in a more random pattern, are smaller, and have a lower tensile strength than type I fibers. Greater amounts of type III collagen are found in tendons undergoing repair and those that are aging and degenerating. From a structural perspective, it is not surprising that damaged and aging rotator cuff tendons are more susceptible to injury.

We stress the importance of recognizing the integral relations among glenohumeral impingement, joint instability, and rotator cuff lesions. Rotator cuff ruptures are often rooted in prior impingement syndromes, with a continuum of causality progressing from mild impingement to complete rotator cuff rupture. Thus, separating the discussion of these two types of injury is, to some extent, artificial. A comprehensive discussion of shoulder pathology should not involve either one of these conditions at the exclusion of the other.

TABLE 7.4

STRUCTURAL AND FUNCTIONAL CHARACTERISTICS OF THE ROTATOR CUFF MUSCLES

	Subscapularis	Supraspinatus	Infraspinatus	Teres minor
Origin	Ventral scapula	Superior scapula	Dorsal scapula	Dorsolateral scapula
Insertion (on humerus)	Lesser tuberosity	Greater tuberosity	Greater tuberosity	Greater tuberosity
Innervation	Subscapular n. (C5-C8)	Suprascapular n. (C4-C6)	Suprascapular n. (C4-C6)	Axillary n. (C5-C6)
Movement function (at glenohumeral joint)	Internal rotator	Abductor	External rotator	External rotator

n = nerve

Glenohumeral Impingement

An impingement syndrome occurs when increased pressure within a confined anatomical space deleteriously affects the enclosed tissues. With respect to the glenohumeral joint, impingement syndrome is an ill-defined term and can refer to either of two major glenohumeral impingement types: subacromial impingement and internal impingement.

Types

Subacromial impingement refers to shoulder abduction that results in suprahumeral structures (most notably the distal supraspinatus tendon, subacromial bursae, and proximal tendon of the long head of the biceps brachii) being forcibly pressed against the anterior surface of the acromion and the coracoacromial ligament (which together form the coracoacromial arch) (figure 7.7a). In addition to affecting suprahumeral structures, subacromial impingement may result in glenohumeral articular cartilage lesions (Guntern et al. 2003).

Subacromial contact pressures are elevated in patients with impingement syndrome; maximal contact pressure develops with the arm in a hyperabducted position or with the arm adducted across the patient's chest with the arm internally rotated (Nordt et al. 1999). Acromioplasty has proved effective in reducing subacromial contact pressure by cutting or shaving bone to flatten the acromion.

Walch and colleagues (1992) described another form of impingement (internal impingement) where the supraspinatus tendon contacts the posterior–superior rim of the glenoid fossa. This mechanism may be significant in the development of rotator cuff pathologies (Edelson and Teitz 2000). Internal impingement often happens in throwing, when the shoulder is abducted and externally rotated (e.g., cocking phase in an overhead throw), but it is not limited to throwers (McFarland et al. 1999). Internal impingement may involve undersurface (articular-sided) tears of the supraspinatus or infraspinatus tendons, posterosuperior labral fraying, anterior labral fraying, and osteochondral lesions (Giaroli et al. 2005; Paley et al. 2000). Caution is warranted in attributing all undersurface rotator cuff lesions to internal impingement, because other mechanisms may be responsible (Budoff et al. 2003).

Etiology

Impingement pathologies fall into two broad age-based categories. Impingement in those younger than 35 years usually happens to participants in sports (e.g., swimming, water polo, baseball, or football) or occupations (e.g., carpenter or painter) involving extensive overhead movements. Older individuals are more likely to suffer from the effects of degenerative processes that lead to bone spur formation, capsular thinning, decreased tissue perfusion, and muscular atrophy.

Repeated abduction places large stresses on the musculotendinous and capsuloligamentous structures and eventually leads to tissue micro-trauma. Continued mechanical loading further weakens the tissues and hastens their failure. Tissue failure, in turn, contributes to glenohumeral instability and greater joint movement. This

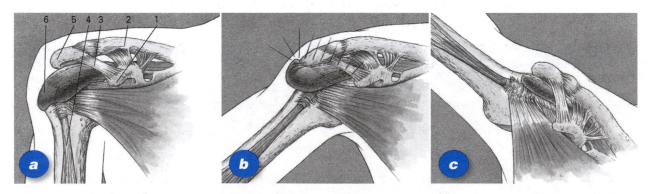

Figure 7.7 Glenohumeral impingement. *(a)* Arm at rest: 1, coracoid process; 2, clavicle; 3, coracoacromial ligament; 4, biceps brachii tendon; 5, acromion process; 6, subacromial bursa. *(b)* Arm abducted 60° to 120° with bursa compressed between the acromion and the rotator cuff tendons. *(c)* Arm abducted greater than 120° with decreased pressure on the bursa.

increases the chance of humeral *subluxation* that further aggravates the impingement condition. Thus, the person is trapped in an unfortunate loop of joint deterioration and compromised function.

A well-established association exists between (1) rotator cuff muscle weakness and subacromial impingement (e.g., Reddy et al. 2000) and (2) altered glenohumeral and scapulothoracic kinematics (e.g., Halder et al. 2001; Yamaguchi et al. 2000). What remains unresolved is whether rotator cuff weakness hastens the impingement or conversely whether the impingement ultimately weakens the muscles.

Acromial structure also is often cited as a factor in glenohumeral impingement. The most commonly used system for classifying acromial morphology is one first presented by Bigliani and Levine (1997), which specifies three shapes as shown in figure 7.8. One group (Farley et al.

1994) suggested a fourth acromion type (inferiorly convex).

Research using alternative methods of classifying acromial morphology has found a stronger relation between arch geometry and shoulder dysfunction (Prato et al. 1998; Tuite et al. 1995; Vaz et al. 2000). Nonetheless, the exact relation between acromial shape and rotator cuff pathology remains unresolved.

Other morphological features have been implicated in impingement and rotator cuff lesions. These include thickening of the coracoacromial ligament (Farley et al. 1994; Ogata and Uhthoff 1990; Soslowsky et al. 1996), anterior tilt of the acromion (Prato et al. 1998), age-related changes in acromial morphology (Wang and Shapiro 1997), and **os acromiale** (Hutchinson and Veenstra 1993).

In light of the evidence, there is likely no single mechanism of impingement injury, rather a variety of factors specific to each individual's morphological characteristics and history of joint loading.

Risk factors for glenohumeral impingement syndrome are the same as those associated with other cumulative trauma disorders of the shoulder (e.g., bicipital tendinitis, subacromial bursitis, or **thoracic outlet syndrome).** These factors include awkward or static postures, heavy work, direct load bearing, repetitive arm movements, working with hands above shoulder height, and fatigue resulting from lack of rest.

Impingement syndrome affects special populations as well. Wheelchair athletes, for example, suffer a high incidence of rotator cuff impingement. Muscle imbalance is suggested as a causal mechanism. The typical pattern of imbalance in wheelchair athletes differs from that in overhead athletes such as baseball pitchers, swimmers, and water polo players. Muscle imbalance in the overhead athlete appears as a relative weakness in the abductors and external rotators. Wheelchair athletes, in contrast, typically exhibit relative weakness in the adductors and overall rotator strength deficiency. The resulting abductor dominance exaggerates superior movement of the humeral head and leads to impingement in the subacromial space (Burnham et al. 1993).

Jobe and Pink (1993) proposed an injury classification based on age-based differences. Group

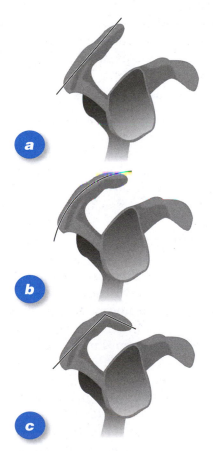

Figure 7.8 Variation in acromion shapes (lateral view). *(a)* Type I, flat. *(b)* Type II, curved. *(c)* Type III, hooked.

Reprinted, by permission, from LU. Bigliani, D.S. Morrison and E.W. April, 1986, "The morphology of the acromion and rotator cuff impingement," *Orthopaedic Transactions* 10: 288.

I injuries are characterized by isolated impingement with no joint instability and are usually found in older (>35) recreational athletes. Group II injuries result from overuse, typically in young (18-35) overhead athletes, and present primarily as glenohumeral instability with secondary impingement. Group III injuries also are common to young overhead athletes and are closely associated with group II. They are differentiated from group II by the presence of generalized ligamentous laxity at the elbow, knee, and fingers.

The mechanisms underlying rotator cuff impingement are the subject of ongoing debate. In a broad sense, the mechanisms can be intrinsic (inflammatory changes within the cuff), extrinsic (forces acting outside the rotator cuff), or both.

Rotator Cuff Rupture

Rupture of musculotendinous structures in the rotator cuff is typically the result of a chain of events that begins with minor inflammation that progresses with continued overuse to advanced inflammation, microtearing of tissue, and partial or complete rupture. Compromised tissue integrity and muscle fatigue contribute to altered movement mechanics, and these modified movements further stress the involved tissues and hasten their eventual failure. The supraspinatus is the most commonly injured muscle in the rotator cuff group (Goldberg et al. 2001). Less frequently, other cuff muscles suffer damage. Supraspinatus injury, in particular, is associated with repeated, and often violent, overhead movement patterns (e.g., throwing, striking, hammering, or painting).

Many of the complex movements at the shoulder stress the muscles of the rotator cuff group. The throwing motion, in particular, places exceptional loads on the shoulder. As a result of these loads, the rotator cuff is especially susceptible to injury. The entire rotator cuff synergistically resists distraction forces that tend to pull the humeral head from the glenoid. Injury or fatigue to any of these muscles leads to altered throwing mechanics and increases the chance of additional tissue damage.

Intrinsic mechanisms have long been implicated in impingement-related rotator cuff lesions. The theory behind these mechanisms posits that the degenerative process is inherent to the supraspinatus itself, largely attributable to compromised blood flow resulting from impingement pressures and regions of relative avascularity in the supraspinatus near its humeral attachment. Convincing evidence supports this concept. First, rotator cuff lesions are often seen in the absence of any extrinsic involvement. Second, studies have shown that degeneration first occurs on the articular surface of the tendon rather than on the bursal side. Support for the role of hypovascularity in impingement pathologies, however, is not universal.

The extrinsic approach implicates external factors such as compression of the rotator cuff tendon by structures external to the tendon, as originally suggested by Neer (1972). Other extrinsic factors that may hasten glenohumeral impingement and resultant rotator cuff tears include faulty posture, capsular tightness, altered joint kinematics and structural pathologies in the coracoacromial arch (Michener et al. 2003), and bone spurs in the joint spaces that serve as *stress risers,* focusing forces on the supraspinatus tendon and producing a functionally smaller supraspinatus outlet (Fu et al. 1991).

Kinematic Patterns and Cuff Lesions

The specific muscles with pathological involvement depend on the shoulder movement pattern. Burkhart (1993) identified four distinct patterns of rotator cuff kinematics associated with specific cuff lesions. These patterns are summarized in table 7.5. All of the patterns involve injury to the supraspinatus.

Burkhart (1993) reported that type I (stable fulcrum kinematics) lesions showed tears of the supraspinatus and part of the infraspinatus but not to a level that disrupted essential force couples. Patients had normal motion and near-normal strength levels. Patients with type II lesions (unstable fulcrum kinematics, posterior cuff tear pattern) exhibited massive tears of the superior and posterior portions of the rotator cuff, which resulted in an uncoupling of the essential force couples and led to an unstable fulcrum for glenohumeral motion. Type II patients could perform little more than a shoulder shrug.

Type III (captured fulcrum kinematics) and type IV (unstable fulcrum kinematics, subscapularis tear pattern) lesions both involved tears of the subscapularis. The less severe type III patterns

TABLE 7.5

KINEMATIC PATTERNS RELATED TO ROTATOR CUFF TEAR LOCATION

Kinematic pattern	Tear location
I. Stable fulcrum kinematics	Supraspinatus + part of infraspinatus
II. Unstable fulcrum kinematics (posterior cuff tear pattern)	Supraspinatus + all of posterior cuff (infraspinatus and teres minor)
III. Captured fulcrum kinematics	Supraspinatus + major posterior cuff + at least 50% subscapularis
IV. Unstable fulcrum kinematics (subscapularis tear pattern)	Supraspinatus + complete subscapularis

Adapted, by permission, from S.S. Burkhart, 1993, "Arthroscopic debridement and decompression for selected rotator cuff tears: Clinical results, pathomechanics, and patient selection based on biomechanical parameters," *Orthopedic Clinics of North America* 24(1): 115.

had partial subscapularis tears (accompanying superior and posterior damage). Muscle damage in type III patients prevented the humeral head from centering in the glenoid fossa, and the humerus subluxated superiorly and formed a captured acromiohumeral fulcrum that restricted humeral elevation. Type IV lesions involved tears of the supraspinatus and subscapularis, with the posterior cuff muscles remaining intact. This was a reversal of the type II pattern in which an unstable glenohumeral fulcrum attributable to the force couple imbalances was created by the muscle tears. Shoulder elevation in the type IV patients was poor (Burkhart 1993).

The last category, type IV, included injuries resulting from a traumatic event such as a fall or direct trauma. Such macrotrauma may be associated with contact sports (Blevins 1997) such as American professional football, where the most common mechanism for full-thickness rotator cuff tears is a fall onto the shoulder (Foulk et al. 2002).

Acromion Types and Cuff Tears

A hooked (type III) acromion (figure 7.8c) has been associated with higher incidence of rotator cuff tears. Evidence of this association is mixed; some studies have reported a positive relation between acromial curvature and rotator cuff lesions (Bigliani and Levine 1997; Toivonen et al. 1995), whereas others have failed to find a significant relation (Banas et al. 1995; Farley et al. 1994). The primary difficulty with finding a significant relation may be the fair to poor interobserver reliability in determining acromion type (Bright

et al. 1997; Haygood et al. 1994; Jacobson et al. 1995; Zuckerman et al. 1997).

Treatment of Rotator Cuff Pathologies

Proper treatment of rotator cuff pathologies is essential to restoring upper-extremity function. Conservative (i.e., nonsurgical) treatment typically is the first course of action and often proves effective. In more severe cases, surgical intervention may be indicated.

Conservative Treatment

The choice of treatment for subacromial impingement syndrome and other rotator cuff pathologies is difficult because of the multifactorial nature of the condition (Michener et al. 2003). Nonoperative (conservative) treatment of impingement and rotator cuff pathology begins with avoidance of aggravating activities and includes use of stretching to increase range of motion, nonsteroidal medications to control inflammation, occasional corticosteroid injections, and physical modalities. These treatments continue until pain is reduced. As pain decreases, a progressive strength training program can begin. Such a treatment program results in patient satisfaction about 50% of the time (Ruotolo and Nottage 2002).

Several factors have been identified that predict a poor prognosis for conservative treatment. These include rotator cuff tears greater than 1 cm, severe muscle weakness, and a history of symptoms lasting more than 1 year (Bartolozzi et al. 1994). Itoi and Tabata (1992) also found a 12-month history of symptoms to be a significantly poor prognostic indicator. Hawkins and Dunlop (1995) reported sleep-interrupting pain as a poor prognostic factor for conservative treatment.

Surgical Treatment

Approximately a century has passed since Codman (1911) reported his surgical repair of a completely ruptured supraspinatus tendon. Since that time, remarkable progress has been made in our understanding of rotator cuff morphology, shoulder pathomechanics, and treatment techniques.

In terms of pain relief and return of muscle strength, surgical repair of rotator cuff tears has a higher success rate than conservative approaches (Ruotolo and Nottage 2002; Wittenberg et al. 2001). Operative (surgical) treatment of a rotator cuff tear is indicated for four groups:

- An active 20- to 30-year-old patient with an acute cuff tear, accompanied by serious functional deficit
- A 30- to 50-year-old patient with an acute tear secondary to a particular event
- A high-level competitive athlete, especially who participates in overhead or throwing activities
- A patient who does not respond to conservative treatment (Mantone et al. 2000)

Although a detailed account of the many surgical techniques used to repair rotator cuff tears is beyond the scope of our discussion, mention is warranted of the pioneering work of Charles Neer, whose recognition of the anterior acromion's role in the pathogenesis of rotator cuff lesions laid the foundation for all subsequent surgical techniques. The goals of Neer's approach to cuff repair were to close the cuff defect, eliminate impingement, preserve the deltoid, and prevent stiffness (Neer 1990). "His surgical technique of an open superior approach, acromioplasty, coracoacromial ligament excision, tendon mobilization, and tendon repair to bone remains the 'gold standard' to which all contemporary methods of surgical treatment of rotator cuff tears must be compared" (Williams et al. 2004, p. 2765).

Whatever the surgical technique used, one must be mindful of the biomechanical aspects of surgical repair. As Burkhart (2000, p. 89) noted,

"Much of the history of rotator cuff repair has been checkered by ill-advised attempts to simply cover the hole in the cuff. By ignoring shoulder mechanics, many of these methods can actually make the shoulder worse. . . . Meticulous attention to detail at every step is critical. A loose suture or a poorly placed anchor can mean loss of integrity of the entire construct. In orthopaedic surgery as in structural engineering, structural integrity is built one step at a time."

Among the important issues requiring consideration are indications for and timing of surgical repair, surgical method (i.e., open, mini-open, or arthroscopic), the need for acromioplasty or coracoacromial ligament excision, and management of irreparable cuff tears (Williams et al. 2004).

All surgical methods have proved successful. Open rotator cuff repair has a long history of success. More recently, mini-open and arthroscopic surgical techniques have also gained favor, often with comparable or better results compared with open repair techniques (Buess et al. 2005; Rebuzzi et al. 2005). Recent comparisons have found no difference in results between arthroscopic and mini-open rotator cuff repair (Sauerbrey et al. 2005; Warner et al. 2005; Youm et al. 2005).

One of the key factors in successful cuff repair is early operative intervention when tears are smaller, tendon degeneration is minimized, and the risk of rerupture is reduced (Williams et al. 2004). Early diagnosis and repair of rotator cuff tears may also limit deterioration of the biceps tendon (Chen et al. 2005).

In cases of massive and irreparable cuff tears, a decision must be made about the surgical approach. Two common options are muscle transfer and **debridement.** Muscle transfer procedures typically use the pectoralis major for anterosuperior lesions (e.g., Wirth and Rockwood 1997) or the atissimus dorsi or teres major for posterosuperior cuff tears (e.g., Warner and Parsons 2001).

In cases where surgery is not feasible or successful, tissue engineering may hold promise. Funakoshi and colleagues (2005) demonstrated the feasibility of rotator cuff regeneration using tissue engineering techniques.

Postoperative Care and Rehabilitation

Following surgery, a well-designed rehabilitation program is essential for full recovery and return to normal activities. Each rehabilitation program should be individualized, with examples of elements listed here. Initially, the rotator cuff repair must be protected for 5 to 6 weeks. Immediately postsurgery, the patient begins exercises aimed at maintaining active elbow, wrist, and hand movements. Shoulder shrugs and scapular adduction exercises are included early on, along with pendulum exercises and gentle isometric exercises with a neutrally positioned arm. Three to four weeks postsurgery, passive mobilization can be added. Active-assisted elevation can be added at 6 weeks, with other progressive resistance exercises added as allowed. Three months postsurgery, patients can return to most typical daily activities but no strenuous movements (e.g., lifting heavy objects or ballistic movements). Although tendon healing may be nearly complete by 3 months, return to full strength may take up to 1 year (Millstein and Snyder 2003).

Rotator Cuff Pathology Prevention

Given the prevalence of both symptomatic and asymptomatic impingement and rotator cuff lesions, prevention is a primary issue, especially for those at risk. A sound prevention program should include two elements: a physical conditioning program and recognition and modification of any underlying pathomechanics. Any injury prevention approach focused solely on improving physical condition may ultimately prove ineffective if underlying pathomechanics (e.g., improper movement mechanics) are not addressed as well. Moreover, these preventive steps must be supplemented with early recognition and intervention strategies.

The physical conditioning program should involve muscles that stabilize and move the glenohumeral and scapulothoracic joints, with emphasis on improving the strength and endurance of the rotator cuff muscles. Special attention should be given to strengthening the subscapularis. Range of motion improvement also is indicated at both the glenohumeral joint (taking care not to overstretch the anterior joint capsule) and scapulothoracic joint. The person's activity should also be monitored to reduce movements that might aggravate the condition. Technique correction may help prevent injury (Jobe 1997).

Labral Pathologies

The glenohumeral *labrum* is a fibrocartilage rim that encircles the articular surface of the scapular glenoid fossa. Inferiorly, the labrum appears as a rounded fibrous structure continuous with the hyaline articular cartilage. Superiorly, the labrum is more meniscus-like with loose attachment to

OF SPECIAL INTEREST

In elderly persons, the rotator cuff can degenerate with accompanying *osteophyte* formation, thinning of the joint capsule, decreased blood perfusion, muscular atrophy, and eventual muscle tears. These pathologies may result in significant pain and loss of function. Studies, however, have also found rotator cuff tear pathologies in asymptomatic individuals. Templehof and colleagues (1999) reported evidence of rotator cuff tears in 23% of 411 asymptomatic adults (≥50 years old), with a clear pattern of more tears with increasing age. In the oldest group (age >80 years), 51% of patients had cuff tears in the absence of symptoms. Other studies have reported similar results (Milgrom et al. 1995; Sher et al. 1995; Worland et al. 2003). In a cadaveric population, Lehman and colleagues (1995) found an increase in full thickness rotator cuff tears with increasing age; 30% of cadavers more than 60 years of age showed evidence of rotator cuff tears.

These studies show that rotator cuff tears "must to a certain extent be regarded as 'normal' degenerative attrition, not necessarily causing pain and functional impairment" (Tempelhof et al. 1999, p. 296).

the glenoid (Nam and Snyder 2003). The labrum's vascular supply arises from capsular or periosteal vessels and is more pronounced peripherally than centrally. Compared with other labral areas, the superior and anterior portions have reduced vascularity (Cooper et al. 1992).

Labral injuries can happen acutely or chronically and are caused by a variety of mechanisms, including compression attributable to falls, traction (tension) from lifting, throwing in overhead sports, and dislocation–subluxation.

Anterior shoulder (glenohumeral) dislocation can cause an avulsion of the anteroinferior glenoid labrum at the attachment site of the inferior glenohumeral ligament complex (IGHL) in what is termed a **Bankart lesion** (figure 7.9a). The Bankart lesion, first described by British orthopedist Arthur Bankart (1923), is invariably accompanied by joint capsule disruption and IGHL stretching. IGHL compromise contributes to recurrent anterior glenohumeral instability.

In 1990, Snyder and colleagues coined the term **SLAP lesion** (superior labrum–anterior to posterior) to describe superior labral injuries. These investigators identified four types of SLAP lesions:

- Type I: Fraying of the superior labrum with no detachment at the biceps insertion (figure 7.9b)
- Type II: Detachment of the superior labrum and biceps tendon (figure 7.9c)
- Type III: Bucket-handle tear of the superior labrum with an intact biceps (figure 7.9d)
- Type IV: Bucket-handle tear of the superior labrum with tearing of the biceps tendon (figure 7.9e)

The prevalence of lesion type varies considerably among studies. For example, the prevalence of type I lesions in three studies ranged from 9.5% to 74%, whereas prevalence of type II lesions ranged from 21% to 55% (Handelberg et al. 1998; Kim et al. 2003; Snyder et al. 1995).

Morgan and colleagues (1998) subdivided type II SLAP lesions into three subtypes according to the location of the labral–biceps tendon detachment: anterior, posterior, and combined anterior–posterior (figure 7.9, f-h). Type II SLAP lesions have been associated with the so-called dead arm

syndrome, which limits a thrower's velocity and control, largely attributable to the pain and uneasiness associated with labral injury (Burkhart and Morgan 2001; Burkhart et al. 2000).

Andrews and colleagues (1985) first described superior labral tears near the origin of the long head of the biceps brachii in a population of overhead throwing athletes. The authors postulated that tension in the biceps tendon during throwing pulled the labrum from its attachment.

Several mechanisms have been suggested for SLAP lesions. Andrews and colleagues (1985) theorized a traction (tension) mechanism. Snyder and colleagues (1990, 1995) promoted a compression mechanism as seen in a fall or collision. Maffet and colleagues (1995) suggested a traction or pulling mechanism. Others (Jobe 1995; Walch et al. 1992) implicated contact with the rotator cuff muscles (e.g., supraspinatus) when the throwing arm is in a cocked position. Burkhart and Morgan (1998) and Morgan and colleagues (1998) described a mechanism involving peel-back of the superior labrum when the biceps attachment is twisted during abduction and external rotation.

The reported frequency of SLAP lesion mechanisms varies, likely due to differences in the populations being assessed, including their age, physical condition, activity patterns, and associated injuries (e.g., rotator cuff pathology). Available information suggests that compression is likely the most common mechanism (table 7.6).

Upper-Arm Injuries

The upper arm (also called *arm*) spans the shoulder and elbow joints and contains the humerus, which is surrounded by two muscle compartments. The anterior compartment contains the biceps brachii, brachialis, and coracobrachialis; the posterior compartment houses only the triceps brachii, a muscle complex with three heads: the long head, medial head, and lateral head (figure 7.2). The humerus articulates proximally with the scapula and distally with the radius and ulna at the elbow joint.

Humeral Fracture

Fractures of the humerus account for 7.5% of all fractures, with an estimated 67,000 annually in the

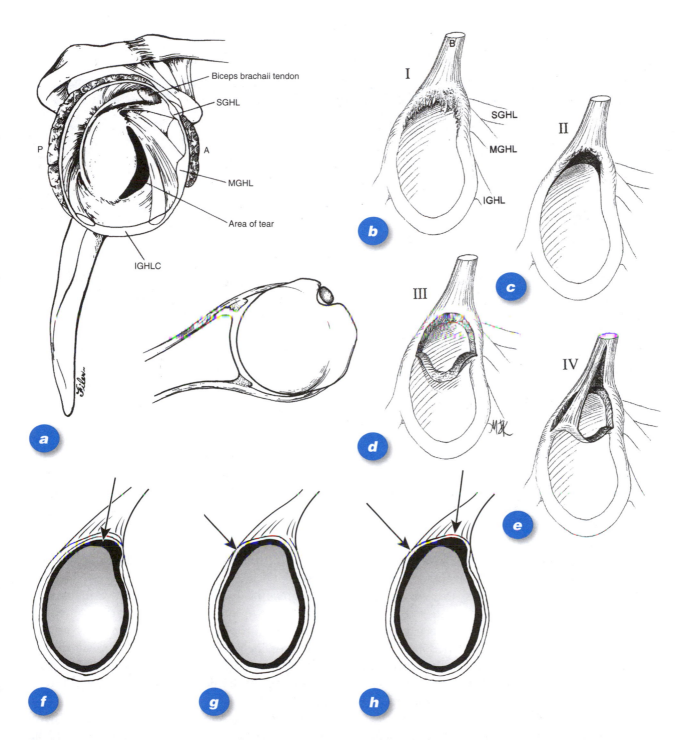

Figure 7.9 Labral injuries. *(a)* Bankart lesion involving an avulsion of the anteroinferior glenoid labrum at the attachment site of the inferior glenohumeral ligament (IGHL) complex. A = anterior; MGHL = middle glenohumeral ligament; P = posterior; SGHL = superior glenohumeral ligament. *(b-e)* SLAP lesion involving damage to the superior labrum—anterior to posterior. *(b)* Type I: Fraying of the superior labrum with no detachment at the biceps insertion. *(c)* Type II: Detachment of the superior labrum and biceps tendon. *(d)* Type III: Bucket-handle tear of the superior labrum with an intact biceps. *(e)* Type IV: Bucket-handle tear of the superior labrum with tearing of the biceps tendon. *(f-h)* Subtypes of type II SLAP lesions. *(f)* Anterior. *(g)* Posterior. *(h)* Combined anterior–posterior.

(a-e) Reprinted, by permission, from J.J. Perry and L.D. Higgins, 2001, Shoulder injuries. In *Sports injuries: Mechanisms, prevention, treatment,* edited by F.H. Fu & D.A. Stone (Baltimore, MD: Lippincott, Williams & Wilkins), 1039.

(g-h) Reprinted from *Arthroscopy,* Vol. 14, C.D. Morgan et al., Type II SLAP lesions: three subtypes and their relatinships to superior instability and rotator cuff tears, pp. 553-565, Copyright 1998, with permission from Elsevier.

TABLE 7.6

MECHANISMS OF SLAP LESION INJURIES

	Snyder et al. (1995) (N = 140)	Handelberg et al. (1998) (N = 32)
Compression (fall or compression)	43 (31)	9 (28)
Traction (lifting weight)	23 (16)	7 (22)
Dislocation–subluxation	27 (19)	7 (22)
Throwing (overhead sports)	16 (12)	8 (25)
Nonspecific onset	31 (22)	1 (3)

Values are n (%).

United States (Praemer et al. 1999). Most humeral fractures (60%) happen in the middle third of the diaphysis (shaft fractures), with 30% occurring in the proximal humerus and the remaining 10% in the distal portion (Tytherleigh-Strong et al. 1998).

Humeral fractures result from either direct or indirect trauma. Direct injuries typically are high energy and exhibit extensive comminution (i.e., bone fragmentation) and soft tissue disruption, whereas indirect trauma involves less energy and minimal bony displacement.

The pattern of bone fracture varies: Compressive forces lead to proximal and distal end disruption, bending results in transverse shaft fracture, torsional loads create spiral fractures, and combined torsion and bending theoretically cause oblique fracture or butterfly fragmentation. A commonly used classification system for humeral shaft fractures is presented in figure 7.10.

Humeral fracture typically results from direct trauma (e.g., fall on an outstretched arm, motor vehicle crash, or direct loading of the arm). Fractures also occur, although rarely, in response to violent muscular contractions, as first reported by Wilkinson (1895).

The pattern of fracture depends on the magnitude, location, and direction of applied forces, with segment movement determined by the action of muscles in the area. For example, when fracture occurs proximal to the pectoralis major attachment site, the distal segment is displaced medially by the pectoralis major, whereas the proximal segment is abducted and internally rotated by the rotator cuff (figure 7.11a). If the fracture site is between the attachments of the pectoralis major and the deltoid, the distal segment is abducted through action of the deltoid, whereas the proximal segment is pulled medially by the pectoralis major, latissimus dorsi, and teres major (figure 7.11b). In fractures distal to the deltoid attachment, the proximal segment is abducted and flexed, with the distal segment displaced superiorly (figure 7.11c).

In most injuries, age plays a prominent role in determining the nature and extent of tissue damage, and the same is true of humeral fractures. Humeral fracture in elderly persons commonly results from falls (Tytherleigh-Strong et al. 1998), whereas in young people the usual culprits are direct impact or vigorous throwing. Humeral fractures have been documented as a result of throwing objects as varied as baseballs, javelins, and hand grenades (e.g., Callaghan et al. 2004; Kaplan et al. 1998). Various theories have been proposed to explain throwing-related fractures, including factors of antagonistic muscle action, violent uncoordinated muscle action, poor throwing mechanics, excessive torsional forces, and fatigue.

A report on a series of 12 spontaneous humeral fractures in baseball players (average age 36) who played regularly in an over-30 league noted that the injured players had been inactive for years before joining the league. The prolonged layoff period may have contributed to disuse atrophy in the humerus and may have predisposed the athletes to both sudden and stress fractures. Four risk factors for fracture were identified: age, prolonged absence from pitching activity, lack of

Essence: All diaphyseal fractures are divided into 3 types according to the contact between the two main fragments after reduction:

A: contact >90% = simple fracture
B: some contact = wedge fracture
C: no contact = complex fracture

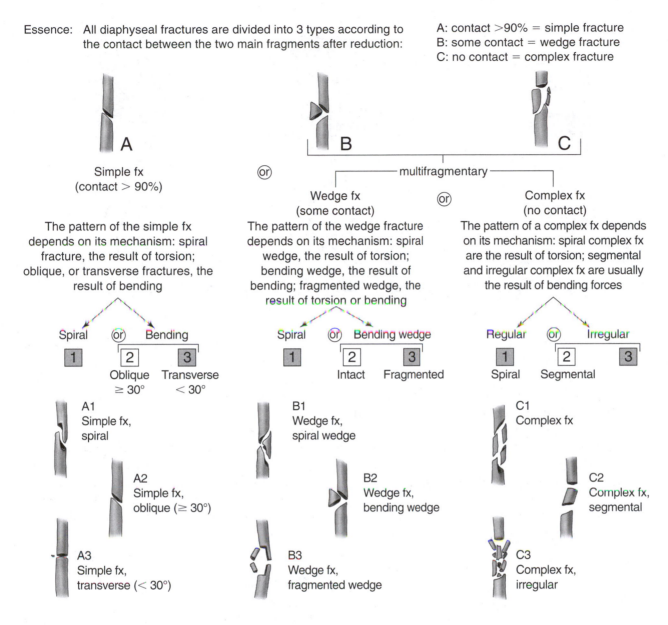

Figure 7.10 Classification system for humeral shaft fractures.

Reprinted from *Skeletal trauma*, 3rd Edition B.D. Browner et al., p. 1485, Copyright 2003, with permission from Elsevier.

regular exercise, and prodromal (precursory) arm pain (Branch et al. 1992). With fracture, as with many injuries, a person's condition and activity pattern play an important role in determining his or her susceptibility to injury.

Biceps Tendon Injuries

The proximal tendon of the long head of the biceps brachii (LHBB) has attachments at both the supraglenoid tubercle (also **biceps anchor)** and the superior rim of the glenoid labrum. The biceps anchor provides primary restraint for the LHBB tendon, with secondary restraint given by the labral attachment (Healey et al. 2001). From this attachment the biceps tendon courses through the rotator cuff interval in the bicipital groove of the humerus. Distally, the biceps brachii attaches to the radial tuberosity.

The primary action of the biceps is to flex the elbow (with lesser involvement at the glenohumeral joint). The proximal biceps tendon is predisposed to injury because of its intimate involvement with the action of the rotator cuff.

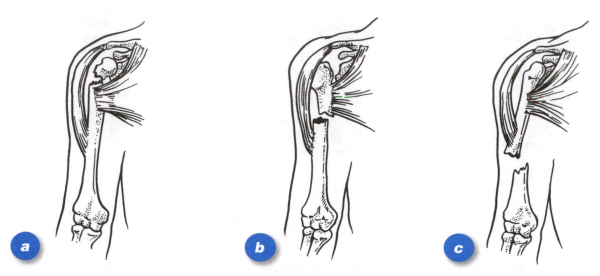

Figure 7.11 *(a)* Fracture proximal to the pectoralis major insertion. *(b)* Fracture distal to the pectoralis major insertion and proximal to the deltoid insertion. *(c)* Fracture distal to the deltoid insertion.

Injuries to the biceps tendon include acute and chronic tenosynovitis, subluxation, dislocation, and rupture.

The cause of biceps injury is unclear, in large part because of the complex interaction of impingement and instability pathologies and considerable anatomical variability. Biceps peritenonitis (tenosynovitis) is commonly associated with repeated overhead tasks such as throwing. Acutely, peritenonitis manifests as swelling and inflammation. With repeated insult, chronic peritenonitis progresses to include tendon fraying, synovial proliferation, fibrosis, and eventual tendon rupture or dislocation (Ptasznik and Hennessy 1995).

The biceps tendon typically dislocates medially, often in conjunction with rotator cuff lesions and acute trauma. The mechanisms of biceps dislocation include abduction with external rotation, falls onto an outstretched hand, direct lateral impact, hyperextension, and anterior glenohumeral dislocation. Anatomical features, most notably the depth and angulation of the bicipital groove, have been implicated as predisposing factors in biceps tendon dislocation. The tendon is more likely to dislodge from a shallow or low-angled groove.

Given the integral relation between rotator cuff and biceps tendon pathologies, it is not surprising that the mechanisms that encourage cuff lesions (e.g., impingement) are strongly associated with biceps tendon degeneration as well. Rupture of

the biceps tendon may be a logical consequence of progressive tissue degradation. Biceps tendon rupture has been associated with distal traction and active biceps contraction, as when a person is actively pulling and his or her arm is suddenly jerked distally.

The structural congruity of the glenoid labrum and biceps tendon often results in combined injuries at the bicipital–labral junction (e.g., SLAP lesion). Although the exact mechanism of injury is unresolved, both biceps tendon traction and humeral head traction may be involved.

Elbow Injuries

The elbow joint is structurally classified as a synovial hinge joint formed by dual articulations of the capitulum of the humerus (7.12*a*) with the head of the radius (7.12*c*) and the trochlea of the humerus (7.12, *a* and *b*) with the trochlear notch of the ulna (figure 7.12*c*). The radius and ulna articulate at the proximal radioulnar joint (7.12, *c* and *d*). Normal elbow motion is confined to uniplanar flexion–extension, with forearm pronation–supination produced by the combined rotations of the proximal and distal radioulnar joints. As a synovial joint, the elbow is surrounded by a thin fibrous capsule that runs from its proximal humeral attachment to become continuous distally with the synovial capsule of the proximal radioulnar joint.

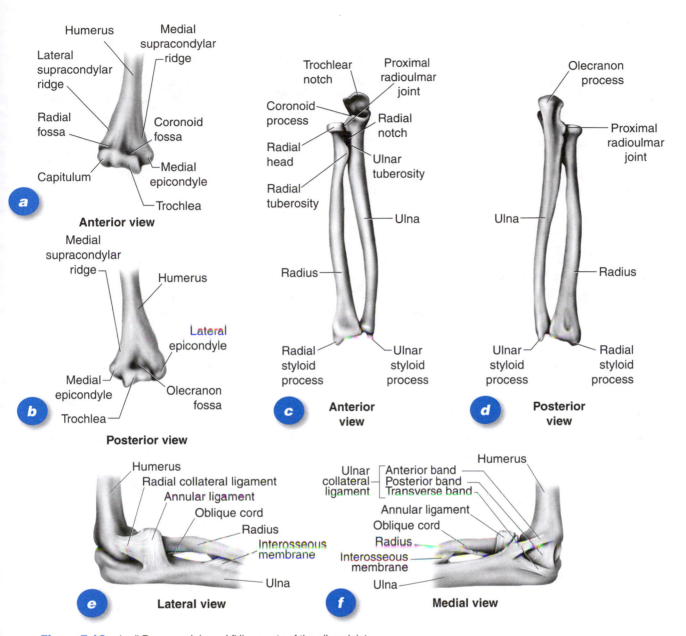

Figure 7.12 *(a-d)* Bones and *(e and f)* ligaments of the elbow joint.

(a-d) Reprinted, by permission, from R.S. Behnke, 2005, *Kinetic anatomy*, 2nd ed. (Champaign, IL: Human Kinetics), 62. *(e, f)* Reprinted, by permission, from R.S. Behnke, 2005, *Kinetic anatomy*, 2nd ed. (Champaign, IL: Human Kinetics), 65.

The elbow is reinforced by the lateral (radial) collateral ligament (LCL) complex, which extends from the lateral epicondyle of the humerus to the annular ligament of the radius (7.12*e*). It is also reinforced by the medial (ulnar) collateral ligament (MCL) complex (7.12*f*), which connects the medial epicondyle with the coronoid process and olecranon of the ulna (figure 7.12, *c, d,* and *f*). The LCL complex consists of three ligaments (radial collateral ligament, annular ligament, and lateral ulnohumeral ligament) and provides primary restraint against varus loading (i.e., lat-

erally directed force on the medial aspect of the elbow). The MCL contains three distinct bundles (anterior, posterior, transverse). The anterior and posterior bundles resist valgus loading (i.e., medially directed force on the lateral aspect of the elbow), whereas the transverse bundle plays a minimal role in joint stabilization. Muscles controlling elbow motion are shown in figure 7.2 and summarized in table 7.7.

Before considering specific injuries, we need to dispense with several pervasive wastebasket terms used to describe elbow injuries. The most

TABLE 7.7

MUSCLES OF THE ELBOW

Muscle	Action
Anconeus	Extends the forearm
Biceps brachii	Flexes the forearm; also supinates the forearm and flexes the arm
Brachialis	Flexes the forearm
Brachioradialis	Flexes the forearm
Triceps brachii	Extends the forearm; long head also extends the arm

common of these is the ubiquitous *tennis elbow,* a term with varied meanings, ranging from a general descriptor of any pain in and around the elbow to the specific designation of lateral epicondylitis (see next section). In the latter case, *tennis elbow* is doubly confusing because it can be interpreted to mean that tennis is solely responsible for epicondylitis or that epicondylitis is the only elbow injury seen in tennis players. Neither of these suppositions is correct. In fact, only a small minority (5%) of those diagnosed with lateral epicondylitis attribute it to tennis (Peterson and Renström 2001).

As with previously noted terms of this type, we refrain from using the colloquial *tennis elbow* and its cousins, *Little League elbow, golfer's elbow,* and *climber's elbow* (Safran 1995) and instead focus on specific clinical descriptors.

Significant injuries to the elbow include epicondylitis, tendinitis, myotendinous strain, osteochondritis dissecans, osteochondrosis, dislocation, bursitis, ligamentous sprain, and fractures of the humerus, ulna, and radius. Many of these injuries are common in athletes and are specific to athletic tasks (table 7.8).

Epicondylitis

Most elbow injuries are overuse conditions characterized by progressive tissue degeneration. As with most chronic injuries, repeated loading produces tissue microtrauma before the condition becomes symptomatic. Even in asymptomatic individuals, evidence of intracytoplasmic calcification, collagen fiber splitting and kinking, and abnormal fiber cross-links has been reported (Kannus and Jozsa 1991). The causes of these his-

topathologies have not been clearly established but may be mechanical or vascular in nature.

Continued loading worsens the microscopic damage and eventually leads to symptomatic tissue involvement in the form of initial inflammation, inflexibility, and tissue weakness. At the elbow, these events often manifest as epicondylitis involving soft tissue attachments on humeral epicondyles. Task specificity determines whether the medial or lateral epicondyle is involved.

Nirschl (1988) described four stages in the progression of epicondylitis:

Stage 1: Inflammation not associated with pathologic tissue alterations

Stage 2: Pathologic tissue alterations characterized by disrupted collagen architecture in the form of a fibroblastic and immature vascular response *(tendinosis)* in the relative absence of inflammatory cells

Stage 3: Tendinosis with tissue structural failure (e.g., microtearing)

Stage 4: Continued structural failure with fibrosis or calcification

The term *epicondylitis* may be misleading, because there is scant evidence of inflammatory cells present at the affected site, and there is significant disagreement about several other aspects of the disorder (Ciccotti and Charlton 2001, p. 77; Nirschl and Ashman 2003). Alternative designations for epicondylitis have been suggested, including *tendinosis* (Kraushaar and Nirschl 1999), *angiofibroblastic tendinosis* (Nirschl and Ashman 2003), and *epicondralgia* (Hotchkiss 2000). Nonetheless, the term *epicondylitis* is so embedded in the literature that its extinction

TABLE 7.8

SPORT-SPECIFIC ELBOW INJURIES

Sport	Injury
Archery	Extensor muscle fatigue, lateral epicondylitis of bow arm
Baseball	Valgus stress of pitching, medial traction, lateral compression, posterior abutment
Basketball	Posterior compartment syndrome with follow-through on jump shot
Bowling	Flexor–pronator soreness
Canoeing, kayaking	Distal bicipital tendinitis
Football	Valgus stress while throwing a pass, hyperextension and dislocation, olecranon bursitis with direct trauma
Golf	Medial epicondylitis on downswing with trailing arm, lateral epicondylitis at impact with leading arm on backhand
Gymnastics	Radiocapitellar overload, posterior impingement, dislocation, olecranon bursitis with direct trauma
Javelin	Valgus extension overload of throwing, medial traction, posterior abutment, lateral compression
Racket sports	Lateral epicondylitis with backhand
Rock climbing	Brachialis or distal biceps tendinitis
Shot put	Posterior impingement with follow-through
Volleyball	Valgus stress at impact of spiking
Waterskiing	Valgus extension overload of posterior compartment
Weight training	Ulnar collateral ligament sprain, ulnar nerve irritation

Adapted, by permission, from M.R. Safran, 1995, "Elbow injuries in athletes: A review," *Clinical Orthopaedics & Related Research* 310: 258.

seems unlikely. Whatever name is used, Whaley and Baker remind us that "the disease itself is not an inflammatory process, but rather a degenerative process" (2004, p. 688).

Lateral Epicondylitis

Lateral epicondylitis is characterized by pain on the lateral aspect of the elbow and is most often attributed to pathology at the proximal attachment of the extensor carpi radialis brevis (ECRB) (Nirschl and Pettrone 1979).

Lateral epicondylitis is prevalent in tennis players (hence the term *tennis elbow)*, with between 40% and 50% of players experiencing this injury at some time during their years of playing. The injury is most common in players between 30 and 50 years old. The suspected causal mechanisms include faulty stroke mechanics, off-center ball contact, grip tightness, and racket vibration. Repeated impact of the racket and ball stresses the muscles that stabilize and control movement of the wrist. These stresses can result from both concentric and eccentric muscle actions.

The mechanics of the backhand stroke in particular have been associated with the incidence of lateral epicondylitis (Priest et al. 1980). Electromyographic studies have shown high levels of wrist extensor activity, especially in the ECRB, during the backhand stroke (Giangarra et al. 1993; Morris et al. 1989). When a player (usually a novice) leads with the elbow, greater forces are generated in the wrist extensors. These loads are transferred through the active and stiffened musculature to the proximal attachment on the lateral humerus. Use of a two-handed backhand stroke has been associated with a lower incidence of lateral epicondylitis because the simplified and coordinated action of trunk rotation and arm movement imposes fewer mechanical demands on the musculoskeletal system.

Lateral epicondylitis is not exclusive to tennis. Other striking sports such as racquetball and golf are implicated, as are occupations involving repetitive motions of the wrist and elbow (e.g., carpentry and surgery). Research has shown that pinching and grasping of the fingers and hand always produce flexor moments at the wrist and that extensor moments are generated to maintain equilibrium. Overuse of the extensor mechanism in repeated pinching and grasping, such as in chronic work with hand tools or writing, increases the susceptibility to lateral epicondylitis (Snijders et al. 1987).

Medial Epicondylitis

Medial epicondylitis occurs infrequently compared with lateral epicondylitis—lateral epicondylitis is 7 to 10 times more common (Leach and Miller 1987). In tennis, medial epicondylitis results from excessive loading during the forehand and service strokes. These motions, especially by advanced players, involve forcible extension of the wrist. The eccentric action of the wrist flexor muscles in controlling wrist extension places considerable stress on these muscles and their attachments on the medial aspect of the humerus.

Medial epicondylitis occurs more often in throwers whose movement patterns include a high-velocity valgus extension mechanism (e.g., baseball or javelin) (Wilson et al. 1983). Valgus loading during throwing, especially during the late-cocking and acceleration phases, produces high tensile forces on the elbow's medial aspect.

Repeated valgus loading can presage medial epicondylitis (Grana 2001). Repetitive valgus loading also can cause other injuries, including damage to the ulnar collateral ligament (UCL), the flexor–pronator musculotendinous unit, and the ulnar nerve (Safran 2004).

Valgus-Extension Loading Injuries

Numerous studies have examined the kinematics, kinetics, and muscle involvement at the elbow during the overhead throwing motion. The throwing motion is divided into five phases: windup, cocking, acceleration, deceleration, and follow-through. A sixth phase, stride, is sometimes included between windup and cocking.

Several studies (e.g., Fleisig et al. 1995) have quantified elbow kinetics during throwing and reported large and potentially injurious forces and moments. Near the end of the cocking phase when the elbow is approaching terminal extension, the elbow experiences valgus loading that is resisted by a varus torque produced by musculotendinous and periarticular tissues (figure 7.13). Varus torques at this point have been estimated as ranging from 64 to 120 N·m, with predicted joint force between the radius and humerus of approximately 500 N (Fleisig et al. 1995; Werner et al. 1993).

The coupling of valgus torque and elbow extension produces a so-called **valgus-extension overload mechanism** that can lead to medial elbow injuries, including epicondylitis, ulnar collateral ligament rupture, avulsion fracture, and nerve damage. In addition, the valgus stress in extension causes impingement of the medial aspect of the olecranon on the olecranon fossa

NOT-SO-FUNNY BONE

The so-called funny bone is neither a bone nor particularly funny. The name is derived from the transient sensation of numbness and tingling experienced when you strike the posteromedial aspect of the elbow. In this area, the *ulnar nerve* passes by the posterior aspect of the humeral epicondyle on its way from the shoulder to the forearm and hand. Violent compression on the ulnar nerve against the humerus causes a temporary blockage of the flow of nerve impulses. This creates the "funny" sensation you feel when this nerve is struck.

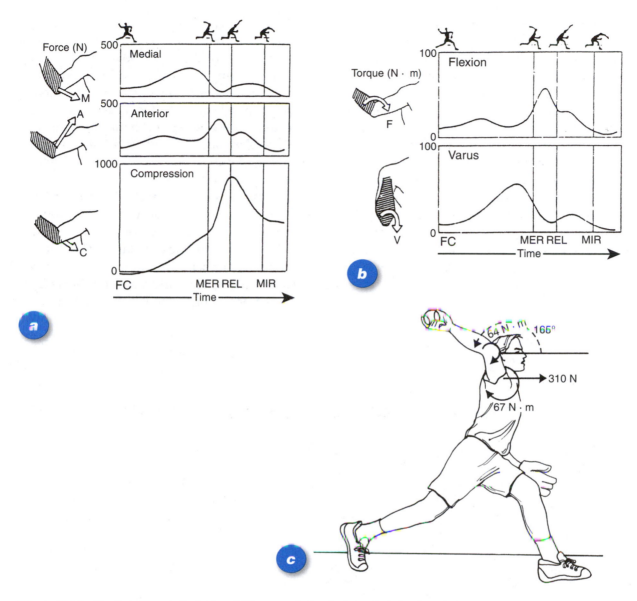

Figure 7.13 Kinetics of baseball pitching. *(a)* Force applied to the forearm at the elbow in the medial (M), anterior (A), and compression (C) directions. The instants of front foot contact (FC), maximum external rotation (MER), ball release (REL), and maximum internal rotation (MIR) torque are shown. *(b)* Torques applied to the forearm at the elbow in the flexion (F) and varus (V) directions. *(c)* Shortly before maximum external rotation, the arm was externally rotated 165° and the elbow was flexed 95°. At this instant there were 67 N·m of internal rotation torque, 310 N of anterior force at the shoulder, and 64 N·m of varus torque at the elbow.

Adapted, by permission, from G.S. Fleisig et al., 1995, "Kinetics of baseball pitching with implications about injury mechanisms," *The American Journal of Sports Medicine* 23(2): 236-268.

and radial head impingement on the capitulum. Repeated impingement may lead to inflammation, chondromalacia, osteophyte formation, and olecranon stress fracture (Ahmad and ElAttrache 2004).

In throwing, as in other dynamic movements, the largest loads are developed during eccentric actions that resist and control high-velocity motion. Repeated application of these high loads may lead to progressive degeneration and eventual tissue failure. Although this process is common, it is not inevitable. Some athletes and workers are able to load their tissues repeatedly with minimal, if any, symptomatic response. Why does a particular loading history cause injury in one person but not in another? The answer likely lies in the fact that injury is a multifactorial puzzle and that some people are more susceptible to injury than others. To define this susceptibility, Meeuwisse (1994) proposed a model that incorporates both intrinsic (e.g., age, strength, flexibility, previous injury) and extrinsic (e.g., biomechanics of movement skills, equipment, environmental conditions, schedule, inherent demands) factors.

Throwing, by its repetitive nature, provides an instructive example of chronic injury development and the interactive nature of these intrinsic and extrinsic factors. Kibler (1995) characterized progressive degeneration as a negative feedback cycle in which five complexes interact to create a downward spiral, or vicious cycle, leading to tissue failure. The five complexes are (Kibler 1995)

1. tissue overload,
2. clinical symptoms,
3. tissue injury,
4. functional biomechanical deficit, and
5. subclinical adaptation.

Applying this model to the overhead throwing motion, for example, we see how the complexes interactively contribute to elbow injury. During the late cocking phase, the valgus extension mechanism causes *tissue overload* on the medial aspect of the elbow in the form of asymptomatic microtrauma. Repeated mechanical insults lead to *clinical symptoms* of point tenderness over the medial epicondyle and *tissue injury* to flexor–pronator group attachments and ulnar collateral ligaments. The elbow then suffers *functional biomechanical deficits* in the form of flexor–pronator inflexibility and weakness, coupled with elbow inflexibility and muscle imbalance. The thrower then alters his or her technique *(subclinical adaptation)* in an attempt to compensate for the deficits. These compensations further overload the tissues, and the insidious cycle continues until ultimate tissue failure. Appropriate rehabilitation must address all five complexes to break the negative feedback cycle.

TENDON TRANSPLANTATION SURGERY

In 1974, holding a 13-3 record, Los Angeles Dodgers pitcher Tommy John was en route to one of his best seasons ever. Then John ruptured the ulnar collateral ligament (UCL) in his pitching arm, and his career appeared to be over. He reportedly asked team physician Frank Jobe to "make up something" to salvage his pitching arm. And Dr. Jobe did just that. In what has become known as *Tommy John surgery*, Dr. Jobe reconstructed John's UCL using a free tendon graft from John's nonpitching arm. John jokingly said, "When they operated, I told them to put in a Koufax fastball. They did—but it was Mrs. Koufax's."

At the time, no one knew what the outcome would be. John missed the 1975 baseball season and returned in 1976 to test his repaired arm. He passed the test with flying colors. Postsurgery, Tommy John pitched for an additional 13 years and amassed 164 wins. These wins, combined with 124 preinjury victories, left John with 288 career triumphs when he finally retired in 1989. Postinjury, John was a three-time All-Star (1978, 1979, 1980) and finished second in the Cy Young Award voting in 1977 and 1979.

Dr. Jobe continued to perform and refine his UCL reconstruction techniques (Jobe et al. 1986) and share his procedure with other professional colleagues. Jobe and colleagues reported good to excellent results in 80% of patients (Conway et al. 1992). The most notable result, however, remains that of Tommy John. John's 288 career wins may be forgotten, but his name will not. The name Tommy John will always be associated with the pioneering surgery that saved his pitching career. As John said, "I'd never have been able to win 288 games without the surgery. We're going to be linked forever."

As for John's perspective of the surgery and his baseball career, "You know what I'm most proud of? . . . I pitched 13 years after the procedure and I never missed a start. I had not one iota of trouble. I'd like people to remember that about me, too."

Elbow Dislocation

Given its relative stability, it is not surprising that the elbow is more than three times less likely than the shoulder to become dislocated (Praemer et al. 1992). Elbow dislocation, nonetheless, is not uncommon. Joint luxations are rarely isolated; they often are accompanied by extensive soft tissue damage as the bones are forcibly displaced. Elbow dislocations follow this pattern and typically involve complete rupture or avulsion of both the ulnar and radial collateral ligaments. Elbow dislocation is most prevalent in young individuals and in conjunction with sport activities.

The elbow's bony configuration provides exceptional resistance to anterior dislocation. In fact, the vast majority of dislocations happen posteriorly (Rettig 2002) and can be directed straight–posterior, posteromedial, or posterolateral (figure 7.14). Dislocation tends to be posterolateral (Nestor et al. 1992; O'Driscoll et al. 1991) to allow the coronoid to pass inferiorly to the trochlea (O'Driscoll 2000). The propensity for posterior dislocation is also seen in children. In one study (Rasool 2004) of 33 elbow dislocations in children (all caused by falls), 91% were posterior.

The most common mechanism for elbow dislocation involves axial force applied to an extended or hyperextended elbow. This force effectively levers the ulna out of the trochlea and causes capsular and ligament rupture that then allows

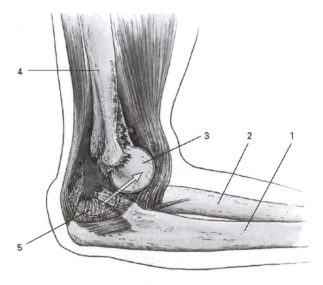

Figure 7.14 Posterior elbow dislocation. 1, ulna; 2, radius; 3, distal humerus; 4, humerus; 5, collateral ligaments.

From *Sports injuries: Their prevention and treatment*, 3rd ed., L. Peterson and P. Renstrom, Copyright 2001, Human Kinetics. Reproduced by permission of Taylor & Francis books UK.

joint dislocation (Hotchkiss 1996). During the forcible hyperextension, the joint's alignment also creates valgus stresses that can lead to rupture of the medial collateral ligament and sometimes rupture at the origin of the flexor–pronator group on the medial aspect of the humerus.

O'Driscoll and colleagues (1991, 1992) described **posterolateral rotatory instability (PLRI),** which results from combined axial compression, valgus instability, and supination. PLRI progresses through four stages of severity:

Stage I: Disruption of the ulnar portion of the lateral collateral ligament complex with resultant posterolateral subluxation and spontaneous reduction

Stage II: Continued disruption anteriorly and posteriorly with accompanying partial dislocation

Stage III A: Disruption of all soft tissues except for the anterior bundle of the medial collateral ligament; complete posterior dislocation

Stage III B: Complete rupture of the medial collateral complex with gross valgus and varus instability

Elbow Fractures

Fractures can occur to any of the three bones (humerus, ulna, or radius) of the elbow. The involvement of each depends on the nature, magnitude, location, and direction of the applied forces.

- Humeral fractures can involve various areas of the distal humerus including the supracondylar, intercondylar (Y and T fractures), condylar, epicondylar, and articular regions (figure 7.15a).

- Ulnar fractures commonly involve the olecranon and result either directly from violent impact to the posterior aspect of the elbow or indirectly from falls that load the elbow joint (figure 7.15b).

- Coronoid fractures happen during joint dislocation when the trochlea shears off the tip of the coronoid process of the ulna (figure 7.15c). Coronoid fracture in conjunction with radial head fracture especially compromises elbow joint stability (Cohen 2004).

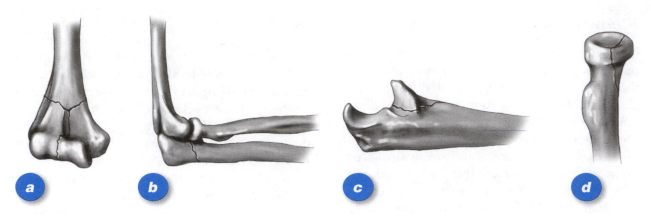

Figure 7.15 Elbow fractures. *(a)* Fracture of the distal humerus. *(b)* Ulnar fracture. *(c)* Radial head fracture. *(d)* Coronoid fracture.

■ Radial head fractures most often result from either longitudinal loading of the radius as a result of a fall or accompanying elbow dislocation (figure 7.15*d*). Radioulnar fractures are more fully described in the following section.

Forearm Injuries

The forearm spans the elbow and wrist joints and consists of two longitudinally aligned bones, the radius (lateral) and ulna (medial), surrounded by numerous muscles that function at the elbow, radioulnar, and wrist joints and whose distal tendons continue to the hand to control finger movements. The forearm muscles are divided into two major groups: the flexor–pronator group and extensor–supinator group. Specific muscles in each of these groups are presented in table 7.9.

The radius and ulna articulate with one another at the proximal radioulnar joint, which is a synovial pivot joint between the head of the radius and the ulnar radial notch, and at the distal radioulnar joint, which is a synovial pivot joint articulating the head of the ulna with the ulnar notch of the radius. Coordinated action of these two joints creates the forearm motions of pronation and supination. In pronation, the radius rolls over a relatively fixed ulna. The reverse occurs in supination, when the radius returns to its anatomical position (figure 7.16).

Diaphyseal Fractures of the Radius and Ulna

Fractures of the radius and ulna can occur in isolation or combination, with numerous suggested injury mechanisms. Identification of the injury mechanism is important because it suggests the location and type of fracture. Combined injury to both the radius and ulna typically involves direct, high-energy trauma, such as motor vehicle accidents or gunshot wounds, or the lower-energy impact of a fall.

Isolated fractures of the upper two-thirds of the radius (proximal fractures) are rare because of the protection offered by the overlying musculature. Forces sufficient to cause radial fracture in this region most often fracture the ulnar shaft as well. Fracture at the junction between the middle and distal thirds of the radius is more common and is often associated with injury to the distal radioulnar joint. Fractures of the distal third of the radius are termed **Galeazzi fractures** (after Galeazzi 1934) and result most often from a direct blow on the dorsolateral side of the wrist or a fall on an outstretched arm that axially loads a usually hyperpronated forearm (Jupiter and Kellam 2003).

Isolated fracture of the ulna arises from various mechanisms. Injury from direct trauma is colloquially termed a **nightstick fracture,** referring to the situation in which a person, in response to an impending overhead blow, raises his or her arm and exposes the medial surface to impact.

Another mechanism of isolated ulnar fracture involves dislocation of the radial epiphysis. This mechanism was originally suggested in 1814 by Giovanni Monteggia, who described fracture of the proximal ulna associated with anterior dislocation of the radial head.

Bado (1967) proposed the term **Monteggia lesion** and enlarged the scope of the injury to include all ulnar fractures resulting from radial

TABLE 7.9

MUSCLES OF THE FOREARM

FLEXOR–PRONATOR GROUP

Superficial group	Pronator teres
	Flexor carpi radialis
	Palmaris longus
	Flexor carpi ulnaris
	Flexor digitorum superficialis
Deep group	Flexor digitorum profundus
	Flexor pollicis longus
	Pronator quadratus

EXTENSOR–SUPINATOR GROUP

Extensors, abductors, and adductors of the wrist	Extensor carpi radialis longus
	Extensor carpi radialis brevis
	Extensor carpi ulnaris
Extensors of the four medial digits	Extensor digitorum
	Extensor indicis
	Extensor digiti minimi
Extensors and abductors of the thumb	Abductor pollicis longus
	Extensor pollicis brevis
	Extensor pollicis longus

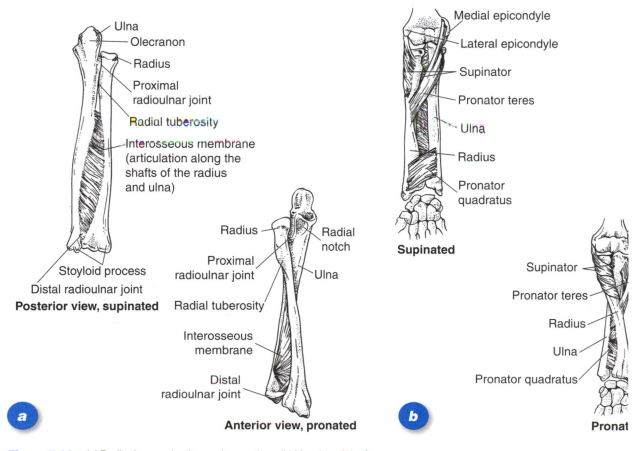

Figure 7.16 (a) Radioulnar supination and pronation. (b) Muscles of the forearm.

epiphyseal luxation. Bado suggested a classification system consisting of four types (I-IV) of injury, as depicted in figure 7.17.

▪ Type I (anterior dislocation) Monteggia fractures are the most common, accounting for about 70% of Monteggia lesions (Wilkins 2002). Several mechanisms have been proposed for type I injuries. In the 1940s, Smith (1947) and Speed and Boyd (1940) suggested that a direct blow to the posterior forearm caused a fracture in the ulnar diaphysis, followed by forcing of the radial head into anterior dislocation. In 1949, Evans described a hyperpronation mechanism from forced pronation that causes ulnar fracture and eventual anterior displacement of the radial head. The current, most accepted mechanism is the hyperextension mechanism proposed by Tompkins (1971). In this case, falling on an outstretched arm hyperextends the elbow. The biceps brachii strongly resists the hyperextension and dislocates the radial head. The compressive load is then transferred to the ulnar diaphysis, which fails in tension in the form of a complete oblique or greenstick fracture (Wilkins 2002).

▪ Type II (posterior dislocation) fractures result from posterior elbow dislocation in which the ulnar shaft fractures before ulnar collateral ligament failure. In this mechanism, a longitudinal force is directed up the forearm with the elbow flexed. This results in failure of the posterior ulnar cortex and posterior radial head dislocation (Penrose 1951). Four subtypes of type II injuries have been identified based on the location and loading mechanism (bending, shearing, or compression) of the ulnar fracture (Jupiter et al. 1991).

▪ Type III (lateral dislocation) lesions are the second most common type, accounting for nearly 25% of Monteggia lesions. Wright (1963)

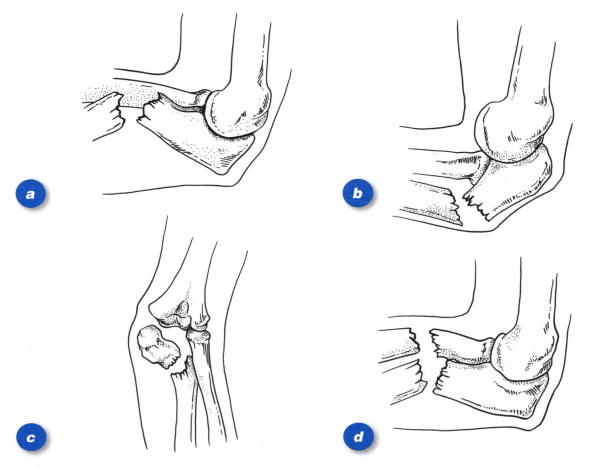

Figure 7.17 Classification of Monteggia fractures. *(a)* Type I, anterior dislocation of radial head and fracture of ulnar diaphysis with anterior angulation. *(b)* Type II, posterior or posterolateral dislocation of radial head and fracture of ulnar diaphysis with posterior angulation. *(c)* Type III, lateral or anterolateral dislocation of radial head and fracture of ulnar metaphysis. *(d)* Type IV, anterior dislocation of radial head and fracture of proximal third of radius and ulna.

described a varus load applied to an extended elbow that results in ulnar failure followed by lateral or anterolateral radial head displacement.

■ Type IV (anterior dislocation with radial head fracture) lesions are rare, accounting for about 1% of all Monteggia lesions. Type IV injuries have a similar mechanism to type I, with the added presence of radial head fracture.

Given the importance of the forearm in mediating movement between the elbow and fingers, clinicians must accurately assess injury mechanisms, make an informed diagnosis, and plan appropriate treatment.

Fracture of the Distal Radius

Fractures of the distal radius are common, accounting for up to one-sixth of all fractures. The most common mechanism is falling on an outstretched arm (Rettig and Raskin 2000). Young males and elderly females greatly outnumber their opposite-gender counterparts in the incidence of fracture.

The literature is replete with eponyms for fractures of the distal radius. These include *Colles'*, *Smith's*, and *Barton's fractures*, each with their own distinguishing characteristics. Here, we use the clinical descriptions with parenthetical reference to corresponding eponymic designations.

Many classification systems have been proposed to describe distal radial fractures. One useful system groups these fractures according to their mechanisms of injury rather than by radiological characteristics (Jupiter and Fernandez 1996). Five types are included: type I, bending fractures; type II, shearing fractures of the joint surface; type III, compression fractures of the joint surface; type IV, avulsion fractures; and type V, combined fractures (figure 7.18).

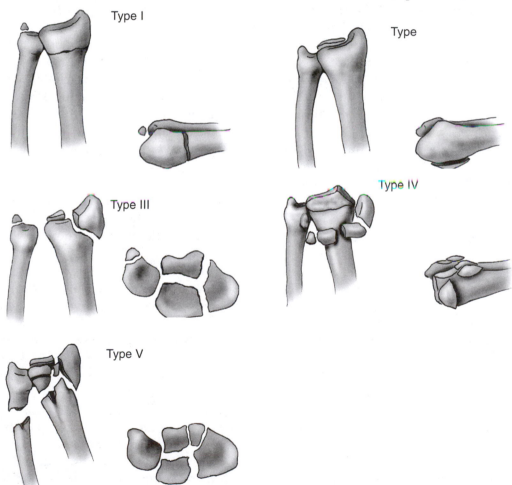

Figure 7.18 Fractures of the distal radius.

Adapted from D.L. Fernandez and J.B. Jupiter, 1996, *Fractures of the distal radius* (New York, NY: Springer-Verlag), 49, with kind permission of Springer Science and Business Media.

Type I bending fractures result from landing on an outstretched arm. Axial compressive loads cause bending of the radius with a fracture pattern (Colles' fracture) showing volar (anterior or palmar) metaphyseal cortex failure in tension and varying degrees of comminution on the dorsal (posterior) surface. These type I injuries constitute the majority of distal radial fractures. An opposite bending results from a fall on either a flexed wrist or an outstretched and supinated arm. The compressive loading then results in tensile failure (Smith's fracture) on the dorsal (posterior) aspect of the metaphysis and compressive comminution on the volar aspect. This fracture pattern also is seen when a slightly flexed clenched fist hits a rigid surface.

High-energy loading, particularly in young individuals, produces type II shearing fracture (Barton's fracture) in which the volar lip of the radial articular surface is sheared off. In type III fractures, high compressive loads (such as those generated in landing from a high fall) cause intra-articular fractures of the joint surface with disruption of the subchondral and cancellous bone.

Mechanical loading that creates high stresses on the osteoligamentous attachments, as in exaggerated torsion, produces avulsion (type IV) fractures. The most complex mechanism of distal radial fracture is type V, or combined fracture, which usually results from high-energy injuries including combinations of bending, compression, shearing, or avulsion mechanisms.

Ulnar Variance

The relative lengths of the ulna and radius play an important role in forearm and wrist mechanics. This length difference is referred to as **ulnar variance (UV).** (Note: the term *ulnar variance* should not be confused with the term *ulnar deviation*, which describes frontal plane wrist–hand movement toward the ulnar side of the forearm.)

If the radius and ulna are the same length, UV is zero. When the ulna is longer than the radius, there exists a **positive ulnar variance.** Conversely, a **negative ulnar variance** exists when radial length exceeds ulnar length. Measures of UV vary, with reported means ranging from –0.84 mm to 0.2 mm (Kristensen et al. 1986; Nakamura et al. 1991). Caution is warranted in interpreting mean values because UV varies by age, ethnicity, and possibly gender. In addition, UV is determined by genetics, elbow pathology, and loading history (De Smet 1994). Loading history is most relevant to our discussion of injury mechanisms. Although the wrist is not designed to function as a load-bearing joint, certain activities (e.g., gymnastics) expose the wrist to considerable loads. These compressive loads are transmitted through the carpals to the radius and ulna, with the radius accepting approximately 80% of the load. In cases of repetitive compressive loading in the skeletally immature individual, this loading differential may dictate premature closure of the radial growth plate. Continued growth of the ulna then would create an acquired ulnar variance.

Gymnasts are especially prone to UV. The dual risk factors of early onset (beginning training at a relatively young age) and repetitive upper-extremity load bearing account for the prevalence of wrist lesions in this population. Aged-matched elite gymnasts had a positive UV of 0.46 mm (De Smet et al. 1994) compared with –1.1 mm in non-elite gymnasts (DiFiori et al. 1997) and –2.3 mm in nongymnasts (Hafner et al. 1989). More positive UV in gymnasts than would be predicted based on nongymnasts suggests that the high loads placed on gymnasts' wrists may inhibit growth of the distal radius, stimulate growth of the ulna, or both (DiFiori et al. 1997). Despite the plausibility of inhibited distal radial growth, conclusive evidence remains elusive (Caine et al. 1997).

Certain gymnastics skills place the athlete's wrist at particular risk. The back handspring, for example, loads the wrist with forces up to 2.4 times body weight, and the radius accepts most of the load (Koh et al. 1992). For men, the pommel horse is the main culprit. Joint forces of up to 2 times body weight and loading rates of up to 219 times the body weight per second have been reported (Markolf et al. 1990). The gymnast's wrist assumes a load-bearing role for which it is ill designed. The consequences manifest in an ulnar impaction syndrome, characterized by progressive degeneration of the triangular fibrocartilage complex and the ulnar carpus.

How much joint loading is too much? Because of the multifactorial nature of the problem, the question remains problematic and unresolved. Some guidance is provided by a study that examined factors associated with wrist pain in young gymnasts. Training intensity, relative to the age of the participant and the age when training began, seems to be a critical determinant of wrist pain development (DiFiori et al. 1996). Although wrist pain is common in gymnasts, UV is not associated with wrist pain or injury (based on radiographic evidence) of the distal radial physis (DiFiori et al. 2002).

Wrist and Hand Injuries

The wrist is not a single joint but rather a group of articulations that include the distal radioulnar, radiocarpal, and intercarpal joints. The hand contains numerous articulations, namely the carpometacarpal (CM), metacarpophalangeal (MP), and interphalangeal (IP) joints (figure 7.19). All of these are synovial joints. Structurally, the MP joints are condyloid, whereas the IP joints are hinge joints. Both MP and IP joints are strengthened by palmar and collateral ligaments.

Muscles in the forearm primarily control wrist and finger motion with assistance from intrinsic muscles of the hand. The distal tendons of most flexor muscles in the forearm pass along the ventral aspect of the wrist, where they are held firmly in place by the flexor retinaculum, a thick and relatively inextensible fascial sheath. These tendons, along with neurovascular structures, pass through the so-called carpal tunnel formed by the carpal bones and the flexor retinaculum (figure 7.20). The distal tendons of the extensors are secured similarly between the carpal bones and the extensor retinaculum.

Carpal Tunnel Syndrome

Injuries resulting from repeated tissue stress are collectively known as cumulative trauma disorders (CTDs), also called *repetitive strain injury, chronic microtrauma, overuse syndrome,* and *cumulative trauma syndrome.* These chronic injuries have been increasing at an alarming rate, especially in occupational settings, and account for more than 60% of all new nonfatal occupational illnesses (National Safety Council 2004). CTDs are prevalent in many occupations and prove costly in both economic and human terms.

One of the most debilitating chronic disorders is **carpal tunnel syndrome (CTS),** a condition first reported by Paget in 1854 (Lo et al. 2002), which is characterized by swelling within the carpal tunnel that creates a compressive neuropathy affecting the median nerve (figure 7.20). Like other entrapment syndromes, CTS involves

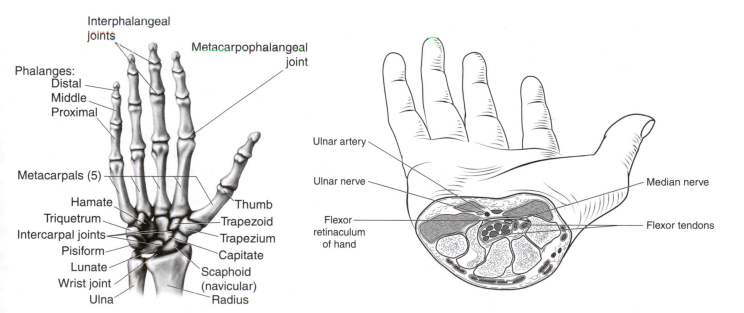

Figure 7.19 Bones and articulations of the wrist and hand.
Reprinted, by permission, from R.S. Behnke, 2005, *Kinetic anatomy,* 2nd ed. (Champaign, IL: Human Kinetics), 77.

Figure 7.20 Structure of the carpal tunnel. Neurovascular structures, including the median nerve, pass through the carpal tunnel bounded by the flexor retinaculum and the carpal bones.

increased pressure within a confined space. The inextensible borders formed by the carpal bones and the flexor retinaculum preclude an increase in tunnel size. Inflammation and edema in response to repeated loading compress neurovascular tissues and compromise their function. Of greatest consequence is compression of the median nerve, which results in sensory symptoms of numbness, tingling, burning, and pain in the wrist and radial 3 1/2 fingers. Sensory deficits from CTS are more pronounced than reductions in motor function.

Symptoms of CTS are associated with specific movement patterns (e.g., wrist flexion–extension) and tasks (e.g., assembly work, typing, playing a musical instrument, polishing, sanding, scrubbing, or hammering). Carpal tunnel syndrome has also been documented in workers across a diverse range of jobs, including keyboard operators, sheet metal workers, supermarket checkers, sheep shearers, fish-processing workers, and sign language interpreters. Making a causal connection in all cases warrants caution, however, because discrimination can be tenuous between work-related and non-work-related cases of CTS. The prevalence of self-reported CTS in the general population is estimated at about 1.5%, with confirmed cases less frequent. CTS is more common among women than men in a ratio of about 2:1 (Phalen 1966; Praemer et al. 1999).

The etiology of CTS is complex, because many mechanisms and risk factors play a contributing role. Fundamentally, the pathophysiology of CTS involves a combination of ischemia and mechanical trauma (Viikari-Juntura and Silverstein 1999; Werner and Andary 2002).

Accurate diagnosis of CTS can be difficult initially because repetitive stress is implicated in other hand and wrist pathologies, such as tendinitis. Once diagnosed, CTS remains mechanically problematic because there are multiple risk factors, including forceful exertions, repetitive or prolonged activities, awkward postures, localized contact stresses, vibration, and cold temperatures.

Silverstein and colleagues (1987) evaluated 652 workers to assess the role of force and repetition on the prevalence of CTS in workers. These authors found the lowest prevalence (0.6%) in workers with low-force, low-repetition jobs and the highest occurrence (5.6%) in workers in high-force, highly repetitive jobs. Silverstein and colleagues concluded that of the two factors, high repetitiveness appeared to be a greater risk factor than high force. Although studies such as this shed some light on the relations among CTS risk factors, unraveling the interrelations and relative contributions of all risk factors remains a challenge.

Carpal Fractures

Wrist fractures encompass osseous injury to the radius and ulna (discussed earlier) and fractures to any of the eight carpal bones. The most common mechanism of injury is a compressive load applied to a hyperextended (dorsiflexed) wrist (figure 7.21). Other, less common mechanisms have been implicated in carpal fracture. These mechanisms include hyperflexion and rotational loading against a fixed object or surface.

The vast majority of carpal fractures occur when axial loads are transmitted through a hyperextended wrist. With the wrist in this position, compressive forces are transmitted through the carpals to the distal radioulnar complex. Predictably, certain carpals are more likely to suffer injury than others. Mitigating influences include the degree of radial or ulnar deviation, amount of energy absorbed, point of application and direction of the applied forces, and relative strength of the bones and ligaments.

Given the 80:20% load distribution between the radius and ulna, respectively, the scaphoid and lunate are most likely to be fractured because of their articulations with the radius. Scaphoid fractures alone account for 60% to 70% of all carpal fractures (Botte and Gelberman 1987). Scaphoid fractures are most likely to occur when the wrist is hyperextended past 95° and the radial portion of the palm accepts most of the load (Dias 2002; Weber and Chao 1978). Typically, the palmar aspect of the scaphoid then fails in tension with the dorsal aspect failing in compression (Ruby and Cassidy 2003). Less commonly, scaphoid fracture can result from a compressive force applied to a neutral or slightly flexed wrist in what has been termed a *puncher's fracture* (Horii et al. 1994).

Lunate fractures are second in frequency, but rare, and typically result from compressive forces

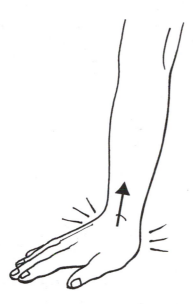

Figure 7.21 Mechanism of injury attributable to compressive load applied to a hyperextended wrist at impact.

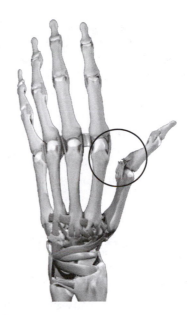

Figure 7.22 Abductor mechanism causing rupture of the ulnar collateral ligament of the thumb.
Copyright Primal Pictures Ltd.

applied through the capitate, which push the lunate ulnarly and exacerbate the dorsal rotation caused by wrist hyperextension. The forces resisting these movements tend to cause transverse lunate fractures.

Fractures of the other carpals are rare, with injury mechanisms ranging from a direct blow to impingement caused by hyperflexion or hyperextension.

Thumb Injuries

The thumb is essential to our prehensile abilities, because it is the digit that opposes the four fingers. Injury to the thumb can severely impair a person's manual dexterity. We describe three common thumb injuries (sprain, fracture, and neural lesion) to illustrate injury mechanisms that can significantly impair thumb function.

The most common sprain in the hand damages the ulnar collateral ligament of the first metacarpophalangeal joint (figure 7.22). This injury, colloquially referred to as *gamekeeper's thumb* or *skier's thumb,* can involve chronic tensile loading (stretching) of the ligament or acute loading that result in any level of sprain, including complete rupture. This injury most commonly occurs when a skier falls onto an outstretched hand with the thumb in an abducted position. The ski pole handle effectively holds the thumb in abduction

as the compressive load of the fall is accepted by the hand. The forceful abduction places excessive tensile loads on the ulnar collateral ligament and hastens its failure.

Ulnar collateral sprain also can result from hyperextension of the first metacarpophalangeal joint, such as when a collision occurs between two athletes. A softball player tagging an opponent who is sliding into second base may have her thumb forcibly hyperextended by the contact force between her thumb and the foot or leg of the incoming base runner.

Fracture subluxation of the trapeziometacarpal joint **(Bennett's fracture)** is an intra-articular fracture of the first metacarpal (thumb) resulting from axial force applied when the metacarpal bone is in flexion. Many circumstances can cause this injury, but it is classically observed as a result of a poorly delivered punch in a boxing match or fist fight.

Our third injury is a neural lesion characterized by perineural fibrosis of the ulnar digital nerve of the thumb. Known as *bowler's thumb,* this injury most commonly involves symptoms of **paresthesia** (tingling) and, to a lesser extent, tenderness and **hyperesthesia** (pathological sensitivity). The injury mechanism is repeated blunt trauma to the thumb's ulnar digital nerve caused by the gripping and release of a bowling ball. This condition can be treated conservatively or prevented by redrilling the bowling ball or modifying the grip mechanics to lessen the repetitive trauma to the nerve.

Metacarpal and Phalangeal Conditions

The pattern of fracture and dislocation involving the metacarpals and phalanges directly depends on the circumstances of injury (direct impact, crushing, indirect trauma) and the nature of the force application (e.g., magnitude, location, direction). Among the implicated injury mechanisms are direct trauma caused by an implement or fall, forcible hyperextension or hyperflexion, twisting forces, violent distraction, forced leverage, crushing, or a combination of these mechanisms.

■ CHAPTER REVIEW

Key Points

- Impairment to any structural component of the upper limb compromises our ability to effectively manipulate things within our reach or propel objects.

- The upper extremity is designed for mobility, and loss of dexterity can have profound effects on our ability to efficiently perform grasping or manipulative tasks.

- Cumulative trauma and repetitive strain injuries are major burdens for today's workplace and the health care system.

- Acute or chronic injuries to any of the upper-extremity joints can significantly reduce a person's quality of life—involving activities of daily living, work, and recreation.

Questions to Consider

1. This chapter's "A Closer Look" examined rotator cuff pathologies in detail. Select another injury presented in the text and write your own "A Closer Look" for that injury.

2. Explain, using specific examples, how upper-extremity injuries may compromise our ability to effectively manipulate items within our reach or propel objects.

3. Chapter 3 discussed a mobility-stability continuum which states that greater joint stability (i.e., resistance to dislocation) usually is associated with lesser mobility, and conversely that greater joint mobility is associated with lesser stability. Upper-extremity joints are among the most mobile joints in the body and hence are relatively unstable. Describe, using specific examples, how this lack of stability may increase injury risk.

4. The text presents carpal tunnel syndrome as an example of a cumulative trauma disorder (CTD). Describe in detail other examples of CTD in the upper extremity.

Suggested Readings

Andrews, J.R., and K.E. Wilk, eds. 1994. *The Athlete's Shoulder*. New York: Churchill Livingstone.

Browner, B.D., J.B. Jupiter, A.M. Levine, and P.G. Trafton. 2003. *Skeletal Trauma* (3rd ed.). Philadelphia: Saunders.

Bucholz, R.W., ed. 2005. *Rockwood, Green, and Wilkins' Fractures* (6th ed.). Philadelphia: Lippincott Williams & Wilkins.

Bulstrode, C., J. Buckwalter, A. Carr, L. Marsh, J. Fairbank, J. Wilson-MacDonald, and G. Bowden, eds. 2002. *Oxford Textbook of Orthopedics and Trauma*. Oxford, UK: Oxford University Press.

Burkhead, W.Z., Jr., ed. 1996. *Rotator Cuff Disorders*. Baltimore: Williams & Wilkins.

Fu, F.H., and D.A. Stone. 2001. *Sports Injuries: Mechanisms, Prevention, Treatment* (2nd ed.). Philadelphia: Lippincott Williams & Wilkins.

Gilula, L.A., ed. 1992. *The Traumatized Hand and Wrist*. Philadelphia: Saunders.

Moore, E.E., D.V. Feliciano, and K.L. Mattox, eds. 2003. *Trauma* (5th ed.). New York: McGraw-Hill.

Morrey, B.F. ed. 1993. *The Elbow and Its Disorders* (2nd ed.). Philadelphia: Saunders.

Nicholas, J.A., and E.B. Hershman, eds. 1995. *The Upper Extremity in Sports Medicine* (2nd ed.). St. Louis: Mosby-Year Book.

Richards, R.R. 1995. *Soft Tissue Reconstruction of the Upper Extremity*. New York: Churchill Livingstone.

Woo, S.L.-Y., and J.A. Buckwalter, eds. 1988. *Injury and Repair of the Musculoskeletal Soft Tissues*. Park Ridge, IL: American Academy of Orthopaedic Surgeons.

CHAPTER 8

HEAD, NECK, AND TRUNK INJURIES

The doctors X-rayed my
head and found nothing.

Dizzy Dean, Hall of Fame baseball pitcher, after being
hit on the head by a ball during the 1934 World Series

OBJECTIVES

To describe the relevant head, neck, and trunk anatomy involved in musculoskeletal injury

To identify and explain the mechanisms involved in musculoskeletal injuries to the head, neck, and spine

Of all the body's regions, the head, neck, and trunk are of paramount importance in controlling life-sustaining functions. Trauma to structures in these regions (e.g., brain, spinal cord, heart) poses the greatest danger to our physical well-being. In the preceding two chapters we discussed injuries to the extremities that result in disability but are rarely fatal. Injuries to the head, neck, and trunk, in contrast, have real and immediate potential for fatality. Understanding the mechanisms responsible for these injuries can assist with their proper diagnosis, treatment, and prevention.

In this chapter, the section titled "Concussion: A Closer Look" provides a detailed exploration of concussion injuries. In this section (p. 251) we expand our discussion beyond the mechanisms of injury to include detailed descriptions of neural tissue structure and mechanics and explanations of clinical evaluation, treatment, and rehabilitation.

Head Injuries

The head includes the skull, brain, meninges, cranial nerves, sense organs, and superior aspects of the digestive system. Structures in the head are protected by an intricate collection of 29 bones. The brain and its protective meningeal covering are contained in a cranial vault composed of eight cranial bones: frontal, occipital, ethmoid, and sphenoid bones, and paired temporal and parietal bones (figure 8.1). The anterior and anterolateral aspects of the head are formed by 14 facial bones: paired nasal, maxillae (upper jaw), zygomatic (cheek), lacrimal, palatine, and inferior nasal conchae bones and singular mandible (lower jaw)

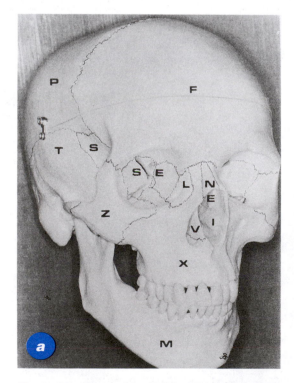

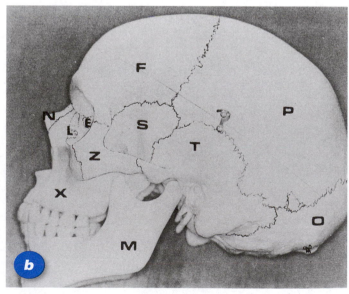

Figure 8.1 Bones of the skull from *(a)* frontal and *(b)* lateral views: ethmoid (E), frontal (F), inferior nasal concha (I), lacrimal (L), mandible (M), nasal (N), occipital (O), parietal (P), sphenoid (S), temporal (T), vomer (V), maxilla (X), zygomatic (Z).

Reprinted with permission from *Health and Neck Injury* (P-276) © 1994 SAE International.

and vomer bones. Of the seven remaining bones of the head, six are housed in the ear. These bones are collectively referred to as auditory ossicles (paired incus, stapes, and malleus bones). The final bone is the hyoid, which is suspended from the styloid process of the temporal bone by ligaments and muscles.

The brain is composed of the cerebrum, cerebellum, and brain stem (figure 8.2). The cerebrum, the largest and most superior of the brain structures, appears as two hemispheres of highly convoluted nervous tissue. Each hemisphere is divided into regions, or lobes (frontal, parietal, temporal, occipital), each with unique and often complementary roles in sensorimotor processing. The cerebrum and spinal cord are covered by three layers of protective tissue known as **meninges.** The outermost layer is the **dura mater,** a tough connective tissue that in the cranium splits into two sheets. The outer (endosteal) sheet adheres tightly to the bone and is separated from the inner sheet in some areas by cavities known as venous sinuses. The middle meningeal layer **(arachnoid)** appears as a weblike membrane and is separated from the innermost layer **(pia mater)** by the subarachnoid space. The subarachnoid space contains **cerebrospinal fluid (CSF),** which circulates around the brain and spinal cord and provides supportive, protective, and nutritive functions.

Located just inferior to the cerebrum is the brain stem. The three structures of the brain

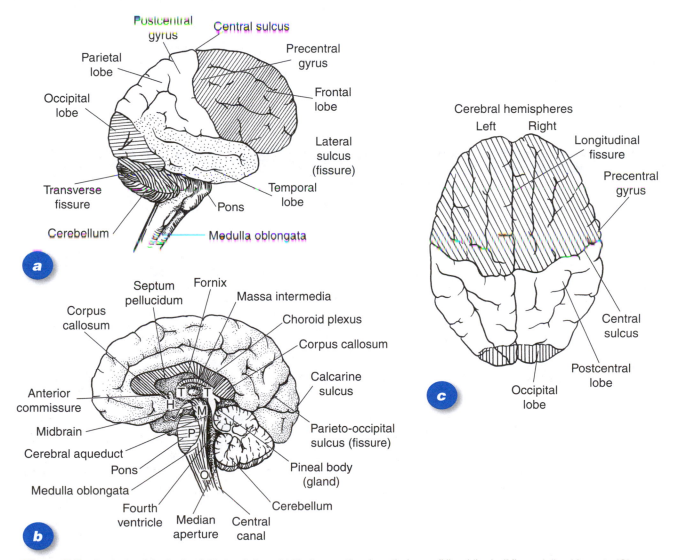

Figure 8.2 Anatomy of the brain. *(a)* Lateral view. *(b)* Median section: hypothalamus (H), midbrain (M), medulla oblongata (O), pons (P), thalamus (T). *(c)* Superior view.

stem (midbrain, pons, and medulla oblongata) serve as pathways for most sensory and motor information passing between the cerebrum and spinal cord and also house numerous vital reflex centers essential for life-supporting functions, such as heart rate, respiration, blood pressure, and levels of consciousness. The brain stem also contains the nuclei for most of the 12 cranial nerves (table 8.1).

The second largest brain structure, the cerebellum, is located inferior to the cerebrum and posterior to the brain stem. Among its functions are coordination of the subconscious movements of skeletal muscles, movement error detection, maintenance of equilibrium and posture, and prediction of the future position of the body during a particular movement. The cerebellum also plays a role in emotional development by modulating sensations of pleasure and anger.

Principles of Head Injury

Head injuries range from mild to life threatening. The severity of injury depends, in part, on several principles based on the head's unique anatomical and neurological constituents. We introduce several terms and discuss principles of head injuries here before exploring specific head injuries in detail.

Traumatic Brain Injury

Traumatic brain injury (TBI) is a term that covers numerous conditions arising from impact or acceleration–deceleration of the head. Specific TBI injuries include focal lesions (e.g., scalp, skull, and dural lesions), contusions, lacerations, intracranial hematoma, brain damage from increased intracranial pressure (ICP), and a variety of diffuse brain injuries (e.g., diffuse axonal injury) (Gennarelli and Graham 2005).

TABLE 8.1

CRANIAL NERVES

Nerve	Name	Sensory functions	Motor functions
I	Olfactory	Olfaction	None
II	Optic	Vision	None
III	Oculomotor	Proprioception	Movement of eyeball and eyelid, lens accommodation, pupil constriction
IV	Trochlear	Proprioception	Movement of eyeball
V	Trigeminal	Proprioception, sensations of touch, pain, and temperature for ophthalmic, maxillary, and mandibular regions	Mastication
VI	Abducens	Proprioception	Movement of eyeball
VII	Facial	Proprioception, taste	Facial expression, saliva and tear secretion
VIII	Vestibulocochlear	Balance, hearing	None
IX	Glossopharyngeal	Taste, blood pressure regulation, proprioception	Secretion of saliva
X	Vagus	Sensations from many organs, proprioception	Visceral muscle movement, speech, swallowing, decreases in heart rate
XI	Accessory	Proprioception	Swallowing, head movements
XII	Hypoglossal	Proprioception	Tongue movement during speech and swallowing

TBI is a major public health issue which, because of the absence of visible injury, has been characterized by the U.S. Centers for Disease Control and Prevention (CDCP) as a "silent epidemic" (Langlois et al. 2004). TBI statistics are staggering. In the United States, more than 1.4 million TBIs happen annually, resulting in more than 1.1 million emergency department visits, 235,000 hospitalizations, and 50,000 deaths. Each year in the United States, 80,000 to 90,000 people suffer the onset of TBI-related long-term disability (Thurman et al. 1999), and estimates indicate that more than 5,000,000 Americans currently live with TBI-related disabilities (National Center for Injury Prevention and Control 2001). And these statistics underestimate the true scope of the problem, because many cases of TBI go undiagnosed or unreported.

Mechanisms

Head injuries sometimes are epidemiologically associated with their causal events (e.g., motor vehicle crashes). Although some authors refer to these events as the mechanism of injury, we prefer to view them as the causal circumstances associated with the injury and reserve the term *mechanism* for the physical processes directly causing an injury. The prevalence of causal events is shown in figure 8.3. These statistics, although interesting, are of limited usefulness for determin-

ing risk or understanding specific factors that cause head injury.

Head injuries occur in response to the sudden application of forces to the head or its connected structures. Numerous interrelated factors combine to determine the exact mechanism of injury. These factors include the type of force and its magnitude, location, direction, duration, and rate.

The forces causing head injuries are characterized as direct or indirect. Direct (contact) loading results from impact, as exemplified by a boxer's punch. Indirect (inertial) loading occurs when forces are transmitted to the head through adjacent structures such as the neck (e.g., whiplash mechanism). Whether direct or indirect, an applied force either accelerates or decelerates the head. Forces applied to a stationary head will tend to accelerate its mass, whereas forces acting in opposition to the head's motion will decelerate it (figure 8.4). A forceful blow to the head typifies an acceleration mechanism. The deceleration mechanism is involved when the head's motion is abruptly stopped by an unyielding surface. These acceleration and deceleration mechanisms often are implicated in brain injury caused by head trauma.

The effects of forces also are categorized by the type of head motion that occurs in response to loading. Forces directed through the head's center of mass cause linear translation of the head (figure 8.5*a*), whereas forces acting eccentrically (off center) from the center of mass result in rotation of the head in any or all of the three primary planes (figure 8.5*b*). In many situations the applied forces cause combined translational and rotational, or general, motion of the skull and its contents.

The mechanical properties of the head's constituent tissues play an important role in determining the location and severity of injury. The skull forms a stiff, yet slightly compressible container housing the brain and its covering structures. The brain, in contrast, is more compliant. When describing potential injury mechanisms, we must consider the different mechanical responses and relative mobility of these tissues. When a load is applied to the head, the acceleration–deceleration profiles of each structural component dictate a complex and not completely understood response.

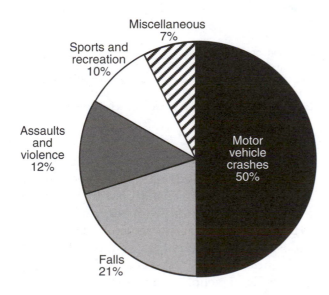

Figure 8.3 Causes of head injury.

Adapted, by permission, from D.V. McGehee, 1996, *Head injury and postconcussive syndrome* (New York, NY: Churchill Livingstone), 58.

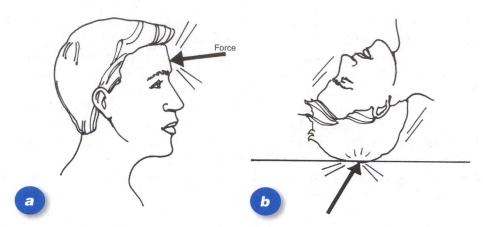

Figure 8.4 *(a)* Force applied to a stationary head. *(b)* Impact force created by contact of a moving head with an unyielding surface.

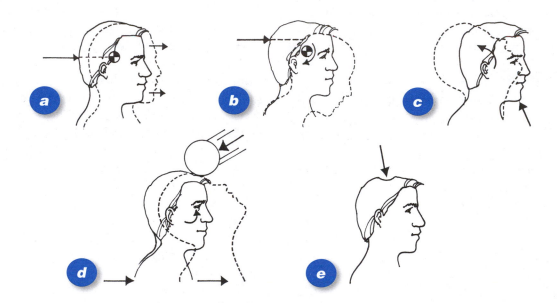

Figure 8.5 Effect of forces applied to the head. (Solid line = original position; dashed line = position following force application). *(a)* Force applied through the head s center of mass causes linear translation of the head. *(b)* Force applied eccentrically (off center) causes rotation of the head. *(c)* Upper-cut punch to a boxer s chin causes rotation of the head. *(d)* Simultaneous impact by a soccer ball to the head and contact force to the back produce both rotation of the head and translation of the body. This combination accentuates the hyperextension mechanism in the cervical spine. *(e)* Blunt trauma to the superior aspect of the head causes a depressed skull fracture.

The brain and other intracranial tissues develop internal stresses in response to loading and usually are deformed or strained. Each of the three principal stresses and strains (compression, tension, shear) can be present. Injury occurs when the capacity of the tissue to withstand the applied load is exceeded. Tissue damage may be restricted to a limited area **(focal injury)** or may pervade a large region of neural tissue **(diffuse injury).**

In terms of structural damage, head injuries are categorized either as closed head injuries or penetrating injuries. Closed head injuries (e.g., concussion, cerebral edema, diffuse axonal injury) result from rapid translational or rotational acceleration of the head with attendant damage to the brain or its covering structures but with no exposure to the external environment. Penetrating injuries, in contrast, occur when an object (e.g., bullet, javelin, arrow) directly penetrates the skull and its neurovascular contents.

The factors just described don't necessarily occur in isolation; rather two or more of them may be used to characterize any given injury. For example, the unfortunate recipient of an upper-cut punch in boxing might experience an acceleration injury that creates violent hyperextension (figure 8.5c). In another example, a soccer player who, while heading the ball, is simultaneously contacted in the back by an opponent (figure 8.5d) would experience the combined effects of forward linear translation of the trunk and hyperextension of the neck caused by the impacting ball and forward movement of the trunk.

We offer three final observations before considering specific head injuries:

- Head injuries often are insidious, in that evidence at the time of primary injury may provide little or no indication of associated secondary injuries that will subsequently develop.

- Injury to superficial structures does not necessarily and invariably result in intracranial damage to the brain and its coverings. Evidence of extensive superficial damage (e.g., copious bleeding), for example, does not always predict cerebral injury. Conversely, extensive brain injury can occur in the absence of superficial damage. In most cases, the overall severity of a head injury is best judged by the degree of internal damage to neural structures because these are responsible for effective cognitive, sensory, and motor function.

- In head injuries, it is not unusual for intracranial damage to occur some time after the actual instant of injury. Computed tomography scans taken immediately after an injury may indicate no intracranial contusion or hemorrhage. Evidence of injury may not appear for several days, so continued monitoring of head injury patients is strongly advised.

Skull Fracture

Fractures of the cranium usually result from blunt trauma as the head hits a surface or object (e.g., falls or motor vehicle accidents) or from a direct blow to the head (e.g., by a baseball bat). Unusual cases have also been reported, such as the person who suffered a compound depressed skull frac-

ture from being assaulted by an attacker using a stiletto-heeled shoe (Stables et al. 2005).

The study of skull biomechanics has a long history. Pioneering work by Gurdjian and colleagues in the 1940 and 1950s (e.g., Gurdjian and Webster 1946; Gurdjian et al. 1947, 1949, 1953) and Hodgson and colleagues in the 1960 and 1970s (e.g., Hodgson 1967; Hodgson and Thomas 1971, 1972, 1973) provided valuable information on skull strength and fracture characteristics.

More recent research has added to our understanding of skull mechanics. Yoganandan and colleagues (1995), for example, tested cadaver skulls and reported failure loads ranging from 4.5 to 14.1 kN (1,011-3,168 lb). The skulls had higher mean failure loads (11.9 kN or 2,674 lb) when loaded dynamically than when loaded quasistatically (6.4 kN or 1,438 lb). Skull strength varies regionally, with highest strength in the occipital (posterior) region, followed by the lateral (side) regions and the frontal region (Yoganandan and Pintar 2004).

Deleterious consequences of the fracture itself are typically minimal. The greater concern is that skull fractures are commonly, although not always, associated with damage to underlying intracranial structures. These injuries include cerebral contusions, intracranial hemorrhage, and in cases of exposure to contaminants (as with scalp laceration or exposure to the nasal cavity and paranasal sinuses) infection of the cerebrospinal fluid (meningitis or cerebritis). Skull fracture is strongly predictive of intracranial hemorrhage. The absence of fracture, however, should not be interpreted as precluding brain injury because approximately 20% of fatal head injuries show no evidence of skull fracture (Gennarelli and Graham 2005).

Skull fractures can occur along the convexity, or vault, of the skull or through the skull base. The mechanism causing convexity fractures is typically low-velocity, blunt trauma, whereas the mechanism in basilar fractures is usually high-velocity acceleration and deceleration. In severe cases, the compressive force of blunt trauma can cause depression of a skull fragment into the subcranial space in what is termed a depressed skull fracture (figure 8.5e). These fractures usually cause cerebral contusion and intracranial bleeding involving the protective layers (meninges) of the brain. The extensive meningeal vasculature increases the likelihood of hemorrhage between the layers, which can be any of the following types:

▪ **Subdural hematoma (SDH)**—Blood accumulation in the subdural space between the outer protective layer (dura mater) of the brain and the middle layer (arachnoid). SDH is the most common form of intracranial mass lesion and can result from a traumatic event (e.g., rapid acceleration–deceleration or blunt force trauma) as acute subdural hematoma or over an extended period of days or weeks as a result of minor head trauma, often in elderly persons. This latter form is termed *chronic subdural hematoma*.

▪ **Subarachnoid hematoma (SAH)**—Blood accumulation in the subarachnoid space between the middle (arachnoid) and inner (pia mater) layers covering the brain. SAH typically happens as a result of head trauma but also can result from a cerebral **aneurysm.**

Blood may also accumulate between the cranial vault and the outer protective layer (dura mater). This lesion is termed an **epidural hematoma.** Blood accumulation caused by hematoma often raises **intracranial pressure (ICP)** on the delicate brain tissues. Unchecked increases in ICP can cause permanent brain damage and eventual death.

Cerebral Contusion

Not surprisingly, head impact often damages the cerebral cortex. What is surprising is the variety of mechanisms that can damage the same areas, as explored in the following section. This damage frequently manifests as a cerebral contusion with secondary brain swelling.

Coup or Contrecoup?

Explaining the observed location of cerebral contusion is no simple task. Contusion can occur directly beneath the site of impact **(coup lesion)** or opposite to the impact location **(contrecoup lesion)** (figure 8.6). The mechanisms for each type of injury are complex and have been the

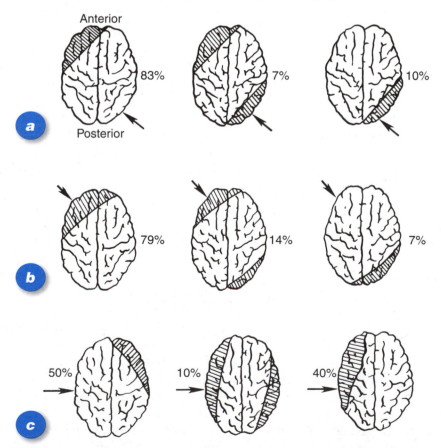

Figure 8.6 Location of coup and contrecoup lesions in 63 cases in which patients died as a result of head injuries. (Arrows indicate location of impact force. Shaded areas indicate areas of cerebral lesions.) *(a)* Occipital impacts. *(b)* Frontal impacts. *(c)* Lateral impacts.

Adapted, by permission, from K. Sano et al., 1967, "Mechanisms and dynamics of closed head injuries (preliminary report)," *Neurologia Medico-Chirurgica* 9: 22.

BOXING: HEAVYWEIGHT ISSUES

The sport of boxing presents complex medical and social issues. No other sport generates more controversy, condemnation, and passion than boxing. On purely medical grounds, the discontinuance of boxing easily could be justified. The sport, however, does not exist in a vacuum; it is surrounded by complex sociological, psychological, moral, and financial considerations. Although some vehemently argue that boxing should be banned, the sport's historical and contemporary complexity make it more amenable to compromise than to abolition.

From a physical perspective, boxing raises many interesting issues. Boxers participate in perhaps the most vigorous of all sports, one that requires an efficient combination of strength, speed, agility, coordination, and stamina. To a greater degree than most other athletes, boxers subject themselves to unprotected physical abuse. Boxing-related injuries range from minor cuts and abrasions to more severe damage, such as head and neck trauma. At particular risk of injury are the hand and wrist (e.g., carpometacarpal fracture, first metacarpophalangeal ulnar collateral ligament tear), face (e.g., ocular injuries, lacerations, facial fractures), and brain (e.g., concussion, subdural hematoma, intracerebral hemorrhage, diffuse axonal injury). In their most severe form, boxing injuries can lead to sudden death. Although other sports (e.g., college football, scuba diving, motorcycle racing, hang gliding, skydiving, and horse racing) have higher fatality rates, boxing nonetheless accounted for more than 650 deaths worldwide in the 20th century.

Perhaps even more devastating than the acute injuries just noted are the cumulative neurological injuries that develop after repeated mechanical insult. These injuries have been described in both common terms (e.g., punch drunk syndrome, dementia pugilistica) and medical terms such as *chronic traumatic boxer's encephalopathy* (Gross 2001), *chronic traumatic brain injury* (CTBI) (Clausen et al. 2005; Jordan 2000), and *chronic traumatic encephalopathy* (McCrory 2002).

The existence of chronic sequelae in some boxers is well documented. Evidence suggests that some boxers may be predisposed to severe chronic brain injury, whereas others can receive innumerable blows over many years and exhibit limited, if any, discernible neurological deficits. Approximately 20% of professional boxers show evidence of CTBI, whereas CTBI in amateur boxers happens infrequently (Jordan 2000). A recent study found no evidence of neuropsychological deterioration in a small group of amateur boxers tracked for 9 years (Porter 2003).

Among the risk factors for CTBI are increased exposure, poor boxing performance, increased sparring, and apolipoprotein (ApoE) genotype (Jordan 2000). Research suggests that some boxers may have a genetic predisposition to CTBI, based on a genetic variation in apolipoprotein E epsilon-4 (Jordan et al. 1997).

On a positive note, Clausen and colleagues (2005) predicted that the incidence of CTBI in boxing may diminish in the future because of two historical developments in boxing:

- Reduced exposure to repetitive head trauma (attributable to the reduction since the 1930s in the average length of a pro career, from 19 to 5 years, and a reduction in average number of career bouts, from 336 to 13)

- Improved medical monitoring via preparticipation physical examinations and periodic neuroimaging assessments

Boxer receiving a punch.

subject of considerable debate in the medical literature.

Early accounts set the stage for this ongoing debate. Holbourn (1943, p. 438) noted that "the idea that the brain is loose inside the skull, and that when the head is struck it rattles about like a die in a box, thereby causing coup and contrecoup injuries, is erroneous." Courville (1942) reported that force wave transmission through the brain accounted for contrecoup lesions and explained the localization of injury to "anatomic conformation of the skull and the relation of the cranial cage to the brain" (Courville 1942, p. 20). These early comments set the stage for decades of observation (i.e., autopsy findings), theorizing, and debate.

Several still relevant principles, summarized by Dawson and colleagues (1980), have emerged:

- Contrecoup contusions result from the impact of a moving head against an unyielding surface and occur most commonly in the temporal and frontal regions (figure 8.7) and only rarely in the occipital area.

- Blunt force applied to a resting yet movable head results in coup injury at the point of impact and only rarely causes contrecoup injury.

- With few exceptions, coup lesions are not seen in the rapid deceleration of a moving head, and contrecoup injuries do not result from a direct blow to a resting head.

- Contusions resulting from transient displacement of skull fracture margins are not associated with coup–contrecoup mechanisms.

- Cerebral lacerations, rather than contusions, are more likely in falls from considerable heights and the crushing of the head between an applied force and a firm surface that shatters the skull.

More recent descriptions of the mechanisms of coup and contrecoup lesions have appeared

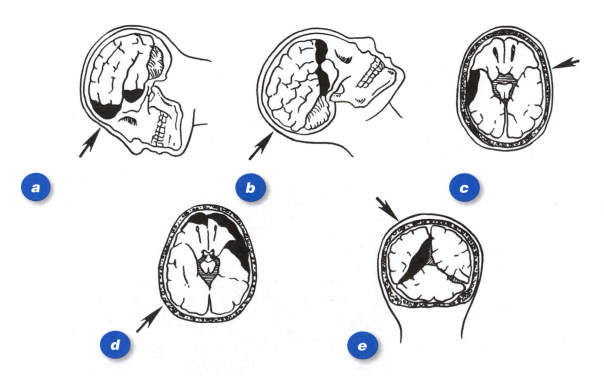

Figure 8.7 Mechanisms of cerebral contusion. Arrows show application point and direction of applied force. Shaded areas indicate location of contusion. (a) Frontal impact with frontotemporal contusion. (b) Occipital impact with frontotemporal contusion. (c) Lateral impact with contralateral temporal lobe contusion. (d) Temporo-occipital impact with contralateral frontotemporal contusion. (e) Vertex impact with diffuse medial temporo-occipital contusion.

in the literature. Adams and Victor (1993, p. 755), for example, stated that "inertia of the malleable brain, which causes it to be flung against the side of the skull that was struck, to be pulled away from the contralateral side, and to rotate against bony promontories with the cranial cavity explain[s] these coup–contrecoup contusions."

Valsamis (1994, pp. 176-177) noted that "the rates of acceleration and deceleration will determine whether contact will be made at the initiation or cessation of skull movement. Thus, fast acceleration and relatively slow deceleration will result in a coup lesion. Relatively slow acceleration coupled with rapid deceleration will result in a contrecoup lesion. If both the components are rapid, a 'coup–contrecoup' lesion will be produced, and, if both components are relatively slow, no contusion will result."

The importance of head rotation, rather than translation, in the development of shearing loads in cerebral tissue has been appreciated for many decades. Holbourn (1943), for example, noted that the shear strains produced by linear acceleration of the head are small compared with those developed in response to rotational accelerations. This fact was experimentally verified by Ommaya and colleagues (1971, p. 515), who discounted the role of linear motion ("pure head translation has never been demonstrated as an injury producing factor for the brain") and submitted that "skull distortion and head rotation . . . explains a greater number of observations on coup and contrecoup injuries." Although much has been learned about the mechanisms of coup and contrecoup head injuries, many questions remain regarding this complex set of cerebral contusion injuries.

Brain Swelling

Contusions and other traumatic brain lesions often are accompanied secondarily by brain swelling (edema), a serious and potentially fatal condition in which the contents of the cranial vault increase in volume. This growth in volume increases intracranial pressure (ICP), which in turn can result in compromised neurovascular function, cerebral ischemia, or herniation into adjacent intracranial spaces.

Brain swelling may be attributable to cerebral **hyperemia** (increased blood volume) or **cerebral edema,** a specialized condition characterized by increased tissue fluid content. Five types of cerebral edema have been identified: vasogenic, hydrostatic, cytotoxic, hypoosmotic, and interstitial (Miller 1993). Vascular injury (e.g., contusion) may produce vasogenic edema, while increased transmural vascular pressure can induce hydrostatic cerebral edema. When cellular membranes are disrupted, cells can become hypoxic (cytotoxic edema). With very low sodium levels in the circulatory system, hypoosmotic edema can result, and interstitial cerebral edema can arise from too much fluid in the cerebral ventricles.

Concussion: A Closer Look

Cerebral concussion is one of the most common head injuries. Many definitions have been proposed for concussion. Although they vary somewhat in their exact diction, they contain the elements of the classic characterization of concussion as "traumatic paralysis of neural function in the absence of lesions" (Denny-Brown and Russell 1941, p. 159). Concussion is characterized by some as *mild traumatic brain injury* (MTBI), although this can be misleading because MTBI implies some form of brain injury, whereas concussion is usually described as involving neural dysfunction without structural brain damage. Also, some physicians and researchers clinically define MTBI to include (in addition to concussion) amnesia, loss of consciousness, skull fracture, intracranial injury of unspecified nature, and unspecified head injury (Bazarian et al. 2005).

Definition and Grading

In 2001, an international group of experts formed the Concussion in Sport Group (CISG) and convened the First International Symposium on Concussion in Sport in Vienna. They proposed the following definition (Aubry et al. 2002, p. 6):

Concussion is defined as a complex patho-physiological process affecting the brain, induced by traumatic biomechanical forces. Several common features that incorporate clinical, pathological, and biomechanical injury constructs that may be used in defining the nature of a concussive head injury include:

1. Concussion may be caused by a direct blow to the head, face, neck, or elsewhere on the body with an "impulsive" force transmitted to the head.

2. Concussion typically results in the rapid onset of short-lived impairment of neurological function that resolves spontaneously.

3. Concussion may result in neuropathological changes but the acute clinical symptoms largely reflect a functional disturbance rather than structural injury.

4. Concussion results in a graded set of clinical syndromes that may or may not involve loss of consciousness. Resolution of the clinical and cognitive symptoms typically follows a sequential course.

5. Concussion is typically associated with grossly normal structural neuroimaging studies.

TABLE 8.2

GLASGOW COMA SCALE

Eye opening	Spontaneous	4
	To speech	3
	To pain	2
	None	1
Best motor response	Obeys	6
	Localizes	5
	Withdraws	4
	Abnormal flexion	3
	Extends	2
	None	1
Verbal response	Oriented	5
	Confused conversation	4
	Inappropriate words	3
	Incomprehensible sounds	2
	None	1
Total EMV score	Eye + motor + verbal	3-15

Adapted, by permission, from G. Teasdale and B. Jennett, 1974, "Assessment of coma and impaired consciousness. A practical scale" *Lancet* 2(7872): 81-84.

Symptoms of concussion may include headache, dizziness, nausea, visual disturbances, auditory problems (e.g., tinnitus), and irritability. Physical manifestations of concussion may include loss of consciousness, confusion, amnesia, compromised balance and coordination, convulsions, seizure, unsteady gait, slowness in answering questions or following directions, poor concentration, display of inappropriate emotions, vomiting, vacant stare, slurred speech, personality changes, and inappropriate playing behavior (McCrory et al. 2005). Although the symptoms are transient by definition, the time course of recovery varies considerably, ranging from several seconds to much longer.

Various scales have been devised to measure the degree of neural dysfunction from head injury. Historically, the most commonly used system has been the Glasgow Coma Scale (GCS), which measures a patient's auditory, motor, and visual response to stimulation and determines the level of brain dysfunction on a 13-point scale ranging from 3 to 15 (table 8.2). A summed GCS score of 13 to 15 indicates mild brain injury, 9 to 12 moderate injury, and 3 to 8 severe injury. Several other concussion rating systems have been increasingly used in recent years. These systems are summarized in table 8.3. Although they have common elements, these systems vary in details.

In 2004, the CISG met again in Prague and developed a summary document for use by doctors, therapists, health professionals, coaches, and others involved in athlete care at all levels (McCrory et al. 2005). This group reconfirmed the Vienna conference (2001) definition of concussion and recommended the abandonment of a grading system approach in favor of simply drawing a distinction between *simple concussion* and *complex concussion* for management of sport-related concussions. This group characterized a simple concussion as an injury that progressively resolves without complications in 7 to 10 days, with rest indicated until symptoms resolve, followed by a graded program of exertion before return to sport. They described complex concussion as involving persistent symptoms, specific sequelae (e.g., convulsions), prolonged loss of consciousness (>1 min), or prolonged cognitive impairment.

TABLE 8.3

CONCUSSION RATING SYSTEMS

RATING SYSTEM	SIGNS AND SYMPTOMS		
	Grade I	Grade II	Grade III
American Academy of Neurology	No loss of consciousness Transient confusion Concussion symptoms resolve in less than 15 min	No loss of consciousness Transient confusion Concussion symptoms or mental status abnormalities on examination resolve in more than 15 min	Any loss of consciousness either brief (seconds) or prolonged (minutes)
American College of Sports Medicine Guidelines	None or transient retrograde amnesia None to slight mental confusion No loss of coordination Transient dizziness Rapid recovery	Retrograde amnesia; memory may return Slight to moderate mental confusion Moderate dizziness Transitory tinnitus Slow recovery	Sustained retrograde amnesia; anterograde is possible with intracranial hemorrhage Severe mental confusion Profound loss of coordination Obvious motor impairment Prolonged tinnitus Delayed recovery
Cantu Concussion Rating Guidelines	No loss of consciousness Concussion symptoms resolving in less than 15 min Posttraumatic amnesia for less than 30 min	Loss of consciousness for less than 5 min Posttraumatic amnesia for more than 30 min but less than 24 hr	Loss of consciousness for more than 5 min Posttraumatic amnesia for more than 24 hr
Colorado Medical Society Concussion Rating Guidelines	No loss of consciousness Transient confusion No amnesia	No loss of consciousness Transient confusion Amnesia	Loss of consciousness

From C. Starkey and J.L. Ryan: Evaluation of orthopaedic and athletic injuries, 2nd ed. F.A. Davis, Philadelphia, 2001, p. 631 with permission.

Epidemiology

Accurate estimates of concussion incidence are elusive, attributable in part to difficulty with definitive diagnosis and nonreporting. After collecting data about emergency department admissions, Bazarian and colleagues (2005) estimated an annual concussion incidence of 127.8 in 100,000. These emergency department data do not, however, include concussions seen in nonemergency venues (e.g., personal physician's office) or possible concussions for which no medical attention is sought.

The primary causes of concussion are predictable: falls, motor vehicle trauma, accidental head impact, assault, and sport and recreational activities. Arguably, the best estimates of concussion incidence are in sport settings.

The CDCP estimates more than 300,000 sport-related concussions annually. Concussion risk varies across sports. In some sports (e.g., ice hockey), concussion is the most common injury (Flik et al. 2005). One study reported concussion rates in high school athletes ranging from 9.36 per 100,000 athlete-exposures in cheer leading to 33.09 per 100,000 athlete-exposures in football (Schulz et al. 2004). Pellman and colleagues (2004) reported 0.41 concussions per National Football League (NFL) game, with two-thirds resulting from impact by another player's helmet.

Kraus and Chu (2005, pp. 23-24) advised skepticism about research literature concerning brain injury, and their advice is clearly appropriate when applied to concussion:

The current brain injury research literature should be read cautiously because of the wide differences in the research methods and interpretation of clinically based, as opposed to epidemiologically based, data. This is especially important in the consideration of the definition of brain trauma and the ways in which injury severity is measured. The results of these methodological inconsistencies (point of interpretation) make cross-study comparisons extremely difficult, if not impossible.

Despite these difficulties in estimating the scope of concussive injury, the available statistics make clear that concussion is a widespread problem with significant medical and financial consequences.

Tissue Structure and Function

The traditional definition of concussion describes neural dysfunction in the absence of structural damage. The events accompanying concussion are more physiological than structural. One of the hallmarks of concussion-related pathophysiology is impairment of cerebral autoregulation. The duration of impaired autoregulation may be correlated with brain injury outcome. Among the physiological changes that accompany concussion are **hyperglycolysis,** increased intracellular calcium and extracellular potassium, and reduced cerebral blood flow. Recovery of neurochemical and metabolic function in mild traumatic brain injury may take up to 2 weeks (Freeman et al. 2005).

In cases of more severe traumatic brain injury, concussion may be accompanied by neural lesions. Brain tissue, because of its incompressibility, handles compressive loads well but is much less resistant to shearing loads caused by head rotation. Interestingly, the estimated rotational acceleration needed to produce concussion is lower for adults (4500 rad/s²) than for infants (10,000 rad/s²). This contrast points to differences in mechanical response between an adult skull and brain and those of a developing child. These differences are attributable to contrasting constitutive material properties, structural geometries, age-dependent physiological responses to mechanical stress, and structural properties of the head (Ommaya et al. 2002). Brains of various sizes have dissimilar injury thresholds and age-related mechanical properties (Goldsmith and Plunkett 2004).

Mechanisms of Injury

Concussion results from a change in momentum of the head and thus most often involves direct impact or acceleration–deceleration mechanisms. In the vast majority of cases, direct impact to the head is implicated. A forceful blow, however, is not required. Rapid acceleration without direct impact also can cause concussion. Whether induced by direct impact or not, rotational (rather than linear) accelerations predominate and are associated with stronger forces. As noted earlier, the importance of rotational acceleration as a mechanism of concussive injury was originally suggested by Holbourn (1943) and has been reiterated by many researchers, notably Ommaya and Gennarelli (1974), who suggested that rotational accelerations to the head cause diffuse and widespread injury, whereas translational accelerations mainly cause focal injuries.

Various mechanisms have been proposed to explain cerebral concussion. Some of these mechanisms are similar to those described earlier for cerebral contusion. Proposed concussion mechanisms include shear strains caused by rotation (Holbourn 1943), coup–contrecoup cavitation caused by displacement between the skull and brain (Gross 1958), linear acceleration causing shear stress or distortion in the brain stem (Gurdjian et al. 1955), and sequential centripetal disruption beginning at the brain surface and extending inward to deeper brain structures (Ommaya and Gennarelli 1974). Despite extensive research over the past half-century, controversy remains over the exact mechanisms causing concussion (Zhang et al. 2001).

Treatment, Rehabilitation, and Prevention

Acute concussion management must include immediate medical evaluation, followed by supervision for several hours to monitor for potential deterioration. The vast majority of concussions (90%) resolve progressively without complication in 7 to 10 days with cognitive and physical rest. Rest is followed by a graded program of exertion prior to return to full activity or sport participation. An appropriate return-to-play protocol is a progression of activity: no activity (until asymptomatic), light activity, increased intensity activity, and finally, after full medical clearance, full activity and game play.

In the 10% of cases that involve complex concussions with persistent symptoms, further medical management is required. This management typically includes formal neuropsychological testing and assessment tools (e.g., Sport Concussion Assessment Tool) and may also include functional neuroimaging, pharmacological treatment, and involvement of a psychologist with expertise in dealing with postconcussive treatment.

Concussion prevention has primarily focused on sport and recreational activities in the form of protective equipment and rule formulation and enforcement. Helmets are commonly used in many sports (e.g., football, ice hockey) and recreational pursuits (e.g., skiing, snowboarding, cycling). A helmet's hard outer shell spreads impact forces over a large area to protect against focal injuries (e.g., scalp lacerations); the helmet's inner lining helps dissipate kinetic energy via an energy-absorptive mechanism (Bailes and Cantu 2001).

Mouth guards provide another means of protection from concussion. Mouth guards sepa-rate the mandibular condyle from the condylar fossa and reduce the energy transfer from forces applied to the jaw to the cranium.

Implementation and enforcement of rules can also help reduce injuries. For example, rules instituted in the 1970s that banned spearing (i.e., using the helmet as a battering device) in American football resulted in a precipitous decrease in cervical spine injuries and concussions.

Injuries such as concussions cannot be eliminated, but advances in equipment design and rule enforcement can help reduce the risk of injury and its physical, financial, and emotional costs.

Diffuse Axonal Injury

One of the distinguishing characteristics of mild cerebral concussion is an absence of detectable pathology. In cases of more severe injury, neural structures suffer damage. These lesions may be in the form of contusion, laceration, hemorrhage, or axonal damage. Axonal lesions have been described as diffuse degeneration of the white matter, diffuse white-matter shearing injury, and inner cerebral trauma. The most common current

OF SPECIAL INTEREST

On rare occasions following a concussion, a subsequent blow that directly impacts or indirectly moves the head can result in dramatic escalation of concussive symptoms. This so-called **second impact syndrome (SIS)** "occurs when an athlete who has sustained an initial head injury, most often a concussion, sustains a second head injury before symptoms associated with the first have fully cleared" (Cantu 1998, p. 37). SIS can result in loss of cerebrovascular regulation, vascular congestion, pupil dilation, increased intracranial pressure (ICP), cerebral edema, respiratory failure, and possible death.

Children and adolescents are at particular risk for SIS. Many reported cases of SIS involve high school athletes. Age, type of sport, and prior history of concussion are potent risk factors for SIS (Cobb and Battin 2004). In a population of high school and collegiate football players, the risk of concussion is nearly six times greater in athletes with a history of concussion compared with players with no history (Zemper 2003).

Given its rarity, clinical evidence for SIS is largely anecdotal. Some researchers, in fact, have questioned whether SIS exists. McCrory (2001), for example, contended that the evidence for SIS is not compelling and that observed neurovascular declines are more likely attributable to diffuse cerebral swelling. Despite the controversy, prudence dictates that individuals with a concussion be restricted from activities in which they might suffer further head impact until concussive symptoms are completely resolved.

designation is **diffuse axonal injury (DAI),** a term that although pervasive may itself be a misnomer, because DAI is not a diffuse injury to the entire brain; DAI rather affects discrete areas of the brain (Meythaler et al. 2001). Some researchers and clinicians now use the term **traumatic axonal injury (TAI)** or **diffuse traumatic axonal injury (DTAI)** to describe these injuries (Gennarelli and Graham 2005).

Many cases of DAI arise from high-speed motor vehicle crashes that involve rapid acceleration and deceleration. DAI also has been reported in football, soccer, and ice hockey players involved in high-speed collisions (Powell and Barber-Foss 1999; Tegner and Lorentzon 1996). Recent reports have associated DAI with shaken-baby syndrome, in which violent shaking of a baby elicits acceleration and deceleration forces purportedly high enough to cause brain trauma (e.g., Case et al. 2001; Duhaime et al. 1998). This relation between violent shaking of an infant with brain injury, however, remains controversial. A recent review concluded that "it is now evident that shaken baby syndrome evolved as a result of a faulty application of scientific reasoning and a lack of appreciation of mechanisms of injury" (Uscinski 2006, p. 57).

Maxwell and colleagues (1993) suggested two mechanisms of injury in nonimpact head injuries: (1) initial axonal shearing and sealing of fragmented axonal membranes, and (2) axonal perturbation that leads to axonal swelling and disconnection. Current research supports the first mechanism (cytoplasmic shear of the intra-axonal cytoskeleton) as the primary cause of injury (Meythaler et al. 2001).

In cases of severe trauma, the mechanical insult may directly disrupt axonal structure and involve the second mechanism immediately. In milder injuries, immediate mechanical disruption of the axon may not occur; axonal damage in these cases may develop over time subsequent to the traumatic event. Delayed and sequential events lead to axonal failure and may result in axonal severance into proximal and distal segments. In these cases, the distal segment predictably experiences *Wallerian degeneration* (see chapter 5) and becomes deafferented (i.e., separated from its sensory components). Injury to a sufficient number of axons produces profound neural dysfunction.

Accurate diagnosis of DAI, especially in its mildest forms, remains problematic because its detection requires careful microscopic examination. Computed tomography scans may miss DAI because it sometimes occurs in the absence of intracranial hemorrhage, elevated intracranial pressure, or cerebral contusion. DAI is graded according to the localization of its lesions (Adams et al. 1989). Grade 1 (mild) lesions exhibit axonal injury in the white matter of the cerebral hemispheres, the corpus callosum (band of nerve fibers connecting the two cerebral hemispheres), the brain stem, and, less often, the cerebellum. Grade 2 (moderate) lesions show added focal lesions in the corpus callosum, whereas grade 3 (severe) lesions show focal lesions in the brain stem. The importance of axonal damage is highlighted by the close association between axonal lesions and prolonged traumatic coma, the sequelae of axonal damage, and the effects of secondary complications such as *hypoxia, ischemia,* and metabolic pathologies.

Cerebral tissue is relatively incompressible and thus not readily susceptible to injury under compressive loading; however, the tissue has limited resistance to shearing loads. Shearing strains arising from angular acceleration of the head are accepted as being responsible for most cases of DAI. More specifically, the plane of angular acceleration largely determines the extent of injury. In studies on nonhuman primates, angular acceleration in the sagittal plane resulted in only grade I lesions. Similar levels of angular acceleration in the transverse (horizontal) plane typically caused grade II injuries. The most severe grade III lesions were associated with accelerations in the frontal (coronal) plane.

Gennarelli and colleagues (1982) used a primate model to explore the effect of angular acceleration *direction* on duration of coma, degree of neurological impairment, and level of DAI. They concluded that "axonal damage produced by coronal head acceleration is a major cause of prolonged traumatic coma and its sequelae" (p. 564).

Our understanding of head injury in general, and DAI in particular, has improved remarkably in the last two decades. Sahuquillo and Poca (2002) presented several important concepts that have changed the accepted view of head injury and its diagnosis, treatment, and biomechanics:

▪ Secondary ischemic damage is highly prevalent in the brains of patients suffering fatal head injuries. Efforts to minimize deleterious ischemic effects greatly improve chances of patient survival and recovery.

▪ DAI is not a single event but rather a dynamic process initiated at impact and continuing for a variable time postinjury. Until as recently as the 1990s, DAI was considered by most to be a primary injury. That view has changed. "Experimental evidence has shown beyond any reasonable doubt that DAI is a complex physiopathological entity where primary immediate damage can coexist with a simultaneously evolving secondary process in which axons that initially are structurally intact may progress towards secondary disconnection or axotomy" (Sahuquillo and Poca 2002, pp. 53-54).

▪ Ommaya's **centripetal theory** is important in explaining lesions after head injury. Drawing on his synthesis of various studies, Ommaya concluded that data "suggested that the distribution of damaging diffuse strains induced by inertial loading would decrease in magnitude from the surface to the center of the approximately spheroidal brain mass" (Ommaya 1995, p. 530). The centripetal theory predicted that deep structures (e.g., brain stem) could not be injured in isolation. This prediction was in conflict with the long-held view that traumatic unconsciousness was an isolated primary brainstem injury. Ommaya consistently found accompanying cerebral lesions with brainstem injuries.

Penetrating Injuries

Penetrating injuries occur when structural damage results in exposure of the intracranial space to the external environment. The term is sometimes used in this broad sense to describe injuries from any cause, including motor vehicle collisions, occupational and sports injuries, and an object penetrating the skull. More commonly, penetrating injuries are limited to those in which an object pierces the cranium and exposes the contents of the cranial vault. These injuries can be divided into missile (or penetrating) injuries, caused by bullets or shrapnel fragments, and nonmissile (or perforation) injuries, resulting from penetration by weapons (e.g., knives) or other implements (e.g., javelin).

In recent decades, cases of nonmissile injury have been relatively few, with most reports being case studies or anecdotal accounts. The variety of penetrating objects makes a curious list, including nails, keys, sewing needles, car antennas, arrows, and javelins. One remarkable and infamous case of a penetrating injury is described in the sidebar on page 258.

The vast majority of penetrating head injuries are caused by gunshot wounds. Increases in shooting-related injuries (hastened by escalated drug use, gang involvement, domestic abuse, and interpersonal violence) have created an epidemic health problem.

Ballistics is the science dealing with the motion of projectiles, or missiles. **Wound ballistics** deals with the interaction of penetrating projectiles and living body tissues (Fackler 1998). In the case of bullets or shrapnel fragments, ballistic principles govern the path and mechanical characteristics that set the stage for head injury upon penetration.

With respect to ballistic injuries, projectile flight can be divided into three phases (Volgas et al. 2005):

▪ **Internal ballistics**—Effect of bullet design, weapon design, and materials on the projectile while in the barrel of the weapon.

▪ **External ballistics**—Effect of external factors (e.g., wind, velocity, drag, gravity) on projectile flight from the barrel to the target.

▪ **Terminal ballistics**—Projectile behavior in tissues.

Internal and external ballistics determine the initial conditions for terminal ballistics. The entering projectile can cause tissue damage via three mechanisms: direct cutting or tearing of tissues, creation of a *permanent cavity*, and creation of a *temporary cavity* (Volgas et al. 2005). The permanent cavity is the path along which the projectile moves through the tissue, and in the absence of fragmentation, the cavity is usually small (figure 8.8). A temporary cavity may form from waves emanating perpendicularly from the projectile's path. The amount of temporary cavitation depends on projectile speed, tissue density, and whether the projectile tumbles or fragments. The temporary cavity creates a vacuum that draws

THE CURIOUS CASE OF PHINEAS GAGE

Most of today's head-penetration injuries result from gunshot wounds. In contrast to these all-too-common contemporary tragedies stands the peculiar case of Phineas Gage, a 25-year-old railroad construction foreman in mid-19th century Vermont. As colorfully described by Damasio (1994, pp. 3-10), Gage was working late one hot summer afternoon placing explosives to blast stone from the planned path of the Rutland & Burlington Railroad. The explosive powder, which had been placed in a hole drilled in the stone, detonated prematurely and launched the iron rod Gage was using to tamp the powder into the hole. The rod, which weighed more than 13 lb (5.9 kg) and measured 1 1/4 inches (3.2 cm) in diameter and more than 3 1/2 feet (1.1 m) in length, penetrated Gage's head. As graphically described by Damasio (p. 4), "The explosion is so brutal. . . . The bang is unusual, and the rock is intact. The iron enters Gage's left cheek, pierces the base of the skull, traverses the front of his brain, and exits at high speed through the top of the head. The rod has landed more than a hundred feet away, covered in blood and brains. Phineas Gage has been thrown to the ground. He is stunned, in the afternoon glow, silent but awake." Miraculously, Gage survived and was "able to talk and walk and remain coherent immediately afterward" (p. 5).

Gage was pronounced fully cured (at least physically) within 2 months of the accident and had no difficulty walking, touching, hearing, or speaking. It seems that the areas of his brain responsible for language, perception, and motor function had survived the accident relatively unaffected. Gage, however, suffered from devastating and progressive alterations in his personality; his social reasoning skills were forever changed.

The implications of Gage's response to injury are profound. "Gage's story hinted at an amazing fact: Somehow, there were systems in the human brain dedicated more to reasoning than to anything else, and in particular to the personal and social dimensions of reasoning. The observance of previously acquired social convention and ethical rules could be lost as a result of brain damage, even when neither basic intellect nor language seemed compromised" (Damasio 1994, p. 10).

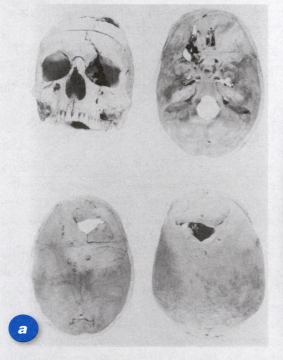

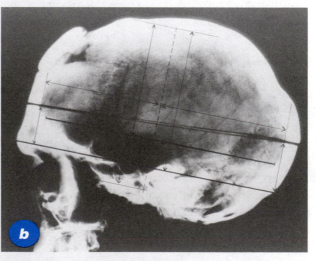

(a) Photographs of several views of the skull of Phineas Gage. (b) An X ray of his skull.

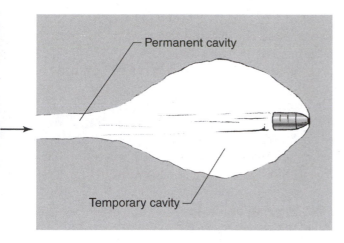

Figure 8.8 Cavitation caused by a projectile penetrating tissue.

tissue and other material into the cavity space. This is followed by tissue rebound, which creates a second temporary cavity as tissues collide and rebound off each other in the vacuum space (Volgas et al. 2005).

Projectiles, by virtue of their motion, possess kinetic energy: $E_k = 1/2 \cdot m \cdot v^2$ (equation 3.20), where m = mass and v = linear velocity. The destructive potential of a bullet is determined, in part, by the magnitude of its kinetic energy at the moment of impact. Clearly, the larger and faster the bullet, the greater is its ability to injure. Of the two components constituting kinetic energy, velocity exerts a more potent influence because it is squared. Doubling a bullet's mass, for example, will double its kinetic energy. But if its velocity is increased twofold, a bullet's kinetic energy quadruples.

In mechanical terms, the destructive energy absorbed (E_a) by the tissues of the head is the amount of kinetic energy lost between bullet impact and exit:

$$E_a = E_i - E_e = 1/2 \cdot m \cdot (v_i^2 - v_e^2) \qquad (8.1)$$

where E_i = kinetic energy at impact, E_e = kinetic energy at exit, m = mass, v_i = impact velocity, and v_e = exit velocity. In cases where the bullet does not exit the skull, all the kinetic energy is absorbed intracranially. Frequently, the bullet has sufficient energy to traverse the entire brain but not enough to pierce the skull a second time and exit the cranium. In these instances, the bullet may ricochet off the interior surface of skull, burrow back into the brain, and cause further damage.

Focusing solely on energy dissipation, however, can be misleading, because many other factors are involved in determining the type and extent of tissue disruption. Two projectiles with the same kinetic energy on entry can cause vastly different injury patterns (Fackler 1996). Velocity and energy measures only indicate the *potential* for injury and are not predictive of the nature and extent of injury (Santucci and Chang 2004). The actual characteristics of tissue disruption are determined by factors in addition to velocity and energy. These factors include the following (Bartlett 2003):

- Projectile stability and entrance profile (e.g., projectile yaw and spin).
- Caliber, construction, configuration, and shape of projectile.
- Distance and path of projectile in the body.
- Biologic characteristics of the tissues (e.g., strength, elasticity, density). Denser tissues, for example, provide greater retarding resistance and contribute to greater energy loss. A bullet penetrating the skull would lose more energy in the brain tissue than would a bullet penetrating an air-filled lung.
- Mechanism of tissue damage (e.g., stretching, crushing, tearing).
- Projectile deformation and fragmentation. Bullets designed and constructed to deform or fragment after impact enhance kinetic energy loss and increase the severity of tissue damage.

All of the foregoing factors interact to determine the complex profile of energy transfer as energy dissipates from the missile and is absorbed by the tissue surrounding the missile's path. The regrettable trend toward firearms designed to use smaller, faster, and more deformable bullets portends an increase in destructive potential and inevitably greater levels of injury and catastrophic death.

Facial Fracture

Facial fractures are of particular concern because of the close proximity of the facial bones to vital neural and sensory structures. A survey of the literature, for example, found that neurological injury associated with facial fractures was as high as 76% (Haug et al. 1994). The mechanism of injury in the vast majority of cases is forceful blunt trauma. Evidence suggests that impact velocity, rather than impact force, correlates most highly with the severity of facial fracture (Rhee et al. 2001).

The impacting object takes a variety of forms. Collisions with a part of another person's body (e.g., head, shoulder, elbow), an implement (e.g., hockey stick, baseball bat), a projectile (e.g., cricket ball, baseball, golf ball), or an unyielding surface (e.g., vehicle steering wheel) have all been reported to cause facial fractures.

Most studies of facial fractures report automobile crashes as the leading cause. The high energy of vehicular collisions greatly increases the likelihood of accompanying injuries. The role of vehicular crashes in facial fracture is a worldwide concern. Studies report that crashes account for 45% of facial fractures in Brazil (Brasileiro and Passeri 2006), 70% in Canada (Hogg et al. 2000), 45% in Germany (Kühne et al. 2007), and 56% in Nigeria (Adebayo et al. 2003).

A recent increase in facial fractures resulting from violent assault suggests a disturbing trend. Lim and colleagues (1993), for example, reported that in Australia assault was the leading etiological factor, accounting for 51.2% of 839 facial fractures reviewed. Similarly, Alvi and colleagues (2003) reported that 41% of facial fractures treated at a trauma center in Chicago were caused by assault, followed by automobile accidents at 26.5%.

Evidence clearly shows that seat belt use has greatly reduced the incidence of chest injury in frontal car crashes. At impact, unrestrained drivers are launched chest first into the steering wheel. In contrast, the driver who is restrained by a lap belt (no shoulder strap) only is more likely to sustain head injury as the cranium is thrown toward the steering wheel (figure 8.9). Drivers who wear shoulder-restraint systems have fewer head injuries than unrestrained drivers, but may have more abdominal injuries since the body is rapidly decelerated by the restraint straps. A higher prevalence of gastrointestinal injury, in particular, has been reported for restrained adults (Wotherspoon et al. 2001) and children (Sokolove et al. 2005).

At first glance, the increasing use of air bag restraints would appear an ideal answer to this problem. In most cases, air bags prevent or reduce the severity of head injury. The air bag should not be considered a panacea, however, because substantial injury can still occur when crash speeds are insufficient to trigger air bag deployment. In addition, serious harm, or even death, can be caused by the explosive deployment of air bags into people of short stature and small body mass (e.g., children and small adults) and those seated close to the steering column.

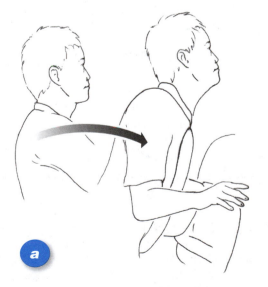 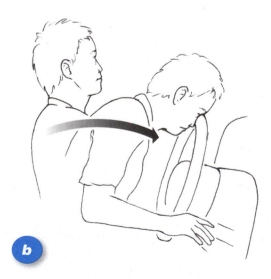

Figure 8.9   (a) Unrestrained driver thrown chest-first into the steering column. (b) The torso of a driver restrained by a lap-only seat belt rotates forward and results in head impact with the steering wheel assembly.

An interesting but unresolved controversy exists concerning whether facial structures influence the severity of neurological injury. On one hand, facial structures that experience impact may increase the likelihood of neural damage by transmitting force to the neural tissues that underlie the craniofacial vault. On the other hand, midfacial bones may actually transmit impact force to the cranium rather than to the soft tissues beneath it, lessening the damage to neural structures.

Because facial injury typically results from direct trauma, the risk of fracture depends largely on the strength of the bony tissue at the impact site. Several researchers have been prominent in reporting facial bone strength. Among these are Hodgson (1967), Nahum and colleagues (1968), Schneider and Nahum (1972), Hopper and colleagues (1994), and Yoganandan and colleagues (1993). Hampson (1995) summarized craniofacial bone tolerances for the zygoma, zygomatic arch, mandible, maxilla, frontal bone, and nose. These data are depicted in figure 8.10.

Reaching valid conclusions regarding facial fracture tolerances is difficult because of the limited number of specimens typically used in impact studies, differences in testing protocols, and specimen age. Nonetheless, Allsop and Kennett (2001) identified some trends:

- Embalming has minimal effect on fracture characteristics.
- Fracture characteristics are not significantly affected by bone mineral content.
- Force pulse duration does not substantially affect fracture force.
- Fracture force is not affected by rate of force onset and strain rate.
- Initial fracture may happen at less than maximal force.
- Skull bone thickness significantly affects fracture force.
- Impactor area of the testing apparatus significantly affects fracture force.

Facial fractures are particularly problematic because they often cross disciplinary boundaries. Treatment of injuries associated with facial fractures frequently requires the combined expertise of orthopedists, neurologists, dentists, ophthalmologists, and other medical specialists.

Neck Injuries

The neck provides the structural link between the head and trunk and contains components of many of the body's principal systems. Many essential structures emanate from the root of

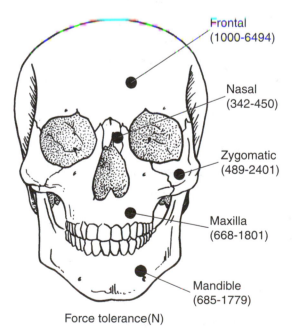

Force tolerance(N)

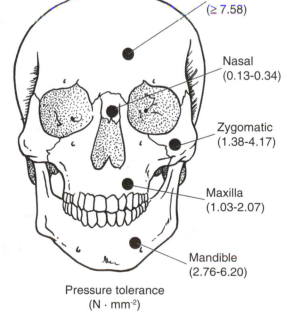

Pressure tolerance
(N · mm⁻²)

Figure 8.10 Force and pressure tolerances for facial bones.

Adapted from *Journal of Biomechanics*, Vol. 28, D. Hampson, Facial injury: A review of biomechanical studies and test procedures for facial injury assessment, pp. 1-7, Copyright 1995, with permission from Elsevier.

the neck. Among these are important vascular (common carotid arteries, subclavian arteries and veins, brachiocephalic trunk and veins), respiratory (trachea, larynx), digestive (esophagus), nervous (sympathetic trunk, phrenic nerve, vagus nerve), and endocrine (thyroid and parathyroid glands) structures. The skeletal portion of the neck includes the cervical vertebrae, held in place by a strong system of ligaments (figure 8.11a). Among the more prominent muscles of the neck are the sternocleidomastoid and trapezius (figure 8.11, b and c).

The vertebral column (spine) is a group of 33 vertebrae extending from the base of the skull to its inferior termination at the coccyx (tailbone). The spine is divided into five regions (figure 8.12): cervical (7 vertebrae), thoracic (12 vertebrae), lumbar (5 vertebrae), sacral (5 fused vertebrae), coccygeal (4 fused vertebrae). Vertebrae in the cervical, thoracic, and lumbar regions are separated by an **intervertebral disc** that is composed of a gelatinous inner mass **(nucleus pulposus)** surrounded by a layered fibrocartilage network

(annulus fibrosus). Each vertebra consists of a body, vertebral arch, and processes arising from the arch (figure 8.13). The size and orientation of these structural elements differ between regions (figure 8.14). Just posterior to the vertebral body is an open passage (vertebral foramen) that houses the spinal cord. Other passages (intervertebral foramina) between adjacent vertebrae allow the exit of spinal nerve roots on both sides of the vertebral column.

Of the many cervical structures susceptible to injury, the spinal cord is the one with the greatest catastrophic potential. The consequences of spinal cord damage range from mild damage, as in *neurapraxia* (transient loss of nerve conduction without structural degeneration), to severe paralysis or death. In cases of severe injury, the level of spinal cord involvement is critical in determining the type and extent of sensorimotor deficit. Injury at the C3-C4 level, for example, may result in complete paralysis of the trunk and extremities and loss of unassisted respiration. Injury at C5-C6 may allow limited arm

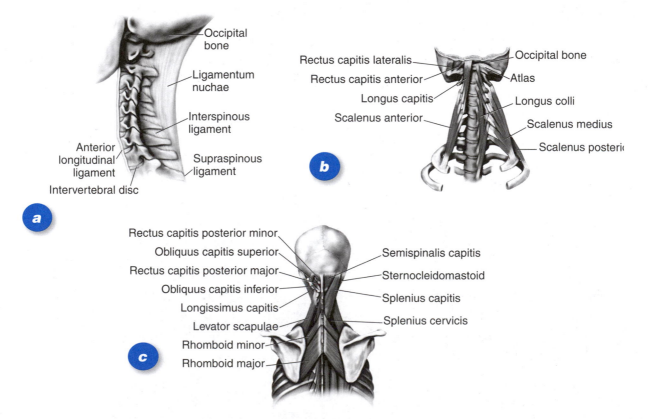

Figure 8.11 *(a)* Cervical vertebrae, with several of their supporting ligaments shown. *(b)* Anterior musculature of the neck. *(c)* Posterior musculature of the neck.

(a) Reprinted, by permission, from R.S. Behnke, 2005, *Kinetic anatomy*, 2nd ed. (Champaign, IL: Human Kinetics), 126. *(b, c)* Reprinted, by permission, from R.S. Behnke, 2005, *Kinetic anatomy*, 2nd ed. (Champaign, IL: Human Kinetics), 130.

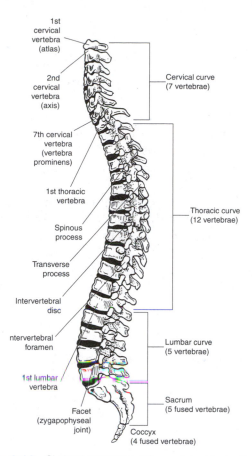

Figure 8.12 Skeletal structures of the vertebral (spinal) column.

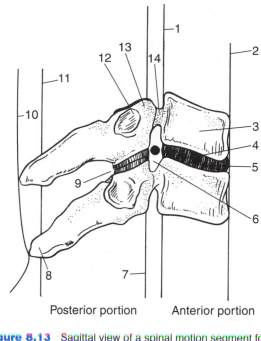

Posterior portion   Anterior portion

Figure 8.13 Sagittal view of a spinal motion segment formed by two adjacent vertebral bodies and the intervening disc. Structures or locations of structures pictured (only the locations of ligaments, and not their structures, are indicated): 1, posterior longitudinal ligament; 2, anterior longitudinal ligament; 3, vertebral body; 4, cartilaginous end plate; 5, intervertebral disc; 6, intervertebral foramen with nerve root; 7, ligamentum flavum; 8, spinous process; 9, intervertebral joint formed by the superior and inferior facets (joint capsules not shown); 10, supraspinous ligament; 11, interspinous ligament; 12, transverse process (intertransverse ligament not shown); 13, vertebral arch; 14, vertebral canal (spinal cord not shown).

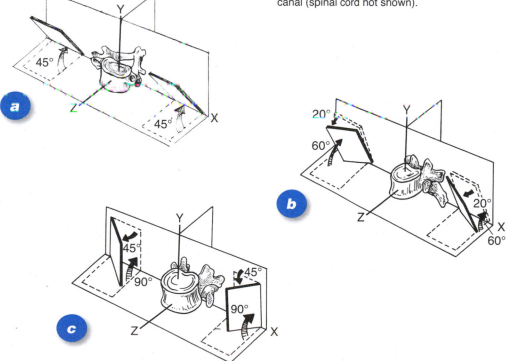

Figure 8.14 Orientation of the vertebral facet joints. *(a)* Cervical spine. The vertebral facets in the cervical spine are inclined 45° above the horizontal plane and are parallel with the frontal plane. *(b)* Thoracic spine. The facets in the thoracic region are inclined 60° above the horizontal deviate 20° behind the frontal plane. *(c)* Lumbar spine. Facets in the lumbar spine are inclined 90° above the horizontal plane and deviate 45° behind the frontal plane. These regional changes in facet orientation play an essential role in determining movement potential between adjacent vertebrae in each region. Angle values are rough estimates. Actual values vary within regions of the spine and among individuals.

Reprinted, by permission, from A.A. White and M.M. Panjabi, 1990, *Clinical biomechanics of the spine*, 2nd ed. (Philadelphia, PA: Lippincott, Williams & Wilkins), 30.

movement, whereas lower-level injury at C7-T1 may spare upper-extremity muscle function and limit paralysis to the lower extremities.

Cervical Trauma

The complex structure and intricate motion of the cervical region present special challenges in identifying and describing mechanisms of cervical injury. The sometimes confusing mixture of engineering and medical terminology used to describe cervical mechanics further complicates the task.

Classification of cervical injury mechanisms requires great care and precision because (1) the overall motion of the head relative to the trunk may not be indicative of local motion between adjacent segments, (2) small deviations (<1 cm) in the point of force application or in head position can change the injury mechanism from compression–flexion to compression–extension, and (3) observed head motions may occur after the instant of injury and thus not reflect the true injury mechanism (McElhaney et al. 2001).

Although various classification systems have been proposed, we favor a revised system (McElhaney et al. 2001) based on the principal loadings applied to the cervical spine (figure 8.15 and table 8.4). Using this system, we present several example injuries and their mechanisms.

Sporting activities provide instructive examples of cervical spine injury mechanisms. It was once thought that cervical injury in football was attributable to either a hyperflexion mechanism or a "guillotine" effect in which the posterior rim of the helmet acted as a pivot during cervical hyperextension. These mechanisms have been discounted as primary causes of cervical injury.

The most common cause is accepted as a **compression–flexion mechanism** (also **flexion– compression mechanism).** When the neck is flexed slightly, the normal cervical lordosis disappears and the cervical vertebrae are aligned axially. In this position, the cervical spine becomes a segmented, straightened column that lacks the curvature required for energy-absorbing bending (Cusick and Yoganandan 2002). This position, termed the *stiffest axis*, puts the spine at greater risk of structural failure (Pintar et al. 1995). The cervical structures in this position must absorb all of the loading energy (figure 8.16). When this energy exceeds the capacity of the cervical structures, failure of the intervertebral discs, the body and processes of the vertebrae, or spinous ligaments may occur. Resulting **wedge fractures** and **burst fractures** are predictably present. Disruption of supporting structures permits further flexion or rotation of the cervical spine and associated vertebral dislocation. This dislocation carries the risk of impinging the spinal cord or spinal nerves. The compression–flexion mechanism is most common in diving and also happens in sports such as football, ice hockey, and surfing.

Cervical injury also results from a **tension– extension mechanism** in which the head is forcibly hyperextended by posterior impact with forcible resistance applied to the chin (figure 8.17*a),* inertial forces resulting from posterior impact (figure 8.17*b),* or forces applied inferiorly to the posterior aspect of the head (figure 8.17*c).* This

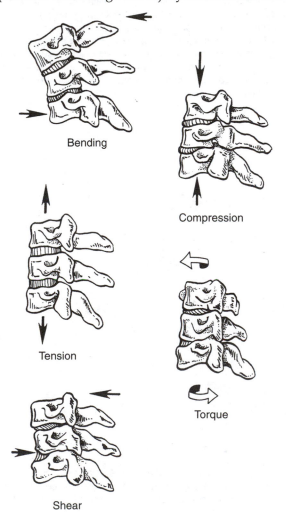

Figure 8.15 Mechanisms of neck loading.

Bending

Compression

Tension

Torque

Shear

TABLE 8.4

REVISED CLASSIFICATION OF CERVICAL SPINE INJURIES

Classification	Injury
Compression	Jefferson fracture
	Multipart atlas fracture
	Multipart vertebral body fracture
Compression–flexion	Burst fracture
	Wedge compression fracture
Compression–extension	Posterior element fractures
Tension	Occipitoatlantal dislocation
Tension–extension	Hangman's fracture
Flexion moment	Hyperflexion sprain
	Bilateral facet dislocation
	Unilateral facet dislocation
Extension moment	Hangman's fracture
	Anterior longitudinal ligamentous rupture
	Disc rupture
	Horizontal fracture of vertebral body
Torsion	Atlantoaxial rotary dislocation
	Atlantoaxial unilateral facet dislocation
Horizontal shear	Transverse ligament rupture
Multiple mechanisms	Odontoid fracture
	Teardrop fracture
	Clay shoveler's fracture

Reprinted from J.H. McElhaney et al., 2001, Biomechanical aspects of cervical trauma. In *Accidental injury: Biomechanics and prevention*, 2nd ed., edited by A.M. Nahum and J.W. Melvin (New York, NY: Springer Verlag), 332, with kind permission of Springer Science and Business Media.

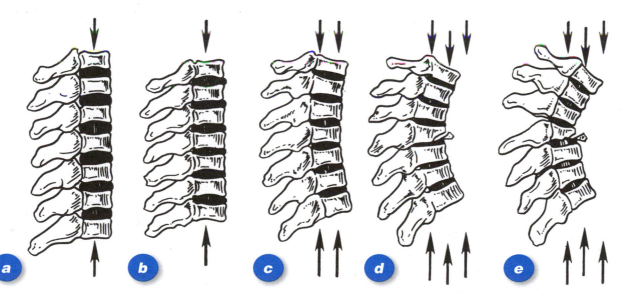

Figure 8.16 *(a)* With the neck slightly flexed (approximately 30°), the cervical spine is straightened and functions as a segmented column. *(b)* Axial compressive forces applied to a segmented column initially compress the column. Increased loading causes *(c)* angular deformation, *(d)* buckling, and *(e)* eventual fracture, subluxation, or dislocation.

Reprinted, by permission, from J.S. Torg et al., 1990, "The epidemiologic, pathologic, biomechanical, and cinematographic analysis of football-induced cervical spine trauma," *The American Journal of Sports Medicine* 18(1): 52-53.

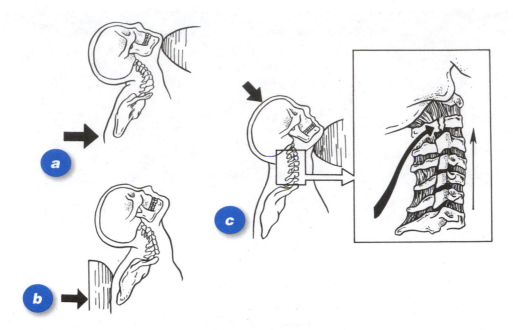

Figure 8.17 Tension–extension injury mechanisms. Cervical hyperextension caused by *(a)* posterior impact with forcible resistance on the chin, *(b)* inertial forces from posterior impact, and *(c)* forces applied inferiorly to the posterior aspect of the head with forcible resistance applied to the chin.

Reprinted with permission from *Health and Neck Injury* (P-276) © 1994 SAE International.

mechanism creates tension stresses in the anterior cervical structures and can involve disruption of the anterior longitudinal ligament, intervertebral disc, or horizontal fracture of the vertebral body. High-energy loading can also result in posterior vertebral displacement and risk of spinal cord injury.

Some cervical injuries result from multiple mechanisms. For example, fracture at the antero-inferior corner of a cervical vertebral body, termed a **teardrop fracture** (figure 8.18), has been attributed to many mechanisms, including axial compression, compression–flexion, tension–extension, and compression–extension loading (McElhaney et al. 2001).

The most common cause of cervical fracture is vehicular trauma. Because of the complexities involved in motor vehicle accidents (MVAs), detailed assessment of cervical biomechanics is difficult. Despite these difficulties, there is agreement on some issues. Yoganandan and colleagues (1989) examined MVA-related cervical injuries and concluded the following:

■ Cervical injuries focus at the occiput axis and C5-C6 regions of the upper and lower cervical spine, respectively.

■ Fatal spinal injury from MVAs most commonly happens at the craniocervical junction and upper cervical spine (O-C1-C2).

■ MVA survivors tend to have lower cervical spine injury more often than upper cervical involvement.

■ Craniofacial injury and cervical spine trauma are closely related, suggesting that occupant restraint systems that limit head and face impact can reduce the incidence of serious cervical spine injury.

Identification of the specific mechanisms of cervical spine injury often is problematic, largely because of the complexity of cervical anatomy, alignment, and loading. Nonetheless, "the influence of cervical alignment and curvature in association with the direction of force application on the potential vertebral component compromise is an important consideration in clarifying the causative forces of specific fracture patterns" (Cusick and Yoganandan 2002, pp. 17-18).

Spinal Cord Injury

Cervical **spinal cord injury (SCI)** occurs in many activities. The reported incidence of SCI for

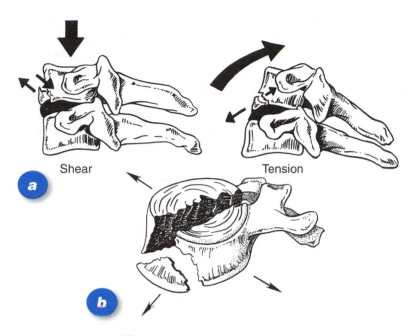

Shear   Tension

Figure 8.18 Vertebral teardrop fracture. *(a)* Bone fragment fracture at the anteroinferior border of the vertebral body resulting from compressive loading (left) that results in shearing at the fragment interface (dark arrows) or from spinal extension (right) that creates tensile loading at the fragment interface. *(b)* A three-part, biplanar teardrop fracture with an anteroinferior corner fracture fragment and a sagittal fracture through the vertebral body.

specific activities depends on the location and circumstances of the population being studied. Motor vehicle collisions are the most common culprit, accounting for up to half of all SCIs. For example, Karacan and colleagues (2000), in a nationwide study in Turkey, reported motor vehicle accidents as the leading cause of SCI (48.8%), followed by falls (36.5%). Societal conditions, however, may influence the statistics. In one study of 616 patients (20-29 years old) in the Republic of South Africa, gunshot wounds were the leading cause of SCI (36%), ranking ahead of motor vehicle accidents (25%) (Hart and Williams 1994). Spinal cord injuries also have been reported attributable to falls from heights, work-related tasks, and sporting and recreational activities. Despite their relative infrequency, sports-related SCIs often achieve notoriety in the media, usually in cases of American football injuries that result in paralysis.

Spinal cord injury, although catastrophic, is relatively rare. More commonly, cervical injury manifests as a temporary sensorimotor lesion caused by pinching of cervical nerve roots or the brachial plexus. These so-called *burners* or *sting-*

ers result in burning pain, with numbness and temporary weakness in the affected arm.

Watkins and Watkins (2001) described two mechanisms associated with this injury. The first involves an off-center axial load applied while the neck is extended and laterally flexed. In this position, the spinal canal and foramina narrow and allow the bony borders to impinge on the exiting nerve roots. This mechanism is most commonly seen in American football players during the impacts of blocking, tackling, and ground contact.

In the second mechanism, the shoulder on the involved side is depressed and the head is forcibly pushed away contralaterally. This motion violently stretches the nerve roots and associated brachial plexus and results in transient neurological symptoms. In mild burners, sensorimotor function returns within several minutes and recovery is complete within a week or two. More severe injury can result in motor loss to muscles (e.g., biceps brachii, deltoid) that persists for weeks or even months.

More severe SCI usually results from vertebral fracture, dislocation, or both. Bony fragments can impinge on and injure the spinal cord. The level of cervical spinal cord injury (C1-C7) is critical in determining the degree of neuromuscular dysfunction and motor deficit (Gardner 2002; Watkins and Watkins 2001). The following summarizes function and deficits according to cervical level.

- **C1-C3**—Limited movement of head and neck. Complete paralysis of trunk and extremities **(quadriplegia).** Patient requires ventilatory support.

- **C3-C4**—Usual head and neck control. C4 may shrug shoulders. Patient requires initial ventilatory assistance.

- **C5**—Head, neck, and shoulder control. Patient can bend elbows and supinate arm to turn palms up.

- **C6**—Head, neck, shoulder, and elbow flexion control. Patient can turn palms up and down and can extend wrist.

- **C7**—Added ability to extend elbow.

- **C8**—Added strength and some finger control, but lack of fine precision hand movements.

Whiplash-Related Injuries

Of all cervical disorders, whiplash-related injuries are among the most common and most misunderstood. The very definition of whiplash is controversial. In one context whiplash describes an injury mechanism, whereas in another it refers to a clinical syndrome. This latter usage is inappropriate because it lacks specificity.

In 1995, the Quebec Task Force on Whiplash-Associated Disorders defined **whiplash** as "an acceleration–deceleration mechanism of energy transfer to the neck which may result from rear-end or side impact, predominately in motor vehicle accidents, and from other mishaps. The energy transfer may result in bony or soft tissue injuries (whiplash injury), which may in turn lead to a wide variety of clinical manifestations (whiplash-associated disorders)" (Cassidy et al. 1995, p. 22).

In recent decades many studies have been conducted to determine the mechanisms of whiplash. Although these studies have been of variable quality (Kwan and Friel 2003; McClune et al. 2002) and have raised concerns about sampling, experimental design, and data interpretation, we have a much better understanding of whiplash mechanisms than we did not long ago.

Whiplash typically has been characterized as involving a hyperextension mechanism. In this view, at the instant of impact, the head remains stationary (according to Newton's first law) while the vehicle is violently pushed forward. When the occupant's trunk and shoulders are accelerated anteriorly, the head is forced into hyperextension. Once its inertia is overcome, the head is thrown (whiplashed) forward into flexion.

Recent work has shown that the hyperextension model is too simplistic and does not adequately describe the complex motion of the cervical spine during whiplash. "The critical revision brought about by modern research into whiplash is that it is not a cantilever movement that is injurious; i.e., it is not an extension–flexion movement of the head, as was commonly believed previously. Rather, within less than 150 ms after impact, the cervical spine is compressed. During this period the cervical spine buckles; upper cervical segments are flexed while lower segments extend around abnormally located axes of rotation" (Bogduk and Yoganandan 2001, p. 272). The simultaneous upper cervical flexion and lower cervical extension results in an S-shaped neck curvature as shown in figure 8.19 (50 ms and 75 ms). The S-shaped configuration happens within 75 ms after impact and then gives way to a C-shaped hyperextension curvature (Grauer et al. 1997). Both the lower cervical spine and upper cervical spine are at risk of injury from the rear-impact mechanism (Panjabi, Pearson et al. 2004).

Although usually considered a sagittal plane injury caused by a rear-end impact, whiplash can also result from lateral or frontal forces that exact their own unique pattern of injury. In addition, motion of the neck sometimes is not confined to a single plane. If a driver is looking to the side at the instant of impact, for example, the injury mechanism involves a combination of hyperextension and rotation. In this case the rotation is enhanced by the impact forces prior to cervical hyperextension, and cervical structures are prestretched and more predisposed to severe injury.

At first glance, whiplash might appear a simple injury mechanism. However, "in an individual accident there is likely to be a complex interaction between different forces depending upon the speed and direction of impact and the attitude of the head and neck" (Barnsley et al. 1994, p. 288). Many structures can be injured in whiplash accidents: structures in the brain, temporomandibular joint, muscles, spinous ligaments, intervertebral discs, vertebral bodies, and facet (zygapophyseal) joints (e.g., Davis 2000; Ito et al. 2004; Panjabi, Ito et al. 2004; Pearson et al. 2004). Various mechanisms and potential injury sites are shown in figure 8.20. Whiplash-associated disorders can manifest as both clinical and psychosocial symptoms (Eck et al. 2001). Possible symptoms are listed in table 8.5.

In summary, not all victims of whiplash sustain injuries. "Whether or not a victim sustains an injury is a function of multiple factors: the magnitude of the impact, their posture at the time, their anatomy, and the material strength of the components of their cervical spine" (Bogduk and Yoganandan 2001, p. 272).

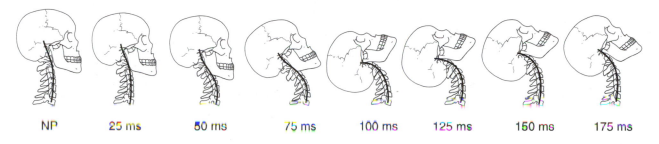

NP   25 ms   50 ms   75 ms   100 ms   125 ms   150 ms   175 ms

Figure 8.19 S-shaped neck curvature resulting from simultaneous upper cervical flexion and lower cervical extension. NP = neutral position.

Reprinted, by permission, from J.N. Grauer et al., 1997, "Whiplash produces an S-shaped curvature with hyperextension at lower levels," *Spine* 22(21): 2492.

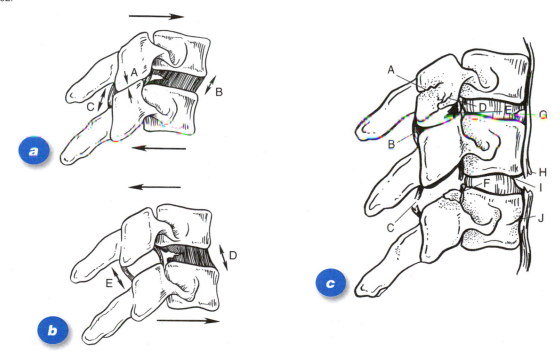

Figure 8.20 Potential mechanisms and injury sites for whiplash-related injuries. Shear forces affecting a spinal motion segment. *(a)* Translation of the superior vertebral body anteriorly relative to the inferior. This movement stresses the articular surfaces of the zygapophyseal joint (A), the anterior annulus fibrosus (B), and the zygapophyseal joint capsule (C). *(b)* Translation of the superior vertebral body posteriorly relative to the inferior body, which stresses the intervertebral disc (D) and the zygapophyseal joint capsules (E). *(c)* Common lesions affecting the cervical spine following whiplash injury. A, articular pillar fracture; B, hemarthrosis (hemorrhage into a joint) of the zygapophyseal joint; C, rupture or tear of the zygapophyseal joint capsule; D, fracture of the subchondral plate; E, contusion of the intra-articular meniscus of the zygapophyseal joint; F, fracture involving the articular surface; G, tear of the annulus fibrosus; H, tear of the anterior longitudinal ligament; I, end-plate avulsion fracture; J, vertebral body fracture.

TABLE 8.5

POSSIBLE CLINICAL AND PSYCHOSOCIAL SYMPTOMS RELATED TO WHIPLASH INJURY

Clinical symptoms	Psychosocial symptoms
Neck pain and stiffness	Depression
Headache	Anger
Shoulder pain and stiffness	Frustration
Vertigo	Anxiety
Dizziness	Family stress
Fatigue	Occupational stress
Temporomandibular joint symptoms	Hypochondriasis
	Compensation neurosis
Arm pain	Drug dependency
Paresthesias	Posttraumatic stress syndrome
Weakness	Sleep disturbances
Visual disturbances	Litigation
Tinnitus	Social isolation
Dysphasia	
Back pain	

Reprinted from *American Journal of Medicine*, Vol. 110, J.C. Eck, S.D. Hodges, S. Craig Humphreys, Whiplash: A Review of a Commonly Misunderstood Injury, pp. 651-656, Copyright 2001, with permission from Elsevier.

Cervical Spondylosis

The etiology of chronic cervical conditions, such as cervical spondylosis, stands in marked contrast with the potentially catastrophic traumatic injuries just described. Although their onset is less dramatic, chronic injuries nonetheless can cause considerable dysfunction. Cervical spondylosis is a term used to describe degenerative changes in the cervical intervertebral discs and surrounding structures, including osteophytosis of the vertebral bodies, ligamentous instability, and osseous hypertrophy of the laminal arches and facets (Lestini and Wiesel 1989). Cervical spondylosis most often affects the C3-C7 vertebrae (McCormack and Weinstein 1996).

As part of the normal aging process, discs lose vertical height and become less extensible, attributable in large part to reduced water content and degradation of the disc substance. Disc degeneration is accompanied by osteophyte (bony outgrowth) formation and increased stresses on articular cartilage. These structural alterations increase the risk of spinal stenosis (narrowed canal), impingement on neural tissue, and impaired blood perfusion of the spinal cord. Symptoms associated with cervical spondylosis include *paresthesia*, neck and arm pain, weakness, and sensory loss. Recent advances in imaging technologies (e.g., magnetic resonance imaging) have improved diagnostic accuracy. Considerable debate continues, however, on the advisability of surgical intervention in treating the lesions associated with cervical spondylosis.

Disturbance or disease of the spinal cord (myelopathy) associated with cervical spondylosis is a well-recognized clinical entity. Mechanical factors that may cause cervical spondylotic myelopathy include a narrowing of the spinal canal, kyphotic conditions causing cervical flexion, spinal cord compression and related ischemia, and ligamentous ossification. Despite

REAR-END COLLISIONS

Over the past half-century, hundreds of research studies have sought to detail the mechanisms involved in rear-end collisions. These studies have involved use of live subjects (in low-speed rear-end impacts), cadaveric simulations, accelerometry, electromyography, and mathematical modeling. As a result of these studies, we have a better understanding of rear-impact dynamics, but controversy remains. The experts do agree on one point—cervical dynamics during rear-impact scenarios are complex and not entirely understood (e.g., Luan et al. 2000).

Pioneering work by Severy (1955) showed that rear-end collisions cause a sequential acceleration of the vehicle, the occupant's trunk and shoulders, and the occupant's head. As the vehicle is impacted (e.g., in an automobile rear-end collision), it accelerates first, reaching a peak acceleration of almost 5 g, that is, five times the acceleration of gravity (A in figure). The vehicle occupant's shoulders reach their peak acceleration of about 7 g 100 ms later (B in figure). Finally, the occupant's head reaches its peak acceleration of greater than 12 g at 250 ms after initial impact (C in figure). This sequential progression of peak acceleration is evidence of both momentum and energy transfers.

Response of the cervical spine depends on impact awareness, muscle involvement, and direction of impact (Kumar et al. 2005). In an unaware vehicle occupant, muscles are recruited late during the whiplash episode. Muscle recruitment and tension development may not happen until 200 to 250 ms after impact. Given that much of the critical cervical motion occurs during the first 200 ms, muscle involvement may only play a role in the late stages of whiplash. Injury may have already happened before the muscles become involved (Bogduk and Yoganandan 2001).

On a positive note, epidemiological evidence suggests that many victims of rear-end collision do not sustain injuries, and most of those who are injured show no long-lasting effects. In one study, 18% of patients had injury-related symptoms 2 years postinjury—82% were asymptomatic (Radanov et al. 1995).

In addition to impact awareness, muscle involvement, and direction of impact, many other factors determine injury risk in rear-end impacts: vehicle mass, velocity, and ability to withstand crashes; road conditions; use of restraint systems; and the passenger's or driver's body and head position at impact, neck rotation, gender, history of neck injury, and age.

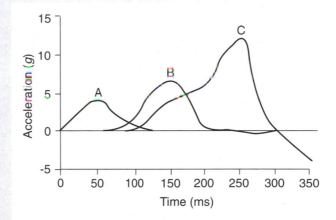

Idealized acceleration curves of (A) an impacted vehicle, (B) an occupant's shoulders, and (C) an occupant's head.

Reprinted from *Pain*, Vol. 58, L. Barnsley, S. Lord, N. Bogduk, Whiplash injury, pp. 283-307, Copyright 1994, with permission from the International Association for the Study of Pain.

the often insidious onset of cervical spondylosis, its sequelae have significant potential to cause severe neuromuscular dysfunction.

Trunk Injuries

The trunk (also called *truncus* or *torso*) extends from the base of the neck down to the pelvic girdle. As the largest body region, the trunk accounts for 45% to 50% of the body's mass and contains such vital organs as the heart (and its major vessels), spinal cord, lungs, stomach and intestines, kidneys, liver, and spleen. The sternum, ribs, and vertebrae of the axial skeleton protect these important organs.

Trunk musculature serves both movement and protective functions. The principal muscles of the anterior trunk are the pectoralis major, serratus anterior, rectus abdominis, external obliques, internal obliques, and transversus abdominis (figure 8.21, *a-d*). Important posterior trunk muscles include the trapezius, latissimus dorsi, rhomboids (major and minor), and erector spinae (figure 8.21, *e* and *f*).

Vertebral Fracture

Vertebral fractures, often associated with osteoporosis, are a major health concern, affecting approximately 25% of postmenopausal women (Melton 1997). Risk factors associated with vertebral compression fractures in the elderly are classified as either *modifiable* or *nonmodifiable*. Among the potentially modifiable risk factors are alcohol and tobacco use, osteoporosis, estrogen deficiency, early menopause, physical frailty, impaired eyesight, lack of physical activity, low body weight, and deficiencies in dietary calcium and vitamin D. Nonmodifiable risk factors include age (advanced), gender (female), dementia, race (Caucasian), fall risk, and adult fracture history (Old and Calvert 2004).

Traumatic vertebral fractures are of particular concern because of their close proximity to the spinal cord. In displaced spinal fractures, bone fragments may be forced into the spinal canal and impinge on the cord. This impingement can cause severe neural damage and attendant paralysis or even death. Vertebral fractures usually are caused by axial compressive loads and occur most com-

monly in the cervical and thoracolumbar regions. Vertebrae in the thoracolumbar region (variably defined to include vertebrae between T11 and L3) are especially susceptible to fracture because of the spine's relatively neutral alignment (minimal curvature) in this region and because this region is a transition zone between the relatively rigid thoracic region and the more flexible lumbar region.

Roaf (1960) and Holdsworth (1963, 1970) provided early descriptions of the mechanism of compressive vertebral fracture. Holdsworth (1963) postulated a two-column model of the spine, consisting of an anterior column (region between the anterior longitudinal ligament and the posterior longitudinal ligament) and a posterior column (posterior bony complex bounded by the posterior longitudinal ligament and the posterior ligamentous complex that includes the supraspinous ligament, interspinous ligament, capsule, and ligamentum flavum). Holdsworth's two-column model was modified by Denis (1983), who proposed a three-column model (figure 8.22) to explain the pattern of thoracolumbar injuries (table 8.6).

In describing the mechanism of compressive vertebral fracture, Holdsworth (1970) coined the term *burst fracture,* noting that when a severe compression force is applied to either cervical or lumbar vertebrae when they are aligned in a straight (noncurved) row, or column, the body of the vertebra can shatter from within (i.e., explode or burst).

The danger of spinal cord lesion attributable to fragment displacement depends on the rate of loading. In controlled conditions, with constant energy input and force direction, spine motion segments impacted at high loading rates fracture with significant encroachment into the spinal canal. Low loading rates, in contrast, produce minimal intrusion (Tran et al. 1995).

The level of instability created by burst fractures has been the subject of some controversy, especially in cases with an absence of neurological deficit associated with the fracture. Panjabi and colleagues (1994) found multiaxial instabilities, especially in response to axial rotation and lateral bending. Their results suggest that treatment of burst fractures requires caution and that their fixation and stabilization should be approached conservatively.

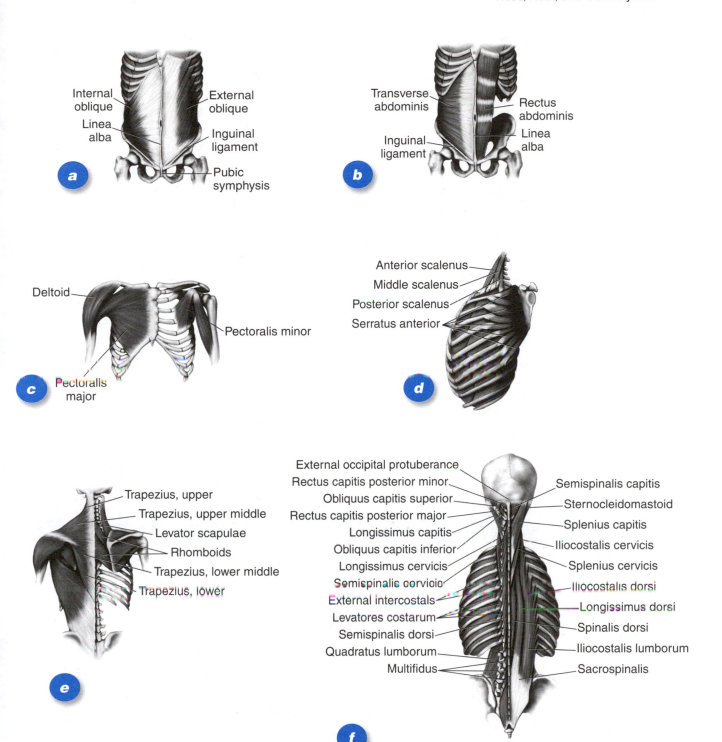

Figure 8.21 Musculature of the trunk. *(a-d)* Anterior views. *(e-f)* Posterior views. Deep muscles of the back, including the three subdivisions of the erector spinae group (iliocostalis, spinalis, longissimus) are shown in *(f)*.

Adapted, by permission, from R.S. Behnke, 2005, *Kinetic anatomy*, 2nd ed. (Champaign, IL: Human Kinetics), 132.

The degree of disc degeneration confounds the mechanism of burst fracture. A finite-element computer model (Shirado et al. 1992) showed that in motion segments with normal discs, the highest stresses were found under the nucleus and at the superoposterior portions of the trabecular body (figure 8.23). In a severely degenerated disc, no stresses were seen under the nucleus and little stress at the middle of the end plate. Based on the computer model of thoracolumbar loading, the

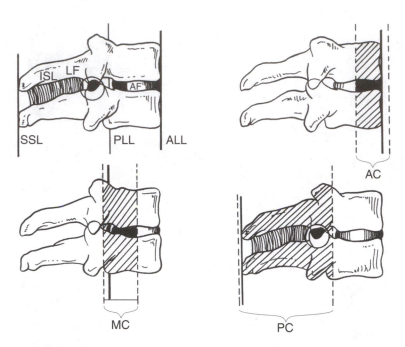

Figure 8.22 Three-column model of the spine: AC, anterior column (upper right); MC, middle column (lower left); PC, posterior column (lower right). Vertebral structures: SSL, supraspinous ligament; PLL, posterior longitudinal ligament; ALL, anterior longitudinal ligament; AF, annulus fibrosus; LF, ligamentum flavum; ISL, interspinous ligament.

Reprinted, by permission, from F. Denis, 1983, "The three column spine and its significance in the classification of acute thoracolumbar spinal injuries," *Spine* 8(8): 818.

TABLE 8.6

FAILURE MECHANISMS USING A THREE-COLUMN MODEL OF THE SPINE

	COLUMN		
Type of fracture	Anterior	Middle	Posterior
Compression	Compression	None	None or distraction (severe)
Burst	Compression	Compression	None
Seat belt type	None or compression	Distraction	Distraction
Fracture dislocation	Compression rotation shear	Distraction rotation shear	Distraction rotation shear

Reprinted, by permission, from F. Denis, 1983, "The three column spine and its significance in the classification of acute thoracolumbar spinal injuries," *Spine* 8(8): 818.

researchers concluded that with healthy discs, rapid axial compressive loading induces a burst fracture with bone fragments retropulsed into the spinal canal. In people with severely degenerated discs (e.g., elderly people with osteoporosis), burst fractures would be much less likely and, should they occur, would involve the anterior column and thus not have associated neurological deficits (Shirado et al. 1992).

Spinal Deformities

Injury, disease, and congenital predisposition all can cause deformities of the spinal column that take the form of abnormal structural alignment or alteration of spinal curvatures. These deformities are not injuries, per se, but because they often result in abnormal force distribution patterns and pathological tissue adaptations, they may indirectly lead to or exacerbate other musculoskeletal injuries and thus deserve mention.

There are three primary types of spinal deformity: scoliosis, kyphosis, and lordosis (figure 8.24). These deformities are classified by their magnitude, location, direction, and cause, and they can occur in isolation or in combination. Spinal deformities have long been associated with cardiopulmonary dysfunction. Hippocrates, for example, noted that

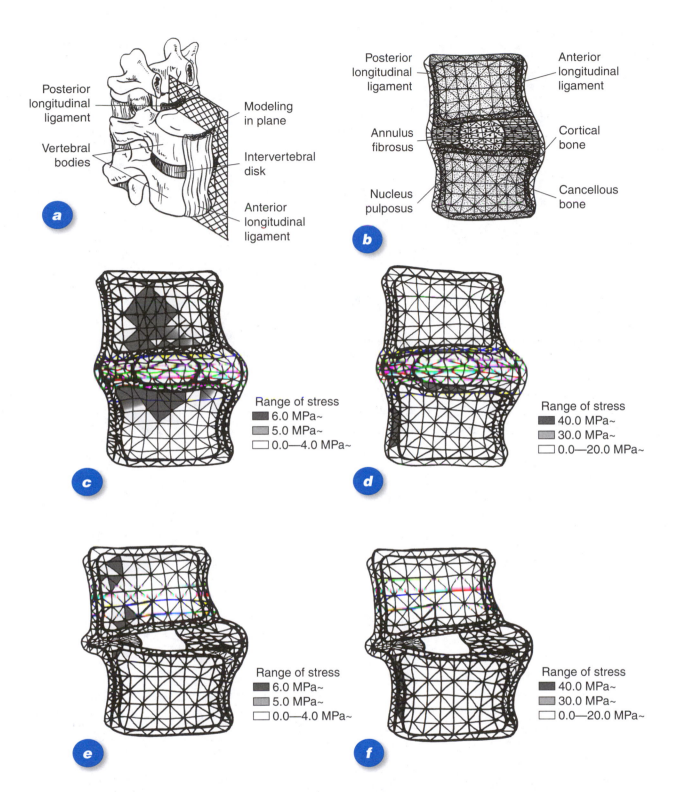

Figure 8.23 *(a)* Median (midsagittal) plane of modeling. *(b)* Finite-element plane model. *(c)* Stress distribution of one motion segment with a healthy disc under axial compression showing the highest trabecular bone stresses under the nucleus pulposus and at the superoposterior regions and *(d)* the highest cortical shell stresses at the middle of the end plate and posterior wall cortex. *(e)* Stress distribution in a model of a severely degenerated disc shows no stresses in the trabecular bone under the nucleus and *(f)* highest stresses in the cortical bone of posterior wall.

Reprinted, by permission, from O. Shirado et al., 1992, "Influences of disc degeneration on mechanism of thoracolumbar burst fractures," *Spine* 17(3):286-292

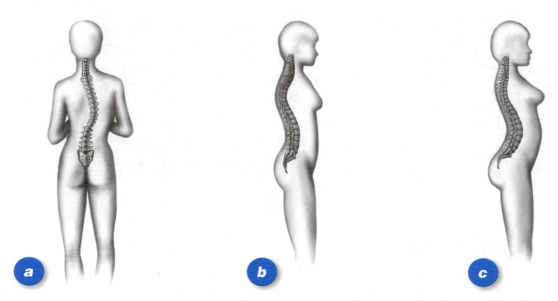

Figure 8.24 Spinal deformities: *(a)* Scoliosis. *(b)* Kyphosis. *(c)* Lordosis.
Reprinted, by permission, from W.C. Whiting and S. Rugg, 2006, *Dynatomy: Dynamic human anatomy* (Champaign, IL: Human Kinetics), 65.

hunchbacks (those with kyphosis) had difficulty breathing, and that patients afflicted with scoliosis commonly exhibited **dyspnea** (shortness of breath) (Padman 1995).

Scoliosis

Scoliosis is defined as a lateral (frontal plane) spinal curvature often associated with twisting of the spine (figure 8.25). Mild spinal deviations are well tolerated and usually asymptomatic. Severe deformities, in contrast, can markedly compromise cardiopulmonary processes. Scoliotic curvatures exceeding 90° greatly increase the risk of cardiorespiratory failure through the cumulative effect of decreased lung and chest wall compliance, poor blood oxygenation **(hypoxemia),** increased work of breathing, reduced respiratory drive, enlarged heart **(cardiomegaly),** and pulmonary arterial hypertension (Padman 1995).

A minority of scoliosis cases (15-20%) are attributable to known causes, including maldevelopmental (congenital) spinal deformities, acquired pathological lesions, neuromuscular disorders, metabolic disorders, and inherited connective tissue diseases (Stehbens 2003). The majority of cases, however, are idiopathic. Various mechanisms and theories have been suggested to explain idiopathic scoliosis. There is support for the role of hereditary (genetic) factors in the pathogenesis of idiopathic scoliosis (e.g., Miller

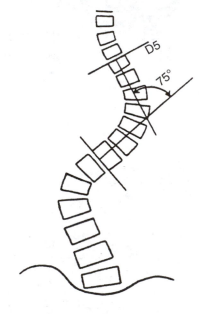

Figure 8.25 Measure of scoliotic spinal curvature: the **Cobb angle,** defined as the angle (75° in the example shown here from Cobb s original work) between the two lines that perpendicularly bisect the lines through the surfaces of the vertebrae at each end of the curvature.

2000). Others favor the involvement of cumulative, repetitive, and asymmetrical biomechanical stresses (e.g., Stehbens 2003). Alternative areas of research include exploration of connective tissue abnormalities, neurological and growth abnormalities, muscle structure, and secondary involvement of melatonin deficiency. The answer remains elusive. "The consensus is that the etiology is multifactorial. With time, continued research will lead to the identification of the various factors involved in the causation of this disorder, which affects so many children and adolescents" (Lowe et al. 2000, p. 1166).

Treatment options for scoliosis are either nonoperative or operative. Nonoperative interventions include electrical stimulation, biofeedback, manipulation, and bracing. Of these, bracing (figure 8.26a) has proved most successful (Parent et al. 2005). A recent review highlights the growing body of evidence that exercise-based approaches can be used to prevent the progression and reverse the signs and symptoms of spinal deformity (Hawes 2003).

For severe scoliotic curvatures, the preferred treatment is operative vertebral fusion in which adjacent vertebrae are fused to forestall further progression of the deformity (figure 8.26b).

The importance of early diagnosis and intervention should not be understated, because in many cases scoliotic deformity is progressive. Lack of early intervention can result in severe, even life-threatening deformities later in life. Although the causal mechanisms of scoliosis are often unknown, the mechanics of treatment in the form of braces or implanted spinal rods are well established and proven effective.

Kyphosis

Kyphosis is a sagittal-plane spinal deformity characterized by excessive flexion, usually in the thoracic region, where it produces a hunchback posture. (Note: *Kyphosis* refers to a forward curvature that is concave anteriorly. The thoracic and sacral regions have a natural kyphosis. Thus, hunchback actually involves a *hyper*kyphosis, or exaggerated kyphotic curvature. Clinically, the term *kyphosis* often is used, as here, to describe this hyperkyphotic condition.)

Kyphosis tends to be more severe in women than in men and is more prevalent with advancing age in both genders. Elderly, postmenopausal women are at particular risk, largely because of the strong association between kyphosis and osteoporosis (Bradford 1995). Kyphosis in these women is readily evident by the presence of a characteristic hump in the upper thoracic region.

The degree of kyphosis is measured by the curvature of the spine in the sagittal plane (using a method similar to the Cobb angle, described in figure 8.25) or alternatively by the sum of the wedge angles between adjacent vertebrae (figure 8.27). The wedge angles of thoracic vertebrae T4-T12 increase exponentially as a function of age, up to 70 years (Puche et al. 1995). This vertebral wedging and its attendant kyphosis affect rib mobility and pulmonary function. Specifically, the thoracic angle in kyphosis has a significant negative correlation with inspiratory capacity, vital capacity, and lateral expansion of the thorax (Culham et al. 1994).

The best treatment for kyphosis may lie in prevention. Women with satisfactory exercise habits have a significantly lower index of kyphosis,

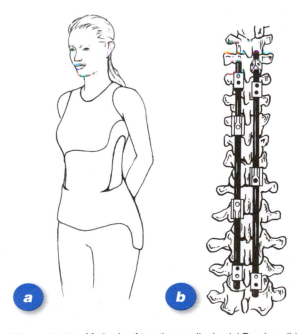

Figure 8.26 Methods of treating scoliosis. *(a)* Bracing. *(b)* Surgical implantation of Harrington instrumentation (rods) to stabilize the spine.

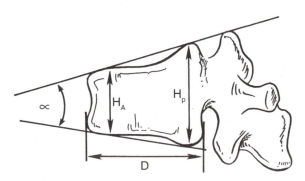

Figure 8.27 Vertebral wedging as measured by the angle α. The amount of wedging can also be measured by calculating the vertebral body height ratio (H_A/H_P). D = depth of vertebral body.

suggesting that physical conditioning programs aimed at proper postural maintenance may delay or prevent the onset of kyphosis associated with aging (Cutler et al. 1993).

Children may exhibit a special type of kyphosis, known as **Scheuermann's kyphosis,** in which structural changes are seen in the end plates of the growing vertebral bodies. Scheuermann's kyphosis is idiopathic, and mild cases are readily controlled by appropriate back extension exercises, with symptoms subsiding on completion of bone growth. The condition is orthopedically significant because it may be progressive. Pathological progression increases the severity of deformity, which may require bracing or surgery (Wegner and Frick 1999).

Lordosis

Lordosis is an abnormal extension deformity, usually seen in the lumbar region, that produces a hollow, or swayback, condition. (Note: *Lordosis* refers to a backward curvature that is concave posteriorly. The cervical and lumbar regions have a natural lordosis. Thus, swayback actually involves a *hyper*lordosis, or exaggerated lordotic curvature. Clinically, the term *lordosis* often is used, as here, to describe this hyperlordotic condition.)

Lordosis is more common in women than men and in persons with a higher **body mass index** (Murrie et al. 2003). Whether lordosis varies with age remains open to debate. Tuzun and colleagues (1999) reported an increase in lordosis with age; Amonoo-Kuofi (1992) found a tendency for decreased lordosis with age; and Murrie and colleagues (2003) found no change in lordosis in older persons.

Forward tilting of the pelvis accentuates lumbar lordosis. This tilting increases the lumbosacral (L5-S1) angle above its normal 30° orientation (figure 8.28), increases shear loading on the intervertebral discs and surrounding structures, and decreases the compressive load on the disc.

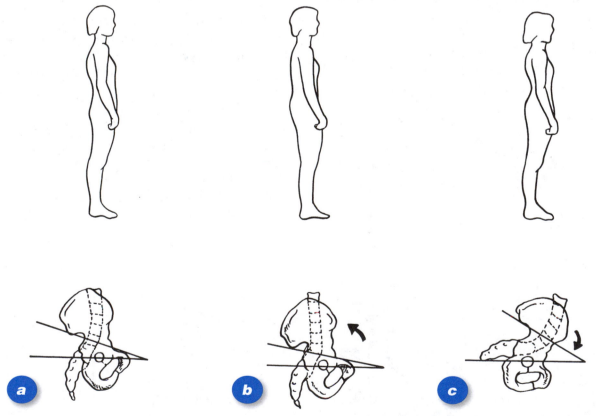

Figure 8.28 Effect of pelvic tilting on the lumbosacral (L5-S1) angle. *(a)* Normal standing creates a lumbosacral angle of approximately 30°. *(b)* Tilting the pelvis backward decreases the lumbosacral angle (<30°) and flattens the lumbar spine. *(c)* Tilting the pelvis forward increases the lumbosacral angle (>30°) and exaggerates the lumbar lordosis.

Reprinted, by permission, from M. Lindh, 1989, Biomechanics of the lumbar spine. In *Basic biomechanics of the musculoskeletal system*, 2nd ed., edited by M. Nordin and V.H. Frankel (Philadelphia, PA: Lea & Febiger), 192.

CHILDREN AND BACKPACKS

Up to 80% of adults will experience low back pain in their lifetime. In recent years, back pain in children has been on the rise, with age-group prevalence approaching 50%. Many theorize that the increase in youth-related back pain is attributable to use of heavy backpacks. Most schoolchildren wear backpacks every day, often carrying excessive weight. Most studies report average daily loads between 10% and 20% of body weight (BW), but loads as high as 46.2% BW have been reported (Negrini et al. 1999). In this study, the authors also found that more than one third of the children carried in excess of 30% BW at least once a week.

Many studies have associated heavy backpack use with back pain (e.g., Negrini and Carabalona 2002; Sheir-Neiss et al. 2003), although this opinion is not universal, because some investigators have questioned the relation between backpack weight and back pain (e.g., Cardon and Balague 2004; Wall et al. 2003).

Maximal recommended load limits of 10% to 15% BW are common (e.g., American Academy of Pediatrics 2007; American Occupational Therapy Association 2005) and justified (Brackley and Stevenson 2004). Backpack weight, however, is only one factor involved in injury risk. Other factors such as load distribution and carrying technique also affect musculoskeletal loading. Most backpacks are designed with two shoulder straps to distribute the load evenly. Some children, especially in the United States, carry two-strap backpacks over only one shoulder, or they use one-strap backpacks. To minimize the risk of injury, they should carry their backpacks on two shoulders rather than one (Cottalorda et al. 2003).

The American Occupational Therapy Association (2005) provided sound guidelines for proper backpack use by children:

- Don't allow children to carry more than 15% of their body weight.
- Load heaviest items closest to the child's back.
- Make sure that the child carries only necessary items.
- Unload heavy backpacks by having the child hand carry a book or other item.
- Have the child wear both shoulder straps.
- Choose a backpack with well-padded shoulder straps.
- Adjust shoulder straps to fit pack snugly against child's back.
- Have the child wear the waist belt if provided.
- Ensure that the bottom of the pack rests in the curve of the low back.
- Choose an age- and size-appropriate backpack with enough room for necessary items.

The backpack–back pain relation is based on *chronic* backpack use; *acute* backpack injuries usually do not involve back injuries. A study of 247 children with acute backpack injuries found that injury location most commonly involved the head and face (22%), followed by the hand (14%), wrist and elbow (13%), shoulder (12%), and foot and ankle (12%). Back injuries ranked sixth at 11%. Acute injuries were more than twice as likely to happen because of tripping over the backpack as by wearing it (Wiersema et al. 2003).

Although much has been learned in recent years, more research is recommended. Research is needed to establish the association between backpack use and injury; how load, backpack design, and personal characteristics (e.g., physical fitness) affect backpack carrying adaptations (Brackley and Stevenson 2004); gait changes attributable to the increased demand imposed by heavy backpacks (Chow et al. 2005); and the involvement of personal physical and psychological factors in the association between backpack load and back pain (Negrini and Carabalona 2002).

Lower-extremity joint pathologies can affect lumbar lordosis. Offierski and MacNab (1983) described a "hip-spine syndrome" in which concurrent pathologies exist at the hip and spine, and they cautioned that failure to recognize the concurrent pathologies may result in misdiagnosis and potential treatment errors. Murata and colleagues (2003) reported what they termed a *knee–spine syndrome,* in which degenerative changes in the knee caused symptoms in the lumbar spine.

Spondylolysis and Spondylolisthesis

Low back pain arises from myriad causes, including structural abnormalities, chronic overuse, and trauma. Two specific conditions that especially afflict young and athletic populations are spondylolysis and spondylolisthesis. These conditions affect the bony structure of the vertebrae, especially at the L4-L5 and L5-S1 levels. **Spondylolysis** is a defect in the area of the lamina between the superior and inferior articular facets known as the **pars interarticularis** (figure 8.29*a*). Radiographic evidence suggests that spondylolysis affects 6% of the adult population (Herman et al. 2003). **Spondylolisthesis** is translational motion, or slippage, between adjacent vertebral bodies (figure 8.29*b*). Most individuals with either condition remain asymptomatic.

Although they have been described for many years, spondylolysis and spondylolisthesis are a source of controversy. Recent advances in imaging techniques have permitted a better understanding of the pathogenesis of both conditions.

Classification of spondylolysis and spondylolisthesis according to their etiological and anatomical characteristics commonly identifies five types: dysplastic, isthmic, degenerative, traumatic, and pathological (Wiltse et al. 1976). The involved anatomical structures and pathogenesis for each type are presented in table 8.7. An alternative classification system based on clinical presentation and spinal morphology has been proposed that may be more appropriate for children and adolescents (Herman and Pizutillo 2005) (see Spondylolysis and Spondylolisthesis Classifications for Children and Adolescents).

Of greatest concern to young athletes is the situation in which repeated loading of the pars region causes microfractures and eventual bone failure. Among the mechanisms responsible for these pars defect failures are repetitive spinal flexion, combined flexion and extension, forcible hyperextension, and lumbar spine rotation.

In spondylolisthesis, slippage occurs between adjacent vertebrae. A recent study using a finite-element model showed that moments created by lateral bending and torsion are associated with the greatest vertebral slippage (Natarajan et al. 2003). The process involved in slippage differs

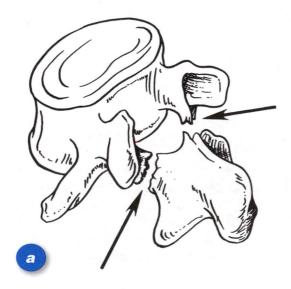

Figure 8.29 *(a)* Spondylolysis showing bilateral fracture of the pars interarticularis (arrows). *(b)* Spondylolisthesis exhibiting slippage (arrow) between the L5 and S1 vertebrae.

TABLE 8.7

CLASSIFICATION OF SPONDYLOLYSIS AND SPONDYLOLISTHESIS

Type	Name	Anatomical involvement	Etiology
I	Dysplastic	Neural arch dysplasia	Hereditary or congenital facet orientation anomalies, particularly L5-S1 facets and supporting structure
II	Isthmic	Pars interarticularis abnormality	Succession of microfractures; mechanical, hormonal, hereditary causes
III	Degenerative	Degenerative disc, facet, and ligamentous disease	Advanced pan-column degenerative changes
IV	Traumatic	Traumatic column instability with delayed translational changes	Trauma
V	Pathological	Pars and other components	Noncongenital or acquired (e.g., infection or neoplasm)

Adapted, by permission, from C.B. Stillerman, J.H. Schneider and J.P. Gruen, 1993, "Evaluation and management of spondylolysis and spondylolisthesis," *Clinical Neurosurgery* 40: 385.

SPONDYLOLYSIS AND SPONDYLOLISTHESIS CLASSIFICATIONS FOR CHILDREN AND ADOLESCENTS

New classification

I. Dysplastic

II. Developmental

III. Traumatic

   Acute

   Chronic
   -stress reaction
   -stress fracture
   -spondylolytic defect (nonunion of pars)

IV. Pathologic

Adapted, by permission, from M.J. Herman and P.D. Pizzutillo, 2005, "Spondylolysis and spondylolisthesis in the child and adolescent: A new classification," *Clinical Orthopaedics and Related Research* 434: 49.

between young and old individuals (typically in women older than 50 years). In older populations, spondylolisthesis occurs most frequently at L4-L5, attributable in part to degenerative lesions associated with arthritis of the facet joints and to relative instability at this level compared with L5-S1. This instability may be caused by a developmental predisposition to more sagittally oriented facet joints at the L4-L5 level (Grobler et al. 1993).

Not surprisingly, the populations most at risk for both spondylolysis and spondylolisthesis are those whose training exposes them to repeated, high, compressive spinal loading, especially in combination with flexion–extension and rotational positions and movements. These include gymnasts, weightlifters, wrestlers, and divers. Spondylolysis is exacerbated by the stresses imposed on the vertebral laminae by the body's lumbar lordosis.

In young athletes, the mechanism allowing spondylolisthesis differs from that observed in adults. In patients between 9 and 18 years old, Ikata and colleagues (1996) found end-plate lesions in all cases of vertebral slip between L5 and S1 exceeding 5%. The implicated mechanism was slippage between the osseous and cartilaginous end plates secondary to spondylolysis. The likelihood of progression (continuing slippage) depends on the type, stability, and degree of slippage and the slip angle (Bradford 1995). With continued slippage, the superior vertebra (L5) moves anteriorly relative to the inferior articular surface (S1), the slip angle (figure 8.30a) changes, and the degree of slippage (as measured by percentage slip as shown in figure 8.30b) increases. There seems to be considerable slowing of slip progression with each decade (Beutler et al. 2003).

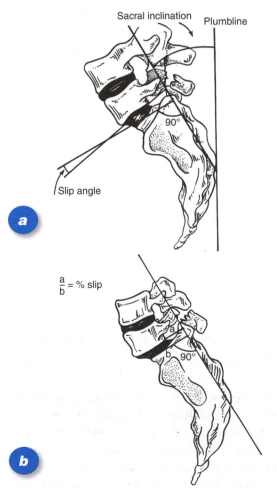

Figure 8.30 Schematic of sacral inclination and *(a)* slip angle and *(b)* percentage slip.

Adapted, by permission, from J.T. Stinson, 1993, "Spondylolysis and spondylolisthesis in the athlete," *Clinics in Sports Medicine* 12(3): 520.

Intervertebral Disc Pathologies

Lumbosacral injuries can involve any of the many structures comprising the spinal column and involve three mechanisms: (1) spinal compression or weight bearing; (2) torsional loading, which results in various patterns of shearing in the transverse (horizontal) plane; and (3) tensile stresses resulting from excessive spinal motion (Watkins and Williams 2001). We present one class of lumbosacral injury (intervertebral disc pathologies) whose injury mechanisms are typical of other lumbosacral injuries.

Normal activities load the intervertebral discs in complex ways. The combined effects of spinal flexion–extension, lateral bending, and rotation exert high forces on the discs and their supporting structures. These forces are highest in the lumbar region, largely attributable to the compressive forces imposed by the weight of superior body segments.

Because disc morphology dictates mechanical response, we need to understand some details of disc anatomy. The intervertebral disc is a viscoelastic structure consisting of two distinct structural elements, the *annulus fibrosus* and *nucleus pulposus*. The disc is separated from the vertebra by a thin layer of hyaline cartilage (cartilaginous end plate). The nucleus pulposus is a gelatinous mass consisting of fine fibers embedded in a mucoprotein gel, with water content ranging from 70% to 90%. Water and proteoglycan contents are highest in the young and decrease with age. The mechanical consequences of these losses include decreases in disc height, elasticity, energy storage ability, and load-carrying capacity. The lumbar nucleus pulposus occupies 30% to 50% of the total disc area in cross section and is located slightly posteriorly (rather than centrally) between adjacent vertebral bodies.

The annulus fibrosus is composed of fibrocartilage and consists of concentric bands of annular fibers that surround the nucleus pulposus and form the outer boundary of the disc. The collagen fibers of adjacent annular bands run in opposite directions (figure 8.31). This crisscrossed fiber orientation allows the annulus to accommodate multidirectional torsional and bending loads.

When subjected to compressive loading, the disc components respond differently. In an unloaded state, the nucleus pulposus exhibits an intrinsic pressure of 10 N/cm² attributable to pre-loading provided by the longitudinal ligaments and the ligamenta flava. In a loaded state, the nucleus pulposus accepts 1.5 times the externally applied load, whereas the annulus experiences only 0.5 times the compressive load (figure 8.32). Because of the relative incompressibility of the

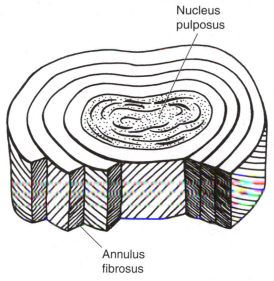

Figure 8.31 Concentric rings of the annulus fibrosus surrounding the centrally located nucleus pulposus, with alternating fiber angulation.

nucleus pulposus, the load is transmitted (see *Poisson's effect,* chapter 3) as a tensile load to the fibers of the annulus fibrosus. These forces radiate circumferentially in what is described as a *hoop effect.* The resulting tensile stresses may be four to five times greater than the externally applied compressive load (Nachemson 1975).

Many mechanisms have been suggested as responsible for lumbar disc pathologies. In a classic work, Charnley (1955) presented a theoretical framework of potential pathoanatomical mechanisms responsible for intervertebral disc pathologies and their relation to **lumbago** and **sciatica**. He wrote, "I make no apology for putting forward some decidedly bold and unorthodox statements in the management of an acute case of lumbago or sciatica; only by stimulating some clear thinking on this fascinating subject can theories be put to the test and progress made" (Charnley 1955, p. 344). Much of what Charnley characterized as "bold and unorthodox" more than 50 years ago is now accepted.

Some of the mechanisms and causal factors involved in intervertebral disc pathologies are summarized in table 8.8 We examine one of these factors (type IV, bulging disc) in detail here. In this mechanism, the nucleus pulposus

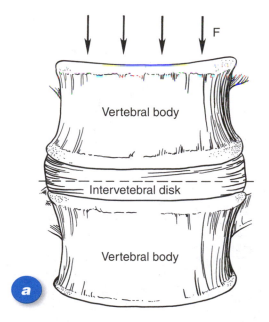

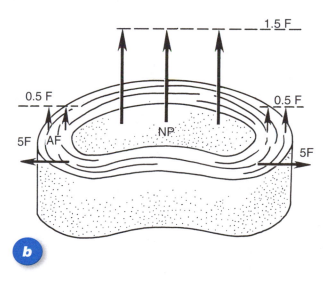

Figure 8.32 Uneven stress distribution across the lumbar intervertebral disc. *(a)* A uniform compressive load *(F)* applied through the vertebral body creates *(b)* an axial stress of 1.5 *F* (per unit area) in the nucleus pulposus (NP). The annulus fibrosus (AF), in contrast, generates an axial stress of only 0.5 *F*. The orthogonal stress in the annulus (perpendicular to the applied load) can reach levels up to five times the applied force (5 *F*).

Reprinted, by permission, from A. Nachemson, 1975, "Towards a better understanding of low-back pain: A review of the mechanics of the lumbar disc," *Rheumatology* 14(3): 129-143.

TABLE 8.8

CAUSAL FACTORS IN INTERVERTEBRAL DISC PATHOLOGIES

Type	Name (description)
I	Acute back sprain (injury to annular fibers, other posterior ligaments, or musculotendinous structures)
II	Fluid ingestion (increased fluid uptake in the nucleus pulposus)
III	Posterolateral annulus disruption (annular disruption with consequent stimulation or sensory innervation by mechanical, chemical, or inflammatory irritants)
IV	Bulging disc (protrusion of the nucleus pulposus with impingement on neural structures)
V	Sequestered fragment (wandering disc material)
VI	Displaced sequestered fragment (displacement of sequestrum into the spinal canal or intervertebral foramen)
VII	Degenerating disc (progressive degeneration of the annulus fibrosus)

Adapted, by permission, from A.A. White and M.M. Panjabi, 1990, *Clinical biomechanics of the spine*, 2nd ed. (Philadelphia, PA: Lippincott, Williams & Wilkins), 391-395.

is displaced from its normal position within the annulus fibrosus. Rotational body movements produce shearing stress in the annular fibers and can lead to circumferential and radial tears. The resulting weakness in the annular layers reduces the ability of the annulus to contain the nucleus pulposus. Compressive loads then squeeze the nucleus pulposus into the area of annular weakness. Sudden disc migration occurs acutely and typically is precipitated by a hyperflexion mechanism, often in conjunction with lateral bending or rotation. This mechanism creates tensile stresses on the posterolateral aspect of the annulus fibrosus (figure 8.33), which, when combined with a compressive load, results in disc movement. Adams and Hutton (1982) suggested that sudden disc migration occurs most often in the lower lumbar region (L4-L5 or L5-S1) and is associated with disc degeneration.

The hyperflexion mechanism is not implicated in most cases of gradual disc herniation

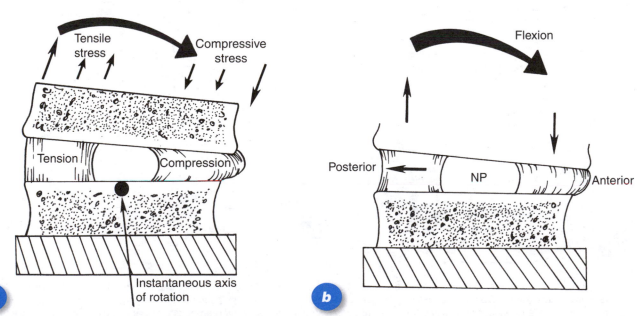

Figure 8.33 Intervertebral disc stress in response to bending. *(a)* In bending, one side of the disc experiences compression while the other side undergoes tension. The compressive and tensile stresses are at a maximum at the outer borders of the disc and decrease toward the center of the disc. *(b)* Forward flexion of the spine tends to squeeze the nucleus pulposus (NP) posteriorly.

Adapted, by permission, from A.A. White and M.M. Panjabi, 1990, *Clinical biomechanics of the spine*, 2nd ed. (Philadelphia, PA: Lippincott, Williams & Wilkins), 15.

(i.e., where there is no identifiable precipitating event). These injuries have been associated with a weakened and degenerative annulus and other loading mechanisms (e.g., bending and twisting). Disc injury typically follows a progression from disc bulging (figure 8.34 *a-b*) to sequestration. The nucleus pulposus puts such pressure on the annulus fibrosus that if there is a weak spot in the walls of the inner annular ring, that ring will rupture. In that way the nucleus pulposus can slowly make its way through successive layers of the annulus fibrosus and form a herniation, which can appear as either a protrusion (figure 8.34*c*) or extrusion (figure 8.34*d*). Eventually herniation can lead to tearing of the outer layer of the annulus fibrosus, at which point the nucleus pulposus begins to leak out of the disc altogether, which can result in sequestration (figure 8.34*e*).

Descriptors of disc pathology are many and varied. To standardize terminology, the North American Spine Society attempted to develop a nomenclature for lumbar disc pathology. These efforts posed challenges, because "standardization of language is difficult, especially among those who have expert knowledge of the subject and clear understanding of what their own words mean. The difficulties must be overcome because deleterious effects ensue when we do not understand what one another's words mean" (Fardon and Milette 2001, p. E93).

Some key terms, with their recommended definitions (Fardon and Milette 2001), are presented here and illustrated in figure 8.34.

- **bulging disc**—A disc in which the contour of the outer annulus fibrosis extends (in the horizontal plane) beyond the edges of the disc space, typically more than 50% (180°) of the disc circumference and less than 3 mm beyond the vertebral body edge (figure 8.34*a-b*). (Note: Disc bulging is not considered a herniation.)

- **herniation**—Localized (i.e., less than 50%, or 180°, of the disc circumference) displacement of disc material beyond the normal margins of the intervertebral disc space.

- **protrusion**—A herniated disc in which the distance between the edges of the disc material beyond the disc space is less than the distance between the edges of the base in the same plane (figure 8.34*c*).

- **extrusion**—A herniated disc in which any one distance between the edges of the disc material beyond the disc space is greater than the distance between the edges of the base in the same plane (figure 8.34*d*).

- **sequestration**—An extruded disc in which part of the disc tissue is displaced beyond the outer annulus and has no connection by disc tissue with the original disc (i.e., a displaced fragment) (figure 8.34*e*).

Whatever the causal mechanism, the herniated disc impinges on adjacent structures. The amount of progressive herniation determines the degree of impingement. Because many disc herniation injuries occur in the posterolateral direction,

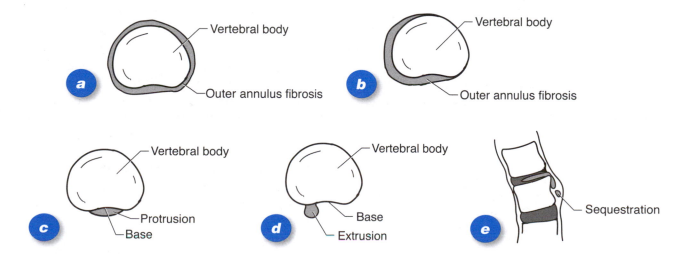

Figure 8.34 Descriptors of intervertebral disc pathology (see text for explanation). (*a*) Symmetrical bulging disc. (*b*) Asymmetrical bulging disc. (*c*) Protrusion. (*d*) Extrusion. (*e*) Sequestration.

LOW BACK PAIN

By any measure, low back pain is the most costly musculoskeletal disorder in industrial societies. Up to 80% of the population will suffer from low back pain in their lifetimes. For some people the pain is merely a temporary annoyance. For others, however, the pain associated with low back pathologies can be completely debilitating.

The pain associated with low back dysfunction arises from chemical or mechanical irritation of pain-sensitive nerve endings in structures of the lumbar spine. Chemical irritation is associated with the biochemical events of inflammatory diseases or subsequent to tissue damage. Mechanical irritation, in contrast, can result from stretching of connective tissues such as ligaments, periostea, tendons, or joint capsules. Compression of spinal nerves by a herniated intervertebral disc, damage to the disc itself, local muscle spasms, and zygapophyseal joint pathology can also cause low back pain. Whatever its source, low back pain either can be felt locally in the lumbar region or may be referred to the buttocks, lower extremities, or, less commonly, to the abdominal wall of the groin.

The efficacy of various treatment strategies remains controversial, largely because an estimated 85% of low back pain cases remain unspecifically diagnosed. Conservative treatment (e.g., ice, rest, gentle activity), manipulative therapies, and therapeutic exercise interventions all have their proponents. Only rarely is surgical intervention indicated. The limited number of well-controlled, randomized studies addressing the issues related to low back pain treatment leaves this area open to continuing debate and controversy. As noted by McGill (2007, p. 99), "Clearly, [low back disorder] causality is often extremely complex with all sorts of factors interacting."

the affected structure often is a nerve root. This results in mechanical, and possibly chemical or inflammatory, irritation of the nerve root with resultant pain in the back, buttocks, thigh, lower leg, and possibly even the foot.

Low back pain, whether from disc pathology or other causes, will afflict up to 80% of individuals at some time during their lives and is the most common occupationally related disability. Given the serious and pervasive nature of this musculoskeletal condition, knowledge of its causal mechanisms is essential in prescribing treatment programs and designing injury prevention strategies.

Concluding Thoughts

An understanding of injury mechanisms in all body regions can assist health professionals in effectively diagnosing and treating patients who suffer these unfortunate injuries and can facilitate the efforts of researchers and medical professionals to reduce the risk of injury and develop effective injury prevention programs. Although our understanding of musculoskeletal injury has increased considerably in recent decades, much remains to be discovered. Among the challenges requiring further investigation are those related to the in vivo responses of biological tissues to mechanical loading, the relations between load volume (i.e., intensity, duration, frequency) and tissue response in real-world situations (e.g., ergonomic, sport), computer modeling of injury dynamics, and effective education programs across all strata of society. Our continuing efforts to understand the biomechanics of musculoskeletal injury are well worth the effort and resources expended. Injury prevention holds the key. After all, the best injury is the one that never happens.

■ CHAPTER REVIEW

Key Points

- Of all the body's regions, the head, neck, and trunk have the greatest potential for catastrophic injury. Injuries to vital structures in these regions can readily compromise essential body functions and, in all too many cases, result in paralysis or death.

- Head injuries happen in response to the sudden application of direct or indirect forces to the head or its connected structures.

- Understanding the mechanisms responsible for head, neck, and trunk injuries can assist with their proper diagnosis, treatment, and prevention.

- The integrated spine comprises a normally balanced system of tissues, alignment, and curvatures that can be disrupted acutely with excessive forces or chronically with repetitive strains.

Questions to Consider

1. This chapter's "A Closer Look" examined concussion injury in detail. Select another injury presented in the text and write your own "A Closer Look" for that injury.

2. Explain, using specific examples, why head, neck, and trunk injuries have the greatest potential for causing catastrophic injury.

3. The sport of boxing is a complex social phenomenon and the subject of controversy. Imagine that you have been selected to participate in a debate addressing the question, "Should boxing be banned?" Prepare a list of arguments to support the abolition of boxing. Prepare another list of arguments to support the retention of boxing.

4. Consider that you have been invited to given a presentation to a group of upper-level elementary school children (grades 4-6) on the topic of backpack safety. Prepare an outline of your presentation based on sound biomechanical principles.

5. Low back pain is a common condition that affects up to 80% of the adult population at some point in their lives. What elements would you include in a biomechanically sound program for adults to prevent low back pain?

Suggested Readings

Becker, D.P., and S.K. Gudeman, eds. 1989. *Textbook of Head Injury*. Philadelphia: Saunders.

Browner, B.D., J.B. Jupiter, A.M. Levine, and P.G. Trafton. 2003. *Skeletal Trauma* (3rd ed.). Philadelphia: Saunders.

Bulstrode, C., J. Buckwalter, A. Carr, L. Marsh, J. Fairbank, J. Wilson-MacDonald, and G. Bowden, eds. 2002. *Oxford Textbook of Orthopedics and Trauma*. Oxford, UK: Oxford University Press.

Cooper, P.R., and J. Golfinos, eds. 2000. *Head Injury* (4th ed.). New York: McGraw-Hill.

Freivalds, A. 2004. *Biomechanics of the Upper Limbs: Mechanics, Modeling, and Musculoskeletal Injuries*. Boca Raton, FL: CRC Press.

Fu, F.H., and D.A. Stone. 2001. *Sports Injuries: Mechanisms, Prevention, Treatment* (2nd ed.). Philadelphia: Lippincott Williams & Wilkins.

Hoerner, E.F., ed. 1993. *Head and Neck Injuries in Sports*. Philadelphia: American Society for Testing and Materials.

King, A.I. 2000. Fundamentals of impact biomechanics: Part I—Biomechanics of the head, neck and thorax. *Annual Review of Biomedical Engineering* 2: 55-81.

Levine, R.S., ed. 1994. *Head and Neck Injury*. Warrendale, PA: Society of Automotive Engineers.

Lonstein, J.E., D.S. Bradford, R.B. Winter, and J.W. Ogilvie, eds. 1995. *Moe's Textbook of Scoliosis and Other Spinal Deformities* (3rd ed.). Philadelphia: Saunders.

McGill, S. 2007. *Low Back Disorders: Evidence-Based Prevention and Rehabilitation* (2nd ed.). Champaign, IL: Human Kinetics.

Moore, E.E., D.V. Feliciano, K.L. Mattox, eds. 2003. *Trauma* (5th ed.). New York: McGraw-Hill.

Reilly, P.L., and R. Bullock, eds. 2005. *Head Injury: Pathophysiology and Management* (2nd ed.). Oxford, UK: Oxford University Press.

Rizzo, M., and D. Tranel, eds. 1996. *Head Injury and Postconcussive Syndrome*. New York: Churchill Livingstone.

Rook, J.L., ed. 2003. *Whiplash Injuries*. Philadelphia: Butterworth-Heinemann.

Torg, J.S., ed. 1991. *Athletic Injuries to the Head, Neck, and Face* (2nd ed.). St. Louis: Mosby-Year Book.

Vinken, P.J., G.W. Bruyn, and H.L. Klawans, eds. 1990. *Head Injury*. New York: Elsevier.

White, A.A., and M.M. Panjabi. 1990. *Clinical Biomechanics of the Spine* (2nd ed.). Baltimore: Williams & Wilkins.

GLOSSARY

abrasion—A scraping away of the superficial skin layer, usually by friction or an abnormal mechanical process.

acceleration—Measure of the change in a body's velocity. Change in velocity divided by change in time.

accident—Unforeseen or unplanned event or circumstance, possibly resulting from carelessness or ignorance and potentially leading to an unexpected injury.

acromioplasty—Excision of the acromion process to relieve pressure in the subacromial space.

actin—Contractile protein that forms thin myofilament in muscle fibers that acts as the binding site during muscular contraction.

action potential—Quick increase **and** decrease in electrical activity traveling along the membrane of a cell.

active stabilization—Contribution of muscle action to joint stability.

acute injury—Injury resulting from a single or a few overload episodes.

acute muscular strain—Overstretching a passive muscle or dynamically overloading an active muscle.

adaptation—Modification of an organism, or the parts of an organism, that makes it fit for existence within the confines of its environment.

adhesive capsulitis—Condition characterized by pain, stiffness, and limited range of motion (e.g., *frozen shoulder*).

adipose tissue—Loose connective tissue that appears as an aggregate of fat cells surrounded by areolar tissue.

aggrecan—Large aggregating proteoglycan (protein modified with carbohydrates) forming the major structural component of cartilage.

allograft—Tissue graft between two individuals of the same species.

all-or-none principle—Principle stating that all fibers within a given motor unit contract or none contract.

amenorrhea—Absence of menstrual cycles.

analytical epidemiology—Use of research strategies to reveal the determinants or underlying causes of disease and injury.

anastomoses—Connections of branches or parts (e.g., convergence of blood vessels).

anatomical position—Standard body reference position, with body erect, head facing forward, and arms hanging straight down with palms facing forward.

aneurysm—Abnormal blood-filled cavity in a blood vessel or organ.

angular acceleration—Change in angular velocity divided by the change in time.

angular displacement—Amount of rotation.

angular impulse—Product of moment $\times$ time; also = torque $\times$ time.

angular momentum—Quantity of angular motion measured by the product of mass moment of inertia $\times$ angular velocity.

angular motion—Form of movement in which a body rotates about an axis of rotation; also *rotational motion*.

angular power—Measure of angular work per unit time. Also equals (torque) $\times$ (angular velocity).

angular velocity—Angular displacement divided by the change in time.

angular work—Mechanical measure of (torque) $\times$ (angular displacement).

anisotropic—Material exhibiting a direction-dependent tissue response to an applied force.

anlage—An initial clustering of embryonic cells from which a body part or organ develops; also referred to as the *primordium*.

annulus fibrosus—Series of fibrocartilage rings surrounding the nucleus pulposus of an intervertebral disc.

anterior drawer mechanism—Action causing anterior translation (e.g., of tibia relative to femur).

anteversion—Forward rotation or displacement relative to a reference plane.

anthropometric—Relating to the study of comparative measurements of the human body (e.g., height, weight, body composition, segment mass, and shape).

anthropometry—Study of comparative measurements of the human body (e.g., height, weight, body composition, muscle mass, and shape).

aponeuroses—Fibrous or membranous sheets connecting a muscle and the part it moves.

applied force—Muscle force; also *effort force*.

appositional growth—Growth of cells that proceeds in the cartilage layers immediately beneath the perichondrium.

arachnoid—Thin tissue membrane comprising the middle meningeal layer between the dura mater and pia mater.

Archimedes' principle—Buoyancy principle stating that the magnitude of the buoyant force is equal to the weight of the displaced liquid.

area moment of inertia—Measure of a body's resistance to bending.

areolar tissue—Weak connective tissue with spaces or holes where only fluid extracellular matrix exists that saturates almost all areas of the body.

ARF—activation–resorption–formation: series of events involved in bone remodeling.

arthritis—Joint inflammation.

arthrology—Study of joints and joint motion.

arthroplasty—Joint replacement through surgical procedures.

articular cartilage—Smooth, shiny layer of hyaline cartilage covering the joint surfaces of articulating bones.

articulation—Junction of two or more bones at their sites of contact; also *joint*.

aseptic necrosis—Cell death in the absence of infection.

atherosclerosis—Vascular disease characterized by plaque buildup in the arterial walls.

atrophy—Decrease in size.

autograft—Graft involving tissue from the same person (also *autologous graft*).

autologous graft—See *autograft*.

avascular necrosis—Cell death caused by an absent or deficient blood supply.

avulsion fracture—Fracture in which a piece of bone is pulled away at the insertion site with the osteotendinous or osteoligamentous attachment still intact.

axis of rotation—Imaginary line about which joint rotation occurs.

axis—Point about which rotation occurs; also *fulcrum* or *pivot*.

axonotmesis—Axonal nerve damage that does not completely sever the surrounding endoneurial sheath.

ballistics—(1) Study of the firing characteristics of a firearm. (2) Study of projectile motion.

Bankart lesion—Avulsion of the anteroinferior glenoid labrum at the attachment site of the inferior glenohumeral ligament complex.

Barton's fracture—Fracture of the distal radius in which the volar lip of the radial articular surface is sheared off.

basement membrane—Thin extracellular supporting layer that separates a layer of epithelial cells from the underlying layer.

beam—Structure that is relatively long and slender.

bending moment—Sum of the moments acting on a beam.

bending—Result of any force acting perpendicular to the longitudinal axis of a beam.

Bennett's fracture—Fracture subluxation of the trapeziometacarpal joint.

biaxial loading—Simultaneous loading along two axes.

biceps anchor—Attachment of the proximal tendon of the long head of the biceps brachii at the supraglenoid tubercle.

bilaminar embryonic disc—Fundamental cellular mass consisting of two cell layers (ectoderm and endoderm).

biomechanics—Area of science related to the application of mechanical principles to biology.

biotribology—Study of the friction, lubrication, and wear of diarthrodial joints.

biphasic response—Two-phase response seen in viscoelastic tissues, with an immediate first phase followed by a delayed second phase response.

blastema—Mass of living cells capable of growth and differentiation.

blastocyst—Mass of about 64 cells arranged in a hollow ball shape.

blister—Fluid accumulation under or within the epidermis caused by heat or by chemical or mechanical means.

body mass index—Measure of body composition as a ratio of body weight to height (kg/m^2).

body—Any collection of matter.

bone mineral content (BMC)—Measure of total mineral in bone (measured in grams).

bone mineral density (BMD)—Measure of the mineral content in a volume of bone, measured either as areal BMD (g/cm^2) or volumetric BMD (g/cm^3).

boosted lubrication—Form of lubrication that incorporates elastohydrodynamic and squeeze-film types of lubrication, which in turn boost the effectiveness of the lubricating fluid film.

boundary lubrication—Form of lubrication where a layer of molecules adheres to each of the two surfaces that are gliding past each other.

brittle—Characteristic of a material that experiences minimal deformation when a load is applied to it before failure (e.g., glass).

bucket-handle tear—Tear in the central part of a C-shaped piece of cartilage (e.g., meniscus).

buoyant force—Equal and opposite force exerted by a liquid against the weight of a body, allowing the body to float.

burst fracture—Shattering of a vertebral body attributable to a severe compressive axial loading.

butterfly fracture—Fracture with more than one fracture line.

canaliculi—Small channels or passageways found in bone.

cancellous bone—See *trabecular bone*.

cantilever bending—Type of loading in which a force offset from the longitudinal axis creates both compression and bending.

capsular ligament—Ligament seen as a thickening of a joint capsule structure.

cardiac muscle—Myocardial tissue that is striated in appearance and is involuntary.

cardiomegaly—Enlargement of the heart.

carpal tunnel syndrome (CTS)—Overuse injury in which contents of the carpal tunnel (e.g., blood vessels, flexor tendons, and median nerve) are subjected to increased compressive forces, with symptoms including tingling, numbness, and pain.

causal association—Relationship between a risk factor and a disease or injury outcome in which the risk factor has been shown to contribute to the outcome in question.

cavitation—Creation of a cavity, or space, within a tissue or body.

center of gravity—Point at which gravity has the same effect as it does on the distributed mass—acts as a balance point.

center of mass—Point about which a body's mass is equally distributed.

centripetal theory—Theory suggesting that the distribution of damaging diffuse strains induced by inertial loading would decrease in magnitude from the surface to the center of brain mass.

cerebral edema—Brain swelling caused by the accumulation of fluid in the brain, caused by trauma, tumor, hypoxia, or exposure to toxic substances.

cerebritis—Inflammation of cerebral tissue.

cerebrospinal fluid (CSF)—Liquid that circulates through the ventricles to the spaces between the meninges about the brain and spinal cord and that primarily maintains uniform pressure within the brain and spinal cord.

cervical spondylotic myelopathy—Condition characterized by degeneration of the cervical intervertebral discs, followed by changes in the bones and soft tissues of the cervical spine.

chemotaxis—Cellular attraction caused by chemical action.

chondral fracture—Isolated damage to articular cartilage; also *flap tear*.

chondrocyte—Mature cartilage cell.

chondromalacia patella—Softening of the articular cartilage on the retropatellar surface.

chondromalacia—End stage of cartilage degeneration (softening), characterized by cartilage fibrillation, loss of elastic support, and disruption of collagen framework.

chondron—Structural unit including a chondrocyte and its lacuna.

chronic injury—Injury caused by repeated insults that may lead progressively to degenerative conditions that set the stage for an acute injury; also *cumulative trauma disorder, overuse injury, repetitive stress syndrome*.

chronological age—Calendar age, determined by calendar years.

circumferential fibrocartilage—Circular ring of fibrocartilage without a center that acts as a spacer in the hip and shoulder joints.

closed fracture—Fracture that remains within the body's internal environment; also known as *simple fracture*.

closure—Completion of the endochondral ossification process evidenced by disappearance of the epiphyseal growth plate.

Cobb angle—Angle between the two lines that perpendicularly bisect the lines through the surfaces of the vertebrae at each end of the curvature.

coefficient of restitution—Ratio between the relative postcollision velocity and the relative precollision velocity.

collagen—Most abundant protein, constituting more than 30% of total protein in the human body. Present in varying amounts in all types of connective tissue.

collagenous—Containing an insoluble fibrous protein (collagen) occurring in vertebrates as the primary constituent of connective tissue fibrils and in bones.

Colles' fracture—Fracture of the distal radius with a fracture pattern showing volar (anterior) metaphyseal cortex failure in tension and varying degrees of comminution on the dorsal (posterior) surface.

collision—Forceful impact between two or more bodies.

comminuted fracture—Fracture in which the configuration consists of more than one fracture line, typically in the form of a multifragmented, shattered bone.

compact bone—Bone with high density (low porosity); also *cortical bone*.

compartment syndrome—Condition caused by the expansion of enclosed tissue within a confined anatomical space; results in increased pressure that affects blood circulation and nerve conduction.

compensatory injury—Secondary injury caused by movement compensation in response to a primary injury.

complete fracture—Bone fracture that traverses the entire bone.

complex glycoprotein—Major protein found in the extracellular matrix.

compliance—Opposite or inverse of stiffness.

compression–flexion mechanism—Injury mechanism that combines simultaneous cervical flexion with forceful axial spinal compression; also *flexion–compression mechanism*.

compression—Force that tends to push ends of a body together.

compressive stress—Internal resistance to being pushed together.

concave—Arched or curved inward, or inner surface of a structure.

concavity compression—Stability created when a convex object is pressed into a concave surface.

concentric—Active muscle shortening.

concurrent force system—Force system in which all forces originate from or are focused on a common point.

conductivity—Ability of a tissue to conduct an electrical impulse or signal.

congenital dislocation—Abnormal hip joint formation during development that predisposes the hip to subluxation or luxation.

connecting fibrocartilage—Fibrocartilage at joints with limited motion (e.g., intervertebral disc).

connective tissue—Classification of all tissues that are not epithelial, muscle, or nervous tissue, including blood, adipose, cartilage, ligament, bone, and fascia.

conservation of energy—No net gain or loss of energy.

conservation of momentum—No net gain or loss of momentum.

continuum mechanical model—Model that considers structures and materials as continuous matter.

contractility—Ability of a muscle to generate force and shorten.

contralateral—Situated on, appearing on, or affecting the opposite side of the body.

contrecoup lesion—Brain injury at a site opposite from impact.

contributing factors in accident causation—Framework proposed by Sanders and Shaw (1988) for categorizing the factors contributing to injury occurrence.

contusion—Skeletal muscle injury that results from a direct compressive impact; muscle bruise, distinguished by intramuscular hemorrhage.

convex—Arched or curved outward, or outer surface of a structure.

coracoacromial arch—Superior border of the glenohumeral joint formed by the acromion process and the coracoacromial ligament.

cortical bone—Bone with high density (low porosity); also *compact bone*.

countermoment—Moment acting in the opposite direction of an applied moment.

countertorque—Torque acting in the opposite direction of an applied torque.

coup lesion—Brain injury at the site or on the same side as impact.

coxa valga—Angle that is greater than normal between the long axis and the neck of the proximal femur.

coxa vara—Angle that is less than normal between the long axis and the neck of the proximal femur.

creep response—Additional deformation (following the initial deformation) over time when a viscoelastic tissue is exposed to a constant load.

cross-bridge—Protruding head of a myosin filament the binds to an actin filament to produce a power stroke and filament sliding during muscle contraction.

cryotherapy—Modality used to counteract the inflammatory process: cold compress, ice, and cooling sprays.

cumulative trauma disorder (CTD)—Injuries resulting from repeated tissue stress; also *repetitive strain injury, repetitive stress injury, chronic microtrauma, overuse syndrome, cumulative trauma syndrome*.

curvilinear motion—Movement along a curved path.

cyclic loading—(1) Repeated force application that allows for dynamic fluid flow (with nutrients and waste products) in and out of the tissue. (2) Repeated force application above a certain threshold that can cause a material to fatigue and exhibit a decreased ability to withstand applied forces.

dampened response—Retardation of an elastic material in returning to its unloaded shape or configuration.

damping—See *dampened response*.

dashpot—Component of a rheological model that represents tissue viscous properties.

de novo—Over again, anew, or from the beginning.

debridement—Surgical removal of lacerated, dead, or contaminated tissue.

deformable-body model—Model in which segments are allowed to deform (e.g., bend).

deformation to failure—Amount of deformation experienced by a tissue when it reaches failure.

deformational energy—Energy stored in a tissue or body during deformation; also *strain energy*.

deformation—Change in a body's shape or configuration when subjected to an external load.

degenerative joint disease (DJD)—Noninflammatory disorder of synovial joints, particularly those with load-bearing involvement, characterized by deterioration of the hyaline articular cartilage and bone formation on joint surfaces and at the joint margins; also *osteoarthritis*.

delayed-onset muscle soreness (DOMS)—Pain resulting from connective and contractile tissue disruption 24 to 72 hr after vigorous exercise, often associated with eccentric muscle action.

dense irregular connective tissue—Fibrous connective tissue with loosely and randomly interwoven fibers, such as fascia.

dense regular connective tissue—Organized fibrous tissues, such as tendons, ligaments, or aponeuroses.

density—Measure of mass concentration: density = mass/volume.

depressed skull fracture—Depression of a skull fragment into the subcranial space caused by compressive force of blunt trauma.

dermatome—New layer of cells formed after the sclerotome has condensed near the notochord.

descriptive epidemiology—Systematic and ongoing collection, analysis, interpretation, and dissemination of public health information to assess public health status, define public health priorities, and evaluate programs set in place to improve the health of a community.

destabilizing component—Force component acting through a joint axis that tends to destabilize the joint.

deterministic model—Model whose output is fully determined (i.e., does not vary) for a given set of inputs.

diaphysis—Main or midsection of a long bone.

diffuse axonal injury (DAI)—Diffuse degeneration of the white matter in the brain.

diffuse injury—Tissue damage over a wide area.

diffuse traumatic axonal injury (DTAI)—See *diffuse axonal injury*.

direct injury—Fracture at the specific site of force application.

direct solution—See *forward solution approach*.

displaced fracture—Fracture in which the bony fragments are moved relative to their normal (prefracture) position.

displacement—Vector measure of movement from one location to another.

distance—Scalar measure of how far a body has moved along its movement path.

distraction—Action that pulls away or separates joint surfaces.

drag force—Force acting parallel to the direction of fluid flow.

ductile—Characteristic of a compliant material that undergoes considerable deformation before failure.

dura mater—Tough tissue membrane comprising the outer meningeal layer.

dynamic equilibrium—Balanced state of bodies in motion with nonzero accelerations.

dynamic friction—See *kinetic friction*.

dynamic model—Model that characterizes the nonzero acceleration of a body or system.

dyspnea—Shortness of breath.

eccentric—Active muscle lengthening.

ectoderm—Layer of cells associated with the primitive amniotic cavity that gives rise to nervous system structures.

edema—Accumulation of excessive fluid (swelling).

effort force—See *applied force*.

elastic cartilage—Flexible and extensible cartilage found in the external ear, epiglottis, portions of the larynx, and eustachian tube.

elastic deformation—Change in a structure's shape or configuration that is followed by a return to its original configuration when the force is removed.

elastic fibers—Slender and extensible fibers that can be stretched to about 150% of their original length before breaking.

elastic limit—Load or stress level above which tissue becomes plastic and experiences permanent deformation.

elastic modulus—See *modulus of elasticity*.

elasticity—Ability of a material to return to its original shape when a load is removed.

elastic—When referred to as fibers, these are bundles of fibers composed of elastin that can stretch and return to their original length.

elastin—Protein in connective tissue that is elastic and returns to its original shape after stretching. Composed primarily of the amino acids glycine, valine, alanine, and praline and made in a reaction that links several tropoelastin molecules catalyzed by lysyl oxidase.

elastohydrodynamic lubrication—Subtype of fluid-film lubrication where one or both of the surfaces are deformable.

elastoplastic—Characteristic of collisions in which there is energy transfer and loss and there may be plastic deformation.

embryo stage—Period of human development from the time from fertilization to week 8.

embryology—Study of embryos and their development.

endochondral ossification—Process of bone growth in which hyaline cartilage is replaced by bony material.

endoderm—Layer of embryonic cells that gives rise to the gastrointestinal tract.

endomysium—Layer of connective tissue, capillaries, nerves, and lymphatics that surround individual muscle fibers.

endoneurium—Delicate sheath surrounding each axon.

endotendineum—Loose connective tissue that binds tendon fascicles together.

endurance training—Form of training that enhances the ability of the cardiorespiratory and muscular systems to resist fatigue.

energy to failure—Measure of the strain energy stored by the tissue prior to failure.

energy—Capacity or ability to perform work.

epicondylitis—Inflammation or damage of an epicondyle.

epidemiology—Study of the distribution and determinants of disease and injury frequency within a given human population.

epidural hematoma—Blood accumulation between the cranial vault and the outer protective meningeal layer (dura mater).

epimysium—Connective tissue surrounding the entire muscle.

epineurium—Tough covering surrounding the nerve trunk.

epiphysis—Rounded end of a long bone filled with red marrow, which produces red blood cells.

epitendineum—Surface covering of the tendon.

epithelial tissue—Covering (lining) tissue that absorbs, secretes, transports, excretes, and protects underlying organ or tissue.

equilibrium—Balanced condition where the sum of forces and the sum of moments equals zero.

ergonomics—Study of how people interact with their immediate environment in a safe and productive manner.

eversion sprain—See *medial ankle sprain*.

exercise-induced muscle injury—Injury resulting from connective and contractile tissue disruption following exercise.

external ballistics—Effect of external factors (e.g., wind, velocity, drag, gravity) on projectile flight from the barrel to the target.

external mechanics—Mechanical factors that produce and control movement outside the body (e.g., gravity or ground reaction forces).

extracapsular ligament—Ligament extrinsic to the joint capsule.

extracellular bone matrix—Inorganic (mineral) and fluid component of bones.

extracellular matrix—Noncellular component of a tissue.

extrapolation—Prediction or estimation of a value beyond known or measured values.

exudate—Substances that move (exude) from blood vessels into surrounding tissues during an inflammatory response.

failure—Complete tearing or rupture.

fascia—Dense, fibrous, unorganized tissue found in layers or sheaths around organs, blood vessels, bones, cartilage, and dermis of the skin.

fasciotomy—Surgical procedure in which fascia is cut (released) to reduce pressure or tension.

fatigue fracture—Bone fracture resulting from loss of material strength attributable to repeated loading.

fatigue—(1) Inability to continue work. (2) Loss of strength and stiffness in materials subjected to repeated cyclic loads.

female athlete triad—Pathological condition caused by the combination of disordered eating, disrupted hormone levels, and poor bone quality and quantity in young female athletes.

fetal stage—Period of human development from week 9 until birth.

fiber—Threadlike structure composed of many fibrils.

fibroblasts—Principal cells in many fibrous connective tissues that form fibers and components of the extracellular matrix.

fibrocartilage—Cartilage found in areas of the body where friction could pose a problem; it is strong, resilient, and flexible.

fibrocytes—Fibrogenic mesenchymal cells.

fibroelastic tissue—Loose, woven network of fibers that encapsulates most organs.

finite-element model—Form of modeling that involves assembling complex geometrical figures, or elements, to represent a structure.

first law of motion—A body at rest or in uniform motion will remain at rest or in uniform motion, unless acted upon by an external force.

first-class lever—Lever system with the axis between the resistance force and the effort force.

first-cycle effect—Mechanical response of tissue to the first loading cycle, which may differ from subsequent loading cycles.

flap tear—Isolated damage to articular cartilage; also *chondral fracture*.

flexion–compression mechanism—See *compression–flexion mechanism*.

flexural rigidity—Measure of the bending stiffness of bone.

fluid flow—Characteristics of a fluid, whether liquid or gas, that allow it to move and that govern the nature of this movement.

fluid mechanics—Branch of mechanics dealing with the properties and behaviors of gases and liquids.

fluid resistance—Opposition to flow.

fluid-film lubrication—Lubrication mechanism where two nonparallel surfaces move past each other on a thin layer of fluid.

focal injury—Tissue damage limited to a specific location.

force couple—A pair of parallel and oppositely directed forces that tend to cause the same rotation about an axis.

force—Mechanical action or effect applied to a body that tends to produce acceleration.

force-relaxation response—Force decrease seen in a viscoelastic tissue stretched or compressed to a given length and then held at that length.

force–velocity relation—Property of skeletal muscle that shows how its force production capability depends on its contraction velocity.

forward solution—Problem-solving approach that uses known or measured kinetic measures (e.g., force) to calculate associated kinematic measures (e.g., acceleration); also *direct solution*.

four-point bending—Bending mode involving four parallel forces, two outer forces in the same direction with two remaining forces acting in the opposite direction of the outer ones.

fracture—Structural breakage in a hard tissue, such as bone.

free-body diagram (FBD)—Diagram depicting all the forces and moments acting in a particular system.

friction—Resistance created at the interface between two bodies that are in contact with one another and that act in the opposite direction of motion or intended motion.

fulcrum—Point about which rotation occurs; also *axis* or *pivot*.

Galeazzi fracture—Fracture of the distal third of the radius.

gastrulation—Transformation of the bilaminar embryonic disc into a three-layered disc containing the three primary germ layers—ectoderm, mesoderm, and endoderm.

general force system—Force system that is neither linear, parallel, nor concurrent.

general motion—Simultaneous linear and angular (rotational) motion.

glycation—Bonding of a sugar molecule to a protein or lipid molecule without the involvement of enzymes.

glycosaminoglycans—Polysaccharides that are constituents of mucoproteins, glycoproteins, and blood-group substances.

gouty arthritis—Joint inflammation caused by excess production of uric acid, with resulting uric acid (urate) crystals embedded in joint structures with attendant irritation and pain; also *gout*.

graft—Surgically implanted living tissue.

ground reaction force (GRF)—Force exerted *by* the ground that is equal and opposite to a force applied *to* the ground by an impacting object (e.g., foot).

growth plate (physis)—Area of developing tissue (hyaline cartilage) near the end of the long bones.

Haversian canals—Small canals through which blood vessels pass, generally oriented longitudinally in bone.

hematoma—Mass of blood that forms in a tissue or body space as a result of a broken blood vessel.

hematopoiesis—Process of blood cell formation.

hemodynamics—Forces and mechanisms involved with blood circulation.

hemorrhage—Profuse bleeding.

high ankle sprain—Injury to the anterior and posterior tibiofibular ligaments just above the ankle joint.

homeostasis—Maintenance of a relatively stable internal physiological environment.

Hooke's law—Principle that stress and strain are linearly related. The resulting strain is proportional to the developed stress.

hoop effect—Radiating circumferential stress developed perpendicular to an applied compressive load.

hyaline cartilage—Firm, glassy appearing cartilage that lines most of the joint surfaces (articular cartilage), the anterior portions of the ribs, and areas of the respiratory system (trachea, nose, and bronchi).

hydrodynamic lubrication—Subtype of fluid-film lubrication produced by two moving surfaces that are nondeformable.

hydroxyapatite—Complex calcium–phosphate crystal that forms the primary structural element of bone.

hyperemia—Presence of excess blood in a body part or region.

hyperesthesia—Heightened or pathological sensitivity of the skin or other sense to stimulation.

hyperglycolysis—Enhanced breakdown of a carbohydrate (e.g., glucose or glycogen) by enzymatic action.

hyperplasia—Abnormal or unexpected increase in the constituents composing a part (e.g., cells in a tissue).

hypertrophy—Increase in size or bulk in the absence of an increase in number of parts (e.g., adaptive increase in muscle fiber size in response to resistance training).

hypovascular—Having a decreased or low number of blood vessels.

hypoxemia—Decrease in blood perfusion.

hypoxia—Decrease in blood (oxygen) delivery to a tissue.

hysteresis loop—Area enclosed by the loading–unloading paths on a load–deformation curve; measures energy loss during a loading–unloading cycle of a viscoelastic material.

hysteresis—Delay or retardation in a material's response when forces are changed.

idealized force vector—Single vector used to represent many vectors.

idiopathic—Of unknown origin or cause.

iliac-type luxation—Anterosuperior hip dislocation; also *pubic-type luxation*.

iliotibial band syndrome (ITBS)—Inflammatory condition caused by friction as the iliotibial band passes over the lateral femoral epicondyle during repeated flexion–extension cycles of the knee.

impact injury—Injury resulting from a direct compressive force.

impact strength—Ability of bone to resist impact loading.

impingement syndrome—Pressure increase within a confined anatomical space with deleterious effect on the enclosed tissues.

impulse—Mechanical agent that changes momentum, calculated as force × time.

impulse–momentum principle—Mechanical principle stating that the impulse equals the change in momentum.

impulsive force—Force component in an impulse.

in situ—Research approach that examines tissues in their normal place, with some elements of natural environment preserved; artificial testing.

in vitro—Research approach that examines tissues in an artificial environment that allows direct measurements.

in vivo—Research approach that examines tissues in the living body.

incidence—Number of new injuries that occur within a given population at risk over a specified time period.

incomplete fracture—Fracture that only partially traverses the bone structure.

indirect injury—Fracture that is remote from the location of force application.

inertia—Resistance to a change in a body's state of linear motion.

inflammation—Localized physiological response to injury that involves redness, heat, pain, swelling, and sometimes loss of function.

initial-cycles effect—Mechanical response of tissue to the first few loading cycles, which may differ from subsequent loading cycles.

injury—damage to tissue caused by physical trauma.

instantaneous joint center—Axis of rotation location at an instant in time.

insufficiency fracture—Chronic fracture found in persons with no increase in physical activity but with decreased bone density.

integrins—Integral membrane proteins in the plasma membrane of cells that attach the cell to the extracellular matrix and facilitate signal transduction from the extracellular matrix to the cell.

interarticular fibrocartilage—Flattened fibrocartilage plates interposed between joint surfaces and held in position by ligaments and tendons that connect to the edges of the fibrocartilage; also *meniscus*.

internal ballistics—Effect of bullet and weapon design and materials on the projectile while in the barrel of the weapon.

internal impingement—Pinching where the supraspinatus tendon contacts the posterior–superior rim of the glenoid fossa.

internal mechanics—Mechanical factors that produce and control movement inside the body (e.g., forces produced by muscle action and stability provided by ligaments surrounding joints).

interpolation—Estimation of values between known or measured values.

interstitial growth—Growth occurring in young cartilage where the cells divide within the lacunae and form cell nests.

intervertebral disc—Structure between two adjacent vertebrae composed of a nucleus pulposus surrounded by the annulus fibrosus.

intracapsular ligament—Ligament located within a joint capsule.

intracranial pressure (ICP)—Pressure within the cranial vault (skull).

intramembranous ossification—Process of bone growth in which a membrane is replaced by bony material.

inverse dynamics—Problem-solving approach that uses known or measured kinematic measures (e.g., acceleration) to calculate associated kinetic measures (e.g., force).

inverse solution—See *inverse dynamics*.

inversion sprain—See *lateral ankle sprain*.

ipsilateral—Situated on, appearing on, or affecting the same side of the body.

irritability—Ability to respond to a stimulus; also *excitability*.

ischemia—Deficient blood supply caused by decreased flow to an area.

isotropic—Property by which a tissue's mechanical responses (e.g., strength or stiffness) to loading are not dependent on direction.

joint reaction force (JRF)—Net effect of muscle and other forces acting across a joint.

joint stability—Ability of a joint to resist dislocation and maintain an appropriate functional position throughout its range of motion.

joint—Junction of two or more bones at their sites of contact; also *articulation*.

joule—Unit of work and energy (1 J = 1 N·m).

Kelvin–Voight model—Rheological model with its spring and dashpot elements in parallel.

kinematics—Description of movement without regard to the forces involved.

kinetic energy—Energy possessed by a body by virtue of its motion.

kinetic friction—Frictional resistance created while an object is moving (e.g., sliding or rolling) along a surface; also *dynamic friction*.

kinetics—Assessment of movement with respect to the forces involved. The study of forces and their effects.

knee extensor mechanism—Functional unit that extends the knee, including the quadriceps muscle group, quadriceps tendon, patella, and patellar tendon–ligament.

kyphosis—Excessive spinal flexion (e.g., thoracic hunchback).

labrum—U-shaped ring of fibrocartilage around the rim of a joint.

laceration—Jagged tearing of the skin, as by a knife or sharp implement.

lacunae—Small pockets or cavities.

lamellar bone—Thin flat scale, membrane, or layer. Primary bone composed of multiple thin layers (lamellae) of bone matrix and cells organized in parallel with the bone surface.

laminar flow—Flow characterized by a smooth, essentially parallel pattern of fluid movement.

lateral ankle sprain—Injury to the lateral ankle ligaments caused by violent foot and ankle supination.

length–tension relation—Property of skeletal muscle that shows how its force production capability depends on the length of the muscle's contractile and noncontractile structures.

level of dysfunction—Degree of impaired or abnormal functioning.

lever arm—See *moment arm*.

lever—Rigid structure, fixed at a single point, to which two forces are applied at two points.

lift force—Force acting perpendicular to the direction of fluid flow.

ligaments—Cords of regular dense fibrous connective tissue that connect one bone to another.

linear acceleration—Change in linear velocity divided by change in time.

linear displacement—Vector measure of the straight-line distance from starting position to the ending position.

linear force system—Force system in which all forces act along a single line.

linear impulse—Product of force × time.

linear limit—See *proportional limit*.

linear momentum—Quantity of linear motion measured by the product of mass × linear velocity.

linear motion—Motion along a straight (rectilinear) or curved (curvilinear) line.

linear power—Measure of mechanical work per unit time $(P = W/t)$. Also equals (force) × (linear velocity).

linear spring—Component of a rheological model that represents tissue elastic properties.

linear velocity—Linear displacement divided by the change in time.

linear work—Mechanical measure of force × displacement.

load–deformation curve—Curve that shows the relation between the external force (load) and the change in shape or configuration (deformation) of a tissue.

load—Externally applied force.

lordosis—Excessive spinal extension (e.g., lumbar lordosis).

lumbago—Chronic or acute pain in the low back.

luxation—Complete dislocation of joint.

macrophage—Migratory cell that contains small holes or vacuoles that can assimilate foreign material, old red blood cells, and bacteria.

macrotrauma—Trauma that can be seen by the unassisted eye (e.g., tibia fracture or laceration).

magnus force—Force created by the spin of an object that creates a deviation in its normal trajectory.

mass moment of inertia—Measure of resistance to change in the state of a body's movement about a fixed axis.

mass—Quantity of matter (in SI units, measured in kilograms).

mast cells—Migratory cells that are relatively large because of their substantial amount of cytoplasm.

material properties—Qualitative measure of a tissue's mechanical properties, usually expressed as a relative measure (e.g., stress and elastic modulus).

material strength—Ultimate stress.

material—Mechanical quality of a tissue.

Maxwell model—Rheological model with its spring and dashpot elements in series.

mechanical energy—Capacity or ability to perform mechanical work.

mechanical strain—Relative measure of a change in shape or configuration of tissue when subjected to a load.

mechanical work—See *work*.

mechanics—Branch of science that deals with the effects of forces and energy on bodies.

mechanism—Fundamental physical process responsible for a given action, reaction, or result.

mechanotransduction—Process by which tissue cells convert a mechanical stimulus into a biochemical response.

medial ankle sprain—Injury to the medial ankle ligaments caused by violent foot and ankle pronation.

medial tibial stress syndrome (MTSS)—Inflammatory reaction of the deep fascial tibial attachments in response to chronic loads, with pain localized to the posteromedial crest of the tibia.

medullary canal—Central canal oriented along the longitudinal axis of long bones.

mega-pascal (MPa)—Measure of pressure or stress (1 MPa = 1 N/mm^2, or 10^6 Pa).

meninges—Protective membrane layers surrounding the brain and spinal cord (include dura mater, pia mater, and arachnoid).

meningitis—Inflammation of the meninges.

meniscus—Interposed fibrocartilage pad that acts as a shock absorber and a wedge at the joint periphery to improve the structural fit of a joint (plural, *menisci).*

mesenchyme—Progenitor primitive tissue of adult connective tissue (e.g., cartilage, ligaments, fascia, tendons, blood cells, blood vessels, skin, and bone) that is loosely woven and formed from undifferentiated cells of the sclerotome.

mesoderm—Middle of the three primary germ layers of an embryo that is the source of many bodily tissues and structures (e.g., bone, muscle, connective tissue, and dermis).

metaphyseal—Deriving from the metaphysis.

metaphysis—Portion of a long bone between the epiphyses and the diaphysis.

microfibrils—Component of elastic fibers composed of microfibrillar-associated glycoproteins, fibrillin, fibulin, and the elastin receptor.

microstrain—Unit of relative deformation ($\mu\epsilon$) equal to 10^{-6}.

microtrauma—Tissue injury that can only be seen under a microscope (e.g., micro cracks in bone).

migratory cells—Cells that wander or move within a tissue or body system.

modeling—Addition (formation) of new bone, occurring mostly in the growing years.

model—Structural or mathematical representation of one or more of an object's or system's characteristics.

modulus of elasticity—Ratio of stress to strain; slope of the linear portion of a stress–strain curve; also *elastic modulus* and *Young's modulus.*

moment arm—Perpendicular distance from the axis of rotation to the line of force action; also *lever arm* and *torque arm.*

moment of inertia—Resistance to a change in a body's state of angular motion or position.

moment—Effect of a force that tends to rotate or bend a body or segment; also *moment of force.*

momentum—Quantity of motion possessed by a moving body.

Monteggia lesion—Fracture in the proximal ulna with dislocation of the radial head.

morphology—Area of study concerned with the form and structure of plants and animals.

morula—Solid mass of cells resulting from cell division following zygote formation.

motor unit—Single motor neuron and all the muscle fibers it innervates.

multiaxial loading—Multidimensional force application to a body in either two- (biaxial) or three- (triaxial) dimensional space.

muscle—Body tissue capable of developing force in response to stimulation. There are three muscle types: skeletal, cardiac, and smooth.

musculotendinous junction—See *myotendinous junction.*

myelopathy—Disease or disorder of the spinal cord.

myofibrils—Contractile units of muscle.

myofilaments—Individual filaments (actin or myosin) that comprise a myofibril.

myosin—Thick myofilament found in muscle fibers that acts as the cross-bridge during muscular contraction.

myositis ossificans—Deposit of an ossified mass within a muscle.

myotendinous junction—Region where a muscle and tendon connect; also *musculotendinous junction.*

myotome—Developmental tissue that gives rise to the musculature.

negative ulnar variance—Ulnar variance with radius longer than the ulna.

neoplasm—Tumor.

nervous tissue—Body tissue responsible for communication. Develops from the ectoderm and comprises the main parts of the nervous system, including the brain, spinal cord, peripheral nerves, nerve endings, and sense organs.

net moment—Sum of all the moments acting on a body; also *net torque.*

net torque—Sum of all the torques acting on a body; also *net moment.*

neurapraxia—Injury to a nerve that disrupts conduction and causes temporary paralysis but not degeneration, followed by rapid and complete recovery.

neuron—Nerve cell.

neurotmesis—Partial or complete severance of a nerve, with axonal disruption.

neutral axis—Line (in a beam) along which neither compressive nor tensile stress exists.

Newtonian fluid—Fluid in which the stress–strain response is linear.

newton—Unit of force ($1 \text{ N} = 1 \text{ kg} \cdot \text{m} \cdot \text{s}^{-2}$).

nightstick fracture—Fracture of the ulna, named for the mechanism of ulnar fracture in blocking an overhead blow.

nonsteroidal anti-inflammatory drugs (NSAIDs)—Inflammation-reducing drugs that do not contain steroids; can produce positive or negative effects on the body's tissues and later human performance, thus altering injury risk.

normal force—Force acting perpendicular to a surface.

notochord—Primitive spinal cord.

nucleus pulposus—Gelatinous mass in the interior of an intervertebral disc.

obturator-type luxation—Anteroinferior hip dislocation.

oligomenorrhea—Irregular menstrual cycles.

open fracture—Fracture that penetrates the skin and exposes bone to external environment; also known as *compound fracture*.

orthotics—Supports or braces for weak or ineffective joints.

os acromiale—Unfused acromial apophysis.

Osgood-Schlatter disease (OSD)—Traction apophysitis at the tibial tuberosity.

osteoarthritis—Noninflammatory condition that affects synovial joints, especially those involved in weight bearing, characterized by cartilage deterioration and bone outgrowth on joint surfaces and joint margins.

osteoblasts—Mononuclear bone cells that produce new bone material.

osteochondral fracture—Injury to cartilage and its underlying bone.

osteochondritis dissecans—Complete or partial detachment of a fragment of bone and cartilage at a joint.

osteochondrosis—Disease in which an ossification center, especially in the epiphyses of long bones, experiences degeneration followed by calcification.

osteoclasts—Large, multinucleated bone cells that break down, or resorb, bone.

osteocytes—Mature bone cells that are smaller and less active than an osteoblast.

osteoid—Organic portion of the extracellular matrix in bone.

osteonecrosis—Bone death caused by cessation of blood flow.

osteopenia—Mild to moderate bone loss; clinically classified as having a bone mineral density (BMD) rating 1.0 to 2.5 standard deviations below the mean BMD for young, healthy adults.

osteophyte—Bone outgrowth; also *bone spur*.

osteophytosis—Condition characterized by formation of osteophytes.

osteoporosis—Severe bone loss marked by increased risk of fractures, primarily of the hip, spine, and wrist; clinically identified by a bone mineral density (BMD) rating 2.5 standard deviations below the mean BMD for young, healthy adults.

osteosarcoma—Malignant bone tumor.

osteotendinous junction—Region where a bone and tendon connect.

overuse injury—See *chronic injury*.

parallel force system—Force system in which all forces acts parallel to one another.

parallel-elastic component—Component of a muscle model that accounts for muscle elasticity in parallel with the contractile component.

paresthesia—Sensation of tingling or creeping on the skin with no specific cause and usually associated with injury or irritation of a sensory nerve.

pars interarticularis—Area of the vertebral lamina between the superior and inferior articular facets.

pascal (Pa)—Unit of pressure or stress ($1 \text{ Pa} = 1 \text{ N/m}^2$).

passive joint stabilizer—Periarticular structure (e.g., ligaments or joint capsule) that stabilizes a joint.

passive stabilization—Contribution of noncontractile components (e.g., periarticular tissues) to joint stability.

patella alta—Abnormally elevated patella situated high in the intercondylar groove.

patellar ligament—Collagenous connective tissue spanning between the inferior pole of the patella and the tibial tuberosity (also *patellar tendon*).

patellar tendon—See *patellar ligament*.

patellar tracking—Sliding movement of the patella in the intercondylar groove of the femur.

peak height velocity (PHV)—Peak rate of longitudinal bone growth.

perfectly elastic collision—Collision in which bodies rebound away from each other following the collision with no energy or momentum loss.

perfectly plastic (inelastic) collision—Collision in which the bodies stick together and move together with a common velocity after impact with no loss of energy or momentum.

perichondrium—Connective tissue membrane surrounding cartilage.

perimysium—Connective tissue that surrounds a bundle of muscle fibers (fascicle).

perineurium—Connective tissue sheath surrounding a bundle of nerve fibers.

periosteal collar—Bony ring surrounding the primary ossification center in developing bone.

peripheral neuropathy—Structural or functional disorder of the peripheral nervous system.

peritenon—Loose areolar connective tissue enveloping the tendon.

permanent set—Plastic deformation in which a tissue does not return to its preloaded shape or configuration.

pes cavus—Foot deformity characterized by a high longitudinal plantar arch.

pes planus—Foot deformity characterized by a flattened longitudinal plantar arch (flat foot).

phagocyte—Cell (e.g., white blood cell) that engulfs and consumes foreign material (e.g., microorganisms) and debris.

phagocytosis—Bodily defense process by which phagocytes engulf and destroy foreign particles and tissue debris.

physiological age—Age based on physiological quality of the tissues.

physiological range—Normal operational range for a physiological variable or measure.

pia mater—Thin tissue membrane comprising the inner meningeal layer.

pivot—Point about which rotation occurs; also *axis* or *fulcrum*.

planar model—Two-dimensional model of an object or system.

plantar fasciitis—Inflammation of the plantar fascia.

plastic deformation—Permanent change in a tissue's or body's shape or dimensions.

pluripotent—Capable of developmental plasticity, or having multiple developmental potentials.

point mass—Concentration of a body's mass at a single point.

Poisson's effect—Inverse response created transversely to an axial load (e.g., a body subjected to a uniaxial compressive load increases its dimension in the axial direction and decreases its dimension in the transverse direction).

Poisson's ratio—Negative of the strain transverse to a load divided by the strain in line with an applied load.

polar moment of inertia—Resistance to torsional loading about a longitudinal axis.

positive ulnar variance—Ulnar variance with the ulna longer than the radius.

posterolateral rotatory instability (PLRI)—Elbow instability that results from combined axial compression, valgus instability, and supination.

potential energy—Energy created by virtue of a body's position (gravitational potential energy) or deformation (strain energy).

power—Rate of work production.

pressure—Total applied force divided by the area over which the force is applied ($p = F/A$).

prevalence—Number of cases (e.g., injuries) both new and old that exist in a given population at a specific point in time divided by the total population number.

primary bone—Bone composed of multiple thin layers (lamellae) of bone matrix and cells organized in parallel with the bone surface, used to replace existing bone.

primary injury—Immediate injury as a consequence of trauma (e.g., skull fracture or torn medial collateral ligament).

primary ossification center—Region in developing bone where initial ossification happens (in the middiaphyseal region of long bones).

primary osteons—Haversian system with concentrically arranged lamellae.

primary spongiosa—Lattice of calcified cartilage.

pronation (foot–ankle)—Combination of ankle dorsiflexion, subtalar eversion, and external rotation of the foot.

proportional limit—Point at which a tissue's load–deformation or stress–strain response changes from linear to nonlinear; also *linear limit*.

proteoglycan—Protein to which are attached one or more specialized carbohydrate side chains, called glycosaminoglycans.

pubic-type luxation—Anterosuperior hip dislocation; also *iliac-type luxation*.

public health approach—Four-step method for gathering epidemiological information.

puncture—Wound created by a sharp implement that penetrates the skin.

Q angle—Angle measured between a line along the longitudinal midline of the thigh and a line connecting the tibial tuberosity with the centroid of the patella (Q = quadriceps)

quadriceps tendon—Tendon connecting the quadriceps muscle group with the superior pole of the patella.

quadriplegia—Paralysis in all four limbs; also *tetraplegia*.

quasi-static model—Model with negligible acceleration that mimics a static model.

range of motion—Measure of joint mobility.

rate of elastic return—Speed at which a deformed material returns to its original shape or configuration.

reaction force—Equal and opposite force created in response to an applied force.

rectilinear motion—Linear motion in a straight line.

reduction—Replacement or realignment of a body part to its normal position following luxation or subluxation.

relative risk—Risk calculated as the injury incidence in group A divided by the injury incidence in group B.

remodeling—Adaptation of existing bone via resorption and reformation.

repetitive stress syndrome—See *chronic injury*.

resident cells—Fixed (nonmoving) cells.

resilient—Characteristic of elastic materials that quickly return to their original shapes.

resistance force—An externally applied force.

resistance training—Mode of physical conditioning designed to enhance muscular strength; also *strength training*.

reticular tissue—Connective tissue containing reticular fibers and some primitive cells; found near lymph nodes and in bone marrow, liver, and spleen.

reticular—Forming or resembling a network.

retroversion—Rearward rotation or displacement relative to a reference plane.

rheological model—Model that characterizes the fluid-related aspects of a tissue or system.

rheology—Study of the deformation and flow of matter.

rheumatoid arthritis—Joint inflammation of autoimmune origin.

rigid-body mechanics—Approach that models each body segment as a nondeformable member.

rigid-body model—Model that treats each segment as a nondeformable member.

risk factors—Factors that may contribute to the occurrence of injury.

risk—Likelihood of injury or death associated with a particular object, task, or environment.

rotational motion—See *angular motion*.

rotatory component—Component of a force acting perpendicular to a segment that tends to cause rotation.

rupture point—Point at which a tissue reaches failure.

sarcomeres—Contractile units of the myofibril, delimited by Z bands at each end of its length.

scalar—Measure that has magnitude only.

scapulohumeral balance—Muscle action at the glenohumeral joint that maintains the net joint reaction force within the glenoid fossa.

Scheuermann's kyphosis—Spinal disorder (kyphosis) in children typically accompanied by vertebral wedging, end plate irregularities, and narrowing of the intervertebral disc spaces.

sciatica—Pain along the tract of the sciatic nerve, usually in the low back, buttocks, or posterior thigh, caused by compression or inflammation of the sciatic nerve.

sclerotome—Group of mesenchymal cells emerging from the ventromedial part of a mesodermic somite and migrating toward the notochord.

scoliosis—Lateral (frontal plane) deviation of the spine.

screw-home mechanism—Tibiofemoral rotation during the final few degrees of knee extension.

second impact syndrome (SIS)—Second head injury sustained before symptoms associated with a first head injury (most often a concussion) have fully resolved.

second law of motion—Force acting upon a body will produce an acceleration proportional to the force ($F = m \cdot a$).

secondary bone—Bone deposited only during remodeling and replacing preexisting primary bone.

secondary injury—Injury happening some time after an initial trauma or in compensation for a primary injury.

secondary ossification centers—Regions in developing bone where subsequent ossification happens (in the epiphyseal region of long bones).

secondary spongiosa—Trabecular lamellar bone formed after the resorption of the primary spongiosa.

second-class lever—Lever system with the resistance force between the axis and the effort force.

sequelae—Aftereffects.

series-elastic component—Component of a muscle model that accounts for muscle elasticity in series with the contractile component.

Sharpey's fibers—Compact bundles of perforating smaller collagen fibers that are tightly embedded in bone to reinforce attachment sites.

shear modulus of elasticity—Ratio of shear stress to shear strain.

shear stress—Internal resistance developed in response to a shear load.

shear—Force that tends to produce horizontal sliding of one layer over another or angulation within a structure.

shoulder separation—Acromioclavicular sprain.

simulation—Process of using a validated model to address questions related to a system and its operation.

skeletal muscle—Muscle tissue responsible for maintaining posture and producing movement.

SLAP lesion—Injury to the superior labrum anterior–posterior region of the glenoid labrum.

sliding filament theory—Theory used to describe how a sarcomere produces force. The final steps associated with excitation–contraction coupling that describe the interaction between the actin and myosin filaments needed to produce force.

Smith's fracture—Fracture of the distal radius with tensile failure on the dorsal aspect of the metaphysis and compressive comminution on the volar (palmar) aspect.

smooth muscle—Muscle tissue that facilitates substance movement through tracts in the circulatory, respiratory, digestive, urinary, and reproductive systems.

somatotype—Shape or physical classification of the human body.

somites—Cuboidal bodies that form distinct surface elevations and influence the external contours of the embryo.

spatial model—Three-dimensional model of an object or system.

speed—Scalar quantity that measures how fast a body is moving.

spinal cord injury—Damage to the spinal cord, usually as a result of high-energy collision.

spondylolisthesis—Vertebral slippage, usually at L5-S1 or L4-L5.

spondylolysis—Fracture of the pars interarticularis.

spondylosis—Intervertebral disc deterioration.

spongy bone—See *trabecular bone.*

sprain—Ligament injury.

squeeze-film lubrication—Theory of cartilage fluid-film lubrication in which two surfaces move at right angles to each other, as might happen in the knee joint at the instant of heel strike in walking, and fluid is forced out of the cartilage to produce a fluid interface between the two surfaces.

stabilizing component—Force component acting through the joint axis that tends to stabilize the joint.

static equilibrium—Balanced state where net accelerations equal zero.

static friction—Resistance created between two surfaces in the absence of movement.

static model—Model of a system that has zero net acceleration, usually motionless.

stem cells—Cells that can differentiate into a variety of cell types, such as connective tissue cells (e.g., fibroblast, chondroblast, or osteoblast) and nerve cells.

stenosis—Narrowing or constriction of a bodily passage or tract.

stiffness—Measure of the relation between stress and strain (i.e., how much a body deforms in response to a given load).

stochastic model—Model whose output varies as a function of probability algorithms within the model.

strain—Damage to a musculotendinous unit (i.e., muscle–tendon complex).

strain energy density—Relative measure of energy stored during deformation, as measured by the area under the stress–strain curve.

strain energy—Energy stored in a tissue during deformation; also *deformational energy.*

strain to failure—Measure of the mechanical strain to the point of tissue failure.

strain-rate dependent—Tissue characteristic in which its mechanical response depends on the rate at which the tissue is deformed.

stratiform fibrocartilage—Layer of fibrocartilage over bone that reduces friction where tendons act.

stress fracture—Bone fracture caused by chronic loading.

stress reaction—Area of enhanced metabolic activity in bone in response to repeated loading.

stress riser—Force or stress concentration at the site of a material discontinuity.

stress—Internal resistance developed in response to an externally applied load.

stress-relaxation response—Stress decrease seen in a viscoelastic tissue stretched or compressed to a given length and then held at that length.

structural properties—Property based on a bone's structure, such as flexural rigidity, load behaviors, and energy.

structural strength—Absolute load a structure can withstand prior to failure.

subacromical impingement—Pinching of structures (distal tendon of the supraspinatus, proximal tendon of the biceps long head, and subacromial bursa) under the coracoacromial arch.

subarachnoid hematoma (SAH)—Blood accumulation between the arachnoid and the pia mater.

subdural hematoma (SDH)—Blood accumulation between the dura mater and the arachnoid.

subluxation—Partial dislocation of a joint.

supination (foot–ankle)—Combination of ankle plantar flexion, subtalar inversion, and internal rotation of the foot.

synapse—Junction between a neuron and its target structure (e.g., another neuron or sarcolemma of skeletal muscle fiber). The region where a nervous impulse is passed from one neuron to another structure.

syncytium—Multinucleated mass resulting from fusion of cells.

synovial membrane—Connective tissue that lines the cavity of a joint and produces synovial fluid.

synovitis—Inflammation of a synovial membrane.

T tubules—Invaginations of the sarcolemma that pass through the muscle fiber between the myofibrils.

taping—To fasten, tie, bind, cover, or support with tape, as in taping a sprained ankle.

teardrop fracture—Fracture at the anteroinferior corner of a cervical vertebral body.

tendinitis—Inflammation of a tendon.

tendon—White, collagenous flexible band that connects muscle to bone.

tensile stress—Internal resistance to being pulled apart.

tension—Force that tends to pull ends of a body apart.

tension–extension mechanism—Injury mechanism in which the head is forcibly hyperextended by posterior impact with forcible resistance applied to the chin, inertial forces resulting from posterior impact, or forces applied inferiorly to the posterior aspect of the head.

terminal ballistics—Projectile behavior in tissues.

tetanus—See *tetany*.

tetany—Steady skeletal muscle contraction caused by rapid arrival of signals from nerves.

third law of motion—For every action there is an equal and opposite reaction.

third-class lever—Lever system with the effort force between the axis and the resistance force.

thoracic outlet syndrome—Compression of nerves or blood vessels between the neck and shoulders that results in neck and shoulder pain, numbness, tingling, and weakened grip.

three-point bending—Bending mode involving three parallel forces, two outer forces in the same direction with the remaining force in between acting in the opposite direction of the outer ones.

tide mark—Boundary between calcified and noncalcified layers of bone.

time—Measure of the duration of a particular event.

tinnitus—Ringing or whistling sounds in the ear.

tissue—Aggregation of cells and intercellular substance that together perform a specialized function.

titin—Giant filamentous protein essential to the structure, development, and elasticity of muscle.

toe region—Portion of a stress-strain curve at initial loading that exhibits relatively high compliance (low stiffness) prior to the collagen fibers becoming taut and increasing tissue stiffness.

tonically recruited—Repeated recruitment of slow muscle fibers.

torque arm—See *moment arm*.

torque—(1) Effect of a force that tends to cause rotation or twisting about an axis. (2) Mechanical agent creating and controlling angular motion.

torsion—Twisting action applied to a structure.

total mechanical energy (TME)—Sum of linear kinetic energy + angular kinetic energy + positional potential energy.

toughness—Measure of a tissue's ability to absorb mechanical energy.

trabecular bone—Bone with high porosity (low density); also *cancellous* and *spongy bone*.

traction apophysitis—Inflammation of the bone at a tendon or ligament insertion, caused by tensile, or pulling, force.

transfer of energy—Exchange of energy from one body to another.

transfer of momentum—Exchange of momentum from one body to another.

translational motion—See *linear motion*.

traumatic axonal injury (TAI)—See *diffuse axonal injury*.

traumatic brain injury (TBI)—Term that covers numerous conditions arising from impact or acceleration–deceleration of the head.

triaxial loading—Simultaneous loading along three axes.

truss mechanism—Mechanical assemblage of two hinged segments (beams), supported at two points and arranged to transmit vertical forces to those points (e.g., longitudinal arch of the foot).

turbulent flow—Fluid flow that exhibits a chaotic pattern that contains areas of turbulence (eddies) and multidirectional movement patterns.

turf toe—Capsuloligamentous injury to the first metatarsophalangeal joint resulting from hyperflexion or hyperextension.

ulnar variance (UV)—Difference in length between the distal radius and ulna.

ultimate load—Highest load a tissue can withstand prior to failure.

undifferentiated mesenchymal stem cells—Cells that can differentiate into a variety of connective tissue cells (e.g., fibroblast, chondroblast, or osteoblast).

undisplaced fracture—Fracture in which the bony segments remain in their original (prefracture) position.

unfused tetanic contraction—Partial relaxation between muscle twitches.

unhappy triad—Knee injury caused by valgus rotation characterized by concurrent injury to the anterior cruciate ligament (ACL), medial meniscus, and medial collateral ligament.

uniaxial loading—Forces applied along a single line, typically along a primary axis of a structure.

vacuolation—Formation or development of a small cavity or space filled with air or fluid.

valgus-extension overload mechanism—Mechanism of elbow injury seen in overhead throwing that involves simultaneous valgus and extension at the elbow.

valgus—Medial deviation of a joint (e.g., knock-kneed).

varus—Lateral deviation of a joint (e.g., bowlegged).

vector—Measure that has both magnitude and direction.

velocity—Measure of the time rate (change) of displacement, calculated by displacement divided by change in time.

viscoelastic—Characteristic of tissues that can return to their original shape or configuration after a load is removed (elastic) and have strain-rate dependent response to loading (viscous effect).

viscosity—Property of a fluid that enables it to develop and maintain a resistance to flow dependent on the flow's velocity (rate of flow).

Volkmann canals—Canals containing transversely oriented blood vessels in bone.

Wallerian degeneration—Axonal degeneration distal to a site of nerve injury.

watt—Unit of mechanical power (1 W = 1 J/s).

wedge fracture—Fracture in which the front (anterior) edge of the vertebra collapses.

whiplash—Acceleration–deceleration mechanism of energy transfer to the neck.

windlass mechanism—Hyperextension of the toes to increase tension in the plantar fascia.

Wolff's law—Law that describes bone's ability to structurally adapt to the time-averaged forces applied to it.

work—Measure of a force (or torque) acting through a displacement in the direction of the force (or torque).

wound ballistics—Study of the interaction of penetrating projectiles and living body tissues.

woven bone—Immature bone without laminar or osteonal organization.

yaw—Movement from side to side.

yellow elastic ligaments—Parallel elastic fibers that are surrounded by loose connective tissue, found in the vocal cords and the ligamenta flava of the vertebrae.

yield point—Point at which there begins a brief region of relatively large strain for little increase in stress (i.e., region of high compliance).

Young's modulus—See *modulus of elasticity*.

zone of Ranvier—Area at the cortical margins of the growth plate toward the primary ossification center.

zygote—Cell formed by the union of an egg (ovum) and sperm (spermatozoa).

REFERENCES

CHAPTER I

Andersen, M.B., and J.M. Williams. 1999. Athletic injury, psychosocial factors and perceptual changes during stress. *Journal of Sports Science* 17: 735-741.

Caine, D.J., C.G. Caine, and K.J. Linder. 1996. *Epidemiology of Sports Injuries*. Champaign, IL: Human Kinetics.

Committee on Trauma Research. 1985. *Injury in America: A Continuing Public Health Problem*. Washington, DC: National Academy Press.

Ghosh, A.K., A. Bhattacherjee, and N. Chau. 2004. Relationships of working conditions and individual characteristics to occupational injuries: A case-control study in coal miners. *Journal of Occupational Health* 46: 470-478.

Gielen, A.C. 1992. Health education and injury control: integrating approaches. *Health Education Quarterly* 19: 203-218.

Haddon, W., Jr., and S.P. Baker. 1981. Injury control. In *Preventative and Community Medicine*, edited by D. Clark and B. MacMahon. New York: Little, Brown.

Heil, J. 1993. *Psychology of Sport Injury*. Champaign, IL: Human Kinetics.

Ichikawa, M., W. Chadbunchachai, and E. Marui. 2003. Effect of the helmet act for motorcyclists in Thailand. *Accident Analysis and Prevention* 35: 183-189.

Keele, K.D. 1983. *Leonardo da Vinci's Elements of the Science of Man*. New York: Academic Press.

LeVay, D. 1990. *The History of Orthopaedics*. Lancashire, UK: Parthenon.

Meeuwisse, W.H. 1994. Assessing causation in sport injury: A multifactorial model. *Clinical Journal of Sport Medicine* 4: 166-170.

Miller, A.P. 1979. Strains of the posterior calf musculature. *American Journal of Sports Medicine* 7: 172-174.

National Center for Injury Prevention and Control (NCIPC), U.S. Department of Health and Human Services. 2001-2002. Injury maps of the USA. *Injury Fact Book: National Center for Injury Prevention and Control, Atlanta, GA*. Available: www.cdc.gov/ncipc/default.htm.

National Safety Council. 2004. *Injury Facts*. Itasca, IL: National Safety Council.

O'Malley, C.D. 1964. *Andreas Vesalius of Brussels*. Berkeley: University of California Press.

Rang, M. 2000. *The Story of Orthopaedics*. Philadelphia: Saunders.

Rice, D.P., and W. Max. 1996. Annotation: The high cost of injuries in the United States. *American Journal of Public Health* 86: 14-15.

Rosenberg, R.I., D.L. Zirkle, and E.A. Neuwelt. 2005. Program self-evaluation: The evolution of an injury prevention foundation. *Journal of Neuroscience* 102: 847-849.

Runge, J.W. 1993. The cost of injury. *Emergency Medicine Clinics of North America* 11: 241-253.

Sanders, M., and B. Shaw. 1988. *Research to Determine the Contribution of System Factors in the Occurrence of Underground Injury Accidents*. Pittsburgh: Bureau of Mines.

Sanders, M.S., and E.J. McCormick. 1993. *Human Factors in Engineering and Design*. New York: McGraw-Hill.

Suchman, E. 1961. On accident behavior. In *Behavioural Approaches to Accident Research*, edited by Edward A. Suchman. Washington, DC: Association for the Aid to Crippled Children.

Trifiletti, L.B., A.C. Gielen, D.A. Sleet, and K. Hopkins. 2005. Behavioral and social sciences theories and models: Are they used in unintentional injury prevention research? *Health Education Research* 20: 298-307.

Witvrouw, E., L. Danneels, P. Asselman, T. D'Have, and D. Cambier. 2003. Muscle flexibility as a risk factor for developing muscle injuries in male professional soccer players: A prospective study. *American Journal of Sports Medicine* 31: 41-46.

CHAPTER 2

Åstrand, P.-O., K. Rodahl, H.A. Dahl, and S.B. Stromme. 2003. *Textbook of Work Physiology: Physiological Bases of Exercise* (4th ed.). Champaign, IL: Human Kinetics.

Berchuck, M., T.P. Andriacchi, B.R. Bach, and B. Reider. 1990. Gait adaptations by patients who have a deficient anterior cruciate ligament. *Journal of Bone and Joint Surgery* 72A: 871-877.

Buckwalter, J.A., and J. Martin, J. 1995. Degenerative joint disease. *Clinical Symposia* 47: 1-32.

Carter, D.R., G.S. Beaupre, W. Wong, R.L. Smith, T.P. Andriacchi, and D.J. Schurman. 2004. The mechanobiology of articular cartilage development and degeneration. *Clinical Orthopaedics and Related Research* 427S: S69-S77.

Cooper, R.R., J.W. Milgram, and R.A. Robinson. 1966. Morphology of the osteon. *Journal of Bone and Joint Surgery* 48A: 1239-1271.

Cormack, D.H. 1987. *Ham's Histology* (9th ed.). Philadelphia: Lippincott.

Currey, J.D. 2002. *Bones: Structure and Mechanics*. Princeton, NJ: Princeton University Press.

Ellman, H. 1975. Anterior angulation deformity of the radial head. *Journal of Bone and Joint Surgery*, 57A: 776-778.

Eyre, D.R. 2004. Collagens and cartilage matrix homeostasis. *Clinical Orthopaedics and Related Research* 427S: S118-S122.

Fawcett, D.W. and E. Raviola. 1994. *Bloom and Fawcett: A Textbook of Histology* (12th ed.). New York: Chapman and Hall.

Ficat, R.P., J. Arlet, and D.S. Hungerford. 1980. *Ischemia and Necroses of Bone*. Baltimore: Williams & Wilkins.

Garrett, W.E., and T.M. Best. 2000. Anatomy, physiology, and mechanics of skeletal muscle. In *Orthopaedic Basic Science* (2nd ed.), edited by S.R. Simon. Park Ridge, IL: American Academy of Orthopaedic Surgeons.

Keele, K.D. 1983. *Elements of the Science of Man*. New York: Academic Press.

Kim, A.W., A.M. Rosen, V.A. Brander, and T.S. Buchanan. 1995. Selective muscle activation following electrical stimulation of the collateral ligaments of the human knee joint. *Archives of Physical Medicine and Rehabilitation* 76: 750-757.

Langman, J. 1969. *Medical Embryology* (2nd ed.). Baltimore: Williams & Wilkins.

Lo, I.K.Y., G. Thornton, A. Miniaci, C.B. Frank, J.B. Rattner, and R.C. Bray. 2003. Structure and function of diarthroidial joints. In *Operative Arthroscopy* (3rd ed.), edited by J.B. McGinty. Philadelphia: Lippincott Williams & Wilkins.

Marieb, E.N. 2004. *Human Anatomy and Physiology* (6th ed.). San Francisco: Pearson Education.

Martin, R.B., and D.B. Burr. 1989. *Structure, Function, and Adaptation of Compact Bone*. New York: Raven Press.

Martin, R.B., D.B. Burr, and N.A. Sharkey. 1998. *Skeletal Tissue Mechanics*. New York: Springer-Verlag.

Morel, V., A. Mercay, and T.M. Quinn. 2005. Pre-strain decreases cartilage susceptibility to injury by ramp compression in vitro. *Osteoarthritis Cartilage* 13: 964-970.

Nigg, B.M., and W. Herzog. 1999. *Biomechanics of the Musculoskeletal System* (2nd ed.). New York: Wiley.

Nordin, M., and V.H. Frankel. 1989. *Basic Biomechanics of the Musculoskeletal System* (2nd ed.). Philadelphia: Lea & Febiger.

Ogden, J.A. 2000a. Anatomy and physiology of skeletal development. In *Skeletal Injury in the Child* (3rd ed.), edited by J.A. Ogden. New York: Springer.

Ogden, J.A. 2000b. Injury to the growth mechanisms. In *Skeletal Injury in the Child* (3rd ed.), edited by J.A. Ogden. New York: Springer.

Ogden, J.A., and D.P. Grogan. 1987. Prenatal development and growth of the musculoskeletal system. In *The Scientific Basis of Orthopaedics* (2nd ed.), edited by J.A. Albright and R.A. Brand. Norwalk, CT: Appleton-Lange.

Ogden, J.A., D.P. Grogan, and T.R. Light. 1987. In *The Scientific Basis of Orthopaedics* (2nd ed.), edited by J.A. Albright and R.A. Brand. Norwalk, CT: Appleton-Lange.

Parry, D.A.D., and J.M. Squire. 2005. Fibrous proteins: coiled coils, collagen and elastomers. In *Advances in Protein Chemistry* (vol. 70), edited by D.A.D. Parry and J.M. Squire. Amsterdam: Academic Press.

Rosier, R.N, P.R. Reynolds, and R.J. O'Keefe. 2000. Molecular and cell biology in orthopaedics. In *Orthopaedic Basic Science* (2nd ed.), edited by S.R. Simon. Park Ridge, IL: American Academy of Orthopaedic Surgeons.

Sadler, T.W. 2004. *Langman's Medical Embryology* (9th ed.). Baltimore: Lippincott Williams & Wilkins.

Sands, W.A., D.J. Caine, and J. Borms. 2003. *Scientific Aspects of Women's Gymnastics. Series: Medicine and Sport Science* (vol. 45). Basel, Switzerland: Karger.

Shim, S.S., D.H. Copp, and F.P. Patterson. 1967. An indirect method of bone blood-flow measurement based on the clearance of a circulating bone-seeking radioisotope. *Journal of Bone and Joint Surgery* 49A: 693-702.

Silver, F.H., and G. Bradica. 2002. Mechanobiology of cartilage: How do internal and external stresses affect mechanochemical transduction and elastic energy storage? *Biomechanics and Modeling in Mechanobiol*ogy 1: 219-238.

Solomonow, M. 2004. Ligaments: A source of work-related musculoskeletal disorders. *Journal of Electromyography and Kinesiology* 14: 49-60.

Surve, I., M.P. Schwellnus, T. Noakes, and C. Lombard. 1994. A fivefold reduction in the incidence of recurrent ankle sprains in soccer players using the Sport-Stirrup orthosis. *American Journal of Sports Medicine* 22: 601-606.

Tidball, J.G. 1991. Force transmission across muscle cell membranes. *Journal of Biomechanics* 24(Suppl. 1): 43-52.

Tothill, P., and J.N. MacPherson. 1986. The distribution of blood flow to the whole skeleton in dogs, rabbits and rats measured with microspheres. *Clinical Physics and Physiological Measurement* 7: 117-123.

Tskhovrebova, L., and J. Trinick. 2003. Titin: Properties and family relationships. *Nature Reviews: Molecular Cell Biology* 4: 679-689.

Whiting, W.C., and S. Rugg. 2006. *Dynatomy-Dynamic Human Anatomy*. Champaign, IL: Human Kinetics.

Woo, S.L.-Y., K.-N. An, C.B. Frank, G.A. Livesay, C.B. Ma, J. Zeminski, J.S. Wayne, and B.S. Myers. 2000. Anatomy, biology, and biomechanics of tendon and ligament. In *Orthopaedic Basic Science* (2nd ed.), edited by S.R. Simon. Park Ridge, IL: American Academy of Orthopaedic Surgeons.

Zelzer, E., R. Mamiuk, N. Ferrara, R.S. Johnson, E. Schipani, and B.R. Olsen. 2004. VEGFA is necessary for chondrocyte survival during bone development. *Development* 131: 2161-2171.

CHAPTER 3

Allen, L.R., D. Flemming, and T.G. Sanders. 2004. Turf toe: Ligamentous injury of the first metatarsophalangeal joint. *Military Medicine* 169: xix-xxiv.

Andreasson, G., U. Lindenberger, P. Renstrom, and L. Peterson. 1986. Torque developed at simulated sliding between sport shoes and an artificial turf. *American Journal of Sports Medicine* 14: 225-230.

Blemker, S.S., and S.L. Delp. 2005. Three-dimensional representation of complex muscle architectures and geometries. *Annals of Biomedical Engineering* 33: 661-673.

Burstein, A.H., and T.M. Wright. 1994. *Fundamentals of Orthopaedic Biomechanics*. Baltimore: Williams & Wilkins.

Ekstrand, J., and B.M. Nigg. 1989. Surface-related injuries in soccer. *Sports Medicine* 8: 56-62.

Hubbard, M. 1993. Computer simulation in sport and industry. *Journal of Biomechanics* 26(Suppl. 1): 53-61.

Kirking, B., J. Krevolin, C. Townsend, C.W. Colwell, Jr., and D.D. D'Lima. 2006. A multiaxial force-sensing implantable tibial prosthesis. *Journal of Biomechanics* 39: 1744-1751.

Meyers, M.C., and B.S. Barnhill. 2004. Incidence, causes, and severity of high school football injuries on FieldTurf versus natural grass: A 5-year prospective study. *American Journal of Sports Medicine* 32: 1626-1638.

Nigg, B.M., and B. Segesser. 1988. The influence of playing surfaces on the load on the locomotor system and on football and tennis injuries. *Sports Medicine* 5: 375-385.

Nigg, B.M., and M.R. Yeadon. 1987. Biomechanical aspects of playing surfaces. *Journal of Sports Sciences* 5: 117-145.

Robertson, L.S. 1998. *Injury Epidemiology* (2nd ed.). New York: Oxford University Press.

Skovron, M.L., I.M. Levy, and J. Agel. 1990. Living with artificial grass: A knowledge update, Part 2: Epidemiology. *American Journal of Sports Medicine* 18: 510-513.

Stansfield, B.W., A.C. Nicol, J.P. Paul, I.G. Kelly, F. Graichen, and G. Bergmann. 2003. Direct comparison of calculated hip joint contact forces with those measured using instrumented implants: An evaluation of a three-dimensional mathematical model of the lower limb. *Journal of Biomechanics* 36: 929-936.

Winter, D.A. 2005. *Biomechanics and Motor Control of Human Movement* (4th ed.). New York: Wiley.

CHAPTER 4

Akeson, W.H., D. Amiel, M.F. Abel, S.R. Garfin, and S.L. Woo. 1987. Effects of immobilization on joints. *Clinical Orthopaedics and Related Research* 219: 28-37.

Ashizawa, N., R. Fujimura, K. Tokuyama, and M. Suzuki. 1997. A bout of resistance exercise increases urinary calcium independently of osteoblastic activation in men. *Journal of Applied Physiology* 83: 1159-1163.

Åstrand, P.-O., K. Rodahl, H.A. Dahl, and S.B. Stromme. 2003. *Textbook of Work Physiology: Physiological Bases of Exercise* (4th ed.). Champaign, IL: Human Kinetics.

Bailey, D.A., R.A. Faulkner, and H.A. McKay. 1996. Growth, physical activity, and bone mineral acquisition. *Exercise and Sport Sciences Reviews* 24: 233-266.

Baldwin, K.M., and F. Haddad. 2002. Cellular and molecular responses to altered physical activity paradigms. *American Journal of Physical Medicine and Rehabilitation* 81(Suppl.): 40-51.

Bandy, W.D., and K. Dunleavy. 1996. Adaptability of skeletal muscle: Response to increased and decreased use. In *Athletic Injuries and Rehabilitation*, edited by J.E. Zachazewski, D.J. Magee, and W.S. Quillen. Philadelphia: Saunders.

Bass, S.L., P. Eser, and R. Daly. 2005. The effect of exercise and nutrition on the mechanostat. *Journal of Musculoskeletal and Neuronal Interactions* 5: 239-254.

Benjamin, M., and B. Hillen. 2003. Mechanical influences on cells, tissues and organs—"Mechanical morphogenesis." *European Journal of Morphology* 41: 3-7.

Beveridge, W. I. B. 1957. The art of scientific investigation (3rd ed.). London: Heinemann.

Bilanin, J., M. Blanchard, and E. Russek-Cohen. 1989. Lower vertebral bone density in male long distance runners. *Medicine and Science in Sports and Exercise* 21: 66-70.

Bostrom, M.P.G., A. Boskey, J.J. Kaufman, and T.A. Einhorn. 1994. Form and function of bone. In *Orthopaedic Basic Science*, edited by S.R. Simon. Park Ridge, IL: American Academy of Orthopaedic Surgeons.

Bourrin, S., A. Toromanoff, P. Ammann, J.P. Bonjour, and R. Rizzoli. 2000. Dietary protein deficiency induces osteoporosis in aged male rats. *Journal of Bone and Mineral Research* 15: 1555-1563.

Brand-Saberi, B. 2005. Genetic and epigenetic control of skeletal muscle development. *Annals of Anatomy* 187: 199-207.

Brandt, K.D. 1992. The pathogenesis of osteoarthritis. *EULAR Bulletin* 3: 75-81.

Buckingham, M., L. Bajard, T. Chang, P. Daubas, J. Hadchouel, S. Meilhac, D. Montarras, D. Rocancourt, and F. Relaix. 2003. The formation of skeletal muscle: from somite to limb. *Journal of Anatomy* 202: 59-68.

Burr, D.B., C. Milgrom, D. Fyhrie, M. Forwood, M. Nyska, A. Finestone, S. Hoshaw, E. Saiag, and A. Simkin. 1996. In vivo measurement of human tibial strains during vigorous activity. *Bone* 18: 405-410.

Butler, D.L., E.S. Grood, F.R. Noyes, and R.F. Zernicke. 1978. Biomechanics of ligaments and tendons. In *Exercise and Sport Sciences Reviews*, edited by R. Hutton. Baltimore: Williams & Wilkins.

Chakravarthy, M.V., B.S. Davis, and F.W. Booth. 2000. IGF-1 restores satellite cell proliferative potential in immobilized old skeletal muscle. *Journal of Applied Physiology* 89: 1365-1379.

Chen, C.T., M. Bhargava, P.M. Lin, and P.A. Torzilli. 2003. Time, stress, and location dependent chondrocyte death and collagen damage in cyclically loaded articular cartilage. *Journal of Orthopaedic Research* 21: 888-898.

Cureton, K.J., M.A. Collins, D.W. Hill, and F.M. McElhannon, Jr. 1990. Muscle hypertrophy in men and women. *Medicine and Science in Sports and Exercise* 20: 338-344.

Currey, J.D. 1979. Mechanical properties of bone tissues with greatly differing functions. *Journal of Biomechanics* 12: 313-319.

Currey, J.D. 1988. The effect of porosity and mineral content on the Young's modulus of elasticity of compact bone. *Journal of Biomechanics* 21: 131-139.

Currey, J.D. 2002. *Bones: Structure and Mechanics*. Princeton, NJ: Princeton University Press.

Currey, J.D. 2003a. How well are bones designed to resist fracture? *Journal of Bone and Mineral Research* 18: 591-598.

Currey, J.D. 2003b. The many adaptations of bone. *Journal of Biomechanics* 36: 1487-1495.

Currey, J.D. 2005. Bone architecture and fracture. *Current Osteoporosis Reports* 3: 52-56.

Currey, J.D., and R.M. Alexander. 1985. The thickness of walls of tubular bones. *Journal of Zoology* (London) 206: 453-468.

Deprez, X., and P. Fardellone. 2003. Nonpharmacological prevention of osteoporotic fractures. *Joint Bone Spine* 70: 448-457.

Doschak, M.R., and R.F. Zernicke. 2005. Structure, function and adaptation of bone-tendon and bone-ligament complexes. *Journal of Musculoskeletal and Neuronal Interactions* 5: 35-40.

Doty, S.B. 2004. Space flight and bone formation. *Materwiss Wersttech* 35: 951-961.

Dull, P. 2006. Hormone replacement therapy. *Primary Care* 33: 953-963.

Ehrhardt, J., and J. Morgan. 2005. Regenerative capacity of skeletal muscle. *Current Opinion in Neurology* 18: 548-553.

Engelke, K., W. Kemmler, D. Lauber, C. Beeskow, R. Pintag, and W.A. Kalender. 2006. Exercise maintains bone density at spine and hip EFOPS: A 3-year longitudinal study in early postmenopausal women. *Osteoporosis International* 17: 133-142.

Englemark, V.E. 1961. Functionally induced changes in articular cartilage. In *Biomechanical Studies of the Musculoskeletal System*, edited by F.G. Evans. Springfield, IL: Charles C Thomas.

Frank, C.B. 1996. Ligament Injuries: Pathophysiology and healing. In *Athletic Injuries and Rehabilitation*, edited by J.E. Zachazewski, D.J. Magee, and W.S. Quillen. Philadelphia: Saunders.

Garrett, W.E., Jr., and T.M. Best. 1994. Anatomy, physiology, and mechanics of skeletal muscle. In *Orthopaedic Basic Science*, edited by S.R. Simon. Park Ridge, IL: American Academy of Orthopaedic Surgeons.

Giannini, S., M. Nobile, S. Sella, and L. Dalle Carbonare. 2005. Bone disease in primary hypercalciuria. *Critical Reviews in Clinical Laboratory Science* 42: 229-248.

Gregor, R.J., P.V. Komi, R.C. Browning, and M. Jarvinen. 1991. A comparison of the triceps surae and residual muscle moments at the ankle during cycling. *Journal of Biomechanics* 24: 287-297.

Hakkinen, K. 2002. *Strength Training for Sport*. Malden, MA: Blackwell.

Harwood, F.L., and D. Amiel. 1992. Differential metabolic responses of periarticular ligaments and tendon to joint immobilization. *Journal of Applied Physiology* 72: 1687-1691.

Hales, S. 1727. Vegetable staticks: or, an account of some statistical experiments on the sap in vegetables: being an essay towards a natural history of vegetation. London: W. and J. Innys, and T. Woodward.

Hengsberger, S., P. Ammann, B. Legros, R. Rizzoli, and P. Zysset. 2005. Intrinsic bone tissue properties in adult rat vertebrae: modulation by dietary protein. *Bone* 36: 134-141.

Hiney, K.M., B.D. Nielsen, D. Rosenstein, M.W. Orth, and B.P. Marks. 2004. High-density exercise of short duration alters bovine bone density and shape. *Journal of Animal Science* 82: 1612-1620.

Ikai, M., and T. Fukunaga. 1968. Calculation of muscle strength per unit cross-sectional area of human muscle by means of ultrasonic measurements. *Internationale Zeitschrift Fur Angewandte Physiologie* 6: 174-177.

Iwamoto, J., C. Shimamura, T. Takeda, H. Abe, S. Ichimura, Y. Sato, and Y. Toyama. 2004. Effects of treadmill exercise on bone mass, bone metabolism, and calcitrophic hormones in young growing rats. *Journal of Bone and Mineral Research* 22: 26-31.

Jin, M., E.H. Frank, T.M. Quinn, E.B. Hunziker, and A.J. Grodzinsky. 2001. Tissue shear deformation stimulates proteoglycan and protein biosynthesis in bovine cartilage explants. *Archives of Biochemistry and Biophysics* 395: 41-48.

Kadi, F., N. Charifi, C. Denis, J. Lexell, J.L. Andersen, P. Schjerling, S. Olsen, and M. Kjaer. 2005. The behaviour of satellite cells in response to exercise: what have we learned from human studies? *Pflugers Archives* 451: 319-327.

Kaplan, F.S., W.C. Hayes, T.M. Keaveny, A. Boskey, T.A. Einhorn, and J.P. Iannotti. 1994. Form and function of bone. In *Orthopaedic Basic Science*, edited by S.R. Simon. Park Ridge, IL: American Academy of Orthopaedic Surgeons.

Kasashima, Y., R.K. Smith, H.L. Birch, T. Takahashi, K. Kusano, and A.E. Goodship. 2002. Exercise-induced tendon hypertrophy: cross-sectional area changes during growth are influenced by exercise. *Equine Veterinary Journal* 34(Suppl.): 264-268.

Kjaer, M. 2004. Role of extracellular matrix in adaptation of tendon and skeletal muscle to mechanical loading. *Physiological Reviews* 84: 649-698.

Komi, P.V. 2003. *Strength and Power in Sport*. Oxford, UK: Blackwell Scientific.

Korpelainen, R., S. Keinanen-Kiukaanniemi, J. Heikkinen, K. Vaananen, and L. Korpelainen. 2006. Effect of impact exercise on bone mineral density in elderly women with low bone mineral density: A population-based randomized controlled 30-month intervention. *Osteoporosis International* 21: 772-779.

Lanyon, L.E. 1987. Functional strain in bone tissue as an objective and controlling stimulus for adaptive bone remodeling. *Journal of Biomechanics* 20: 1083-1093.

Lieberman, D.E., O.M. Pearson, J.D. Polk, B. Demes, and A.W. Crompton. 2003. Optimization of bone growth and remodeling in response to loading in tapered mammalian limbs. *Journal of Experimental Biology* 206: 3125-3138.

Linnamo, V., A. Pakarinen, P.V. Komi, W.J. Kraemer, and K. Hakkinen. 2005. Acute hormonal responses to submaximal and maximal heavy resistance and explosive exercises in men and women. *Journal of Strength Conditioning Research* 19: 566-571.

Liu, D., H.P. Veit, and D.M. Denbow. 2004. Effects of long-term dietary lipids on mature bone mineral content, collagen, crosslinks, and prostaglandin E2 production in Japanese quail. *Poultry Science* 83: 1876-1883.

Lo, I.K.Y., G. Thornton, A. Miniaci, C.B. Frank, J.B. Rattner, and R.C. Bray. 2003. Structure and function of diarthrodial joints. In *Operative Arthroscopy* (3rd ed.), edited by J.B. McGinty. Philadelphia: Lippincott Williams & Wilkins.

Loitz, B.J., and R.F. Zernicke. 1992. Strenuous exercise-induced remodeling of mature bone: Relationships between in vivo strains and bone mechanics. *Journal of Experimental Biology* 170: 1-18.

Loitz-Ramage, B.J., and R.F. Zernicke. 1996. Bone biology and mechanics. In *Athletic Injuries and Rehabilitation*, edited by J.E. Zachazewski, D.J. Magee, and W.S. Quillen. Philadelphia: Saunders.

Lorentzon, M., D. Mellstrom, and C. Ohlsson. 2005. Association of amount of physical activity with cortical bone size and trabecular volumetric BMD in young adult men: The GOOD study. *Journal of Bone and Mineral Research* 20: 1936-1943.

Macaluso, A., and G. De Vito. 2004. Muscle strength, power, and resistance training in older people. *European Journal of Applied Physiology* 91: 450-472.

MacDougall, J., C. Webber, J. Martin, S. Ormerod, A. Chesley, E. Younglai, C. Gordon, and C. Blimkie. 1992. Relationship among running mileage, bone density, and serum testosterone in male runners. *Journal of Applied Physiology* 73: 1165-1170.

MacDougall, J.D. 2003. Hypertrophy or hyperplasia? In *Strength and Power in Sport: The Encyclopedia of Sports Medicine* (2nd ed.), edited by P. Komi. Oxford, UK: Blackwell.

Maddalozzo, G.G., J.J. Widrick, B.J. Cardinal, K.M. Winters-Stone, M.A. Hoffman, and C.M. Snow. 2007. The effects of hormone replacement therapy and resistance training on spine bone mineral density in early postmenopausal women. *Bone* 40: 1244-1251.

Mankin, H.J., V.C. Mow, J.A. Buckwalter, J.P. Iannotti, and A. Ratcliffe. 1994. Articular cartilage structure, composition and function. In *Orthopaedic Basic Science*, edited by S.R. Simon. Park Ridge, IL: American Academy of Orthopaedic Surgeons.

Marcus, R., C. Cann, P. Madvig, J. Minkoff, M. Goddard, M. Bayer, M. Martin, W. Haskell, and H. Genant. 1985. Menstrual function and bone mass in elite women distance runners. Endocrine and metabolic features. *Annals of Internal Medicine* 102: 158-163.

Martin, R.B., D.B. Burr, and N.A. Sharkey. 1998. *Skeletal Tissue Mechanics*. New York: Springer-Verlag.

Matsuda, J.J., R.F. Zernicke, A.C. Vailas, A. Pedrini-Mille, and J.A. Maynard. 1986. Morphological and mechanical adaptation of immature bone to strenuous exercise. *Journal of Applied Physiology: Respiratory, Environmental, and Exercise Physiology* 60: 2028-2034.

McArdle, W.D., F.I. Katch, and V.L. Katch. 2001. *Exercise Physiology*. Philadelphia: Lippincott Williams & Wilkins.

McCall, G.E., W.C. Byrnes, and S.J. Fleck. 1999. Acute and chronic hormonal responses to resistance training designed to promote muscle hypertrophy. *Canadian Journal of Applied Physiology* 24: 96-107.

Merriam-Webster. 2003. *Merriam-Webster's Collegiate Dictionary* (11th ed.). Springfield, MA: Merriam-Webster.

Mow, V.C., A. Ratcliffe, and A.R. Poole. 1992. Cartilage and diarthrodial joints as paradigms for hierarchical materials and structures. *Biomaterials* 1: 67-97.

Mullner, T., O. Kwasny, R. Reihsner, V. Lohnert, and R. Schabus. 2000. Mechanical properties of a rat patellar tendon stress-shielded in situ. *Archives of Orthopaedic and Trauma Surgery* 120: 70-74.

Oganov, V.S. 2004. Modern analysis of bone loss mechanisms in microgravity. *Journal of Gravitational Physiology* 11: P143-P146.

Ormerod, S., J. MacDougall, and C. Webber. 1988. The effects of different forms of exercise on bone mineral content. *Canadian Journal of Sport Science* 13: 74P.

Palmes, D., H.U. Spiegel, T.O. Schneider, M. Langer, U. Stratmann, T. Budny, and A. Probst. 2002. Achilles tendon healing: Long-term biomechanical effects of postoperative mobilization and immobilization in a new mouse model. *Journal of Orthopaedic Research* 20: 939-946.

Parfitt, A.M. 1984. The cellular basis of bone remodeling: The quantum concept reexamined in light of recent advances in the cell biology of bone. *Calcified Tissue International* 36: S38.

Platt, M.A. 2005. Tendon repair and healing. *Clinics in Podiatric Medicine and Surgery* 22: 553-560.

Pogoda, P., M. Priemel, H.M. Rueger, and M. Amling. 2005. Bone remodeling: new aspects of a key process that controls skeletal maintenance and repair. *Osteoporosis International* 16: S18-S24.

Robinson, T., C. Snow-Harter, D. Taafe, D. Gillis, J. Shaw, and R. Marcus. 1995. Gymnasts exhibit higher bone mass than runners despite prevalence of amenorrhea and oligomenorrhea. *Journal of Bone and Mineral Research* 19: 26-35.

Rubin, C.T., and L.E. Lanyon. 1985. Regulation of bone mass by mechanical strain magnitude. *Calcified Tissue International* 37: 411-417.

Snow-Harter, C., and R. Marcus. 1991. Exercise, bone mineral density, and osteoporosis. *Exercise and Sport Sciences Reviews* 19: 351-388.

Sorrenti, S.J. 2006. Achilles tendon rupture: Effect of early mobilization in rehabilitation after surgical repair. *Foot Ankle International* 27: 407-410.

Specker, B., T. Binkley, and N. Fahrenwald. 2004. Increased periosteal circumference remains present 12 months after an exercise intervention in preschool children. *Bone* 35: 1383-1388.

Staud, R. 2005. Vitamin D: more than just affecting calcium and bone. *Current Rheumatology Reports* 7: 356-364.

Suva, L.J., D. Gaddy, D.S. Perrien, R.L. Thomas, and D.M. Findlay. 2005. Regulation of bone mass by mechanical loading: microarchitecture and genetics. *Current Osteoporosis Report* 3: 46-51.

Tajbakhsh, S. 2003. Stem cells to tissue: Molecular, cellular and anatomical heterogeneity in skeletal muscle. *Current Opinion in Genetics and Development* 13: 413-422.

Taylor, A.H., N.T. Cable, G. Faulkner, M. Hillsdon, M. Narici, and A.K. Van Der Bij. 2004. Physical activity and older adults: A review of health benefits and the effectiveness of interventions. *Journal of Sports Science* 22: 703-725.

Tidball, J.G. 1983. The geometry of actin filament-membrane associations can modify adhesive strength of the myotendinous junction. *Cell Motility and the Cytoskeleton* 3: 439-447.

Tipton, C.M., R.D. Matthes, J.A. Maynard, and R.A. Carey. 1975. The influence of physical activity on ligaments and tendons. *Medicine and Science in Sports* 7: 165-175.

Tuukkanen, J., B. Wallmark, P. Jalovaara, T. Takala, S. Sjogren, and K. Vaananen. 1991. Changes induced in growing rat bone by immobilization and remobilization. *Bone* 12: 113-118.

Walker, J.M. 1996. Cartilage of human joints and related structures. In *Athletic Injuries and Rehabilitation*, edited by J.E. Zachazewski, D.J. Magee and W.S. Quillen. Philadelphia: Saunders.

Williams, P.E., and G. Goldspink. 1981. Longitudinal growth of striated muscle. *Journal of Cell Science* 9: 751-767.

Wilmore, J.H. 1979. The application of science to sport: Physiological profiles of male and female athletes. *Canadian Journal of Applied Sport Science* 4: 102-115.

Woo, S.L.-Y., K.-N. An, C.B. Frank, G.A. Livesay, C.B. Ma, J. Zeminski, J.S. Wayne, and B.S. Myers. 1994. Anatomy, biology, and biomechanics of tendon and ligament. In *Orthopaedic Basic Science*, edited by S.R. Simon. Park Ridge, IL: American Academy of Orthopaedic Surgeons.

Woo, S.L.-Y., M.A. Gomez, T.J. Sites, P.O. Newton, C.A. Orlando, and W.H. Akeson. 1987. The biomechanical and morphological changes in the medial collateral ligament of the rabbit after immobilization and remobilization. *Journal of Bone and Joint Surgery* 69A: 1200-1211.

Woo, S.L., M. Thomas, and S.S. Chan Saw. 2004. Contribution of biomechanics, orthopaedics and rehabilitation: The past, present and future. *Surgeon* 2: 125-136.

Yamada, H. 1973. *Strength of Biological Materials*. Huntington, NY: Krieger.

Zernicke, R.F., J.J. Garhammer, and F.W. Jobe. 1977. Human patellar tendon rupture: A kinetic analysis. *Journal of Bone and Joint Surgery* 59A: 179-183.

Zernicke, R.F., J. McNitt-Gray, C. Otis, B. Loitz, G. Salem, and G. Finerman. 1994. Stress fracture risk assessment among elite collegiate women runners. *Journal of Biomechanics* 27: 978-986.

Zioupos, P., and J.D. Currey. 1994. The extent of microcracking and the morphology of microcracks in damaged bone. *Journal of Materials Science* 29: 978-986.

CHAPTER 5

Akesson, K. 2003. Principles of bone and joint disease control programs—osteoporosis. *Journal of Rheumatology* 67(Suppl.): 21-25.

American Academy of Orthopaedic Surgeons. 1999. Musculoskeletal conditions in the U.S. *AAOS Bulletin* 47: 34-36.

American Academy of Orthopaedic Surgeons. 2007. Total hip replacement.. Available: http://orthoinfo.aaos.org/topic.cfm?topic=A00377.

Anderson, R.N., A.M. Minino, L.A. Fingerhut, M. Warner, and M.A. Heinen. 2004. Deaths: Injuries, 2001. *National Vital Statistical Report* 52: 1-86.

Bunn, T.L., S. Slavova, T.W. Struttmann, and S.R. Browning. 2005. Sleepiness/fatigue and distraction/inattention as factors for fatal versus nonfatal commercial motor vehicle driver injuries. *Accident Analysis and Prevention* 37: 862-869.

Christiansen, C. 1995. Osteoporosis: Diagnosis and management today and tomorrow. *Bone* 17(Suppl): 513S-516S.

Close, G.L., T. Ashton, A. McArdle, and D.P.M. MacLaren. 2005. The emerging role of free radicals in delayed onset muscle soreness and contraction-induced muscle injury. *Comparative Biochemistry and Physiology, Part A* 142: 257-266.

Committee on Trauma Research. 1985. *Injury in America: A Continuing Public Health Problem.* Washington, DC: National Academy Press.

Cooper, R.R., and S. Misol. 1970. Tendon and ligament insertion. A light and electron microscopic study. *Journal of Bone and Joint Surgery* 52A: 1-20.

Croft, P., C. Cooper, C. Wickham, and D. Coggon. 1992. Osteoarthritis of the hip and occupational activity. *Scandinavian Journal of Work and Environmental Health* 18: 59-63.

Croisier, J.L. 2004. Factors associated with recurrent hamstring injuries. *Sports Medicine* 34(10): 681-695.

Crowninshield, R.D. 1990. Computer-assisted prosthetic design. In *Joint Replacement: State of the Art*, edited by R. Coombs, A. Gristina, and D. Hungerford. St. Louis: Mosby-Year Book.

Day, S.M., R.F. Ostrum, E.Y.S. Chao, C.T. Rubin, H.T. Aro, and T.A. Einhorn. 1994. Bone injury, regeneration, and repair. In *Orthopaedic Basic Science*, edited by J.A. Buckwalter, T.A. Einhorn, and S.R. Simon. Park Ridge, IL: American Academy of Orthopaedic Surgeons.

DeSmet, K.A., C. Pattyn, and R. Verdonck. 2002. Early results of primary Birmingham hip resurfacing using a hybrid metal-on-metal couple. *Hip International* 12: 158-162.

Ellender, L., and M.M. Linder. 2005. Sports pharmacology and ergogenic aids. *Primary Care: Clinics in Office Practice* 32: 277-292.

Felson, D.T., J.J. Anderson, A. Naimark, A.M. Walker, and R.F. Meenan. 1988. Obesity and knee osteoarthritis: The Framingham study. *Annals of Internal Medicine* 109(1): 18-24.

Frank, C.B. 1996. Ligament injuries: Pathophysiology and healing. In *Athletic Injuries and Rehabilitation* (2nd ed.), edited by J.E. Zachazewski, D.J. Magee, and W.S. Quillen. Philadelphia: Saunders.

Fulkerson, J.P., C.C. Edwards, and O.D. Chrisman. 1987. Articular cartilage. In *The Scientific Basis of Orthopaedics* (2nd ed.), edited by J.A. Albright and R.A. Brand. Norwalk, CT: Appleton-Lange.

Garrett, W.E., Jr., P.K. Nikolaou, B.M. Ribbeck, R.R. Glisson, and A.V. Seaber. 1988. The effect of muscle architecture on the biomechanical failure properties of skeletal muscle under passive conditions. *American Journal of Sports Medicine* 16: 7-12.

Garrett, W.E., Jr., M.R. Safran, A.V. Seaber, R.R. Glisson, and B.M. Ribbeck. 1987. Biomechanical comparison of stimulated and nonstimulated skeletal muscle pulled to failure. *American Journal of Sports Medicine* 15: 448-454.

Goltzman, D. 2000. *The Osteoporosis Primer.* Cambridge, UK: Cambridge University Press.

Gusmer, P.B., and H.G. Potter. 1995. Imaging of shoulder instability. *Clinics in Sports Medicine* 14: 777-795.

Harris, W.H. 1986. Etiology of osteoarthritis of the hip. *Clinical Orthopaedics and Related Research* 213: 20-33.

Hipp, J.A., and W.C. Hayes. 2003. Biomechanics of fractures. In *Skeletal Trauma: Basic Science, Management, and Reconstruction* (2nd ed.), edited by B.D. Browner, J.B. Jupiter, A.M. Levine, and P.G. Trafton. Philadelphia: Saunders.

Horton, W.E., R. Yagi, D. Laverty, and S. Weiner. 2005. Overview of studies comparing human normal cartilage with minimal and advanced osetoarthritic cartilage. *Clinical and Experimental Rheumatology* 23: 103-112.

Houston, C.S., and L.E. Swischuk. 1980. Varus and valgus—no wonder they are confused. *New England Journal of Medicine* 302: 471-472.

Julin, M.J., and M. Mathews. 1990. Shoulder injuries. In *The Team Physicians Handbook* (1st ed.), edited by M.B. Mellion, W.M. Walsh, and G.L. Shelton. Philadelphia: Hanley and Belfus, Inc.

Lawrence, R.C., M.C. Hochberg, J.L. Kelsey, F.C. McDuffie, T.A. Medsger, Jr., W.R. Felts, and L.E. Shulman. 1989. Estimates of the prevalence of selected arthritic and musculoskeletal diseases in the United States. *Journal of Rheumatology* 16(4): 427-441.

Leadbetter, W.B. 2001. Soft tissue athletic injury. In *Sports Injuries: Mechanisms, Prevention, Treatment* (2nd ed.), edited by F.H. Fu and D.A. Stone. Philadelphia: Lippincott Williams & Wilkins

Mankin, H.J. 1993. Articular cartilage, cartilage injury, and osteoarthritis. In: *The Patellofemoral Joint,* edited by J.M. Fox and W. Del Pizzo. New York: McGraw-Hill.

Marion, D.W. 1994. Head injuries. In *Sports Injuries: Mechanisms, Prevention, Treatment,* edited by F.H. Fu and D.A. Stone. Baltimore: Williams & Wilkins.

McMaster, P.E. 1933. Tendon and muscle ruptures. *Journal of Bone and Joint Surgery* 15: 705-722.

National Safety Council. 2004. *Injury Facts.* Itasca, IL: National Safety Council.

Pacifici, R., and L.V. Avioli. 1993. Effects of aging on bone structure and metabolism. In: *The Osteoporotic Syndrome: Detection, Prevention, and Treatment* (3rd ed.), edited by L.V. Avioli. New York: John Wiley & Sons.

Reginster, J.-Y., E. Abadie, P. Delmas, R. Rizzoli, W. Dere, P. van der Auwere, B. Avouac, M.-L. Brandi, A. Daifotis, A. Diez-Perez, G. Calvo, O. Johnell, J.-M. Kaufman, G. Kreutz, A. Laslop, F. Lekkerkerker, B. Mitlak, P. Nilsson, J. Orloff, M. Smillie, A. Taylor, Y. Tsouderos, D. Ethgen, and B. Flamion. 2006. Recommendations for an update of the current (2001) regulatory requirements for registration of drugs to be used in the treatment of osteoporosis in postmenopausal women and in men. *Osteoporosis International* 17: 1-7.

Rockwood, C.A., D.P. Green, R.W. Bucholz, and J.D. Heckman, eds. 1996. *Rockwood and Green's Fractures in Adults* (4th ed.). Philadelphia: Lippincott-Raven.

Salter, R.B. 1999. *Textbook of Disorders and Injuries of the Musculoskeletal System.* Baltimore: Williams & Wilkins.

Sanders, M., and Albright, J.A. 1987. Bone: Age-related changes and osteoporosis. In *The Scientific Basis of Orthopaedics* (2nd ed.), edited by J.A. Albright and R.A. Brand. Norwalk, CT: Appleton-Lange.

Sanders, M., and E.J. McCormick. 1993. *Human Factors in Engineering and Design* (7th ed.). New York: McGraw-Hill.

Sunderland, S. 1990. The anatomy and physiology of nerve injury. *Muscle Nerve* 13: 771-784.

Tanzer, M., and N. Noiseux. 2004. Osseous abnormalities and early osteoarthritis: The role of hip impingement. *Clinical Orthopaedics and Related Research* 429: 170-77.

Tidball, J.G. 1991. Myotendinous junction injury in relation to junction structure and molecular composition. In *Exercise and Sport Sciences Reviews,* edited by J.O. Holloszy. Baltimore: Williams & Wilkins.

Tidball, J.G., and M. Chan. 1989. Adhesive strength of single muscle cells to basement membrane at myotendinous junction. *Journal of Applied Physiology* 67: 1063-1069.

Tidball, J.G., G. Salem, and R.F. Zernicke. 1993. Site and mechanical conditions for failure of skeletal muscle in experimental strain injuries. *Journal of Applied Physiology* 74: 1280-1286.

Toni, A., S. Paderni, A. Sudanese, E. Guerra, F. Traina, F. Giardina, B. Antonietti, and A. Giunti. 2001. Anatomic cementless total hip arthroplasty with ceramic bearings and modular necks: 3 to 5 years follow-up. *Hip International* 11: 1-17.

Weissmann, G. 1992. Inflammation: Historical perspective. In *Inflammation: Basic Principles and Clinical Correlates* (2nd ed.), edited by J.I. Gallin, I.M. Goldstein, and R. Snyderman. New York: Raven Press.

Woo, S.L.-Y., K.-N. An, S.P. Arnoczky, J.S. Wayne, D.C. Fithian, and B.S. Myers. 1994. Anatomy, biology, and biomechanics of tendon, ligament, and meniscus. In *Orthopaedic Basic Science,* edited by J.A. Buckwalter, T.A. Einhorn, and S.R. Simon. Park Ridge, IL: American Academy of Orthopaedic Surgeons.

CHAPTER 6

Agel, J., E. Arendt, and B. Bershadsky. 2005. Anterior cruciate ligament injury in National Collegiate Athletic Association basketball and soccer. *American Journal of Sports Medicine* 33: 524-531.

Ahmed, A.M., and D.L. Burke. 1983. In-vitro measurement of static pressure distribution in synovial joints: Part I. Tibial surface of the knee. *Journal of Biomechanical Engineering* 105: 216-225.

Ahmed, A.M., D.L. Burke, and A. Yu. 1983. In-vitro measurement of static pressure distribution in synovial joints: Part II. Retropatellar surface. *Journal of Biomechanical Engineering* 105: 226-236.

Ahmed, I.M., M. Lagopoulos, P. McConnell, R.W. Soames, and G.K. Sefton. 1998. Blood supply of the Achilles tendon. *Journal of Orthopaedic Research* 16: 591-596.

Allen, C.R., G.A. Livesay, E.K. Wong, and S.L.-Y. Woo. 1999. Injury and reconstruction of the anterior cruciate ligament and knee osteoarthritis. *Osteoarthritis and Cartilage* 7: 110-121.

Allen, L.R., D. Flemming, and T.G. Sanders. 2004. Turf toe: Ligamentous injury of the first metatarsophalangeal joint. *Military Medicine* 169: xix-xxiv.

Almeida, S.A., K.M. Williams, R.A. Shaffer, and S.K. Brodine. 1999. Epidemiological patterns of musculoskeletal injuries and physical training. *Medicine and Science in Sports and Exercise* 31: 1176-1182.

American Academy of Orthopaedic Surgeons. 2005a. Falls and hip fractures. Available: http://orthoinfo.aaos.org/fact/thr_report.cfm?Thread_ID=77&topcategory=Hip.

American Academy of Orthopaedic Surgeons. 2005b. Available: http://orthoinfo.aaos.org/topic.cfm?topic=A00213.

Andrews, J.R., J.C. Edwards, and Y.E. Satterwhite. 1994. Isolated posterior cruciate ligament injuries. *Clinics in Sports Medicine* 13: 519-530.

Arendt, E., B. Bershadsky, and J. Agel. 2002. Periodicity of noncontact anterior cruciate ligament injuries during the menstrual cycle. *Journal of Gender-Specific Medicine* 5: 19-26.

Arendt, E., and R. Dick. 1995. Knee injury patterns among men and women in collegiate basketball and soccer. NCAA data and review of literature. *American Journal of Sports Medicine* 23: 694-701.

Arndt, A.N., P.V. Komi, G.P. Bruggemann, and J. Lukkariniemi. 1998. Individual muscle contributions to the in vivo Achilles tendon force. *Clinical Biomechanics* 13: 532-541.

Askling, C., J. Karlsson, and A. Thorstensson. 2003. Hamstring injury occurrence in elite soccer players after preseason training with eccentric overload. *Scandinavian Journal of Medicine and Science in Sports* 13: 244-250.

Baker, S.P. 1985. Fall injuries in the elderly. *Clinics in Geriatric Medicine* 1: 501-511.

Batt, M.E. 1995. Shin splints—a review of terminology. *Clinical Journal of Sports Medicine* 5: 53-57.

Beck, B.R. 1998. Tibial stress injuries. An aetiological review for the purposes of guiding management. *Sports Medicine* 26: 265-279.

Behnke, R.S. 2006. *Kinetic Anatomy* (2nd ed.). Champaign, IL: Human Kinetics.

Beiner, J.M., and P. Jokl. 2001. Muscle contusion injuries: current treatment options. *Journal of the American Academy of Orthopaedic Surgeons* 9: 227-237.

Beiner, J.M., and P.J. Jokl. 2002. Muscle contusion injury and myositis ossificans traumatica. *Clinical Orthopaedics and Related Research* 403S: S110-S119.

Bjordal, J., F. Arnly, B. Hannestad, and T. Strand. 1997. Epidemiology of anterior cruciate ligament injuries in soccer. *American Journal of Sports Medicine* 25: 341-345.

Boden, B., G. Dean, J. Feagin, and W. Garrett. 2000. Mechanisms of anterior cruciate ligament injury. *Orthopedics* 23: 573-578.

Bogey, R.A., J. Perry, and A.J. Gitter. 2005. An EMG-to-force processing approach for determining ankle muscle forces during normal human gait. *IEEE Transactions of Neural Systems and Rehabilitation Engineering* 13: 302-310.

Booth, D.W., and B.M. Westers. 1989. The management of athletes with myositis ossificans traumatica. *Canadian Journal of Sport Sciences* 14: 10-16.

Bradley, J., J. Klimkiewicz, M. Rytel, and J. Powell. 2002. Anterior cruciate ligament injuries in the National Football League: Epidemiology and current treatment trends among team physicians. *Arthroscopy* 18: 502-509.

Brechter, J.H., and C.M. Powers. 2002. Patellofemoral stress during walking in persons with and without patellofemoral pain. *Medicine and Science in Sports and Exercise* 34: 1582-1593.

Brechter, J.H., C.M. Powers, M.R. Terk, S.R. Ward, and T.Q. Lee. 2003. Quantification of patellofemoral joint contact area using magnetic resonance imaging. *Magnetic Resonance Imaging* 21: 955-959.

Brien, W.W., S.H. Kuschner, E.W. Brien, and D.A. Wiss. 1995. The management of gunshot wounds to the femur. *Orthopedic Clinics of North America* 26: 133-138.

Brockett, C.L., D.L. Morgan, and U. Proske. 2004. Predicting hamstring strain injury in elite athletes. *Medicine and Science in Sports and Exercise* 36: 379-387.

Burdett, R.G. 1982. Forces predicted at the ankle during running. *Medicine and Science in Sports and Exercise* 14: 308-316.

Butler, D.L., Y. Guan, M.D. Kay, J.F. Cummings, S.M. Feder, and M.S. Levy. 1992. Location-dependent variations in the material properties of the anterior cruciate ligament. *Journal of Biomechanics* 25: 511-518.

Carr, J.B. 2003. Malleolar fractures and soft tissue injuries of the ankle. In *Skeletal Trauma: Basic Science, Management, and Reconstruction* (3rd ed.), edited by B.D. Browner, J.B. Jupiter, A.M. Levine, and P.T. Trafton. Philadelphia: Saunders.

Cascio, B., L. Culp, and A. Cosgarea. 2004. Return to play after anterior cruciate ligament reconstruction. *Clinics in Sports Medicine* 23: 395-408.

Casparian, J.M., M. Luchi, R.E. Moffat, and D. Hinthorn. 2000. Quinolones and tendon ruptures. *Southern Medical Journal* 93: 488-491.

Church, J.S., and W.J.P. Radford. 2001. Isolated compartment syndrome of the tibialis anterior muscle. *Injury, International Journal of the Care of the Injured* 32: 170-171.

Ciullo, J.V., and J.D. Shapiro. 1994. Track and field. In *Sports Injuries: Mechanisms, Prevention, Treatment*, edited by F.H. Fu and D.A. Stone. Baltimore: Williams & Wilkins.

Cooper, C., H. Inskip, P. Croft, L. Campbell, G. Smith, M. McLaren, and D. Coggon. 1998. Individual risk factors for hip osteoarthritis: obesity, hip injury, and physical activity. *American Journal of Epidemiology* 147: 516-522.

Court-Brown, C., and J. McBirnie. 1995. The epidemiology of tibial fractures. *Journal of Bone and Joint Surgery* 77B: 417-421.

Crisco, J.J., P. Jokl, G.T. Heinen, M.D. Connell, and M.M. Panjabi. 1994. A muscle contusion injury model: Biomechanics, physiology, and histology. *American Journal of Sports Medicine* 22: 702-710.

Croisier, J.L., B. Forthomme, M.H. Namurois, M. Vanderthommen, and J.M. Crielaard. 2002. Hamstring muscle strain recurrence and strength performance disorders. *American Journal of Sports Medicine* 30: 199-203.

Cummings, S.R., J.L. Kelsey, M.C. Nevitt, and K.J. O'Dowd. 1985. Epidemiology of osteoporosis and osteoporotic fractures. *Epidemiologic Review* 7: 178-208.

Cummings, S.R., and L.J. Melton, III. 2002. Epidemiology and outcomes of osteoporotic fractures. *The Lancet* 359: 1761-1767.

Dandy, D.J. 2002. General observations on surgery of the anterior cruciate ligament. In *Oxford Textbook of Orthopedics and Trauma*, edited by C. Bulstrode, J. Buckwalter, A. Carr, L. Marsh, J. Fairbank, J. Wilson-MacDonald, and G. Bowden. Oxford, UK: Oxford University Press.

Davidson, C.W., M.J. Meriles, T.J. Wilkinson, J.S. McKie, and N.L. Gilchrist. 2001. Hip fracture mortality and morbidity—Can we do better? *New Zealand Medical Journal* 114: 329-332.

Dayton, P.D., and R.T. Bouche. 1994. Compartment syndromes. In *Musculoskeletal Disorders of the Lower Extremities*, edited by L.M. Oloff. Philadelphia: Saunders.

DeCoster, T.A., and D.R. Swenson. 2002. Femur shaft fractures. In *Oxford Textbook of Orthopedics and Trauma*, edited by C. Bulstrode, J. Buckwalter, A. Carr, L. Marsh, J. Fairbank, J. Wilson-MacDonald, and G. Bowden. Oxford, UK: Oxford University Press.

Demirag, B., C. Ozturk, Z. Yazici, and B. Sarisozen. 2004. The pathophysiology of Osgood-Schlatter disease: A magnetic resonance investigation. *Journal of Pediatric Orthopaedics B* 13: 379-382.

Diaz, J.A., D.A. Fischer, A.C. Rettig, T.J. Davis, and K.D. Shelbourne. 2003. Severe quadriceps muscle contusions in athletes. *American Journal of Sports Medicine* 31: 289-293.

Dienst, M., R.T. Burks, and P.E. Greis. 2002. Anatomy and biomechanics of the anterior cruciate ligament. *Orthopedic Clinics of North America* 33: 605-620.

Dragoo, J.L., R.S. Lee, P. Benhaim, G.A. Finerman, and S.L. Hame. 2003. Relaxin receptors in the human female anterior cruciate ligament. *American Journal of Sports Medicine* 31: 577-584.

Dye, S.F. 2004. Reflections on patellofemoral disorders. In *Patellofemoral Disorders: Diagnosis and Treatment*, edited by R.M. Biedert. West Sussex, UK: Wiley.

Eriksen, H.A., A. Pajala, J. Leppilhati, and J. Risteli. 2002. Increased content of type III collagen at the rupture site of human Achilles tendon. *Journal of Orthopaedic Research* 20: 1352-1357.

Etheridge, B.S., D.P. Beason, R.R. Lopez, J.E. Alonso, G. McGwin, and A.W. Eberhardt. 2005. Effects of trochanteric soft tissues and bone density on fracture of the female pelvis in experimental side impacts. *Annals of Biomedical Engineering* 33: 248-254.

Fanelli, G.C., and C.J. Edson. 2005. Posterior cruciate ligament injuries in trauma patients: Part II. *Arthroscopy* 11: 526-529.

Farrell, K.C., K.D. Reisinger, and M.D. Tillman. 2003. Force and repetition in cycling: Possible implications for iliotibial band friction syndrome. *The Knee* 10: 103-109.

Fredericson, M., A.G. Bergman, K.L. Hoffman, and M.S. Dillingham. 1995. Tibial stress reaction in runners: Correlation of clinical symptoms and scintigraphy with a new magnetic resonance imaging grading system. *American Journal of Sports Medicine* 23: 472-481.

Fukashiro, S., P.V. Komi, M. Jarvinen, and M. Miyashita. 1995. In vivo Achilles tendon loading during jumping in humans. *European Journal of Applied Physiology and Occupational Physiology* 71: 453-458.

Funsten, R.V., P. Kinser, and C.J. Frankel. 1938. Dashboard dislocation of the hip: A report of 20 cases of traumatic dislocations. *Journal of Bone and Joint Surgery* 20A: 124-132.

Garneti, N., C. Holton, and A. Shenolikar. 2005. Bilateral Achilles tendon rupture: A case report. *Accident and Emergency Nursing* 13: 220-223.

Garrett, W.E. 1995. Basic science of musculotendinous injuries. In *The Lower Extremity and Spine in Sports Medicine*, edited by J.A. Nicholas, J.A. and E.B. Hershman. St. Louis: Mosby-Year Book.

Garrett, W.E., F.R. Rich, P.K. Nikolaou, and J.B. Vogler. 1989. Computed tomography of hamstring muscle strains. *Medicine and Science in Sports and Exercise* 21: 506-514.

Garrett, W.E., Jr., M.R. Safran, A.V. Seaber, R.R. Glisson, and B.M. Ribbeck. 1987. Biomechanical comparison of stimulated and nonstimulated skeletal muscle pulled to failure. *American Journal of Sports Medicine* 15: 448-454.

Giddings, V.L., G.S. Beaupre, R.T. Whalen, and D.R. Carter. 2000. Calcaneal loading during walking and running. *Medicine and Science in Sports and Exercise* 32: 627-634.

Gindele, A., D. Schwamborn, K. Tsironis, and G. Benz-Bohm. 2000. Myositis ossificans traumatica in young children: report of three cases and review of the literature. *Pediatric Radiology* 30: 451-459.

Gokcen, E.C., A.R. Burgess, J.H. Siegel, S. Mason-Gonzalez, P.C. Dischinger, and S.M. Ho. 1994. Pelvic fracture mechanism of injury in vehicular trauma patients. *Journal of Trauma* 36: 789-796.

Goldblatt, J., S. Fitzsimmons, E. Balk, and J. Richmond. 2005. Reconstruction of the anterior cruciate ligament: Meta-analysis of patellar tendon versus hamstring tendon autograft. *Arthroscopy* 21: 791-803.

Grabiner, M.D., M.J. Pavol, and T.M. Owings. 2002. Can fall-related hip fractures be prevented by characterizing the biomechanical mechanisms of failed recovery? *Endocrine* 17: 15-20.

Griffin, L.Y., J. Agel, M.J. Albohm, E.A. Arendt, R.W. Dick, W.E. Garrett, J.G. Garrick, T.E. Hewett, L. Huston, M.L. Ireland, R.J. Johnson, W.B. Kibler, S. Lephart, J.L. Lewis, T.N. Lindenfeld, B.R. Mandelbaum, P. Marchak, C.C. Teitz, and E.M. Wojtys. 2000. Noncontact anterior cruciate ligament injuries: risk factors and prevention strategies. *Journal of the American Academy of Orthopaedic Surgeons* 8: 141-150.

Grood, E.S., F.R. Noyes, D.L. Butler, and W.J. Suntary. 1981. Ligamentous and capsular restraints preventing straight medial and lateral laxity in intact human cadaver knees. *Journal of Bone and Joint Surgery* 63A: 1257-1269.

Gulli, B., and D. Templeman. 1994. Compartment syndrome of the lower extremity. *Orthopedic Clinics of North America* 25: 677-684.

Hait, G., J.A. Boswick, and J.J. Stone. 1970. Heterotopic bone formation secondary to trauma (myositis ossificans traumatica). *Journal of Trauma* 10: 405-411.

Hayes, W.C., E.G. Myers, J.N. Morris, T.N. Gerhart, H.S. Yett, and L.A. Lipsitz. 1993. Impact near the hip dominates fracture risk in elderly nursing home residents who fall. *Calcified Tissue International* 52: 192-198.

Hewett, T., G. Myer, and K. Ford. 2005. Reducing knee and anterior cruciate ligament injuries among female athletes: A systematic review of neuromuscular training interventions. *Journal of Knee Surgery* 18: 82-88.

Hierton, C. 1983. Regional blood flow in experimental myositis ossificans. *Acta Orthopaedica Scandinavia* 54: 58-63.

Hirano, A., T. Fukubayashi, T. Ishii, and N. Ochiai. 2002. Magnetic resonance imaging of Osgood-Schlatter disease: The course of the disease. *Skeletal Radiology* 31: 334-342.

Hubbell, J.D., and E. Schwartz. 2005. Anterior cruciate ligament injury. Available: www.emedicine.com/sports/topic9.htm.

Huberti, H.H., and W.C. Hayes. 1984. Patellofemoral contact pressures: The influence of q-angle and tendofemoral contact. *Journal of Bone and Joint Surgery* 66A: 715-724.

Huberti, H.H., W.C. Hayes, J.L. Stone, and G.T. Shybut. 1984. Force ratios in the quadriceps tendon and ligamentum patellae. *Journal of Orthopaedic Research* 2: 49-54.

Hull, M.L., and C.D. Mote. 1980. Leg loading in snow skiing: Computer analyses. *Journal of Biomechanics* 13: 481-491.

Hungerford, D.S., and M. Barry. 1979. Biomechanics of the patellofemoral joint. *Clinical Orthopaedics and Related Research* 144: 9-15.

Hurschler, C., R. Vanderby, Jr., D.A. Martinez, A.C. Vailas, and W.D. Turnipseed. 1994. Mechanical and biochemical analyses of tibial compartment fascia in chronic compartment syndrome. *Annals of Biomedical Engineering* 22: 272-279.

Inman, V.T. 1976. *The Joints of the Ankle*. Baltimore: Williams & Wilkins.

Inoue, M., E. McGurk-Burleson, J.M. Hollis, and S.L.-Y. Woo. 1987. Treatment of the medial collateral ligament injury. *American Journal of Sports Medicine* 15: 15-21.

Ireland, M.L. 2002. The female ACL: Why is it more prone to injury? *Orthopedic Clinics of North America* 33: 637-651.

Ireland, M.L., and S.M. Ott. 2001. Special concerns of the female athlete. In *Sports Injuries: Mechanisms, Prevention, Treatment*, edited by F.H. Fu and D.A. Stone. Philadelphia: Lippincott Williams & Wilkins.

Jarvholm, B., R. Lundstrom, H. Malchau, B. Rehn, and E. Vingard. 2004. Osteoarthritis in the hip and whole-body vibration in heavy vehicles. *International Archives of Occupational and Environmental Health* 77: 424-426.

Jarvinen, T.A., P. Kannus, M. Paavola, T.L. Jarvinen, L. Jozsa, and M. Jarvinen. 2001. Achilles tendon injuries. *Current Opinions in Rheumatology* 13: 150-155.

Johnell, O., and J.A. Kanis. 2004. An estimate of the worldwide prevalence, mortality and disability associated with hip fracture. *Osteoporosis International* 15: 897-902.

Jozsa, L., J.B. Balint, P. Kannus, A. Reffy, and M. Barzo. 1989. Distribution of blood groups in patients with tendon rupture. *Journal of Bone and Joint Surgery* 71B: 272-274.

Kellersmann, R., T.R. Blattert, and A. Weckbach. 2005. Bilateral patellar tendon rupture without predisposing systemic disease or steroid use: A case report and review of the literature. *Archives of Orthopaedic Trauma and Surgery* 125: 127-133.

Khaund, R., and S.H. Flynn. 2005. Iliotibial band syndrome: a common source of knee pain. *American Family Physician* 71: 1545-1550.

Kibler, W.B., and T.J. Chandler. 1994. Racquet sports. In *Sports Injuries: Mechanisms, Prevention, Treatment*, edited by F.H. Fu and D.A. Stone. Baltimore: Williams & Wilkins.

Kibler, W.B., C. Goldberg, and T.J. Chandler. 1991. Functional biomechanical deficits in running athletes with plantar fasciitis. *American Journal of Sports Medicine* 19: 66-71.

King, J.B. 1998. Post-traumatic ectopic calcification in the muscles of athletes: A review. *British Journal of Sports Medicine* 32: 287-290.

Kirk, K.L., T. Kuklo, and W. Klemme. 2000. Iliotibial band friction syndrome. *Orthopedics* 23: 1209-1214.

Komi, P.V., S. Fukashiro, and M. Jarvinen. 1992. Biomechanical loading of Achilles tendon during normal locomotion. *Clinics in Sports Medicine* 11: 521-531.

Koulouris, G., and D. Connell. 2003. Evaluation of the hamstring muscle complex following acute injury. *Skeletal Radiology* 32: 582-589.

Kujala, U.M., M. Jarvinen, A. Natri, M. Lehto, O. Nelimarkka, M. Hurme, L. Virta, and J. Finne. 1992. ABO blood groups and musculoskeletal injuries. *Injury* 23: 131-133.

Kvist, M. 1994. Achilles tendon injuries in athletes. *Sports Medicine* 18: 173-201.

Lam, F., J. Walczak, and A. Franklin. 2001. Traumatic asymmetrical bilateral hip dislocation in an adult. *Emergency Medicine Journal* 18: 506-507.

Lauritzen, J.B. 1997. Hip fractures. Epidemiology, risk factors, falls, energy absorption, hip protectors, and prevention. *Danish Medical Bulletin* 44: 155-168.

Leadbetter, W.B. 2001. Soft tissue athletic injury. In *Sports Injuries: Mechanisms, Prevention, Treatment*, edited by F.H. Fu and D.A. Stone. Philadelphia: Lippincott Williams & Wilkins.

Levin, P.E., and B.D. Browner, B.D. 1991. Dislocations and fracture-dislocations of the hip. In *The Hip and Its Disorders*, edited by M.E. Steinberg. Philadelphia: Saunders.

Levy, A.S., J. Bromberg, and D. Jasper. 1994. Tibia fractures produced from the impact of a baseball bat. *Journal of Orthopaedic Trauma* 8: 154-158.

Lieber, R.L., and J. Friden. 2002. Mechanisms of muscle injury gleaned from animal models. *American Journal of Physical Medicine and Rehabilitation* 81(11 Suppl.): S70-S79.

Lievense, A.M., S.M.A. Bierma-Zeinstra, A.P. Verhagen, M.E. van Baar, J.A.N. Verhaar, and B.W. Koes. 2002. Influence of obesity on the development of osteoarthritis of the hip: A systematic review. *Rheumatology* 41: 1155-1162.

Linko, E., A. Harilainen, A. Malmivaara, and S. Seitsalo. 2005. Surgical versus conservative interventions for anterior cruciate ligament ruptures in adults. *The Cochrane Database of Systematic Reviews* 2: CD001356.

Lofman, O., K. Berglund,, L. Larsson, and G. Toss. 2002. Changes in hip fracture epidemiology: Redistribution between ages, genders and fracture types. *Osteoporosis International* 13: 18-25.

Long, W.T., W. Chang, and E.W. Brien. 2003. Grading system for gunshot injuries to the femoral diaphysis in civilians. *Clinical Orthopedics* 408: 92-100.

Maffulli, N., J.A. Reaper, S.W. Waterston, and T. Ahya. 2000. ABO blood groups and Achilles tendon rupture in the Grampian Region of Scotland. *Clinical Journal of Sports Medicine* 10: 269-271.

Maffulli, N., and J. Wong. 2003. Rupture of the Achilles and patellar tendons. *Clinics in Sports Medicine* 22: 761-776.

Magnusson, H.I., N.E. Westlin, F. Nyqvist, P. Gärdsell, E. Seeman, and M.K. Karlsson. 2001. Abnormally decreased regional bone density in athletes with medial tibial stress syndrome. *American Journal of Sports Medicine* 29: 712-715.

Mahan, K.T., and S.R. Carter. 1992. Multiple ruptures of the tendo Achillis. *Journal of Foot Surgery* 31: 548-559.

Mandelbaum, B.R., H.J. Silvers, D.S. Watanabe, J.F. Knarr, S.D. Thomas, L.Y. Griffin, D.T. Kirkendall, and W. Garrett. 2005. Effectiveness of a neuromuscular and proprioceptive training program in preventing anterior cruciate ligament injuries in female athletes: 2-year follow-up. *American Journal of Sports Medicine* 33: 1003-1010.

Markolf, K.L., D.M. Burchfield, M.M. Shapiro, M.F. Shepard, G.A. Finerman, and J.L. Slauterbeck. 1995. Combined knee loading states that generate high anterior cruciate ligament forces. *Journal of Orthopaedic Research* 13: 930-935.

Markolf, K.L., J.R. Slauterbeck, K.L. Armstrong, M.S. Shapiro, and G.A. Finerman. 1996. Effects of combined knee loadings on posterior cruciate ligament force generation. *Journal of Orthopaedic Research* 14: 633-638.

Markolf, K.L., J.R. Slauterbeck, K.L. Armstrong, M.S. Shapiro, and G.A. Finerman. 1997. A biomechanical study of replacement of the posterior cruciate ligament with a graft. Part II: Forces in the graft compared with forces in the intact ligament. *Journal of Bone and Joint Surgery* 79A: 381-386.

Marks, R., J.P. Allegrante, C.R. MacKenzie, and J.M. Lane. 2003. Hip fractures among the elderly: Causes, consequences and control. *Ageing Research Reviews* 2: 57-93.

Matsumoto, K., H. Sumi, Y. Sumi, and K. Shimizu. 2003. An analysis of hip dislocations among snowboarders and skiers: A 10-year prospective study from 1992 to 2002. *Journal of Trauma* 55: 946-948.

McLean, S., X. Huang, A. Su, A. Van den Bogert. 2004. Sagittal plane biomechanics cannot injure the ACL during sidestep cutting. *Clinical Biomechanics* 19: 828-838.

McLean, S., X. Huang, A. Su, A. Van den Bogert. 2005. Association between lower extremity posture at contact and peak knee valgus moment during sidestepping: Implications for ACL injury. *Clinical Biomechanics* 20: 863-870.

Merriam-Webster. 2005. *Merriam-Webster's Medical Desk Dictionary*. Springfield, MA: Merriam-Webster.

Michelson, J.D., A. Myers, R. Jinnah, Q. Cox, and M. Van Natta. 1995. Epidemiology of hip fractures among the elderly. Risk factors for fracture type. *Clinical Orthopaedics and Related Research* 311: 129-135.

Miyasaka, K.C., D.M. Daniel, M.L. Stone, and P. Hirshman. 1991. The incidence of knee ligament injuries in the general population. *American Journal of Knee Surgery* 4: 3-8.

Monma, H., and T. Sugita. 2001. Is the mechanism of traumatic posterior dislocation of the hip a brake pedal injury rather than a dashboard injury? *Injury, International Journal for Care of the Injured* 32: 221-222.

Moorman, C.T., III, R.F. Warren, E.B. Hershman, J.F. Crowe, H.G. Potter, R. Barnes, S.J. O'Brien, and J.H. Guettler. 2003. Traumatic posterior hip subluxation in American football. *Journal of Bone and Joint Surgery* 85A:1190-1196.

Mubarak, S.J. 1981. *Compartment Syndromes and Volkmann's Contracture*. Philadelphia: W.B. Saunders.

Myer, G., K. Ford, and T. Hewett. 2005. The effects of gender on quadriceps muscle activation strategies during a maneuver that mimics a high ACL injury risk position. *Journal of Electromyographical Kinesiology* 115: 181-189.

Myklebust, G., L. Engebretsen, I. Braekken, A. Skjolberg, O. Olsen, and R. Bahr. 2003. Prevention of anterior cruciate ligament injuries in female team handball players: A prospective intervention study over three seasons. *Clinical Journal of Sports Medicine* 13: 71-78.

Noonan, T.J., and W.E. Garrett. 1992. Injuries at the myotendinous junction. *Clinics in Sports Medicine* 11: 783-806.

Norman, A., and H.D. Dorfman. 1970. Juxtacortical circumscribed myositis ossificans: Evolution and radiographic features. *Radiology* 96: 304-306.

Nowinski, R.J., and C.T. Mehlman. 1998. Hyphenated history: Osgood-Schlatter disease. *American Journal of Orthopaedics* 27: 584-585.

O'Donoghue, D.H. 1984. *Treatment of Injuries to Athletes* (4th ed.). Philadelphia: Saunders.

Orchard, J.W. 2001. Intrinsic and extrinsic risk factors for muscle strains in Australian football. *American Journal of Sports Medicine* 29: 300-303.

Orthopaedic Trauma Association. 1996. Fracture and dislocation compendium. *Journal of Orthopaedic Trauma* 10: v-ix, 31-45.

Owings, T.M., M.J. Pavol, and M.D. Grabiner. 2001. Mechanisms of failed recovery following postural perturbations on a motorized treadmill mimic those associated with an actual forward trip. *Clinical Biomechanics* 16: 813-819.

Papadimitropoulos, E.A., P.C. Coyte, R.G. Josse, and C.E. Greenwood. 1997. Current and projected rates of hip fracture in Canada. *Canadian Medical Association Journal* 157: 1357-1363.

Paul, J., K. Spindler, J. Andrish, R. Parker, M. Secic, and J. Bergfeld. 2003. Jumping versus nonjumping anterior cruciate ligament injuries: A comparison of pathology. *Clinical Journal of Sports Medicine* 13: 1-5.

Pavol, M.J., T.M. Owings, K.T. Foley, and M.D. Grabiner. 1999. Gait characteristics as risk factors for falling from trips induced in older adults. *The Journals of Gerontology Series A Biological Sciences and Medical Sciences* 54: M583-M590.

Pavol, M.J., T.M. Owings, K.T. Foley, and M.D. Grabiner. 2001. Mechanisms leading to a fall from an induced trip in healthy older adults. *The Journals of Gerontology Series A Biological Sciences and Medical Sciences* 56: M428-M437.

Petersen, W., C. Braun, W. Bock, K. Schmidt, A. Weimann, W. Drescher, E. Eiling, R. Stange, T. Fuchs, J. Hedderich, and T. Zantop. 2005. A controlled prospective case control study of a prevention training program in female team handball players: The German experience. *Archives of Orthopaedic Trauma and Surgery* 125: 614-621.

Peterson, L., and P. Renström. 2001. *Sports Injuries: Their Prevention and Treatment*. Champaign, IL: Human Kinetics.

Piziali, R.L., J. Rastegar, D.A. Nagel, and D.J. Schurman. 1980. The contribution of the cruciate ligaments to the load-displacement characteristics of the human knee joint. *Journal of Biomechanical Engineering* 102: 277-283.

Pourcelot, P., M. Defontaine, B. Ravary, M. Lemâtre, and N. Crevier-Denoix. 2005. A non-invasive method of tendon force measurement. *Journal of Biomechanics* 38: 2124-2129.

Powers, C.M. 2003. The influence of altered lower-extremity kinematics on patellofemoral joint dysfunction: A theoretical approach. *Journal of Orthopaedic and Sports Physical Therapy* 33: 639-646.

Proske, U., D.L. Morgan, C.L. Brockett, and P. Percival. 2004. Identifying athletes at risk of hamstring strains and how to protect them. *Clinical Experiments in Pharmacology and Physiology* 31: 546-550.

Quarles, J.D., and R.G. Hosey. 2004. Medial and lateral collateral injuries: Prognosis and treatment. *Primary Care* 31: 957-975.

Robinovitch, S.N., E.T. Hsiao, R. Sandler, J. Cortez, Q. Liu, and G.D. Paiement. 2000. *Exercise and Sport Sciences Reviews* 28: 74-79.

Robinovitch, S.N., T.A. McMahon, and W.C. Hayes. 1995. Force attenuation in trochanteric soft tissues during impact from a fall. *Journal of Orthopaedic Research* 13: 956-962.

Rooser, B., S. Bengtson, and G. Hagglund. 1991. Acute compartment syndrome from anterior thigh muscle contusion: a report of eight cases. *Journal of Orthopaedic Trauma* 5: 57-59.

Rose, P.S., and F.J. Frassica. 2001. Atraumatic bilateral patellar tendon rupture, a case report and review of the literature. *Journal of Bone and Joint Surgery (American)* 83: 1382-1386.

Ross, M.D., and D. Villard. 2003. Disability levels of college-aged men with a history of Osgood-Schlatter disease. *Journal of Strength and Conditioning Research* 17: 659-663.

Rothwell, A.G. 1982. Quadriceps hematoma: A prospective clinical study. *Clinical Orthopaedics and Related Research* 171: 97-103.

Rubenstein, L.Z., and K.R. Josephson. 2002. The epidemiology of falls and syncope. *Clinics in Geriatric Medicine* 18: 141-158.

Safran, M.R., R.S. Benedetti, A.R. Bartolozzi, III, and B.R. Mandelbaum. 1999. Lateral ankle sprains: A comprehensive review. Part 1: Etiology, pathoanatomy, histopathogenesis, and diagnosis. *Medicine and Science in Sports and Exercise* 31: S429-S437.

Salem, G.J., and C.M. Powers. 2001. Patellofemoral joint kinetics during squatting in collegiate women athletes. *Clinical Biomechanics* 16: 424-430.

Salsich, G.B., S.R. Ward, M.R. Terk, and C.M. Powers. 2003. In vivo assessment of patellofemoral joint contact area in individuals who are pain free. *Clinical Orthopaedics and Related Research* 417: 277-284.

Sandler, R., and S. Robinovitch. 2001. An analysis of the effect of lower extremity strength on impact severity during a backward fall. *Journal of Biomechanical Engineering* 123: 590-598.

Sangeorzan, B.J. 1991. Subtalar joint: Morphology and functional anatomy. In *Inman's Joints of the Ankle* (2nd ed.), edited by J.B. Stiehl. Baltimore: Williams & Wilkins.

Schepsis, A.A., M. Fitzgerald, and R. Nicoletta. 2005. Revision surgery for exertional anterior compartment syndrome of the lower leg. *American Journal of Sports Medicine* 33: 1040-1047.

Schepsis, A.A., H. Jones, and A.L. Haas. 2002. Achilles tendon disorders in athletes. *American Journal of Sports Medicine* 30: 287-305.

Scott, S.H., and D.A. Winter. 1990. Internal forces at chronic running injury sites. *Medicine and Science in Sports and Exercise* 22: 357-369.

Seedhom, B.B., and V. Wright. 1974. Functions of the menisci: A preliminary study. *Journal of Bone and Joint Surgery* 56B: 381-382.

Seering, W.P., R.L. Piziali, D.A. Nagel, and D.J. Schurman. 1980. The function of the primary ligaments of the knee in varus-valgus and axial rotation. *Journal of Biomechanics* 13: 785-794.

Sibley, M.B., and F.H. Fu. 2001. Knee injuries. In *Sports Injuries: Mechanisms, Prevention, Treatment*, edited by F.H. Fu and D.A. Stone. Philadelphia: Lippincott Williams & Wilkins.

Siegler, S., J. Block, and C.D. Schneck. 1988. The mechanical characteristics of the collateral ligaments of the human ankle joint. *Foot and Ankle* 8: 234-242.

Siliski, J.M. 2003. Dislocations and soft tissue injuries of the knee. In *Skeletal Trauma: Basic Science, Management, and Reconstruction*, edited by B.D. Browner, J.B. Jupiter, A.M. Levine, and P.T. Trafton. Philadelphia: Saunders.

Speer, K.P., C.E. Spritzer, F.H. Bassett, J.A. Feagin, and W.E. Garrett. 1992. Osseous injury associated with acute tears of the anterior cruciate ligament. *American Journal of Sports Medicine* 20: 382-389.

Speer, K.P., R.F. Warren, T.L. Wickiewicz, L. Horowitz, and L. Henderson. 1995. Observations on the injury mechanism of anterior cruciate ligament tears in skiers. *American Journal of Sports Medicine* 23: 77-81.

Spilker, R.L., P.S. Donzelli, and V.C. Mow. 1992. A transversely isotropic biphasic finite element model of the meniscus. *Journal of Biomechanics* 25: 1027-1045.

Stanitski, C.L. 2002. Sports injuries in children and adolescents. In *Oxford Textbook of Orthopedics and Trauma*, edited by C. Bulstrode, J. Buckwalter, A. Carr, L. Marsh, J. Fairbank, J. Wilson-Mac-Donald, and G. Bowden. Oxford, UK: Oxford University Press.

Staubli, H.U., U. Durrenmatt, B. Porcellini, and W. Rauschning. 1999. Anatomy and surface geometry of the patellofemoral joint in the axial plane. *Journal of Bone and Joint Surgery* 81B: 452-458.

Stedman's Medical Dictionary for the Health Professions and Nursing (5th ed.). 2005. Philadelphia: Lippincott Williams & Wilkins.

Sterett, W.I., and W.B. Krissoff. 1994. Femur fractures in alpine skiing: Classification and mechanisms of injury in 85 cases. *Journal of Orthopaedic Trauma* 8: 310-314.

Stevenson, H., J. Webster, R. Johnson, and B. Beynnon. 1998. Gender differences in knee injury epidemiology among competitive alpine ski racers. *Iowa Orthopaedic Journal* 18: 64-66.

St-Onge, N., Y. Chevalier, N. Hagemeister, M. Van De Putte, and J. De Guise. 2004. Effect of ski binding parameters on knee biomechanics: a three-dimensional computational study. *Medicine and Science in Sports and Exercise* 36: 1218-1225.

Swenson, T.M., and C.D. Harner. 1995. Knee ligament and meniscal injuries: Current concepts. *Orthopaedic Clinics of North America* 26: 529-546.

Tearse, D., J.A. Buckwalter, J.L. Marsh, and E.A. Brandser. 2002. Stress fractures. In *Oxford Textbook of Orthopedics and Trauma*, edited by C. Bulstrode, J. Buckwalter, A. Carr, L. Marsh, J. Fairbank, J. Wilson-MacDonald, and G. Bowden. Oxford, UK: Oxford University Press.

Thelen, D.G., E.S. Chumanov, D.M. Hoerth, T.M. Best, S.C. Swanson, L. Li, M. Young, and B.C. Heiderscheit. 2005. Hamstring muscle kinematics during treadmill sprinting. *Medicine and Science in Sports and Exercise* 37: 108-114.

Thomee, R., P. Renstrom, J. Karlsson, and G. Grimby. 1995. Patellofemoral pain syndrome in young women. I. A clinical analysis of alignment, pain parameters, common symptoms and functional activity level. *Scandinavian Journal of Medicine and Science in Sports* 5: 237-244.

Thompson, W.O., F.L. Thaete, F.H. Fu, and S.F. Dye. 1991. Tibial meniscal dynamics using three-dimensional reconstruction of magnetic resonance images. *American Journal of Sports Medicine* 19: 210-215.

Upadhyay, S.S., A. Moulton, and R.G. Burwell. 1985. Biological factors predisposing to traumatic posterior dislocation of the hip. *Journal of Bone and Joint Surgery* 67B: 232-236.

Urabe, Y., M. Ochi, K. Onari, and Y. Ikuta. 2002. Anterior cruciate ligament injury in recreational alpine skiers: analysis of mechanisms and strategy for prevention. *Journal of Orthopaedic Science* 7: 1-5.

van den Bogert, A.J., M.J. Pavol, and M.D. Grabiner. 2002. Response time is more important than walking speed for the ability of older adults to avoid a fall after a trip. *Journal of Biomechanics* 35: 199-205.

van den Kroonenberg, A.J., W.C. Hayes, and T.A. McMahon. 1995. Dynamic models for sideways falls from standing height. *Journal of Biomechanical Engineering* 117: 309-318.

Vanek, D., Saxena, A., and J.M. Boggs. 2003. Fluoroquinolone therapy and Achilles tendon rupture. *Journal of the American Podiatric Medical Association* 93: 333-335.

Venes, D., ed. 2005. *Taber's Cyclopedic Medical Dictionary*. Philadelphia: Davis.

Verrall, G.M., J.P. Slavotinek, and P.G. Barnes. 2005. The effect of sports specific training on reducing the incidence of hamstring injuries in professional Australian Rules football players. *British Journal of Sports Medicine* 39: 363-368.

Verall, G.M., J.P. Slavotinek, P.G. Barnes, and G.T. Fon. 2003. Diagnostic and prognostic value of clinical findings in 83 athletes with posterior thigh injury: Comparison of clinical findings with magnetic resonance imaging documentation of hamstring muscle strain. *American Journal of Sports Medicine* 31: 969-973.

Verrall, G.M., J.P. Slavotinek, P.G. Barnes, G.T. Fon, and A.J. Spriggins. 2001. Clinical risk factors for hamstring muscle strain injury: A prospective study with correlation of injury by magnetic resonance imaging. *British Journal of Sports Medicine* 35: 435-439.

Verzijl, N., R.A. Bank, J.M. TeKoppele, and J. DeGroot. 2003. Ageing and osteoarthritis: A different perspective. *Current Opinions in Rheumatology* 15: 616-622.

Verzijl, N., J. DeGroot, Z.C. Ben, O. Brau-Benjamin, A. Maroudas, R.A. Bank, J. Mizrahi, C.G. Schalkwijk, S.R. Thorpe, J.W. Baynes, J.W. Bijlsma, F.P. Lafeber, and J.M. TeKoppele. 2002. Crosslinking by advanced glycation end products increases the stiffness of the collagen network in human articular cartilage: A possible mechanism through which age is a risk factor for osteoarthritis. *Arthritis and Rheumatology* 46: 114-123.

Vuori, J.-P., and H.T. Aro. 1993. Lisfranc joint injuries: Trauma mechanisms and associated injuries. *Journal of Trauma* 35: 40-45.

Walker, P., and M. Erkman. 1975. The role of the menisci in force transmission across the knee. *Clinical Orthopaedics* 109: 184-192.

Ward, S.R., and C.M. Powers. 2004. The influence of patella alta on patellofemoral joint stress during normal and fast walking. *Clinical Biomechanics* 19: 1040-1047.

Watson, J.T. 2002. Tibial shaft fractures. In *Oxford Textbook of Orthopedics and Trauma* (Vol. 3), edited by C. Bulstrode, J. Buckwalter, A. Carr, L. Marsh, J. Fairbank, J. Wilson-MacDonald, and G. Bowden. Oxford, UK: Oxford University Press.

Wind, W.M., Jr., J.A. Bergfeld, and R.D. Parker. 2004. Evaluation and treatment of posterior cruciate ligament injuries: revisited. *American Journal of Sports Medicine* 32: 1765-1775.

Winquist, R.A., and S.T. Hansen, Jr. 1980. Comminuted fractures of the femoral shaft treated by intramedullary nailing. *Orthopedic Clinics of North America* 11: 633-648.

Winquist, R.A., S.T. Hansen, Jr., and D.K. Clawson. 1984. Closed intramedullary nailing of femoral fractures: A report of five hundred and twenty cases. *Journal of Bone and Joint Surgery* 66A: 529-539.

Wiss, D.A., W.W. Brien, and V. Becker, Jr. 1991. Interlocking nailing for the treatment of femoral fractures due to gunshot wounds. *Journal of Bone and Joint Surgery* 73A: 598-606.

Wojtys, E.M., L.J. Huston, M.D. Boynton, K.P. Spindler, and T.N. Lindenfeld. 2002. The effect of the menstrual cycle on anterior cruciate ligament injuries in women as determined by hormone levels. *American Journal of Sports Medicine* 30: 182-188.

Woo, S.L., J.M. Hollis, D.J. Adams, R.M. Lyon, and S. Takai. 1991. Tensile properties of the human femur-anterior cruciate ligament-tibia complex. The effects of specimen age and orientation. *American Journal of Sports Medicine* 19: 217-225.

Woo, S. L.-Y., M.A. Knaub, and M. Apreleva. 2001. Biomechanics of ligaments in sports medicine. In *Sports Injuries: Mechanisms, Prevention, Treatment*, edited by F.H. Fu and D.A. Stone. Philadelphia: Lippincott Williams & Wilkins.

Worrell, T.W. 1994. Factors associated with hamstring injuries: An approach to treatment and preventative measures. *Sports Medicine* 17: 338-345.

Yates, B., M.J. Allen, and M.R. Barnes. 2003. Outcome of surgical treatment of medial tibial stress syndrome. *Journal of Bone and Joint Surgery* 85A: 1974-1980.

Yates, B., and S. White. 2004. The incidence and risk factors in the development of medial tibial stress syndrome among naval recruits. *American Journal of Sports Medicine* 32: 772-780.

Yinger, K., B.R. Mandelbaum, and L.C. Almekinders. 2002. Achilles rupture in the athlete: Current science and treatment. *Clinics in Podiatric Medicine and Surgery* 19: 231-250.

Zantop, T., B. Tillmann, and W. Petersen. 2003. Quantitative assessment of blood vessels of the human Achilles tendon: An immunohistochemical cadaver study. *Archives of Orthopaedic Trauma Surgery* 123: 501-504.

Zernicke, R.F. 1981. Biomechanical evaluation of bilateral tibial spiral fractures during skiing: A case study. *Medicine and Science in Sports and Exercise* 13: 243-245.

Zernicke, R.F., J. Garhammer, and F.W. Jobe. 1977. Human patellar-tendon rupture: A kinetic analysis. *Journal of Bone and Joint Surgery* 59A: 179-183.

CHAPTER 7

Ahmad, C.S., and N.S. ElAttrache. 2004. Valgus extension overload syndrome and stress injury of the olecranon. *Clinics in Sports Medicine* 23: 665-676.

Allman, F.L., Jr. 1967. Fractures and ligamentous injuries of the clavicle and its articulation. *Journal of Bone and Joint Surgery* 49A: 774-784.

American Academy of Orthopaedic Surgeons. 2007. Number of physician visits for rotator cuff problems. Available: http://www.aaos.org/Research/stats/Rotator%20Cuff.pdf.

Andrews, J.R., W.G. Carson, Jr., and W.D. McLeod. 1985. Glenoid labrum tears related to the long head of the biceps. *American Journal of Sports Medicine* 13: 337-341.

Bado, J.L. 1967. The Monteggia lesion. *Clinical Orthopaedics and Related Research* 50: 71-86.

Banas, M.P., R.J. Miller, and S. Totterman. 1995. Relationship between the lateral acromion angle and rotator cuff disease. *Journal of Shoulder and Elbow Surgery* 4: 454-461.

Bankart, A.S.B. 1923. Recurrent or habitual dislocation of the shoulder joint. *British Medical Journal* 2: 1132-1133.

Bartolozzi, A., D. Andreychik, and S. Ahmad. 1994. Determinants of outcome in the treatment of rotator cuff disease. *Clinical Orthopaedics and Related Research* 308: 90-97.

Bigliani, L.U., and W.N. Levine. 1997. Subacromial impingement syndrome. *Journal of Bone and Joint Surgery* 79A: 1854-1868.

Bigliani, L.U., D.S. Morrison, and E.W. April. 1986. The morphology of the acromion and its relationship to rotator cuff tears. *Orthopaedic Transactions* 10: 228.

Blevins, F.T. 1997. Rotator cuff pathology in athletes. *Sports Medicine* 24: 205-220.

Botte, M.J., and R.H. Gelberman. 1987. Fractures of the carpus, excluding the scaphoid. *Hand Clinics of North America* 3: 149-161.

Branch, T., C. Partin, P. Chamberland, E. Emeterio, and M. Sabetelle. 1992. Spontaneous fractures of the humerus during pitching: A series of 12 cases. *American Journal of Sports Medicine* 20: 468-470.

Bright, A.S., B. Torpey, D. Magid, T. Codd, and E.G. McFarland. 1997. Reliability of radiographic evaluation for acromial morphology. *Skeletal Radiology* 26: 718-721.

Browner, B.D., J.B. Jupiter, A.M. Levine, and P.G. Trafton. 2003. *Skeletal Trauma* (3rd ed.). Philadelphia: Saunders.

Budoff, J.E., R.P. Nirschl, O.A. Ilahi, and D.M. Rodin. 2003. Internal impingement in the etiology of rotator cuff tendinosis revisited. *Arthroscopy* 19: 810-814.

Buess, E., K.U. Steuber, and B. Waibl. 2005. Open versus arthroscopic rotator cuff repair: A comparative view of 96 cases. *Arthroscopy* 21: 597-604.

Burkhart, S.S. 1993. Arthroscopic debridement and decompression for selected rotator cuff tears. Clinical results, pathomechanics, and patient selection based on biomechanical parameters. *Orthopedic Clinics of North America* 24: 111-123.

Burkhart, S.S. 2000. A stepwise approach to arthroscopic rotator cuff repair based on biomechanical principles. *Arthroscopy* 16: 82-90.

Burkhart, S.S., and C.D. Morgan. 1998. The peelback mechanism: Its role in producing and extending posterior type II SLAP lesions and its effect on SLAP repair rehabilitation. *Arthroscopy* 14: 637-640.

Burkhart, S.S., and C. Morgan. 2001. SLAP lesions in the overhead athlete. *Orthopedic Clinics of North America* 32: 431-441.

Burkhart, S.S., C.D. Morgan, and W.B. Kibler. 2000. Shoulder injuries in overhead athletes: The "dead arm" revisited. *Clinics in Sports Medicine* 19: 125-158.

Burnham, R.S., L. May, E. Nelson, R. Steadward, and D.C. Reid. 1993. Shoulder pain in wheelchair athletes: The role of muscle imbalance. *American Journal of Sports Medicine* 21: 238-242.

Buss, D.D., and J.D. Watts. 2003. Acromioclavicular injuries in the throwing athlete. *Clinics in Sports Medicine* 22: 327-341.

Caine, D., W. Howe, W. Ross, and G. Bergman. 1997. Does repetitive physical loading inhibit radial growth in female gymnasts? *Clinical Journal of Sport Medicine* 7: 302-308.

Callaghan, E.B., D.L. Bennett, G.Y. El-Khoury, and K. Ohashi. 2004. Ball-thrower's fracture of the humerus. *Skeletal Radiology* 33: 355-358.

Chen, C.H., K.Y. Hsu, W.J. Chen, and C.H. Shih. 2005. Incidence and severity of biceps long head tendon lesion in patients with complete rotator cuff tears. *Journal of Trauma* 58: 1189-1193.

Ciccotti, M.G., and W.P.H. Charlton. 2001. Epicondylitis in the athlete. *Clinics in Sports Medicine* 20: 77-93.

Clasper, J. 2002. Frozen shoulder. In *Oxford Textbook of Orthopedics and Trauma*, edited by C. Bulstrode, J. Buckwalter, A. Carr, L. Marsh, J. Fairbank, J. Wilson-MacDonald, and G. Bowden. Oxford, UK: Oxford University Press.

Codman, E.A. 1911. Complete rupture of the supraspinatus tendon: Operative treatment with report of two successful cases. *Boston Medical and Surgical Journal* 164: 708-710.

Codman, E.A. 1934. Tendinitis of the short rotators. In *Ruptures of the Supraspinatus Tendon and Other Lesions on or About the Subacromial Bursa*, edited by E.A. Codman. Boston: Thomas Todd.

Cohen, M.S. 2004. Fractures of the coronoid process. *Hand Clinics* 20: 443-453.

Conway, J.E., F.W. Jobe, R.E. Glousman, and M. Pink. 1992. Medial instability of the elbow in throwing athletes: Treatment by repair or reconstruction of the ulnar collateral ligament. *Journal of Bone and Joint Surgery* 74A: 67-83.

Cooper, D.E., S.P. Arnoczky, S.J. O'Brien, R.F. Warren, E. DeCarlo, and A.A. Allen. 1992. Anatomy, histology, and vascularity of the glenoid labrum: An anatomical study. *Journal of Bone and Joint Surgery* 74A: 46-52.

Cresswell, T.R., and R.B. Smith. 1998. Bilateral anterior shoulder dislocations in bench pressing: an unusual cause. *British Journal of Sports Medicine* 32: 71-72.

De Smet, L. 1994. Ulnar variance: Facts and fiction. Review article. *Acta Orthopaedica Belgica* 60: 1-9.

De Smet, L., A. Clasessens, J. Lefevre, and G. Beunen. 1994. Gymnast wrist: An epidemiologic survey of ulnar variance and stress changes of the radial physis in elite female gymnasts. *American Journal of Sports Medicine* 22: 846-850.

Dias, J. 2002. Scaphoid fractures. In *Oxford Textbook of Orthopedics and Trauma*, edited by C. Bulstrode, J. Buckwalter, A. Carr, L. Marsh, J. Fairbank, J. Wilson-MacDonald, and G. Bowden. Oxford, UK: Oxford University Press.

Dias, R., S. Cutts, and S. Massoud. 2005. Frozen shoulder. *British Medical Journal* 331: 1453-1456.

DiFiori, J.P., J.C. Puffer, B. Aish, and F. Dorey. 2002. Wrist pain, distal radial physeal injury, and ulnar variance in young gymnasts: Does a relationship exist? *American Journal of Sports Medicine* 30: 879-885.

DiFiori, J.P., J.C. Puffer, B.R. Mandelbaum, and F. Dorey. 1997. Distal radial growth plate injury and positive ulnar variance in nonelite gymnasts. *American Journal of Sports Medicine* 25: 763-768.

DiFiori, J.P., J.C. Puffer, B.R. Mandelbaum, and S. Mar. 1996. Factors associated with wrist pain in the young gymnast. *American Journal of Sports Medicine* 24: 9-14.

Duplay, E.S. 1872. De la periarthritis scapulohumerale et des raiderus de l'epaule qui en son la consequence. *Archives of General Medicine* 20: 513-542.

Edelson, G., and C. Teitz. 2000. Internal impingement in the shoulder. *Journal of Shoulder and Elbow Surgery* 9: 308-315.

Emond, M., N. Le Sage, A. Lavoie, and L. Rochette. 2004. Clinical factors predicting fractures associated with an anterior shoulder dislocation. *Academy of Emergency Medicine* 11: 853-858.

Evans, M. 1949. Pronation injuries of the forearm. *Journal of Bone and Joint Surgery* 31B: 578-588.

Farley, T.E., C.H. Neumann, L.S. Steinbach, and S.A. Petersen. 1994. The coracoacromial arch: MR evaluation and correlation with rotator cuff pathology. *Skeletal Radiology* 23: 641-645.

Fleisig, G.S., J.R. Andrews, C.J. Dillman, and R.F. Escamilla. 1995. Kinetics of baseball pitching with implications about injury mechanisms. *American Journal of Sports Medicine* 23: 233-239.

Foulk, D.A., M.P. Darmelio, A.C. Rettig, and G. Misamore. 2002. Full-thickness rotator-cuff tears in professional football players. *American Journal of Orthopedics* 31: 622-624.

Fu, F.H., C.D. Harner, and A.H. Klein. 1991. Shoulder impingement syndrome: A critical review. *Clinical Orthopaedics and Related Research* 269: 162-173.

Fu, F.H., and D.A. Stone. 2001. *Sports Injuries: Mechanisms, Prevention, Treatment* (2nd ed.). Philadelphia: Lippincott Williams & Wilkins.

Funakoshi, T., T. Majima, N. Iwasaki, N. Suenaga, N. Sawaguchi, K. Shimode, A. Minami, K. Harada, and S. Nishimura. 2005. Application of tissue engineering techniques for rotator cuff regeneration using a chitosan-based hyaluronan hybrid fiber scaffold. *American Journal of Sports Medicine* 33: 1193-1201.

Galeazzi, R. 1934. Uber ein besonderes syndrom bei verltzunger im bereich der unterarmknochen. *Archiv Fur Orthopadische und Unfall-Chirurgie* 35: 557-562.

Giangarra, C.E., B. Conroy, F.W. Jobe, M. Pink, and J. Perry. 1993. Electromyographic and cinematographic analysis of elbow function in tennis players using single- and double-handed backhand strokes. *American Journal of Sports Medicine* 21: 394-399.

Giaroli, E.L., N.M. Major, and L.D. Higgins. 2005. MRI of internal impingement of the shoulder. *American Journal of Roentgenology* 185: 925-929.

Goldberg, B.A., R.J. Nowinski, F.A. Matsen, III. 2001. Outcome of nonoperative management of full-thickness rotator cuff tears. *Clinical Orthopaedics and Related Research* 382: 99-107.

Grana, W. 2001. Medial epicondylitis and cubital tunnel syndrome in the throwing athlete. *Clinics in Sports Medicine* 20: 541-548.

Guntern. D.V., C.W. Pfirrmann, M.R. Schmid, M. Zanetti, C.A. Binkert, A.G. Schneeberger, and J. Hodler. 2003. Articular cartilage lesions of the glenohumeral joint: Diagnostic effectiveness of MR arthrography and prevalence in patients with subacromial impingement syndrome. *Radiology* 226: 165-170.

Hafner, R.A., K. Poznanski, and J.M. Donovan. 1989. Ulnar variance in children—standard measurements for evaluation of ulnar shortening in juvenile rheumatoid arthritis, hereditary multiple exostosis and other bone or joint disorders in childhood. *Skeletal Radiology* 18: 513-516.

Halder, A.M., K.D. Zhao, S.W. Odriscoll, B.F. Morrey, and K.N. An. 2001. Dynamic contributions to superior shoulder stability. *Journal of Orthopedic Research* 19: 206-212.

Handelberg, F., S. Willems, M. Shahabpour, J.-P. Huskin, and J. Kuta. 1998. SLAP lesions: A retrospective multicenter study. *Arthroscopy* 14: 856-862.

Harryman, D.T., II, J.A. Sidles, J.M. Clark, K.J. McQuade, T.D. Gibb, and F.A. Matsen, III. 1990. Translation of the humeral head on the glenoid with passive glenohumeral motion. *Journal of Bone and Joint Surgery* 72A: 1334-1343.

Hawkins, R.H., and R. Dunlop. 1995. Nonoperative treatment of rotator cuff tears. *Clinical Orthopedics* 321: 178-188.

Haygood, T.M., C.P. Langlotz, J.B. Kneeland, J.P. Iannotti, G.R. Williams, Jr., and M.K. Dalinka. 1994. Categorization of acromial shape: Interobserver variability with MR imaging and conventional radiography. *American Journal of Roentgenology* 162: 1377-1382.

Healy, J.H., S. Barton, P. Noble, H.W. Kohl, III, and O.A. Ilahi. 2001. Biomechanical evaluation of the origin of the long head of the biceps tendon. *Journal of Arthroscopic and Related Surgery* 17: 378-382.

Horii, E., R. Nakamura, K. Watanabe, and K. Tsunoda. 1994. Scaphoid fracture as a "puncher's" fracture. *Journal of Orthopaedic Trauma* 8: 107-110.

Horvath, F., and L. Kery. 1984. Degenerative deformations of the acromioclavicular joint in elderly. *Archives of Gerontology and Geriatrics* 3: 259-265.

Hotchkiss, R.N. 1996. Fractures and dislocations of the elbow. In: *Rockwood and Green's Fractures in Adults*, edited by C.A. Rockwood, D.P. Green, R.W. Bucholz, and J.D. Heckman. Philadelphia: Lippincott-Raven.

Hotchkiss, R.N. 2000. Epicondylitis—lateral and medial. *Hand Clinics* 16: 505-508.

Hutchinson, M.R., and M.A. Veenstra. 1993. Arthroscopic decompression of shoulder impingement secondary to os acromiale. *Arthroscopy* 9: 28-32.

Itoi, E., and S. Tabata. 1992. Conservative treatment of rotator cuff tears. *Clinical Orthopaedics and Related Research* 275: 165-173.

Jacobson, S.R., K.P. Speer, J.T. Moor, D.H. Janda, S.R. Saddemi, P.B. MacDonald, and W.J. Mallon. 1995. Reliability of radiographic assessment of acromial morphology. *Journal of Shoulder and Elbow Surgery* 4: 449-453.

Jobe, C.M. 1995. Posterior superior glenoid impingement: expanded spectrum. *Arthroscopy* 11: 530-536.

Jobe, C.M. 1997. Superior glenoid impingement. *Orthopedic Clinics of North America* 28: 137-143.

Jobe, F.W., and M. Pink. 1993. Classification and treatment of shoulder dysfunction in the overhead athlete. *Journal of Orthopaedic and Sports Physical Therapy* 18: 427-432.

Jobe, F.W., H. Stark, and S.J. Lombardo. 1986. Reconstruction of the ulnar collateral ligament in athletes. *Journal of Bone and Joint Surgery* 68A: 1158-1163.

Jupiter, J.B., and D.L. Fernandez. 1996. *Fractures of the Distal Radius.* New York: Springer-Verlag.

Jupiter, J.B., and J.F. Kellam. 2003. Diaphyseal fractures of the forearm. In *Skeletal Trauma* (3rd ed.), edited by B.D. Browner, J.B. Jupiter, A.M. Levine, and P.G. Trafton. Philadelphia: Saunders.

Jupiter, J.B., S.J. Leibovic, W. Ribbans, and R.M. Wilk. 1991. Posterior Monteggia lesion. *Journal of Orthopaedic Trauma* 5: 395-402.

Kannus, P., and L. Jozsa. 1991. Histopathologic changes preceding spontaneous rupture of a tendon. *Journal of Bone and Joint Surgery* 73A: 1517-1525.

Kaplan, H., A. Kiral, M. Kuskucu, M.O. Arpacioglu, A. Sarioslu, and O. Rodop. 1998. Report of eight cases of humeral fracture following the throwing of hand grenades. *Archives of Orthopaedic Trauma and Surgery* 117: 50-52.

Kibler, W.B. 1995. Pathophysiology of overload injuries around the elbow. *Clinics in Sports Medicine* 14: 447-457.

Kim, T.K., W.S. Queale, A.J. Cosgarea, and E.G. McFarland. 2003. Clinical features of the different types of SLAP lesions. *Journal of Bone and Joint Surgery* 85A: 66-71.

Koh, T.J., M.D. Grabiner, and G.G. Weiker. 1992. Technique and ground reaction forces in the back handspring. *American Journal of Sports Medicine* 20: 61-66.

Kraushaar, B.S., and R.P. Nirschl. 1999. Tendinosis of the elbow (tennis elbow). Clinical features and findings of histological, immunohistochemical, and electron microscopy studies. *Journal of Bone and Joint Surgery* 81A: 259-278.

Kristensen, S.S., E. Thomassen, and F. Christensen. 1986. Ulnar variance determination. *Journal of Hand Surgery* 11B: 255-257.

Labriola, J.E., T.Q. Lee, R.E. Debski, and P.J. McMahon. 2005. Stability and instability of the glenohumeral joint: The role of shoulder muscles. *Journal of Shoulder and Elbow Surgery* 14(Suppl.): 32S-38S.

Lambert, S.M., and R. Hertel. 2002. Dislocations about the shoulder girdle, scapular fractures, and clavicle fractures. In *Oxford Textbook of Orthopedics and Trauma*, edited by C. Bulstrode, J. Buckwalter, A. Carr, L. Marsh, J. Fairbank, J. Wilson-MacDonald, and G. Bowden. Oxford, UK: Oxford University Press.

Leach, R.E., and J.K. Miller. 1987. Lateral and medial epicondylitis of the elbow. *Clinics in Sports Medicine* 6: 259-272.

Lehman, C., F. Cuomo, F.J. Kummer, and J.D. Zuckerman. 1995. The incidence of full thickness rotator cuff tears in a large cadaveric population. *Bulletin of the Hospital for Joint Diseases* 54: 30-31.

Lippitt, S., and F. Matsen. 1993. Mechanisms of glenohumeral joint stability. *Clinical Orthopaedics and Related Research* 291: 20-28.

Lo, S.L., K. Raskin, H. Lester, and B. Lester. 2002. Carpal tunnel syndrome: A historical perspective. *Hand Clinics* 18: 211-217.

Maffet, M.W., G.M. Gartsman, and B. Moseley. 1995. Superior labrum-biceps tendon complex lesions of the shoulder. *American Journal of Sports Medicine* 23: 93-98.

Malcarney, H.L., and G.A.C. Murrell. 2003. The rotator cuff: biological adaptations to its environment. *Sports Medicine* 33: 993-1002.

Mantone, J.K., W.Z. Burkhead, Jr., and J. Noonan, Jr. 2000. Nonoperative treatment of rotator cuff tears. *Orthopedic Clinics of North America* 31: 295-311.

Markolf, K.L., M.S. Shapiro, B.R. Mandelbaum, and L. Teurlings. 1990. Wrist loading patterns during pommel horse exercises. *Journal of Biomechanics* 23: 1001-1011.

McFarland, E.G., C.Y. Hsu, C. Neira, and O. O'Neil. 1999. Internal impingement of the shoulder: A clinical and arthroscopic analysis. *Journal of Shoulder and Elbow Surgery* 8: 458-460.

Meeuwisse, W.H. 1994. Assessing causation in sport injury: A multifactorial model. *Clinical Journal of Sports Medicine* 4: 166-170.

Mengiardi, B., C.W.A. Pfirrmann, C. Gerber, J. Hodler, and M. Zanetti. 2004. Frozen shoulder: MR arthrographic findings. *Radiology* 233: 486-492.

Michener, L.A., P.W. McClure, and A.R. Karduna. 2003. Anatomical and biomechanical mechanisms of subacromial impingement syndrome. *Clinical Biomechanics* 18: 369-379.

Milgrom, C., M. Schaffler, S. Gilbert, and M. van Holsbeeck. 1995. Rotator-cuff changes in asymptomatic adults. The effect of age, hand dominance and gender. *Journal of Bone and Joint Surgery* 77B: 296-298.

Millstein, E.S., and S.J. Snyder. 2003. Arthroscopic management of partial, full-thickness, and complex rotator cuff tears: Indications, techniques, and complications. *Arthroscopy* 19: 189-199.

Morgan, C.D., S.S. Burkhart, M. Palmeri, and M. Gillespie. 1998. Type II SLAP lesions: Three subtypes and their relationships to superior

instability and rotator cuff tears. *Arthroscopy* 14: 553-565.

Morris, M., F.W. Jobe, J. Perry, M. Pink, and B.S. Healy. 1989. Electromyographic analysis of elbow function in tennis players. *American Journal of Sports Medicine* 17: 241-247.

Nakamura, R., Y. Tanaka, T. Imaeda, and T. Miura. 1991. The influence of age and sex on ulnar variance. *Journal of Hand Surgery* 16B: 84-88.

Nam, E.K., and S.J. Snyder. 2003. The diagnosis and treatment of superior labrum, anterior and posterior (SLAP) lesions. *American Journal of Sports Medicine* 31: 798-810.

National Safety Council. 2004. *Injury Facts, 2004 edition.* Itasca, IL: National Safety Council.

Neer, C.S., II. 1972. Anterior acromioplasty for the chronic impingement syndrome in the shoulder: a preliminary report. *Journal of Bone and Joint Surgery* 54A: 41-50.

Neer, C.S., II. 1990. *Shoulder Reconstruction.* Philadelphia: Saunders.

Nestor, B.J., S.W. O'Driscoll, and B.F. Morrey. 1992. Ligamentous reconstruction for posterolateral instability of the elbow. *Journal of Bone and Joint Surgery* 74A: 1235-1241.

Neviaser, J.S. 1945. Adhesive capsulitis of the shoulder. *Journal of Bone and Joint Surgery* 27A: 211-212.

Nirschl, R.P. 1988. Prevention and treatment of elbow and shoulder injuries in the tennis player. *Clinics in Sports Medicine* 7: 289-308.

Nirschl, R.P., and E.S. Ashman. 2003. Elbow tendinopathy: Tennis elbow. *Clinics in Sports Medicine* 22: 813-836.

Nirschl, R., and F. Pettrone. 1979. Tennis elbow: The surgical treatment of lateral epicondylitis. *Journal of Bone and Joint Surgery* 61A: 832-841.

Nordt, W.E., III, R.B. Garretson, III, and E. Plotkin. 1999. The measurement of subacromical contact pressure in patients with impingement syndrome. *Arthroscopy* 15: 121-125.

O'Driscoll, S.W. 2000. Classification and evaluation of recurrent instability of the elbow. *Clinical Orthopaedics and Related Research* 370: 34-43.

O'Driscoll, S.W., D.F. Bell, and B.F. Morrey. 1991. Posterolateral rotatory instability of the elbow. *Journal of Bone and Joint Surgery* 73A: 440-446.

O'Driscoll, S.W., B.F. Morrey, S. Korinek, and K.N. An. 1992. Elbow subluxation and dislocation: A spectrum of instability. *Clinical Orthopaedics* 280: 186-197.

Ogata, S., and H.K. Uhthoff. 1990. Acromial enthesopathy and rotator cuff tear. A radiologic and histologic postmortem investigation of the coracoacromial arch. *Clinical Orthopaedics and Related Research* 254: 39-48.

Paley, K.J., F.W. Jobe, M.M. Pink, R.S. Kvitne, and N.S. ElAttrache. 2000. Arthroscopic findings in the overhand throwing athlete: Evidence for posterior internal impingement of the rotator cuff. *Arthroscopy* 16: 35-40.

Penrose, J.H. 1951. The Monteggia fracture with posterior dislocation of the radial head. *Journal of Bone and Joint Surgery* 33B: 65-73.

Perry, J.J., and L.D. Higgins. 2001. Shoulder injuries. In *Sports Injuries: Mechanisms, Prevention, Treatment*, edited by F.H. Fu and D.A. Stone. Philadelphia: Lippincott Williams & Wilkins.

Peterson, L., and P. Renström. 2001. *Sports Injuries: Their Prevention and Treatment*. Champaign, IL: Human Kinetics.

Phalen, G.S. 1966. The carpal-tunnel syndrome. *Journal of Bone and Joint Surgery* 48A: 211-218.

Praemer, A., S. Furner, and D.P. Rice. 1992. *Musculoskeletal Conditions in the United States*. Park Ridge, IL: American Academy of Orthopaedic Surgeons.

Praemer, A., S. Furner, and D.P. Rice. 1999. *Musculoskeletal Conditions in the United States*. Park Ridge, IL: American Academy of Orthopaedic Surgeons.

Prato, N., D. Peloso, A. Franconeri, G. Tegaldo, G.B. Ravera, E. Silvestri, and L.E. Derchi. 1998. The anterior tilt of the acromion: Radiographic evaluation and correlation with shoulder diseases. *European Journal of Radiology* 8: 1639-1646.

Priest, J.D., J. Braden, and S.G. Gerberich. 1980. The elbow and tennis. *The Physician and Sportsmedicine* 8: 80-85.

Ptasznik, R., and O. Hennessy. 1995. Abnormalities of the biceps tendon of the shoulder: Sonographic findings. *American Journal of Roentgenology* 164: 409-414.

Rasool, M.N. 2004. Dislocations of the elbow in children. *Journal of Bone and Joint Surgery* 86B: 1050-1058.

Rebuzzi, E., N. Coletti, S. Schiavetti, and F. Giusto. 2005. Arthroscopic rotator cuff repair in patients older than 60 years. *Arthroscopy* 21: 48-54.

Reddy, A.S., K.J. Mohr, M.M. Pink, and F.W. Jobe. 2000. Electromyographic analysis of the deltoid and rotator cuff muscles in persons with sub-acromial impingement. *Journal of Shoulder and Elbow Surgery* 9: 519-523.

Rettig, A.C. 2002. Traumatic elbow injuries in the athlete. *Orthopedic Clinics of North America* 33: 509-522.

Rettig, M.E., and K.B. Raskin. 2000. Acute fractures of the distal radius. *Hand Clinics* 16: 405-415.

Rockwood, C.A., D.P. Green, R.W. Bucholz, and J.D. Heckman, eds. 1996. *Rockwood and Green's Fractures in Adults* (4th ed.). Philadelphia: Lippincott-Raven.

Rockwood, C.A., Jr., G.R. Williams, and D.C. Young. 1996. Injuries to the acromioclavicular joint. In *Rockwood and Green's Fractures in Adults* (4th ed.), edited by C.A. Rockwood, D.P. Green, R.W. Bucholz, and J.D. Heckman. Philadelphia: Lippincott-Raven.

Ruby, L.K., and C. Cassidy. 2003. Fractures and dislocations of the carpus. In *Skeletal Trauma* (3rd ed.), edited by B.D. Browner, J.B. Jupiter, A.M. Levine, and P.G. Trafton. Philadelphia: Saunders.

Ruotolo, C., and W.M. Nottage. 2002. Surgical and nonsurgical management of rotator cuff tears. *Arthroscopy* 18: 527-531.

Safran, M.R. 1995. Elbow injuries in athletes: A review. *Clinical Orthopaedics and Related Research* 310: 257-277.

Safran, M.R. 2004. Ulnar collateral ligament injury in the overhead athlete: Diagnosis and treatment. *Clinics in Sports Medicine* 23: 643-663.

Samilson, R.L., and V. Prieto. 1983. Posterior dislocation of the shoulder in athletes. *Clinics in Sports Medicine* 2: 369-378.

Sauerbrey, A.M., C.L. Getz, M. Piancastelli, J.P. Iannotti, M.L. Ramsey, and G.R. Williams, Jr. 2005. Arthroscopic versus mini-open rotator cuff repair: A comparison of clinical outcome. *Arthroscopy* 21: 1415-1420.

Sher, J.S., J.W. Uribe, A. Posada, B.J. Murphy, and M.B. Zlatkin. 1995. Abnormal findings on magnetic resonance images of asymptomatic shoulders. *Journal of Bone and Joint Surgery* 77B: 10-15.

Silverstein, B.A., L.J. Fine, and T.J. Armstrong. 1987. Occupational factors and carpal tunnel syndrome. *American Journal of Industrial Medicine* 11: 343-358.

Smith, F.M. 1947. Monteggia fractures: an analysis of 25 consecutive fresh injuries. *Surgery, Gynecology and Obstetrics* 85: 630-640.

Snijders, C.J., A.C.W. Volkers, K. Mechelse, and A. Vleeming. 1987. Provocation of epicondylalgia lateralis (tennis elbow) by power grip or pinching. *Medicine and Science in Sports and Exercise* 19: 518-523.

Snyder, S.J., M.P. Banas, and R.P. Karzel. 1995. An analysis of 140 injuries to the superior glenoid labrum. *Journal of Shoulder and Elbow Surgery* 4: 243-248.

Snyder, S.J., R.P. Karzel, W. Del Pizzo, R.D. Ferkel, and M.J. Friedman. 1990. SLAP lesions of the shoulder. *Arthroscopy* 6: 274-279.

Soslowsky, L.J., C.H. An, C.M. DeBano, and J.E. Carpenter. 1996. Coracoacromial ligament: In situ load and viscoelastic properties in rotator cuff disease. *Clinical Orthopaedics and Related Research* 330: 40-44.

Speed J.S., and H.B. Boyd. 1940. Treatment of fractures of ulna with dislocation of head of radius. *Journal of the American Medical Association* 125: 1699-1704.

Speer, K.P. 1995. Anatomy and pathomechanics of shoulder instability. *Clinics in Sports Medicine* 14: 751-760.

Templehof, S., S. Rupp, and R. Seil. 1999. Age-related prevalence of rotator cuff tears in asymptomatic shoulders. *Journal of Shoulder and Elbow Surgery* 8: 296-299.

Toivonen, D.A., M.J. Tuite, and J.F. Orwin. 1995. Acromial structure and tears of the rotator cuff. *Journal of Shoulder and Elbow Surgery* 4: 376-383.

Tompkins, D.G. 1971. The anterior Monteggia fracture. *Journal of Bone and Joint Surgery* 53A: 1109-1114.

Tossy, J.D., N.C. Mead, and H.M. Sigmond. 1963. Acromioclavicular separations: Useful and practical classification for treatment. *Clinical Orthopaedics and Related Research* 28: 111-119.

Tuite, M.J., D.A. Toivonen, J.F. Orwin, and D.H. Wright. 1995. Acromial angle on radiographs of the shoulder: Correlation with the impingement syndrome and rotator cuff tears. *American Journal of Roentgenology* 165: 609-613.

Tytherleigh-Strong, G., N. Walls, and M.M. McQueen. 1998. The epidemiology of humeral shaft fractures. *Journal of Bone and Joint Surgery* 80B: 249-253.

van der Windt, D.A., B.W. Koes, B.A. de Jong, and L.M. Bouter. 1995. Shoulder disorders in general practice: Incidence, patient characteristics, and management. *Annals of the Rheumatic Diseases* 54: 959-964.

Vaz, S., J. Soyer, P. Pries, and J.P. Clarac. 2000. Subacromial impingement: Influence of coracoacromial arch geometry on shoulder function. *Joint Bone Spine* 67: 305-309.

Vecchio, P., R. Kavanagh, B.L. Hazleman, and R.H. King. 1995. Shoulder pain in a community-based rheumatology clinic. *British Journal of Rheumatology* 34: 440-442.

Viikari-Juntura, E., and B. Silverstein. 1999. Role of physical load factors in carpal tunnel syndrome. *Scandinavian Journal of Work and Environmental Health* 25: 163-185.

Walch, G., P. Boileau, E. Noel, and S.T. Donell. 1992. Impingement of the deep surface of the supraspinatus tendon on the posterior glenoid rim: An arthroscopic study. *Journal of Shoulder and Elbow Surgery* 1: 238-245.

Wang, J.C., and M.S. Shapiro. 1997. Changes in acromial morphology with age. *Journal of Shoulder and Elbow Surgery* 6: 55-59.

Warner, J.J., and I.M. Parsons, IV. 2001. Latissimus dorsi tendon transfer: A comparative analysis of primary and salvage reconstruction of massive, irreparable rotator cuff tears. *Journal of Shoulder and Elbow Surgery* 10: 514-521.

Warner, J.J., P. Tetreault, J. Lehtinen, and D. Zurakowski. 2005. Arthroscopic versus mini-open rotator cuff repair: A cohort comparison study. *Arthroscopy* 21: 328-332.

Weber, E.R., and E.Y. Chao. 1978. An experimental approach to the mechanism of scaphoid wrist fractures. *Journal of Hand Surgery* 3: 142-148.

Werner, R.A., and M. Andary. 2002. Carpal tunnel syndrome: Pathophysiology and clinical neurophysiology. *Clinical Neurophysiology* 113: 1373-1381.

Werner, S.L., G.S. Fleisig, C.J. Dillman, and J. Andrews. 1993. Biomechanics of the elbow during baseball pitching. *Journal of Orthopaedic and Sports Physical Therapy* 17: 274-278.

Whaley, A.L., and C.L. Baker. 2004. Lateral epicondylitis. *Clinics in Sports Medicine* 23: 677-691.

Wilkins, K.E. 2002. Changes in the management of Monteggia fractures. *Journal of Pediatric Orthopaedics* 22: 548-554.

Wilkinson, G.T. 1895. Complete transverse fracture of the humerus by muscular action. *Lancet* 2: 733.

Williams, G.R., V.D. Nguyen, and C.A. Rockwood, Jr. 1989. Classification and radiographic analysis of acromioclavicular dislocations. *Applied Radiology* 18: 29-34.

Williams, G.R., C.A. Rockwood, Jr., L.U. Bigliani, J.P. Iannotti, and W. Stanwood. 2004. Rotator cuff tears: Why do we repair them? *Journal of Bone and Joint Surgery* 86A: 2764-2776.

Wilson, F.D., J.R. Andrews, T.A. Blackburn, and G. McCluskey. 1983. Valgus extension overload in the pitching elbow. *American Journal of Sports Medicine* 11: 83-88.

Wirth, M.A., and C.A. Rockwood, Jr. 1997. Operative treatment of irreparable rupture of the subscapularis. *Journal of Bone and Joint Surgery* 79A: 722-731.

Wittenberg, R.H., F. Rubenthaler, T. Wolk, J. Ludwig, R.E. Willburger, and R. Steffen. 2001. Surgical or conservative treatment for chronic rotator cuff calcifying tendinitis—a matched-pair analysis of 100 patients. *Archives of Orthopaedic and Trauma Surgery* 121: 56-59.

Worland, R.L., D. Lee, C.G. Orozco, F. SozaRex, and J. Keenan. 2003. Correlation of age, acromial morphology, and rotator cuff tear pathology diagnosed by ultrasound in asymptomatic patients. *Journal of the Southern Orthopedic Association* 12: 23-26.

Wright, P.R. 1963. Greenstick fracture of the upper end of the ulna with dislocation of the radiohumeral joint or displacement of the superior radial epiphysis. *Journal of Bone and Joint Surgery* 45B: 727-731.

Yamaguchi, K., J.S. Sher, W.K. Anderson, R. Garretson, J.W. Uribe, K. Hechtman, and R.J. Neviaser. 2000. Glenohumeral motion in patients with rotator cuff tears: A comparison of asymptomatic and symptomatic shoulders. *Journal of Shoulder and Elbow Surgery* 9: 6-11.

Youm, T., D.H. Murray, E.N. Kubiak, A.S. Rokito, and J.D. Zuckerman. 2005. Arthroscopic versus mini-open rotator cuff repair: a comparison of clinical outcomes and patient satisfaction. *Journal of Shoulder and Elbow Surgery* 14: 455-459.

Zuckerman, J.D., F.J. Kummer, F. Cuomo, and M. Greller. 1997. Interobserver reliability of acromial morphology classification: an anatomic study. *Journal of Shoulder and Elbow Surgery* 6: 286-287.

CHAPTER 8

Adams, J.H., D. Doyle, I. Ford, T.A. Gennarelli, D.I. Graham, and D.R. McLellan. 1989. Diffuse axonal injury in head injury: definition, diagnosis and grading. *Histopathology* 15: 49-59.

Adams, M.A., and W.C. Hutton. 1982. Prolapsed intervertebral disc. A hyperflexion injury. *Spine* 7: 184-191.

Adams, R.D., and M. Victor. 1993. *Principles of Neurology.* New York: McGraw-Hill.

Adebayo, E.T., O.S. Ajike, and E.O. Adekeye. 2003. Analysis of the pattern of maxillofacial fractures in Kaduna, Nigeria. *British Journal of Oral and Maxillofacial Surgery* 41: 396-400.

Allsop, D., and K. Kennett. 2001. Skull and facial bone trauma. In *Accidental Injury*, edited by A.M. Nahum and J.W. Melvin. New York: Springer.

Alvi, A., T. Doherty, and G. Lewen. 2003. Facial fractures and concomitant injuries in trauma patients. *Laryngoscope* 113: 102-106.

American Academy of Pediatrics. 2007. Backpack safety. Available: http://www.aap.org/advocacy/backpack_safety.pdf.

American Occupational Therapy Association. 2005. Backpack strategies for parents and students. Available: www.promoteot.org/AI_Backpack-Strategies.html.

Amonoo-Kuofi, H.S. 1992. Changes in the lumbosacral angle, sacral inclination and the curvature of the lumbar spine during aging. *Acta Anatomica* 145: 373-377.

Aubry, M., R. Cantu, J. Dvorak, T. Graf-Baumann, K. Johnston, J. Kelly, M. Lovell, P. McCrory, W. Meeuwisse, P. Schamasch, the Concussion in Sport (CIS) Group. 2002. Summary and agreement statement of the first International Conference on Concussion in Sport, Vienna 2001. *British Journal of Sports Medicine* 36: 6-10.

Bailes, J.E., and R.C. Cantu. 2001. Head injury in athletes. *Neurosurgery* 48: 26-46.

Barnsley, L., S. Lord, and N. Bogduk. 1994. Whiplash injury. *Pain* 58: 283-307.

Bartlett, C.S. 2003. Clinical update: gunshot wound ballistics. *Clinical Orthopaedics and Related Research* 408: 28-57.

Bazarian, J.J., J. McClung, M.N. Shah, Y.T. Cheng, W. Flesher, and J. Kraus. 2005. Mild traumatic brain injury in the United States, 1998-2000. *Brain Injury* 19: 85-91.

Beutler, W.J., B.E. Fredrickson, A. Murtland, C.A. Sweeney, W.D. Grant, and D. Baker. 2003. The natural history of spondylolysis and spondylolisthesis: 45-year follow-up evaluation. *Spine* 28: 1027-1035.

Bogduk, N., and N. Yoganandan. 2001. Biomechanics of the cervical spine. Part 3: minor injuries. *Clinical Biomechanics* 16: 267-275.

Brackley, H.M., and J.M. Stevenson. 2004. Are children's backpack weight limits enough? A critical review of the relevant literature. *Spine* 29: 2184-2190.

Bradford, D.S. 1995. Kyphosis in the elderly. In *Moe's Textbook of Scoliosis and Other Spinal Deformities*, edited by J.E. Lonstein, D.S. Bradford, R.B. Winter, and J.W. Ogilvie. Philadelphia: Saunders.

Brasileiro, B.F., and L.A. Passeri. 2006. Epidemiological analysis of maxillofacial fractures in Brazil: a 5-year prospective study. *Oral Surgery Oral Medicine Oral Pathology Oral Radiology and Endodontics* 102: 28-34.

Cantu, R.C. 1998. Second-impact syndrome. *Clinics in Sports Medicine* 17: 37-44.

Cardon, G., and F. Balague. 2004. Backpacks and spinal disorders in school children. *Europa Medicophysica* 40: 15-20.

Case, M.E., M.A. Graham, T.C. Handy, J.M. Jentzen, and J.A. Monteleone. 2001. Position paper on fatal abusive head injuries in infants and young children. *American Journal of Forensic Medicine and Pathology* 22: 112-122.

Cassidy, J.D., J. Duranceau, M.H. Liang, L.R. Salmi, M.L. Skovon, and W.O. Spitzer. 1995. Scientific monograph of the Quebec Task Force on whiplash associated disorders. *Spine* 20: S8-S58.

Charnley, J. 1955. Acute lumbago and sciatica. *British Medical Journal* 4909: 344-346.

Chow, D.H., M.L. Kwok, A.C. Au-Yang, A.D. Holmes, J.C. Cheng, F.Y. Yao, and M.S. Wong. 2005. The effect of backpack load on the gait of normal adolescent girls. *Ergonomics* 48: 642-656.

Clausen, H., P. McCrory, and V. Anderson. 2005. The risk of chronic traumatic brain injury in professional boxing: Change in exposure variables over the past century. *British Journal of Sports Medicine* 39: 661-664.

Cobb, S., and B. Battin. 2004. Second-impact syndrome. *Journal of School Nursing* 20: 262-267.

Courville, C.B. 1942. Coup-contrecoup mechanism of craniocerebral injuries: Some observations. *Archives of Surgery* 45: 19-43.

Cottalorda, J., A. Rahmani, M. Diop, V. Gautheron, E. Ebermeyer, and A. Belli. 2003. Influence of school bag carrying on gait kinetics. *Journal of Pediatric Orthopaedics B* 12: 357-364.

Culham, E.G., H.A. Jimenez, and C.E. King. 1994. Thoracic kyphosis, rib mobility, and lung volumes in normal women and women with osteoporosis. *Spine* 19: 1250-1255.

Cusick, J.F., and N. Yoganandan. 2002. Biomechanics of the cervical spine 4: Major injuries. *Clinical Biomechanics* 17: 1-20.

Cutler, W.B., E. Friedmann, and E. Genovese-Stone. 1993. Prevalence of kyphosis in a healthy sample of pre- and postmenopausal women. *American Journal of Physical Medicine and Rehabilitation* 72: 219-225.

Damasio, A.R. 1994. *Descartes' Error: Emotion, Reason, and the Human Brain*. New York: Grosset/Putnam.

Damasio, H., T. Grabowski, R. Frank, A.M. Galaburda, and A.R. Damasio. 1994. The return of Phineas Gage: Clues about the brain from the skull of a famous patient. *Science* 264: 1102-1105.

Davis, C.G. 2000. Injury threshold: Whiplash-associated disorders. *Journal of Manipulative and Physiological Therapeutics* 23: 420-427.

Dawson, S.L., C.W. Hirsch, F.V. Lucas, and B.A. Sebek. 1980. The contrecoup phenomenon: Reappraisal of a classic problem. *Human Pathology* 11: 155-166.

Denis, F. 1983. The three column spine and its significance in the classification of acute thoracolumbar spinal injuries. *Spine* 8: 817-831.

Denny-Brown, D., and W.R. Russell. 1941. Experimental cerebral concussion. *Brain* 64: 93-164.

Duhaime, A.C., C.W. Christian, L.B. Rorke, and R.A. Zimmerman. 1998. Non-accidental head injury in infants—the "shaken-baby syndrome." *New England Journal of Medicine* 338: 1822-1829.

Eck, J.C., S.D. Hodges, and S.C. Humphreys. 2001. Whiplash: A review of a commonly misunderstood injury. *American Journal of Medicine* 110: 651-656.

Fackler, M.L. 1996. Gunshot wound review. *Annals of Emergency Medicine* 28: 194-203.

Fackler, M.L. 1998. Civilian gunshot wounds and ballistics: Dispelling the myths. *Emergency Medicine Clinics of North America* 16: 17-28.

Fardon, D.F., and P.C. Milette. 2001. Nomenclature and classification of lumbar disc pathology. *Spine* 26: E93-E113.

Flik, K., S. Lyman, and R.G. Marx. 2005. American collegiate men's ice hockey: An analysis of injuries. *American Journal of Sports Medicine* 33: 183-187.

Freeman, J.R., J.T. Barth, D.K. Broshek, and K. Plehn. 2005. Sports injuries. In *Textbook of Traumatic Brain Injury*, edited by J.M. Silver, T.W. McAllister, and S.C. Yudofsky. Washington, DC: American Psychiatric Publishing.

Gardner, B. 2002. Rehabilitation of spinal cord injuries. In *Oxford Textbook of Orthopedics and Trauma*, edited by C. Bulstrode, J. Buckwalter, A. Carr, L. Marsh, J. Fairbank, J. Wilson-MacDonald, and G. Bowden. Oxford, UK: Oxford University Press.

Gennarelli, T.A., and D.I. Graham. 2005. Neuropathology. In *Textbook of Traumatic Brain Injury*, edited by J.M. Silver, T.W. McAllister, and S.C. Yudofsky. Washington, DC: American Psychiatric Publishing.

Gennarelli, T.A., L.E. Thibault, J.H. Adams, D.I. Graham, C.J. Thompson, and R.P. Marcincin. 1982. Diffuse axonal injury and traumatic coma in the primate. *Annals of Neurology* 12: 564-574.

Goldsmith, W., and J. Plunkett. 2004. A biomechanical analysis of the causes of traumatic brain injury in infants and children. *American Journal of Forensic Medicine and Pathology* 25: 89-100.

Grauer, J.N., M.M. Panjabi, J. Cholewicki, K. Nibu, and J. Dvorak. 1997. Whiplash produces an S-shaped curvature of the neck with hyperextension at lower levels. *Spine* 22: 2489-2494.

Grobler, L.J., P.A. Robertson, J.E. Novotny, and M.H. Pope. 1993. Assessment of the role played by lumbar facet joint morphology. *Spine* 18: 80-91.

Gross, A.G. 1958. Impact thresholds of brain concussion. *Journal of Aviation Medicine* 29: 725-732.

Gross, L.B. 2001. Boxing. In *Sports Injuries: Mechanisms, Prevention, Treatment*, edited by F.H. Fu and D.A. Stone. Philadelphia: Lippincott Williams & Wilkins.

Gurdjian, E.S., and J.E. Webster. 1946. Deformation of the skull in head injury studied by stresscoat technique. *Surgery, Gynecology & Obstetrics* 83: 219-233.

Gurdjian, E.S., J.E. Webster, and H.R. Lissner. 1947. The mechanism of production of linear skull fractures. *American Journal of Surgery* 85: 195-210.

Gurdjian, E.S., J.E. Webster, and H.R. Lissner. 1949. Studies on skull fracture with particular reference to engineering factors. *American Journal of Surgery* 87: 736-742.

Gurdjian, E.S., J.E. Webster, and H.R. Lissner. 1953. Observations on prediction of fracture site in head injury. *Radiology* 60: 226-235.

Gurdjian, E.S., J.E. Webster, and H.R. Lissner. 1955. Observations on the mechanism of brain concussion, contusion, and laceration. *Surgery, Gynecology & Obstetrics* 101: 680-690.

Hampson, D. 1995. Facial injury: A review of biomechanical studies and test procedures for facial injury assessment. *Journal of Biomechanics* 28: 1-7.

Hart, C., and E. Williams. 1994. Epidemiology of spinal cord injuries: A reflection of changes in South African society. *Paraplegia* 32: 709-714.

Haug, R.H., J.M. Adams, P.J. Conforti, and M.J. Likavec. 1994. Cranial fractures associated with facial fractures: A review of mechanism, type, and severity of injury. *Journal of Oral and Maxillofacial Surgery* 52: 729-733.

Hawes, M.C. 2003. The use of exercises in the treatment of scoliosis: An evidence-based critical review of the literature. *Pediatric Rehabilitation* 6: 171-182.

Herman, M.J., and P.D. Pizzutillo. 2005. Spondylosis and spondylolisthesis in the child and adolescent. *Clinical Orthopaedics and Related Research* 434: 46-54.

Herman, M.J., P.D. Pizzutillo, and R. Cavalier. 2003. Spondylolysis and spondylolisthesis in the child and adolescent athlete: A new classification. *Orthopedic Clinics of North America* 34: 461-467.

Hodgson, V.R. 1967. Tolerance of the facial bones to impact. *American Journal of Anatomy* 120: 113-122.

Hodgson, V.R., and L.M. Thomas. 1971. *Breaking Strength of the Human Skull vs. Impact Surface Curvature* (HS-800-583). Springfield, VA: U.S. Department of Transportation.

Hodgson, V.R., and L.M. Thomas. 1972. Effect of long-duration impact on head. In *Proceedings of the 16th Stapp Car Crash Conference*, Warrendale, PA: Society of Automotive Engineers.

Hodgson, V.R., and L.M. Thomas. 1973. *Breaking Strength of the Human Skull vs. Impact Surface Curvature* (HS-801-002). Springfield, VA: U.S. Department of Transportation.

Hogg, N.J., T.C. Stewart, J.E. Armstrong, and M.J. Girotti. 2000. Epidemiology of maxillofacial injuries at trauma hospitals in Ontario, Canada, between 1992 and 1997. *Journal of Trauma* 49: 425-432.

Holbourn, A.H.S. 1943. Mechanics of head injuries. *Lancet* 2: 438-441.

Holdsworth, F.W. 1963. Fractures, dislocations, and fracture-dislocations of the spine. *Journal of Bone and Joint Surgery* 45B: 6-20.

Holdsworth, F.W. 1970. Fractures, dislocations, and fracture-dislocations of the spine. *Journal of Bone and Joint Surgery* 52A: 1534-1541.

Hopper, R.H., J.H. McElhaney, and B.S. Myers. 1994. Mandibular and basilar skull fracture tolerance (SAE 942213). In *Proceedings of the 38th Stapp Car Crash Conference*, Warrendale, PA: Society of Automotive Engineers.

Ikata, T., R. Miyake, S. Katoh, T. Morita, and M. Murase. 1996. Pathogenesis of sports-related spondylolisthesis in adolescents. *American Journal of Sports Medicine* 24: 94-98.

Ito, S., P.C. Ivancic, M.M. Panjabi, and B.W. Cunningham. 2004. Soft tissue injury threshold during simulated whiplash. *Spine* 29: 979-987.

Jordan, B.D. 2000. Chronic traumatic brain injury associated with boxing. *Seminars in Neurology* 20: 179-185.

Jordan, B.D., N.R. Relkin, L.D. Ravdin, A.R. Jacobs, A. Bennett, and S. Gandy. 1997. Apolipoprotein E epsilon4 associated with chronic traumatic brain injury in boxing. *Journal of the American Medical Association* 278: 136-140.

Karacan, I., H. Koyuncu, O. Pekel, G. Sumbuloglu, M. Kirnap, H. Dursun, A. Kalkan, A. Cengiz, A. Yalinkilic, H.I. Unalan, K. Nas, S. Orkun, and I. Tekeoglu. 2000. Traumatic spinal cord injuries in Turkey: a nation-wide epidemiological study. *Spinal Cord* 38: 697-701.

Kraus, J.F., and L.D. Chu. 2005. Epidemiology. In *Textbook of Traumatic Brain Injury*, edited by J.M. Silver, T.W. McAllister, and S.C. Yudofsky. Washington, DC: American Psychiatric Publishing.

Kühne, C.A., C. Krueger, M. Homann, C. Mohr, and S. Ruchholtz. 2007. Epidemiology and management in emergency room patients with maxillofacial fractures. *Mund Kiefer und Gesichtschirurgie* 11: 201-208.

Kumar, S., R. Ferrari, and Y. Narayan. 2005. Kinematic and electromyographic response to whiplash loading in low-velocity whiplash impacts—a review. *Clinical Biomechanics* 20: 343-356.

Kwan, O., and J. Friel. 2003. A review and methodologic critique of the literature supporting "chronic whiplash injury": part I—research articles. *Medical Science Monitor* 9: 203-215.

Langlois, J.A., W. Rutland-Brown, and K.E. Thomas. 2004. *Traumatic Brain Injury in the United States: Emergency Department Visits, Hospitalizations, and Deaths*. Atlanta: Centers for Disease Control and Prevention, National Center for Injury Prevention and Control.

Lestini, W.F., and S.W. Wiesel. 1989. The pathogenesis of cervical spondylosis. *Clinical Orthopaedics and Related Research* 239: 69-93.

Lim, L.H., L.K. Lam, M.H. Moore, J.A. Trott, and D.J. David. 1993. Associated injuries in facial fractures: Review of 839 patients. *British Journal of Plastic Surgery* 46: 635-638.

Lindh, M. 1989. Biomechanics of the lumbar spine. In *Basic Biomechanics of the Musculoskeletal System* (2nd ed.), edited by M. Nordin and V.H. Frankel. Philadelphia: Lea & Febiger.

Lowe, T.G., M. Edgar, J.Y. Margulies, N.H. Miller, V.J. Raso, K.A. Reinker, and C.-H. Rivard. 2000. Etiology of idiopathic scoliosis: current trends in research. *Journal of Bone and Joint Surgery* 82A: 1157-1168.

Luan, F., K.H. Yang, B. Deng, P.C. Begeman, S. Tashman, and A.I. King. 2000. Qualitative analysis of neck kinematics during low-speed rear-end impact. *Clinical Biomechanics* 15: 649-657.

Maxwell, W.L., C. Watt, D.I. Graham, and T.A. Gennarelli. 1993. Ultrastructural evidence of axonal shearing as a result of lateral acceleration of the head in non-human primates. *Acta Neuropathologica* 86: 136-144.

McClune, T., A.K. Burton, and G. Waddell. 2002. Whiplash associated disorders: A review of the literature to guide patient information and advice. *Emergency Medicine Journal* 19: 499-506.

McCormack, B.M., and P.R. Weinstein. 1996. Cervical spondylosis: an update. *Western Journal of Medicine* 165: 43-51.

McCrory, P. 2001. Does second impact syndrome exist? *Clinical Journal of Sport Medicine* 11: 144-149.

McCrory, P. 2002. Boxing and the brain. *British Journal of Sports Medicine* 36: 2.

McCrory, P., K. Johnston, W. Meeuwisse, M. Aubry, R. Cantu, J. Dvorak, T. Graf-Baumann, J. Kelly, M. Lovell, and P. Schamasch. 2005. Summary and agreement statement of the 2nd International Conference on Concussion in Sport, Prague 2004. *Clinical Journal of Sports Medicine* 15: 48-55.

McElhaney, J.H., R.W. Nightingale, B.A. Winkelstein, V.C. Chancey, and B.S. Myers. 2001. Biomechanical aspects of cervical trauma. In *Accidental Injury*, edited by A.M. Nahum and J.W. Melvin. New York: Springer.

McGehee, D.V. 1996. Head injury in motor vehicle crashes: Human factors, effects, and prevention. In *Head Injury and Postconcussive Syndrome*, edited by M. Rizzo and D. Tranel. New York: Churchill Livingstone.

McGill, S. 2007. *Low Back Disorders: Evidence-based Prevention and Rehabilitation*. Champaign, IL: Human Kinetics.

Melton, L.J., III. 1997. Epidemiology of spinal osteoporosis. *Spine* 22(24 Suppl.): 2S-11S.

Meythaler, J.M., J.D. Peduzzi, E. Eleftheriou, and T.A. Novack. 2001. Current concepts: Diffuse axonal injury—associated traumatic brain injury. *Archives of Physical Medicine and Rehabilitation* 82: 1461-1471.

Miller, J.D. 1993. Traumatic brain swelling and edema. In *Head Injury* (3rd ed.), edited by P.R. Cooper. Baltimore: Williams & Wilkins.

Miller, N.H. 2000. Genetics of familial idiopathic scoliosis. *Spine* 25: 2416-2418.

Murata, Y., K. Takahashi, M. Yamagata, E. Hanaoka, and H. Moriya. 2003. The knee-spine syndrome. *Journal of Bone and Joint Surgery* 85B: 95-99.

Murrie, V.L., A.K. Dixon, W. Hollingworth, H. Wilson, and T.A.C. Doyle. 2003. Lumbar lordosis: Study of patients with and without low back pain. *Clinical Anatomy* 16: 144-147.

Nachemson, A. 1975. Towards a better understanding of low-back pain: A review of the mechanics of the lumbar disc. *Rheumatology and Rehabilitation* 14: 129-143.

Nahum, A.M., J.D. Gatts, C.W. Gadd, and J.P. Danforth. 1968. Impact tolerance of the skull and face (680785). In *Proceedings of the 12th Stapp Car Crash Conference*, Warrendale, PA: Society of Automotive Engineers.

Natarajan, R.N., R.B. Garretson, III, A. Biyani, T.H. Lim, G.B. Andersson, and H.S. An. 2003. Effects of slip severity and loading directions on the stability of isthmic spondylolisthesis: A finite element model study. *Spine* 28: 1103-1112.

National Center for Injury Prevention and Control. 2001. *Injury Fact Book 2001-02*. Atlanta: Centers for Disease Control and Prevention.

National Safety Council. 2004. *Injury Facts*. Itasca, IL: National Safety Council.

Negrini, S., and R. Carabalona. 2002. Backpacks on! Schoolchildren's perceptions of load, associations with back pain and factors determining the load. *Spine* 27: 187-195.

Negrini, S., R. Carabalona, and P. Sibilla. 1999. Backpack as a daily load for schoolchildren. *Lancet* 354: 1974.

Offierski, C.M., and I. MacNab. 1983. Hip-spine syndrome. *Spine* 8: 316-321.

Old, J.L., and M. Calvert. 2004. Vertebral compression fractures in the elderly. *American Family Physician* 69: 111-116.

Ommaya, A.K. 1995. Head injury mechanisms and the concept of preventive management: A review and critical synthesis. *Journal of Neurotrauma* 12: 527-546.

Ommaya, A.K., and T.A. Gennarelli. 1974. Cerebral concussion and traumatic unconsciousness: Correlations and experimental and clinical observations on blunt head injuries. *Brain* 97: 633-654.

Ommaya, A.K., W. Goldsmith, and L. Thibault. 2002. Biomechanics and neuropathology of adult and paediatric head injury. *British Journal of Neurosurgery* 16: 220-242.

Ommaya, A.K., R.L. Grubb, Jr., and R.A. Naumann. 1971. Coup and contre-coup injury: Observations on the mechanics of visible brain injuries in the rhesus monkey. *Journal of Neurosurgery* 35: 503-516.

Padman, R. 1995. Scoliosis and spine deformities. *Delaware Medical Journal* 67: 528-533.

Panjabi, M.M., S. Ito, A.M. Pearson, and P.C. Ivancic. 2004. Injury mechanisms of the cervical intervertebral disc during simulated whiplash. *Spine* 29: 1217-1225.

Panjabi, M.M., T.R. Oxland, R.-M. Lin, and T.W. McGowen. 1994. Thoracolumbar burst fracture: A biomechanical investigation of its multidirectional flexibility. *Spine* 19: 578-585.

Panjabi, M.M., A.M. Pearson, S. Ito, P.C. Ivancic, and J.-L. Wang. 2004. Cervical spine curvature during simulated whiplash. *Clinical Biomechanics* 19: 1-9.

Parent, S., P.O. Newton, and D.R. Wenger. 2005. Adolescent idiopathic scoliosis: Etiology, anatomy, natural history, and bracing. *Instructional Course Lectures* 54: 529-536.

Pearson, A.M., P.C. Ivancic, S. Ito, and M.M. Panjabi. 2004. Facet joint kinematics and injury mechanisms during simulated whiplash. *Spine* 29: 390-397.

Pellman, E., J. Powell, D. Viano, I. Casson, A. Tucker, H. Feuer, M. Lovell, J. Waeckerle, and D. Robertson. 2004. Concussion in professional football: epidemiological features of game injuries and review of the literature—part 3. *Neurosurgery* 54: 81-94.

Pintar, F.A., N. Yoganandan, L.M. Voo, J.F. Cusick, D.J. Maiman, and A. Sances, Jr. 1995. Dynamic characteristics of the human cervical spine. *SAE Transactions* 104: 3087-3094.

Porter, M.D. 2003. A 9-year controlled prospective neuropsychologic assessment of amateur boxing. *Clinical Journal of Sport Medicine* 13: 339-352.

Powell, J.W., and K.D. Barber-Foss. 1999. Traumatic brain injury in high school athletes. *Journal of the American Medical Association* 282: 958-963.

Puche, R.C., M. Morosano, A. Masoni, N.P. Jimeno, S.M. Bertoluzzo, J.C. Podadera, M.A. Podadera, R. Bocanera, and R. Tozzini. 1995. The natural history of kyphosis in postmenopausal women. *Bone* 17: 239-246.

Radanov, B.P., M. Sturzenegger, and G. Di Stefano. 1995. Long-term outcome after whiplash injury: A 2-year follow-up considering features of injury mechanism and somatic, radiologic, and psychosocial findings. *Medicine* 74: 281-297.

Rhee, J.S., L. Posey, N. Yoganandan, and F. Pintar. 2001. Experimental trauma to the malar eminence: fracture biomechanics and injury patterns. *Otolaryngology Head and Neck Surgery* 125: 351-355.

Roaf, R. 1960. A study of the mechanism of spinal injuries. *Journal of Bone and Joint Surgery* 42B: 810-823.

Sahuquillo, J., and M.A. Poca. 2002. Diffuse axonal injury after head trauma. A review. *Advances and Technical Standards in Neurosurgery* 27: 23-86.

Sano, K., N. Nakamura, K. Hirakawa, H. Masuzawa, and K. Hashizume. 1967. Mechanism and dynamics of closed head injuries (preliminary report). *Neurologia medico-chirurgica (Tokyo)* 9: 21-33.

Santucci, R.A., and Y.-J. Chang. 2004. Ballistics for physicians: Myths about wound ballistics and gunshot injuries. *Journal of Urology* 171: 1408-1414.

Schneider, D.C., and A.M. Nahum. 1972. Impact studies of facial bones and skull (SAE 720965). In *Proceedings of the 16th Stapp Car Crash Conference*, Warrendale, PA: Society of Automotive Engineers.

Schulz, M., S. Marshall, F. Mueller, J. Yang, N. Weaver, W. Kalsbeek, and J. Bowling. 2004. Incidence and risk factors for concussion in high school athlete, North Carolina, 1996-1999. *American Journal of Epidemiology* 160: 937-944.

Severy, D.M., J.H. Mathewson, and C.O. Bechtol. 1955. Controlled automobile rearend collisions, an investigation of related engineering and medical phenomena. *Canadian Services Medical Journal* 11: 727-759.

Sheir-Neiss, G.I., R.W. Kruse, T. Rahman, L.P. Jacobson, and J.A. Pelli. 2003. The association of backpack use and back pain in adolescents. *Spine* 28: 922-930.

Shirado, O., K. Kaneda, S. Tadano, H. Ishikawa, P.C. McAfee, and K.E. Warden. 1992. Influence of disc degeneration on mechanism of thoracolumbar burst fractures. *Spine* 17: 286-292.

Sokolove, P.E., N. Kuppermann, and J.F. Holmes. 2005. Association between the "seat belt sign" and intra-abdominal injury in children with blunt torso trauma. *Academy of Emergency Medicine* 12: 808-813.

Stables, G., G. Quigley, S. Basu, and R. Pillay. 2005. An unusual case of a compound depressed skull fracture after an assault with a stiletto heel. *Emergency Medicine Journal* 22: 303-304.

Starkey, C., and J.L. Ryan. 2001. *Evaluation of Orthopedic and Athletic Injuries* (2nd ed.). Philadelphia, PA: F.A. Davis.

Stehbens, W.E. 2003. Pathogenesis of idiopathic scoliosis revisited. *Experimental and Molecular Pathology* 74: 49-60.

Stillerman, C.B., J.H. Schneider, and J.P. Gruen. 1993. Evaluation and management of spondylolysis and spondylolisthesis. *Clinical Neurosurgery* 40: 384-415.

Stinson, J.T. 1993. Spondylolysis and spondylolisthesis in the athlete. *Clinics in Sports Medicine* 12: 517-528.

Teasdale, G., and B. Jennett. 1974. Assessment of coma and impaired consciousness. A practical scale. *Lancet* 2(7872): 81-84.

Tegner, Y., and R. Lorentzon. 1996. Concussion among Swedish elite ice hockey players. *British Journal of Sports Medicine* 30: 251-255.

Thurman, D.J., C. Alverson, K.A. Dunn, J. Guerrero, and J.E. Sniezek. 1999. Traumatic brain injury in the United States: A public health perspective. *Journal of Head Trauma and Rehabilitation* 14: 602-615.

Torg, J.S., J.J. Vegso, M.J. O'Neill, and B. Sennett. 1990. The epidemiologic, pathologic, biomechanical, and cinematographic analysis of football-induced cervical spine trauma. *American Journal of Sports Medicine* 18: 50-57.

Tran, N.T., N.A. Watson, A.F. Tencer, R.P. Ching, and P.A. Anderson. 1995. Mechanism of the burst fracture in the thoracolumbar spine: The effect of loading rate. *Spine* 20: 1984-1988.

Tuzun, C., I. Yorulmaz, A. Cindas, and S. Vatan. 1999. Low back pain and posture. *Clinical Rheumatology* 18: 308-312.

Uscinski, R.H. 2006. Shaken baby syndrome: An odyssey. *Neurologia medico-chirurgica (Tokyo)* 46: 57-61

Valsamis, M.P. 1994. Pathology of trauma. *Neurosurgery Clinics of North America* 5: 175-183.

Volgas, D.A., J.P. Stannard, and J.E. Alonso. 2005. Ballistics: a primer for the surgeon. *Injury* 36: 373-379.

Wall, E.J., S.L. Foad, and J. Spears. 2003. Backpacks and back pain: Where's the epidemic? *Journal of Pediatric Orthopaedics* 23: 437-439.

Watkins, R.G., and W.G. Watkins, IV. 2001. Cervical spine and spinal cord injuries. In *Sports Injuries: Mechanisms, Prevention, Treatment*, edited by F.H. Fu and D.A. Stone. Philadelphia: Lippincott Williams & Wilkins.

Watkins, R.G., and L.A. Williams. 2001. Lumbar spine injuries. In: *Sports Injuries: Mechanisms, Prevention, Treatment*, edited by F.H. Fu and D.A. Stone. Philadelphia: Lippincott Williams & Wilkins.

Wegner, D.R., and S.L. Frick. 1999. Scheuermann kyphosis. *Spine* 24: 2630-2639.

White, A.A., and M.M. Panjabi. 1990. *Clinical Biomechanics of the Spine* (2nd ed.). Philadelphia: Lippincott.

Wiersema, B.M., E.J. Wall, and S.L. Foad. 2003. Acute backpack injuries in children. *Pediatrics* 111: 163-166.

Wiltse, L.L., P.H. Newman, and I. MacNab. 1976. Classification of spondylolysis and spondylolisthesis. *Clinical Orthopaedics and Related Research* 117: 23-29.

Wotherspoon, S., K. Chu, and A.F. Brown. 2001. Abdominal injury and the seat-belt sign. *Emergency Medicine (Fremantle)* 13: 61-65.

Yoganandan, N., N.M. Haffner, D.J. Maiman, H. Nichols, F.A. Pintar, J. Jentzen, S.S. Weinshel, S.J. Larson, and A. Sances, Jr. 1989. Epidemiology and injury biomechanics of motor vehicle related trauma to the human spine. In *Proceedings of the 33rd Stapp Car Crash Conference* (SAE 892438), Warrendale, PA: Society of Automotive Engineers.

Yoganandan, N., and F.A. Pintar. 2004. Biomechanics of temporo-parietal skull fracture. *Clinical Biomechanics* 19: 225-239.

Yoganandan, N., F. Pintar, J. Reinartz, and A.J. Sances. 1993. Human facial tolerance to steering wheel impact: A biomechanical study. *Journal of Safety Research* 24: 77-85.

Yoganadan, N., F.A. Pintar, A. Sances, Jr., P.R. Walsh, C.L. Ewing, D.J. Thomas, and R.G. Snyder. 1995. Biomechanics of skull fracture. *Journal of Neurotrauma* 12: 659-668.

Zemper, E.D. 2003. Two-year prospective study of relative risk of a second cerebral concussion. *American Journal of Physical Medicine & Rehabilitation* 82: 653-659.

Zhang, L., K.H. Yang, and A.I. King. 2001. Biomechanics of neurotrauma. *Neurological Research* 23: 144-156.

NAME INDEX

A

Adams, M.A. 284
Adams, R.D. 251
Alexander, R.M. 103, 104
Allman, F.L. Jr. 208
Allsop, D. 261
Alvi, A. and colleagues 260
Amonoo-Kuofi, H.S. 278
Andersen, M.B. 11
Andrews, J.R. and colleagues 220
Åstrand, P.-O. and colleagues 45

B

Bailey, D.A. and colleagues 112
Baker, C. L. 227
Baker, S.P. 158
di Bandino Baroncelli, Bernardo 6f
Bandy, W.D. 120
Bankart, Arthur 220
Baroncelli, Bernardo di Bandino 6f
Bazarian, J.J. and colleagues 253
Berchuck, M. and colleagues 39
Best, T.M. 43
Bigliani, L.U. 215
Bircher, 6
Blemker, S.S. 97
Bogey, R.A. 197
Booth, D.W. 162
Boyd, H.B. 234
Brado, J.L. 232
Brandt, K.D. 115
Brechter, J.H. 181
Brechter, J.H. and colleagues 180
Burdett, R.G. 197
Burkhart, S.S. 216, 218, 220

C

Carter, D.R. and colleagues 29
Celsus, Aulus Cornelius 133
Charnley, J. 283
Chu, L.D. 253
Clausen, H. and colleagues 249
Codman, E.A. 209, 218
Connell, D. 165
Courville, C.B. 250
Crabb, W.A. 128
Crisco, J.J. and colleagues 161–162
Currey, J.D. 103, 104, 105, 106

D

Damasio, H. 258
Davidson, C.W. and colleagues 155
Dawson, S.L. and colleagues 250

Delp, S.L. 97
Demirag, B. and colleagues 182
Denis, F. 272, 274f
Dunleavy, K. 120
Dunlop, R. 218
Duplay, E.S. 209
Dye, S.F. 181–182

E

Eriksen, H.A. and colleagues 199
Evans, M. 234

F

Farrell, K.C. and colleagues 184
Flynn, S.H. 184
Fugunaga, T. 120
Fukashiro, S. and colleagues 197
Funakoshi, T. and colleagues 218
Funsten, R.V. 159

G

Gage, Phineas 258
Galen 5, 133
Garrett, W.E. 43
Garrett, W.E. and colleagues 145
Gennarelli, T.A. 254
Gennarelli, T.A. and colleagues 256
Ghosh, A.K. and colleagues 11
Giddings, V.L. and colleagues 197
Gielen, A.C. 13
Goldspink, G. 119
Gurdjian, E.S. and colleagues 247

H

Hampson, D. 261
Handleberg, F. and colleagues 222t
Harris, W.H. 140
Harryman, D.T. and colleagues 210
Hawkins, R.H. 218
Hayes, W.C. and colleagues 157
Heil, J. 12
Hippocrates 4, 7, 274
Hodgson, V.R. 247, 261
Hodgson, V.R. and colleagues 247
Holbourn, A.H.S. 250, 251, 254
Holdsworth, F.W. 272
Hopper, R.H. and colleagues 261
Houston, C.S. 128
Hubbard, M. 95
Hutton, W.C. 284

I

Ikai, M. 120

Ikata, T. and colleagues 282
Inman, V.T. 196
Itoi, E. 217–218

J

Job, Frank 230
Jobe, F.W. 215
Jobe, F.W. and colleagues 230
John, Tommy 230

K

Karacan, I. and colleagues 267
Keith 107
Kennett, K. 261
Khaund, R. 184
Kibler, W.B. 230
Kim, A.W. and colleagues 40
Komi, P.V. and colleagues 197
Koulouris, G. 165
Kraus, J.F. 253
Krissoff, W.B. 164

L

Lanyon, L.E. 108
Lehman, C. and colleagues 219
Leonardo da Vinci 5, 6
Levine, W.N. 215
Levy, A.S. and colleagues 190
Lim, L.H. and colleagues 260
Linko, E. and colleagues 173
Long, W.T. and colleagues 163–164

M

MacDougall, J. 110
MacNab, I. 180
Maffet, M.W. and colleagues 220
Maffulli, N. and colleagues 199
Marcus, R. 111
Markolf, K.L. and colleagues 169
Matsumoto, K. and colleagues 160
Maxwell, W.L. and colleagues 256
McCormick, E.J. 11, 131
McCrory, P. 255
McGill, S. 286
McLean, S. and colleagues 175–176
Meeuwisse, W.H. 230
Monma, H. 159
Monteggia, Giovanni 232
Moorman, C.T. and colleagues 160
Morgan, C.D. 220
Morgan, C.D. and colleagues 220
Murata, Y. and colleagues 280
Murrie, V.L. and colleagues 278

N

Nahum, A.M. and colleagues 261
Neer, Charles 218
Neer, C.S. 216
Neviaser, J.S. 209
Newton, Isaac 65
Nirschl, R.P. 226
Noiseux, N. 140

O

O'Donoghue, D.H. 188
O'Driscoll, S.W. and colleagues 231
Offierski, C.M. 180
Ommaya, A.K. 254, 257
Ommaya, A.K. and colleagues 251

P

Paget 237
Panjabi, M.M. and colleagues 272
Pellman, E. and colleagues 253
Peterson, L. 174–175
Pink, M. 215
Poca, M.A. 256
Pourcelot, P. and colleagues 197
Powers, C.M. 179, 181

R

Renström, P. 174–175
Roaf, R. 272
Robertson, L.S. 68
Robinovitch, S.N. and colleagues 157
Robinson, T. and colleagues 112
Rubin, C.T. 108
Runge, J.W. 11

S

Sahuquillo, J. 256
Salem, G.J. 181
Salter, 30
Salter, Robert B. 128
Sanders, M. 11, 131
Sanders, M.S. 11
Schneider, D.C. 261
Severy, D.M. 271
Shaw, B. 11
Silverstein, B.A. and colleagues 238
Smith, F.M. 234
Snow-Harter, C. 111
Snyder, S.J. and colleagues 220, 222t
Solomonow, M. 40
Speed, J.S. 234
Speer, K.P. and colleagues 173

Sterett, W.I. 164
Stevenson, H.J. and colleagues 175
Suchman, E. 3
Sugita, T. 159
Surve, I. and colleagues 38
Swischuk, L.E. 128

T

Tabata, S. 217–218
Tanzer, M. 140
Templehof, S. and colleagues 219
Thelen, D.G. and colleagues 165
Thomas, L.M. 247
Thomee, R. and colleagues 181
Tidball, J.G. and colleagues 145
Tipton, C.M. and colleagues 118
Tompkins, D.G. 234
Tossey, J.D. and colleagues 208
Tuzun, C. and colleagues 278

V

Valsamis, M.P. 251
Vesalius 5
Victor, M. 251

W

Walch, G. and colleagues 214
Ward, S.R. 181
Watkins, R.G. 267
Watkins, W.G. 267
Westers, B.M. 162
Whaley, A.L 227
White, S. 187
Wilkinson, G.T. 222
Williams, J.M. 11
Williams, P.E. 119
Wilmore, J.H. 120
Wolff, Julius 107
Woo, S.L. and colleagues 171
Wright, P.R. 234–235

Y

Yamada, H. 105
Yates, B. 187, 189
Yoganandan, N. and colleagues 247, 261, 266

Z

Zernicke, R.F. 191
Zernicke, R.F. and colleagues 183
Zioupos, P. 106

SUBJECT INDEX

Note: The italicized *f* and *t* following page numbers refer to figures and tables, respectively.

A

ᾱ (average angular acceleration) 57*f*
ā (average linear acceleration) 57*f*
abdomen 148, 260
abduction/adduction
  overviews 46*t*–48*t*, 75*t*, 76*t*
  arm 63, 64
  forearm 233*ft*
  hip 154
  muscles 156*ft*, 161
acceleration 56–58, 66
accidents 2, 3, 9, 11, 130, 169. *See also* motor vehicle accidents
accommodation 127
acetabular structures 154, 155*f*, 159
Achilles tendon (calcaneal). *See also* tendon
  overviews 112, 115, 191, 192*f*
  injuries 119, 197–199
  risk factors, injury 125, 198, 199
  treatment 118
ACL (anterior cruciate ligament)
  overviews 170–171, 178
  injury 38–39, 171–173
  prevention of injury 175–176
  rehabilitation 174–175
  risk factors, injury 176
  treatment 173–174
acromial structures 205*f*, 208, 214, 215, 217, 218
acromioclavicular (AC) structures
  overviews 47*t*, 141, 204–206
  injuries of 206–209
  injury 206
  treatments 7
actin 41, 42, 118
action potentials 42
acute injuries 125, 127, 129, 143, 185. *See also specific injuries*
adduction. *See* abduction
adipose tissue 22*f*, 23, 24
adolescence. *See* youth
adult bone mass 110
adult cartilage 35
adult processes 107
age/aging. *See also* hip, fractures; youth/adolescence
  overviews 30, 129
  and Achilles tendon 198
  and back 277, 278, 279, 280–282

and bone 106, 108*f*, 135–136
brain injuries and 248, 254, 255
and cartilage 34, 114, 115
elbow injuries and 231
and fracture/strain injuries 157, 164, 222
impingement syndromes and 214, 215–216
and knee injuries 132, 171, 173, 177, 182
and ligaments/tendons 117
and meniscus injuries 176
OA and 141, 160
and OSD 182–183
and physeal injuries 30
shoulder and 206, 209, 212, 213, 215, 219
and ski injuries 164
and strength 120
ulnar variance and 236
α (instantaneous angular acceleration) 57*f*
a (instantaneous linear acceleration) 57*f*
air bags 260
all-or-none principle 44
amphiarthrodial joint 46
amputation 200
angular acceleration (deg/s2 or rad/s2) 57*f*
angular concepts and measures. *See also* moment of inertia
  kinetics 61–64
  measures 57*f*, 68, 69, 70
  motion 54–55*f*, 56, 57*f*, 58
angular displacement (deg or rad) 57*f*
angular displacement (θ) 57*f*
angular position (r,θ) 57*f*
angular velocity (deg/s or rad/s) 57*f*
angulation 80, 81*f*, 137*f*, 171
anisotropic materials 83, 105, 127
ankle. *See also* Achilles tendon; joints
  overviews 47*t*, 190–191, 192–193, 195*f*
  causes of injury to 72
  and knee 179
  plane of action 75*t*
  range of motion 76*t*
  sprains 38–39*f*, 184, 193–197
  stability 190–191, 193, 197
  twisted 124
anlage 27, 28, 29, 35
annular ligament 225
annulus fibrosis 141
anterior cruciate ligament. *See* ACL

anterior talofibular ligament (ATFL) 190, 196
anterior tibial translation (ATT) 171
anteversion 160
anthropometrics 72, 132
anticoagulants 25
antigens 134
antiseptics 6
apolipoprotein (ApoE) 249
aponeuroses 23, 36, 37, 192, 199*f*
applied force (F) 77
arches, foot 191–192, 193*f*, 194*f*, 199
arch geometry 215
Archimedes' principle 73
area moment of inertia (I) 64, 87, 88, 102–103
areolar tissue 22, 23–24, 37, 49*f*
arm (upper arm) 59, 220–224. *See also* humerus; ulna and ulnar structures
arteries 31–32, 74
arthritis 149, 281. *See also* osteoarthritis (OA)
arthrology 46–49
arthroplasty 142
arthroscopy 6, 218
articular cartilage
  overviews 34*f*, 35, 75
  adaptation of 114–115
  biomechanics 112–113
  degeneration of 182
  injuries of 139–141
  of knee 176, 182
  OA in 141, 161
  of shoulder 214, 219
  of spine 270
articular processes 49*t*
articulation 46. *See also* joints
assessment of injury 125–126
atherosclerosis 74
athletes 106, 111–112, 118. *See also* sports; *specific sports*
  back injury 280
  fatigue 130
  meniscus injuries 177
  muscle mass 120
  psychological factors for 12
  rehabilitation 174
  stress fractures 189
  tendon ruptures 182
  wheelchair 215
atrophy 115, 118, 121
automobiles. *See* motor vehicle accidents
average angular acceleration (α) 57*f*

average angular velocity(ω̄) 57
average linear acceleration (ā) 57f
average linear velocity (v̄) 57f
avulsion, ligament 231
avulsion fractures 115, 117, 137, 138, 228
axial loading 137f
axis. *See also* moment of inertia
  anatomical 171
  in angular kinetics 54–55f, 61–66
  in joints generally 77–79
  in knees 166, 179, 180
  longitudinal 82, 83, 87, 89f, 105, 179, 189
  neutral 87, 88f
  spinal 263f, 264, 266, 284f
  stiffest 264
  in subtalar joints 191, 194f
axons 23, 150, 151, 255–257

B
back. *See* spinal structures
backpacks 279
balance. *See* equilibrium
balance point 60
Bankart lesion 220, 221f
baseball
  Dizzy Dean syndrome 127
  elbow injuries 227t, 228
  fractures 222
  kinetics 229f
  shoulder injuries 206, 207t
  sprains 239
baseball bats 64, 190, 197, 198f
basement membranes 20
basketball
  and Achilles tendon 197, 198f
  and ankles 192
  knee injuries 172–173
  and OSD 182
  prevention of injury in 175
  shoulder injuries 207t
  stress fractures 106, 189
behavioral approaches to prevention/control 13
bending
  overviews 86–89, 101–107
  cantilever 65, 87, 89f
  four-point 87, 89f, 101–102
  fractures 137f, 190, 235–236
  injuries 187
biceps
  overviews 205f, 206f, 210f
  brachii 162, 165, 207t, 214, 223, 267
  femoris 148, 156ft, 164, 165
  injuries 204, 223–224
  risk factors for injury 224
biomechanics 3, 15, 191, 197. *See also* kinematics; kinetics
blastocysts 18, 19f
bleeding 134, 149, 247. *See also* hemorrhage
blood 26, 27, 31, 138, 198–199. *See also* bleeding; circulation; hemorrhage; vascularization
body, defined 59
body mass index (BMI) 161

bone. *See also* myositis ossificans; skeletal tissue; trabecular (cancellous) bone; *specific bones*
  overviews 26–30, 108
  adaptation 107–112
  compact (cortical) 27, 101–106
  components 30–32
  density 110
  fragility 31, 106, 117
  healing 138–139
  injuries 134–138
  macrostructure 32–33
  matrix 20, 23, 27f
  primary 32–33
  quality/quantity 158
bone-ligament junction 118
bone mineral content (BMC) 108, 109
bone mineral density (BMD) 108, 135, 160, 187, 188
bony fit 76, 77, 154, 166, 204, 209
bowleggedness (genu varum) 128, 168, 184
boxing 239, 246, 247, 249
brain
  overviews 149, 243–247
  concussion 251–255
  contusion, cerebral 248–251
  DAI 255–257
  injury mechanisms 245–247
  mechanisms of injury 250–251
  penetrating injuries 257–259
  prevention of concussion 255
  risk factors, injury 245, 249, 255
  swelling 251
  treatment of concussion 254–255
brittleness 93, 106
bruises 147, 149, 161–162. *See also* contusions
bucket-handle tear 177, 220, 221f
bullets 163, 258–259

C
calcaneal tendon. *See* Achilles tendon
calcaneofibular ligament (CFL) 190–191, 192f, 196
calcaneus 191, 192f, 193f, 194f, 195f
calcification 27f, 28, 162, 182. *See also* myositis ossificans
calcium 33, 105, 107, 109–110, 148
callus 138, 139
cancellous (trabecular) bone 27, 106–107
carpal structures
  injuries 125, 134, 150, 237–239
  metacarpal (CM) joints 141, 239–240
  risk factors, injury 238
cartilage 30, 33–36, 75, 114–115, 139. *See also specific types of cartilage*
catching 178, 197, 198f
categorizations 125–126
causal association 9, 11
cause-and-effect relations 124
causes of injury/disease 8, 15
cavitation 163
cavities, joint 46
cells
  overviews 20–22, 24–25

bone 26–33
cartilage 29, 33–36
death 135, 150
embryology 18–20
ligament/tendon 39–40
mesenchymal 138
mesenchymal stem 18, 20, 24, 27, 33, 34, 49
muscle 40–45
center of gravity 59, 60, 63
center of mass (COM) 59, 60, 63
childhood. *See* youth/adolescence
chondroblasts 20, 25
chondrocytes 25, 33–34, 115, 139
chondronectin 26
chronic compartment syndrome (CCS) 185
chronic injuries 125, 127, 143
chronic traumatic brain injury (CTBI)
  risk factors 249
chronological age 129
circulation 73, 74, 134, 135, 185, 254. *See also* blood; vascularization
clavicle 29, 204, 208
climbing 226
C2-L5 joints 49t
coefficient of restitution 71
collagen
  overviews 25, 27f, 31, 34f–35
  and aging 114, 117
  fibers 20, 23, 25, 142f, 166f
  and immobilization 118
  in ligaments 40
  in rotator cuff tendons 213
  and ruptures 199
  in tendons 37
  and tensile loads 113, 116
collateral ligaments
  overview 225
  dislocation 231f
  lateral (LCL) 168, 178
  medial (MCL) 168, 178
  ulnar (UCL) 228, 230, 239
collision 69, 70–71, 256, 271. *See instead* motor vehicle accidents; whiplash
coma 252, 256
Committee on Trauma Research (CTR) 10, 13, 15
communion 163
compartment syndrome (CS) 134, 161, 185–187
compensatory injury 127
compliance 81, 82, 129
compression wraps 185
compressive loads (compression)
  overviews 80, 81f, 83, 85
  and axis of rotation 62
  and cartilage 28, 35
  and chondrocytes 29
  concavity 210
  deformation 113
  injuries from 137, 147
  in multiaxial loading 85–86
compressive stress 29, 81, 85f, 87, 90, 105
concave (inner) surfaces 87, 101, 210
concentric action 45, 147, 165

concussions 126t, 251–255
conductivity 23
connective tissues. *See also specific connective tissues*
    overviews 20, 22, 23–24
    constituents 24–26
    mineral content 31
contactility 23
contact/noncontact injuries 172
contraction
    in Poisson's effect 85f, 86f
    of sarcomeres 42f
    skeletal muscle 23
    theories of 41
    unfused tetanic 118
    violent 212, 222
contributory factors 11, 124, 129–132. *See also risk factors*
contusions 147, 149, 161–162
contusions, cerebral 247, 248–251
convex (outer) surfaces 87
coordination 119, 120
coracoacromial structures 205f, 214, 215, 218
coracobrachialis 206f, 207t, 210f
coracoclavicular ligament 204, 205f, 206, 208
coracohumeral ligament 204, 209
coracoid process 204, 205, 205f, 214f
coronoid process 225, 231, 232f
cortical (compact) bone
    overview 27
    biomechanics of 101–106
    and exercise 110
    fracture risk of 137
    and osteoporosis 135
    structure 32f, 33, 83
costovertebral joints 49t
costs of injury 2, 9, 11, 16, 131
countermoment (M2) 63
coxa vara/valga 65, 128
cramps 147–148
crash-test dummies 93–94
creep response 91–92, 113, 115
cross-bridges 42, 43f, 46, 118
cross-links, collagen 25, 41, 117, 160, 226
cross-sectional geometry 102–104, 116–117, 118, 119, 120, 135
cruciate ligaments. *See also* ACL
    overview 168
    posterior (PCL) 168, 169–170, 171
    prevention of injury 175–176
    rehabilitation 174–175
    ruptures 169, 170, 172, 173, 178
    treatment 173–174
crushing forces 137f
cryotherapy 133, 173
C/t ratio (cross-sectional thickness/radius of bone) 104
cuboid 192, 193f, 194f
cuneiforms 191, 192, 193f, 194f
curvilinear motion 54–55f
cycling 73, 184

D

d (linear displacement) 57f

damage 106, 125–126
dampened response 91
dancers 189
dashboard impact 159
dashpot 96
death. *See* fatalities
deformation
    overviews 71, 80–82
    of collagen fibers 113
    from compression/tension 84
    constant 91, 92f
    of elastic materials 91
    energy stored from 69
    to failure 84
    plastic 83, 102
    rate of 92
    study of 95
    from torsion 90
    in twisting 90
    without fracture 106
deg (angular displacement) 57f
degenerative joint disease (DJD) (osteoarthritis) 115, 140–141
deg/s2 (angular acceleration) 57f
deg/s (angular velocity) 57f
delayed-onset muscle soreness (DOMS) 147
deltoid 207t, 210, 210f, 267
deltoid ligaments (DL) 190, 192f, 196, 197, 206f, 208
deltoid muscles 209
deltoid tuberosity 205f
dendrites 23
denervation 150
dense tissue 22, 23, 25, 36
density 134
    and fracture risk 137
deposition, bone 29, 107, 108, 109, 139
dermatome 20
dermis 21f
dermomyotome 21f
design, product or environmental 13
destabilizing component of force 62, 63f
development 108–109, 160
Δ force/Δ length. *See* stiffness
diaphyseal fractures 232–237
diaphysis 28, 102, 190f
diarthrodial joints 46, 49, 73, 75, 113–114. *See also* articular cartilage
diet (nutrition) 121
differentiation, cell 18, 20
digits 233ft
dimensional change 84
direct solution approaches 95
discontinuities 92–93
discs, intervertebral. *See also* neck; spinal structures; whiplash
    overviews 49t, 75, 141, 156f
    biomechanics 62, 283–285
    in deformities 36, 280, 281
    fracture 265t, 272–274
    injuries 141, 280, 285–286
    risk factors, injury 111, 129
    spondylosis 270
disease 130, 134, 135. *See also specific diseases*

dislocating component of force 62
dislocation. *See also under specific bones; specific joints*
    overviews 148, 195
    ancient treatments 5, 7
    biceps 224
    congenital 160
    and osteonecrosis 135
    resistance to 77, 144
    and SLAP lesions 222t
    toe 200
disorganized tissue 36
displaced fragments 138
displacement 56–57f, 138
distracting component of force 62, 176
disuse 112. *See also* immobilization; physical activity
diving 281
division of cells 18, 19f
dorsiflexion 195f, 197, 200
double conydloid joints 166
drivers, vehicle 161
drugs 130–132, 134, 164
ductile materials 93, 106
dummies, crash-test 93–94
dynamic friction 72

E

E (kinetic energy) 68–69
eccentric action 45, 147, 164, 165
economic perspectives 10–11. *See also* costs of injury
ectoderm 18, 19f, 20, 21
edema 26, 133, 134, 161, 251
education 13
effort force (F) 77. *See also* force
Egypt 4
elastic cartilage 36
elastic deformation 71
elastic fibers 20, 23, 25–26
elasticity 71, 82, 90, 91, 95, 105–106
elastic ligaments 40
elastic limit 80, 102
elastic modulus 82. *See also* stiffness (Δ force/Δ length)
elastic return 91
elastin 25
elastohydrodynamic lubrication 114
elbow. *See also* joints; throwing
    overviews 48t, 75, 76t, 77, 224–226t
    biomechanics 55f, 59f, 63f, 78–79f, 228–229, 229f
    dislocation 227t, 231, 232–235
    fracture 231–232
    gender comparisons 120
    overuse injuries 226–230
    risk factors for injury 227, 230
    stability 225, 231
elderly. *See* age/aging
elements 96–97
elongation 80, 81f, 85–86, 116, 117, 150
embryology 18–20
emergencies 10
enabling factors 9
endochondral ossification 28
endoderm 18, 19f, 20, 21f

endomysium 40
endoneurium 150
endosteal structures 103, 135
endotendineum 36
end plates 44, 45, 140f
endplates 278
end plates (spine) 263f, 273, 275f, 278, 282
endurance training 119, 121
energy. See also kinetics
   overviews 15, 29, 68–69, 84
   of bone 102
   and failure 145, 164
   muscle 121
   potential 68, 69
   stored/strain 91, 171
   and toughness 106
engineering 3, 6, 59, 93, 98, 158, 218, 264
engineers 10, 11, 13, 15
entrapment 134, 150, 237–238
environmental factors 130
environmental relations 138
epicondralgia 226
epicondylitis 226–228
epidemiological perspectives 7–10, 15, 68
epimysium 40
epineurium 150
epiphyseal fractures 137, 163
epiphyseal plate 114
epiphyseal regions 28, 29
epitendineum 37
epithelial tissue 20–22
equilibrium 67, 86f, 100, 113, 228
equipment 13, 130, 162, 173, 191, 255
ergonomics 131
error, human 11
estrogen 29–30, 111–112
ethnicity 160, 236
etiological factors 138
exercise 110–112, 114–115, 117–118, 147, 223.
   See also physical activity
experience 132
experiments vs. models 93–94
extensibility 25
extension
   of knee 168, 178
   and knee injuries 177
   of sarcomeres 42f
extensor digitorum tendons 193f
extensors 185, 186f, 187t
   forearm 232, 233t
   knee 169, 179–183
   wrist and hand 237
extent of injury 138
external loads 81f
external mechanics 55. See also joint
   mechanics
extracellular matrix (ECM) 20, 23, 25–26,
   31, 135
extrapolation 95
extrinsic risk factors 9
exudate 133
eye wounds 149

F
F (effort force) 77. See also force
face wounds 149

facial injuries 149, 259–261, 266
factors 8–9, 11–12, 12t
   economic/social/cultural 10
failure, material 83–84, 84f
   overview 92–93
   from bending 89f
   in bending 87
   of collagen fibers 113, 116
   in cyclic loading 106
   and fractures 189
   in passive/stimulated muscle 144
   and shoulder instability 214
   spiral 90
   of tissue 126
falls 120, 157–158, 169, 173
   head injuries 245f
   prevention of 158
fascia 40, 185, 187, 192
   plantar 199
fascicles 150
fascicular disorganization 150
fasciitis 187
fasciotomy 185
fat 109, 110, 120
fatalities
   overviews 2, 10
   from airbags 260
   boxing 249
   from deformable bullets 259
   from head injuries 245, 247, 248, 248f,
     257
   rates of 155
   and second impact syndrome 255
   from spinal injury 266, 272
   and spine 262, 272
   sports 249
fatigue, material 92, 106, 108. See also stress
   fractures
fatigue, mental/physical 11, 130, 147, 161,
   166, 215
fatigue, muscle
   as cause of injury 164, 200, 210, 212, 216,
     222
   of fiber types 43, 44t
   as symptom of injury 270
feet 141
female athlete triad 111
femoral structures. See also femur; hip
   overviews 154–156ft, 161, 167f, 168t
   biceps femoris 148, 164, 165
femur. See also ACL; cruciate ligaments;
   hip
   overviews 65, 154, 161
   bending of 87, 89f
   deformation of 97f
   diagrammed 155f, 167f, 170f, 174f
   fetal 29
   fractures 159, 162–164
   in hip replacement 142
   immobilized 112
   injuries 157, 159, 162
   and joints 46t, 49t
   and lubrication 114
   material properties 105
   osteoarthritic 140f
   structural properties 102

fibroblasts 20, 24, 25, 139, 162, 166f
fibrocartilage 36, 141, 143, 154, 168, 236. See
   also specific structures
fibroelasticity 22, 23–24
fibrous connective tissue 45
fibula 105, 167f, 184, 186f, 189, 192f, 193f,
   195f
finite-element (FE) models 275f
flexibility 14, 36, 132, 164
flexion 168, 178, 179, 179f, 181f
flexion-extension plane 154
flexors 185, 186f, 187, 199f
   forearm 232, 233t
   wrist and hand 237
fluid mechanics 73–74, 90–91. See also
   rheology
fluids. See also blood; edema; swelling;
   viscoelasticity; viscosity
   overviews 25, 26, 31, 33, 95–96
   accumulation of 185
   brain 243
   and cartilage 35
   cerebrospinal 247
   diffusion of 21
   in load responses 113
   and loads 176
   in swelling 133
foot 191–192, 193f
   risk factors, injury 200
football
   Achilles tendon ruptures 197
   ankles and 198f
   dislocations 160
   hamstring sprains 164
   impingement 214
   knee injuries 172
   neck/spinal injuries 264, 267
   shoulder injuries 207t, 217
force. See also kinematics; kinetics; lever
   systems; loading; muscle
   overview 58–59
   direct/indirect 137f
   distracting component of 62, 176
   external/internal 80, 81, 95
   in falls 158
   fluid 73
   impulsive (F) 70
   and injury 136–137, 147
   in models 95
   moment of (torque) (M) 78–79f, 179
   vs. movement 55
   normal (N) 71
   off-angle 127
   per unit area (stress) 102
   transfer 141
Δ force/Δ length. See stiffness
force (F) 66, 70, 77, 79, 103
forearm 232, 233t. See also radius; ulna;
   wrist and hand
form and function 105, 107
fractures. See also stress fractures; specific
   bones; specific joints
   overview 136–139, 163–164
   and age 30, 108
   healing 29, 138–139
   prevention 93

risk factors 129, 135, 137, 138, 222–223
types of 163f, 189, 222–223f, 232, 235–236,
238–239, 267, 272
friction 36, 71–73, 75, 113, 149, 150

G

g (gravity) 58–59, 66f, 69, 112, 257. *See also*
center of gravity
γ (shear strain) 90
γ (strain) 90, 251
gastrocnemius 148, 161, 162, 185, 186f,
187t
gender comparisons. *See also* menopause
overview 129
Achilles tendon ruptures 198
ACL injuries 175–176
bone mass 135f–136
bones 29, 108–109
hip fractures 154, 157
hip OA 160
knee injuries 169
kyphosis 277
OSD 182–183
pelvic fractures 155
proneness to injury 30
shoulder injuries 209
spinal deformities 278
strength 120
ulnar variation 236
genetics/heredity 129, 160, 199, 236, 249,
276, 281
genu varum (bowleggedness) 128, 168,
184
geometry, cross-sectional 102–104, 116–117,
118, 119, 120, 135
glenohumeral (GH) structures. *See* shoulder
glenoid fossa/labrum 204, 209, 210, 214
gluteus muscles 97, 156ft
glycation end product (AGE) 160
glycoproteins 25–26
glycosaminoglycans 26, 114
golf 207t, 226, 228
gracilis muscle/tendon 156ft, 161, 174
grain 105, 106
gravity (g) 58–59, 66f, 69, 112, 257. *See also*
center of gravity
ground reaction force (GRF) 66, 80f
growth 28, 29, 30f, 107, 108–109, 119
growth/growth plates 236
gunshot wounds 163
gymnastics 172, 207t, 236, 281

H

hamstrings 161, 164–166, 174. *See also*
tendon
prevention of injury 166
rehabilitation 164, 166
risk factors, strain 165
hand 141. *See also* wrist and hand
handball 176
head. *See also* brain
overviews 242–247
fracture, facial 259–261
healing 138–139, 141, 143–144, 149
Health Canada 2, 9

health professionals 10, 15–16
heart 22, 23, 74, 120
heat 21, 133
heel pain syndrome 199
helmets 255
hematoma 138, 161, 248
hematopoiesis 27, 134
hemorrhage 147, 149, 161, 247–248. *See
also* bleeding
heparin 25
heredity/genetics 129, 160, 199, 236, 249,
276, 281
high-energy injuries 190
hingelike joints 166
hip. *See also* joints
overviews 36, 46t, 75, 76t, 154
anatomy 155f, 156ft
dislocation 157, 159–160
fracture 65, 135–136, 154–158
joint moments 65
and knee 179
OA of 140–141, 160–161
replacement/resurfacing 142
risk factors, injury 157, 158, 161, 164
stability 77, 154
histamine 25
historical origins 4–7
Holland 9
homeostasis 100, 129
homeostatic balance 113
Homer 4
Hooke's law 82, 83, 101
hoop effect 176
hormone replacement therapy (HRT) 111
hormones 29, 30, 108, 109, 111–112, 175
human error 11
human factors 131
human interactions 130
humerus
overviews 204, 205f, 210
elbow dislocation 231f
fractures 220, 221–223, 224f, 231–232f
material properties 105
risk factors, fracture 222–223
hunchback 276, 277–278
hyaline cartilage 33, 34f–35, 113, 139
hydrodynamic lubrication 114
hydrodynamics 73
hydrophilic tissues 26
hyperplasia 119
hypertrophy, chondrocyte 115
hypertrophy, muscle 119
hypoxia 198
hysteresis 91, 115

I

I (area moment of inertia) 64, 87, 88,
102–103
idealizations (simplifications) 58, 59, 69,
95, 101
iliacus 97, 156ft, 161
iliopsoas tendon 156f, 156t, 157, 161. *See
also* tendon
iliotibial band 183–184
risk factors, injury 184

imbalances, biochemical 149
imbalances, muscle 215
immobility 77, 112
immobilization 115, 117–118. *See also* atrophy; physical activity
immune system 134
impact 60, 105, 147, 161. *See also* collision
impingement syndromes 134, 204, 214–216,
228–229, 270, 272
impulse (force) 70
incidence vs. prevalence 8
inelasticity 71. *See also* elasticity; viscoelasticity
inertia 58. *See also* moment of inertia
infants 130, 160, 256
infections 133–134, 138, 149
inflammation. *See also* arthritis; synovitis
overviews 132, 133–134
in contusion injuries 162
and DOMS 147
in strain 164
synovitis 148
treatment 133
infraorbital wounds 149
infraspinatus 204, 206f, 207t, 210, 213t,
214, 216
infraspinous fossa 205f
initial-cycles effect 92f
injuries. *See also* acute injuries; chronic
injuries; principles of injury; *specific
injuries; specific structures*
overviews 2–3, 125–127
biomechanical overviews 66, 70, 82, 92
causes of 68, 72, 78
contact/noncontact 172
direct/indirect 136, 137
factors in 58
historical perspective 4–7
likelihood of 11, 124
mechanisms of 3, 124–125
nonmusculoskeletal 149–151
Injury in America (Committee on Trauma
Research) 10, 15
inner (concave) surfaces 87, 101, 210
insertion sites. *See also* avulsion fractures
calcaneal 197, 198, 199
direct/indirect 37, 40
electron micrograph of 143f
injuries to 137, 183
of upper arm 213t, 221f, 224f
weakening of 117, 118
instantaneous angular acceleration (α)
57f
instantaneous axis of rotation 166
instantaneous joint center 78, 79f
instantaneous linear acceleration (a) 57f
interdisciplinary approaches 3, 14
internal mechanics 55. *See also* material
mechanics
interphalangeal (IP) joints 47t, 141, 193f
intervertebral joints 75t, 262–263f
irregular tissue 22, 23
ischemia 127, 185
isometric work 45
isotropic materials 105

J

J (joules) 68
javelin 207t, 228
joint capsule 49f, 66–67f, 209, 210, 219, 220, 224, 269f
joint mechanics. *See also specific joints*
    overview 74–75
    joint reaction force (JRF) 79–80
    lever systems 77–78, 79
    mobility and stability 76–77
joints. *See also specific joints*
    overview 46–49
    friction 73
    injuries 148–149
    kinds of 75, 206
    Leonardo da Vinci on 5
    ligaments of 40
    replacement of 142
    sprains 148
joules (J) 68
jumper's knee 182
jumping 70, 172–173, 189, 195, 197
junctions 37, 118. *See also joints; specific junctions*

K

kinematics 55–58, 95, 215, 216, 217t
kinetics. *See also force; moment of force (M); moment of inertia*
    angular 61–64
    collisions 70–71
    energy (E) 68–69, 259
    equilibrium 67
    friction 71–73
    linear 58–61
    modelling 95
    momentum 69–70
    Newton's Laws 65–66
    projectiles 259
    work and power 67–68
knee. *See also* joints; meniscus; patella and patellar structures
    overviews 49f, 75, 76t, 78, 156f, 166–169
    bowlegged/knock-kneed curvatures 128, 168, 184
    collateral ligaments 168, 178
    cruciate ligaments 168, 169–176
    extensors 169, 179–183
    force components on 63f
    and friction 72
    ligaments 167f, 168
    moments and 79f
    and obesity 132
    prevention of injury 175–176
    rehabilitation 174–175
    ruptures 182, 183
    sprains 38, 178
    stability 40, 168, 171, 173, 178
    treatment 173–174, 178
kyphosis 274, 276, 277–278
    prevention 277–278

L

labral structures
    anatomy 219–220, 223
    injuries 140, 204, 214, 220–221f, 224
lacerations 141, 149
lamellar bone 33, 107, 139
laminar flow 73
latissimus dorsi 206f, 207t, 209, 210
laws of motion 65–66, 71, 80f
L-d (load-deformation) 81, 84, 91, 101–102
legs, lower 184–190. *See also specific structures of the lower leg*
length-tension relation 45, 45f
levator scapulae 206f, 210
lever arm (m) 61–62, 65
lever systems 77–78
ligament. *See also specific structures*
    overview 36–40
    biomechanics 115–118
    classification of injuries to 125–126t
    composition 23
    injuries 148
    long vs. short 117
    and stretch 73, 119
likelihood of injury 124, 160, 168. *See also contributory factors*
linear concepts and measures
    displacement (d) 57f
    elasticity 85
    kinetics 58–61
    measures 56–57f, 67–70
    motion 54–56, 58
    power (P) 68
    spring 95–96
    stress-strain 82–83, 96
    velocity (m/s) 57f
load-carrying tissues 45, 176
load-deformation (L-d) 81, 84, 91, 101–102
loading. *See also* bending; force; visco-elasticity
    overview 125
    acute 136
    chronic/repeated 136, 147, 157, 189, 199
    combined 87, 169, 172
    constant 91, 92f
    cyclic 113
    deformation and stiffness 80–81
    direction of 83, 105
    health effects of 149
    history 229–230, 236
    multiaxial 85–86, 101
    rate of 56, 183, 197
    stress and strain 81–82
    and time 56
    and tissue structure 127, 129
    torsion 88–90, 137f
    tortional 90f
    types 82–90
    uniaxial 82–85
    and viscosity 73
load response 113
loads. *See also* compression; failure; fatigue, material; material mechanics; response to loading
    axial 81

calculating 79
cyclic 100, 106
deprivation of 118
long-term 115
maximum 84f, 102, 145
reduced 115
ultimate 115
long bones 29, 31f, 83, 86, 105. *See also specific long bones*
loose tissue 22, 23, 25, 36
lordosis 274, 276f, 278–279
lower extremities 154. *See also specific joints; specific tissues*
    risk factors, injury 112, 175
lower legs 184–190
    risk factors, injury 184
lubrication 35, 49, 113–114. *See also* fluid mechanics
lumbar structures 262, 263, 278–284
luxation. *See* dislocation
lymphatic system 26

M

M. *See* moment of force
macrophages 25, 139, 162
malleolus 193f, 197
marrow 29f, 106–107, 190f
mass, bone 103, 108–109, 120, 135f–136
mass (m) 58–60, 63, 64, 66, 67, 134
material mechanics 80–85
material properties 102
matrix 23, 27f, 28, 113
maturation 108–109
Maxwell model 96
M2 (countermoment) 63
medial tibial stress syndrome (MTSS) 187–189
megapascal (MPa) 81
meniscus 36, 141, 168–169, 173, 176–178
menopause 109, 110
menstruation 111–112
mesoderm 19f, 21f
metacarpophalangeal (MP) joints 48, 237, 239, 249
metaphysis 28, 29, 31, 32, 234f, 236
metatarsals 136, 191, 192, 193f, 194f, 200
metatarsophalangeal (MP) joints 47t, 141, 200
microcracks/damage/strain/trauma
    overviews 106, 108, 127
    elbow 226, 230
    measurement units 82
    and PF 200
    quadriceps 182
    shoulder 214
mild traumatic brain injury (MTBI) 251, 254
mineralization 28–29, 105, 106, 107–109
minerals 31, 33, 110–111
minor injuries 126, 200, 216
ms2 (linear acceleration) 57f
m (linear displacement) 57f
M-line proteins 41, 42f
m (mass) 58–60, 63, 64, 66, 67, 134
mobility/stability 76–77

modeling, bone 107, 108, 110, 275f. See also remodeling
models, biomechanical 93–97, 271, 286
moment arm (m) 61–62, 65
moment of force (M) (torque)
  overview 61–64, 87
  applied torque (T) 90f, 91f
  and joint motion 65, 78, 179
  measuring 95
  in neck 264f
  in skiing 191
  in throwing 229f
moment of inertia 58, 64, 87, 88, 89, 90f, 102–103
momentum (p) 58–61, 69–70
motion. See also movement
  asymmetry 78
  laws of 65–66, 71, 80f
  and moments 78
  rectilinear 54–55f
  rolling 78, 79f
  translational 54
motor units 44–46
motor vehicle accidents (MVA)
  commercial 130
  dislocations 159–160
  fractures 154–155, 162, 190, 247
  head injuries 245f, 253, 256, 260
  models 93
  PCL injuries 169
  risk factors, injury 271
  and SCI 267
  spinal injuries 266, 267, 268–269
motor vehicle drivers 161
movement. See also kinematics; motion
  overview 54–55
  analysis 75
  influence on skeleton 29
  and joint development 49
  modeling 95
  and pain 132
  production of 23
  velocity of 79
MPa (megapascal) 81
m/s (linear velocity) 57f
muscle. See also individual structures
  overview 20, 21f, 40–46
  death of 134
  fiber 41f, 42, 43–44t
  force 43–46, 55, 95, 118–121, 144
  passive vs. stimulated 145–147, 164
  tissue 22, 23, 41f
musculotendinous structures 37, 145. See also myotendinous junction
myofibers 162
myofibrils 40–41, 45f, 46, 119, 164
myofilament 41
myosin 41, 42, 43f, 118
myositis ossificans 147, 162
myotendinous junction (MTJ)
  overviews 37, 121, 141, 142f, 143
  in contusion injuries 161
  injuries 144, 165–166, 197
  strain to failure 146f, 147f

N
N (normal force) 71
N (newtons) 58
N · m (newton-meters) 61–62, 183
National Safety Council 2, 3, 9, 10–11, 131
natural selection 104, 105
navicular structures 191, 192, 193f, 194f, 196f, 237
neck. See also discs, intervertebral; spinal structures; vertebrae; whiplash
  overviews 36, 48t–49t, 261–264f
  cervical trauma 264–266
  spondylosis 270–272
  whiplash injuries 268–270t
nerves
  overviews 33, 150–151
  cranial 244
  median 237, 238
  of neck 262, 267
  pressure on 134
  of rotator cuff 213
  of spine 263, 264, 267, 286
  thumb 239
  ulnar 227, 228
  ulnar digital 239–240
nervous tissue 21–23, 149–151
neural structures. See also axons; brain
  action potential of 42
  adaptations 119
  in compartment syndrome 185
  development 19, 20f, 21t
  dysfunction 251, 252, 254
  injuries 261
neurons 23, 44, 45f, 150
newton-meters (N m) 61–62, 183
newtons (N) 58
nightstick fracture 4, 232
nonlinear behavior 83, 94
nonmusculoskeletal injuries 149–151
normal force (N) 71
normal stresses 81, 85, 87f, 90
notochords 18, 20, 21f
nutrition 109–110, 129, 161

O
obesity 132, 141, 158, 161
organization of tissue 18, 22f
Osgood–Schlatter disease (OSD) 182–183, 184f
ossification 27, 27f, 28, 29, 29f
osteoarthritis (OA) 115, 140–141, 160–161, 163
osteoblasts 20, 25, 27f, 30, 31, 138, 139
osteoclasts 30, 31, 139
osteocytes 25, 30–31, 139, 143
osteoligamentous structures 141, 144, 236
osteons 32f, 33, 106, 108, 135
osteopenia 110, 135
osteophytes 140, 229, 270
osteoporosis
  overview 135–136
  as contributing factor 65, 158
  gender comparisons 272
  prevention 111
  risk factors 129, 136
  and spinal deformities 277

osteotendinous junction 37, 93, 138, 141, 143, 182
outer (convex) surfaces 87
overextension 41, 42f
overhead movement 214, 215, 216, 222t, 224. See also sports; throwing
overload
  overview 125
  and injury 141, 143, 147, 228, 230
  of joints 161, 181, 227
  and myotendinous junction 121
  as prevention 181
  principle of 119
overuse
  overview 125
  Achilles tendon 197
  back 280
  caused by drugs 132
  elbow 226, 228
  hand 228, 237
  knee 181
  leg 188–189
  in OA 141
  shoulder 216

P
P (linear power) 68
p (momentum) 58–61
pain
  overviews 132, 133, 134
  back 283, 286
  from inflammation 149
  Leonardo da Vinci on 5, 6f
  in muscles 147
Pa (pascal) 60–61, 81
pascal (Pa) 60–61, 81
patella and patellar structures
  overviews 167f, 169, 179
  disorders 179–184
  ligaments/tendons 115, 169, 182, 186f (See also ligament; tendon)
  risk factors, injury 182
patellofemoral joint (PFJ) 47t, 166, 169, 180f, 181–182
patellofemoral pain (PFP) 179, 181–182
peak height velocity (PHV) 108
pectineus 156ft, 161
pectoralis 206f, 207t, 209, 210f
peel-back 220
pelvis 46t, 154–155, 278
penetrating forces 137f, 246
pennation 43–44
perichondrium 33, 36
perimysium 40
perineurium 150
periosteal bone deposition 29, 29f, 138–139
periosteum 28, 103, 188, 190f
peritenon 37, 143, 197
permanent set 83, 102
peroneus muscles/tendons 185, 186f, 187t, 192f, 193f
pes cavus/planus 200
phagocytes 25, 133
phalanges 102, 193f, 237, 240
phantom foot injury 173

phosphorous 108
physical activity 110–112, 161. *See also* exercise
physical condition 14, 129
physics 15
physiological age 129
physiological range 125
physiological status 129
physiology 15
physis 28–29, 30, 32
piriformis 156*ft*
planes of action 75, 94–95
plantaris 185, 187*t*
plantar structures 192, 195*f*, 196, 199–200
plastic deformation (plastic strain) 71, 83, 106
Poisson's effect/ratio 84–85, 88, 105–106, 283
polymorphonucleocytes 162
popliteus 187*t*
porosity 105
position, anatomical 75
position, measuring 56–57*f*
"position of no return" 175
posterior talofibular ligament (PTFL) 191, 196
posture 23, 132
potential energy 68, 69
power 68. *See also* force; strength
predictive ability 95
predisposing factors 9, 124, 129, 138
prepubescence 120
presomite embryos 20*f*
pressure 60–61, 81, 134
pressure bandages 5
prevalence vs. incidence 8
prevention 8, 10–11, 12–15, 78, 131. *See also* equipment; *specific injuries; specific structures*
previous injury 130, 166
primary injuries 127
primary ossification center 28
primary spongiosa 28
principles of injury
   contributory factors 129–132
   rehabilitation 132–133
   terminology 125–127
   tissue structure 127–129
probabilities 94
pronation, foot 184, 195*f*, 196, 200
pronation, forearm 232, 233*ft*
proportional limit 83, 102
protein
   overviews 25–26, 31, 40, 41, 42*f*, 113
   and aging 114
   and atrophy 121
   in course of injury 162
   dietary 109–110
   and exercise 115, 118, 119
proteoglycan 25–26, 114, 115
psoas 97, 156*ft*, 161
psychological factors 11–12, 15, 16, 129–130
puberty 108, 120
pubic symphysis 46*t*

pubic-type luxation 160
public health approaches 8
pulling 80
puncture wounds 149
pushing 80

Q
θ (angular displacement) 57*f*
quadriceps 161–162, 179, 180*f*, 182
qualitative descriptions 56
quantification of responses 15, 183
quantitative descriptions 56

R
R (resistance force) 77
race comparisons 135*f*, 272
racquetball 228
rad (angular displacement) 57*f*
radial measures 57*f*
radians 57*f*
radioulnar joint 48*t*, 75*t*, 76*t*
radius (bone)
   overviews 225, 232, 233*f*, 237
   dislocation/fracture 30*f*, 231–236
   tensile stress 105
rad/s2 (angular acceleration) 57*f*, 254
rad/s (angular velocity) 57*f*
range of motion (ROM) 44*f*, 76, 132
Ranvier, zone of 28
reaction force 71
rectilinear motion 54–55*f*
regular tissue 22, 23
rehabilitation 10, 11, 121, 132–133, 164. *See also under specific structures*
reinjury 14
relative risk 9
relaxation
   force-relaxation response 91, 115
   response 91, 92*f*
   of sarcomeres 41*f*, 42*f*
   of tendons 37
remobilization 112, 115, 117*f*–118
remodeling
   overviews 3, 31, 33, 106, 107–108
   after fractures 138–139, 189
   and exercise 110
   in ligaments 144
   and nutrition 109
repair rate 132
reporting percentages 10
research 15. *See also* scientific approaches
resilience 35, 36, 91, 114
resistance
   to dislocation 77
   fluid 73
   frictional 71–72
   impact 105
   initial 91
   internal 81
   to torque 90*f*
   viscous 91
resistance force (R) 77
resistance training 119, 120–121
resorption 108, 109
response to loading 80–82, 96. *See also* loading

reticular fiber/tissue 22, 23, 24, 40
retinaculum 186*f*, 193*f*, 237, 238
retroversion 160
reversible effects 127
rheology 95–96
rhomboids 206*f*, 210
rigid-body models 69, 80, 94, 183
rigidity 31, 102
risk 12
risk factors 8–9, 11, 13. *See also* contributory factors; *specific injuries; specific structures*
rotation 62, 63*f*, 78, 79*f*, 137*f*
rotational acceleration 254. *See also* rad/s2 (angular acceleration)
rotator cuff
   overviews 204, 207*t*, 209–210, 212–213
   and biceps 223
   injuries 214–215, 216–217
   prevention of injury 219
   rehabilitation 218–219
   risk factors, injury 215–216
   treatment 217–218
(r, θ) (angular position) 57*f*
running. *See also* sports
   and Achilles tendon 197
   and bone mineral content (BMC) 110–111
   and fractures 189
   Ground Reaction Force (GRF) 66
   hamstring sprain 165
   and knees 182, 184
   and lower leg 187, 188
   and PF 199
   and tibiofemoral joint 176
ruptures 127, 129, 142. *See also under specific anatomical structures*
   overview 83, 84
   of capillaries 161
   of collagen fibers 116
   of ligaments 39*f*

S
sacroiliac 46*t*
safety 12–14, 15. *See also* prevention
saggital plane 75, 154
sarcomeres 41, 42*f*, 118, 164, 165
sartorius 156*ft*, 161, 186*f*
scapula 204, 205*f*
scapulohumeral balance 210
scar tissue 116, 164
scientific approaches 14–15, 16
sclerotome 20, 21*f*
scoliosis 274, 276–277
seat belts 159, 260, 266, 274*t*
secondary injuries 127
second impact syndrome 255
   risk factors 255
semimembranosus muscles 148, 156*ft*, 164, 165
semitendinosus muscles 148, 156*ft*, 164, 165
semitendinosus tendon 174. *See also* tendon
seniors. *See* age/aging
sensory receptors 38, 40
serratus anterior 206*f*, 210

severity 125–126, 132
shaken baby syndrome 130, 256
shape effects 102
shear
    forces 71, 113
    injuries 28, 90, 254, 255, 256
    loads 37, 80–81, 81f, 150
    strain (γ) 90, 251, 254, 256
    stress (τ) 29, 81, 85–86, 87–88f
shin splints 188
shoulder. *See also* joints; rotator cuff
    overviews 36, 47t, 75, 76t, 204, 207t
    anatomy 205f, 206f
    biomechanics 63, 134
    dislocation 7, 148f, 207t, 209–211, 212, 220
    frozen 209
    glenohumeral (GH) impingement 213–217
    prevention of injury 219
    rehabilitation 218–219
    risk factors, injury 209, 212, 215, 216, 222
    ruptures 204, 208, 213, 216–219, 224
    separation 206–208
    sprains 206, 208
    stability 204, 206, 209–211, 214
    treatment 217–219
simple tissue 20, 22
simulations 93–97
sinovial structures
    overview 46t–49ft
    fluid 35, 49, 75, 112, 113
    joints 75, 113, 139, 140 (*See also specific joints*)
    membrane 49f, 148
sinusoids 31
in situ 100–101
size effects 102
skeletal tissue 22, 23, 27. *See also* bone
skiing
    ACL injuries 172f, 173
    dislocations, shoulder 207t
    fractures, leg 164, 189, 191
    prevention of injuries from 175, 176
    sprains, hand 239
    tears, knee 177
skin 149
skull 242–243, 247–248, 261f. *See also* head
SLAP lesions 220, 221f, 222t, 224
sliding 41, 72, 78, 79f, 80
smoking 161
snowboarding 160, 255
soccer 69, 70, 175, 176, 177, 247
soleus 185, 186f, 187
somatotype 132
somites 19–20, 21f
soreness 14, 147
space flight 112
spasms 147
speed 56. *See also* velocity
spinal structures. *See also* discs, intervertebral; joints; neck; vertebrae
    overviews 75t, 76t, 77f, 262f–263f, 273f
    cervical 66, 264–266, 268–272

deformities 5, 274–280
    dislocation 264, 265t, 267, 274
    and OA 135, 141
    prevention of deformities/injury 277–278, 279
    risk factors, injury 264, 272, 276, 279
    rupture 85, 265, 269f
    spinal cord 18, 21f, 149, 243, 262–263, 266–268, 272
    spondylolysis/spondylolisthesis 280–282
    spondylosis 270, 272
    sprain 265, 284t
    treatments for pain 286
spiral fracture 91, 137, 189, 191, 222
spondylolisthesis/spondylolysis 280–282
spondylosis 270, 272
sports. *See also* jumping; squatting; throwing; *specific sports*
    artificial turf 72
    concussions 252, 253, 255
    contusions 161
    cruciate injuries 169
    cruciate ligament injuries 169, 171
    elbow injuries 227t, 231
    fatalities 249
    head injuries 245f
    impingement syndromes 214
    injuries 159–160, 172, 200, 207t, 217
    and myositis ossificans 161
    neck/spinal injuries 264, 267
    and OA 161
    prevention of injury 255
sports medicine 124–125
sprain 38–39, 144. *See also under specific joints*
sprinting 164–165. *See also* running
squatting 178, 179, 181
stability 25, 76–77, 138, 139, 174. *See also under specific joints*
stabilizing component of force 62, 63f
standard units (SI). *See specific units*
static force relaxation 115
static friction (f) 71, 72f
statistical data
    ACL injuries 171
    classification difficulties 8
    falls 158
    femoral fractures 162
    hip injury 155–156, 160
    injuries 2
    myositis ossificans 162
    pelvic fractures 154
statistical measures 9
stem cells 18, 20, 24, 27
stenosis, spinal 270
sternoclavicular joint 47t, 141, 204
sternomanubrial joints 49t
steroids 131–132, 182
stiffness (Δ force/Δ length). *See also* fatigue, material
    overviews 25, 46, 81–82, 84f, 106
    age/exercise/immobilization effects on 117, 118, 171
    bending 102, 103

    of bones 31, 35, 104, 105
    functional 40
    and injuries/disease 160, 171, 176, 185
    of ligaments/tendons 115, 116–118
    in osteotendinous junction 143
    and torsion 90
    and viscoelasticity 91
stiffness (sensation) 14, 144, 147, 209, 218
s (time) 57f
strain, mechanical
    overviews 81–83, 85f
    energy 69, 91
    energy density 84
    to failure 84, 145, 146f
    functional 108
    in linear spring 95
    maximal 105, 145
    in multiaxial loading 85–86
    in muscle injury 164
    rates of 90–91, 92, 95, 106, 115
    in rheology 95
strain (γ) 90, 251
strain injuries
    overview 142–148
    and contusions 162
    hamstrings 164–166
    prevention of 166
    risk factors 164, 165
strain rate 90, 91, 92, 95, 96, 115
stratified tissue 20, 22
strength. *See also* muscle, force; power; tensile strength
    bending 105
    of bones 31, 32, 35
    of cartilage 36
    of connective tissues 25
    and immobilization 118
    impact 105
    of junctions 37
    material 83, 84f, 93
    maximal 45, 120
    muscle 120
    structural 84f
    of tissues 36, 83
    and trabecular bone 106
strength training 121
stress
    overview 81–84
    in constant deformation 91, 92f
    distribution of 92
    focused 93
    in linear spring 95
    material 87f
    maximum 84f, 87f, 90, 102, 145
    in multiaxial loading 85–86
    relaxation of 91–92
    in rheology 95
    and viscoelasticity 90
stress, psychological 11
stress fractures 106, 112, 127, 136, 189
stress raisers 93
stress reaction 189, 190f
stress-relaxation response 91–92
stress responsivity 11
stress risers 92–93, 165, 216

stress-strain curves 116
stress-strain relation 91
stretching 14. *See also* extensibility
    and cramps 148
    and organization 40
    rate of 73
    and sarcomeres 119
    of tissue 23–24, 36, 91
striated muscle 23
structural properties 102, 105
structural strength 83, 117
subclavian muscles 206*f*
subluxation 148, 160, 215. *See also* disloca-
    tion
subscapular fossa 205*f*
subscapularis 204, 206*f*, 207*t*, 210, 213*t*,
    216–217
subtalar eversion/inversion 197, 200
subtalar joint 47*t*, 191, 193*f*, 194*f*, 195*f*
sugar 109, 110
supination 195–196, 233*ft*
supination, forearm 232
supraglenoid tubercle 205*f*
supraorbital wounds 149
supraspinatus
    overviews 204–207, 213–217
    rupture 218
    SLAP lesions and 220
supraspinous fossa 205*f*
surfaces 70, 72
surgery 5–6, 10, 173–174, 189, 218, 230,
    278
surveillance 8, 9
swayback 278–279
swelling 26, 133, 134, 147, 251
swimming 73, 178, 206, 207*t*, 264. *See also*
    sports
sympathetic nerves 23
symphysis 49*t*
synapses 42
synarthrodial joints 46
syncytiums 23, 31
syndesmosis 197
synovial fluid 75, 113
synovial joints 75, 134, 204. *See also* osteo-
    arthritis (OA)
synovial membrane 148
synovitis 148
synovium 148

T
T (applied torque) 90*f*, 91*f*
τ (shear stress) 29, 81, 85–86, 87–88*f*
t (tangential stress) 85
talocalcaneal joint 191, 192*f*, 194*f*
talocrural joint 193*f*, 195*f*
talofibular ligaments 192*f*
talus 191, 192*f*, 193, 194*f*, 195*f*, 197
Tang dynasty 5
tangential acceleration 57*f*
tangential stress (*t*) 85
taping 38
tapping forces 137*f*
tarsal structures 47*t*, 136, 191, 192, 193*f*,
    194*f*, 200

tears. *See also* ruptures
    overviews 116, 146*f*
    in joints 148
    ligament 144
    location of 145
    meniscus 177
    skin 149
    tendons 138
temporomandibular joint 48*t*, 141
tendinitis (tendonitis) 73, 125, 143, 197
tendon. *See also* collagen; myotendinous
        junction; *specific structures*
    overviews 36–37, 40, 41*f*, 44*f*, 141
    biomechanics 45, 80, 115–118
    experiments 101
    injuries 73, 127, 137, 138, 141–146, 226
    transplantation 230
tendonopathy 197
tennis 173, 207*t*, 227–228
tensile force, distribution of 143
tensile loads (tension)
    overviews 80, 82, 83, 84
    and collagen failure 113
    fractures from 137*f*
    and ligaments/tendons 116
    maximal 118, 145
    in multiaxial loading 85–86
    and nervous tissue 150
    passive 45
    quick applications of 115
tensile strength 105, 116
tensile stress 81, 85*f*, 87, 90, 115
tension, muscle 44, 45
tensor fascia lata 156*ft*, 161
teres minor/major 204, 206*f*, 207*t*, 210,
    213*t*
terminology 8, 125–127
testosterone 29–30, 120
tetany (tetanus) 118, 162
*Textbook of Disorders and Injuries of the Mus-
        culoskeletal System* (Salter) 128
thighs 161–166
thoracolumbar spine 76*t*, 272, 273–274, 275*f*
three-point bending 87, 89*f*, 101–102
throwing. *See also* elbow; overhead move-
        ment; sports
    overviews 69, 70, 228–230
    and injury 164, 197, 210, 214, 220, 222
thumb 233*ft*, 239–240
thyroid 29, 109, 136, 209, 262
tibia. *See also* ACL
    overviews 49, 167*f*, 184*f*, 186*f*, 192*f*, 193*f*,
        195*f*
    fractures 189
    injuries 187–190
    lubrication 114
    stressed 190*f*
    tensile strength 105
    treatment 189
    use/immobilization 110, 112
tibialis muscles 185, 186*f*, 187, 193*f*
tibial translation 173
tibial tubercule 184
tibiofemoral joint 47*t*, 141, 166, 176
tibiofibular ligaments 192*f*, 193, 197

tightness 14
time (s) 57*f*
time (t) 56–57*f*
tissue 20–23, 73, 95–97, 127–129. *See also*
        specific tissues
titin 41, 42*f*
toes 200. *See also* joints
tonically recruited fibers 121
torque. *See* moment of force
torque arm (m) 61–62, 65
torsion 61, 88–90, 137*f*
total mechanical energy (TME) 69
toughness 106
trabecular (cancellous) bone
    overviews 27, 101, 106–107
    and exercise 110, 112
    gender comparisons 108, 111
    and osteoporosis 135
    risk factors, fracture 137
track and field sports 177, 207*t*
training 119, 120–121, 164, 176. *See also*
        exercise
transducers, in vivo force-sensing 80
translational motion 54
transverse ligament 205*f*
transverse plane 75, 84, 154
trapezius 206*f*, 208, 209, 210
trapezoid ligament 205*f*
trauma 5, 6*f*, 150, 161, 162. *See also specific*
        traumas
traumatic brain injury (TBI) 244–247, 254
    risk factors 249
triceps brachii 206*f*, 207*t*, 226
triceps surae 185, 191, 197. *See also* gastroc-
        nemius; soleus
triplanar motion 204
tripping 182
trochanter, greater and lesser 157*f*
trunk 272, 273*f*. *See also* spinal structures
t (time) 57*f*
T tubules 42
tuberosities
    of calcaneous 199
    of humerous 205*f*, 213*t*
    radial/ulnar 223, 225*f*, 233*f*
    tibial 167*f*, 169, 179, 180*f*, 182, 184*f*
tubes 102–104
turbulent flow 73
twisting 61, 88, 90
twitch properties 43, 44, 118, 119, 120,
    121, 165

U
ulna and ulnar structures
    overviews 225, 232, 233*f*, 237
    dislocation/fracture 231–235
    material properties 105
    risk factors, injury 236
    variance (UV) 236–237
ultimate load 83, 84*f*, 115
ultrastructure of bone 32*f*
undisplaced fragments 138
unfused tetanic contraction 118
unhappy triad 173
uniaxial loading 82–85, 84
unstimulated muscle 145, 146*f*

upper arm 59, 220–224. *See also* humerus; ulna and ulnar structures
upper body strength 120
upper extremities 204. *See also specific structures*
  risk factors, injury 236
U.S. National Safety Council 2, 3, 9, 10–11, 131
use 7, 125. *See also* physical activity
uterus 18*f*

V
v̄ (average linear velocity) 57*f*
vacuolation 162
valgus
  overviews 128
  and ACL injuries 175–176
  extension-loading injuries 228–230
  external rotation 173
  loading 168, 171, 172, 178, 225
  in meniscus injuries 177
  moment 170
variability 132
"Varus and Valgus—No Wonder They Are Confused" (Houston and Swischuk) 128
varus loading 225
varus-valgus movement 166, 168, 171
vascularization. *See also* blood; circulation
  and Achilles tendon 198–199
  and death of cells 135
  in elbow injuries 226
  of labrum 220
  of rotator cuff 213
  of supraspinatus 216
  of tendons/ligaments 139
vasoconstrictors/dilators 25, 133
vectors 56, 58, 59
vehicles. *See* motor vehicle accidents
velocity 45, 56–57*f*, 71, 78. *See also* moment of inertia
vertebrae. *See also* neck; whiplash
  overviews 40, 40*t*–49*t*, 62, 262–263*f*
  in deformities 276*f*, 277
  fracture 265*f*, 267*f*, 272–275*f*
  risk factors, injury 264, 272
  and running 111
  spondylosis/spondylysis/spondylolisthesis 270, 280*f*, 282
v (instantaneous linear velocity) 57*f*
viscoelasticity 90–92, 95–97, 106, 107, 115, 138. *See also* elasticity
viscosity 73, 91, 114, 117
vitamin D 109, 272
in vitro/vivo 100–101
in vivo force-sensing transducers 80
volleyball 182, 189, 207*t*
volume 134

W
ω̄ (average angular velocity) 57*f*
W (linear work) 67
walking 176, 197
Wallerian degeneration 150, 256
warm-up 164
warnings 14
water polo 214
watts 68
weakest link 129, 138
weightlifting
  back injuries 281
  and bending 89*f*
  and laws of motion 66, 67*f*
  and OA 141, 161
  and patellar tendon 115, 183
  and shoulders 212, 219, 220, 222
  SLAP lesions and 222*t*
  and spine 281
  work and power during 68
whiplash 125, 266, 268–270*t*, 271
ω (instantaneous angular velocity) 57*f*
Wolff's law 107, 189
work, angular/linear/mechanical 67–68, 69, 105
work-related injuries 131, 141, 178, 184, 189, 214, 228, 237, 238
World Health Organization (WHO) 9, 10, 135
wounds 149, 163, 257, 258, 267
woven bone 32
wrestling 207*t*, 281
wrist and hand. *See also* joints
  overviews 36, 75
  injuries 237–240
  and OA 141
  and osteoporosis 135
  risk factors, injury 236–237, 238
  sprains 239

X
(x, y) (linear position) 57*f*

Y
yield point 83
Young's modulus 82, 84*f*, 95. *See also* elastic modulus; stiffness
youth/adolescence 108, 109, 163. *See also* age/aging
  back pain 279
  deaths 10
  kyphosis 278
  meniscus injuries 176, 177
  spinal deformities 277

Z
Z disc 142*f*, 145, 146*f*
Z-line proteins 41, 42*f*
zone of Ranvier 28

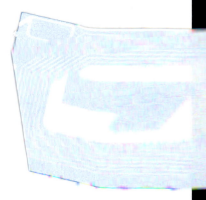

ABOUT THE AUTHORS

William C. Whiting, PhD, is a professor and director of the biomechanics laboratory in the department of kinesiology at California State University at Northridge (CSUN) and an adjunct professor in the department of physiological science at UCLA. He has taught undergraduate and graduate courses in biomechanics and human anatomy for more than 25 years. As an author and researcher, Whiting has written more than 60 research articles, abstracts, and book chapters, along with two other books, *Dynatomy: Dynamic Human Anatomy* and the first edition of this book, both published by Human Kinetics, Inc.

Whiting has served on many editorial boards, including *Journal of Strength and Conditioning Research, Strength and Conditioning Journal, ACSM's Health & Fitness Journal,* and *American Council on Exercise, FitnessMatters.* He also serves as a reviewer for numerous journals, including *Journal of Applied Biomechanics, Medicine & Science in Sports & Exercise, Journal of Strength and Conditioning Research, Strength and Conditioning Journal, Clinical Biomechanics, Clinical Kinesiology,* and *ACSM's Health & Fitness Journal.*

In 2002 Whiting received CSUN's Distinguished Teaching Award. He is a fellow of the American College of Sports Medicine and a member of the American Society of Biomechanics, the International Society of Biomechanics, the National Strength and Conditioning Association, and the American Alliance for Health, Physical Education, Recreation, and Dance. He is also a former president of the Southwest Chapter of the American College of Sports Medicine.

Whiting lives in Glendale, California, with his wife, Marji, and children Trevor, Emmi, and Tad. He enjoys playing basketball, hiking, and reading.

Ronald F. Zernicke, PhD, is a professor at the University of Michigan in the division of kinesiology and the departments of orthopaedic surgery and biomedical engineering. He is also the director of the Bone & Joint Injury Prevention and Rehabilitation Center.

Before moving to Ann Arbor in 2007, Zernicke was professor and chair of the department of kinesiology at UCLA and was Wood professor in joint injury research at the University of Calgary faculties of kinesiology (dean, 1998-2005), medicine, and engineering. At the University of Calgary, Zernicke was the executive director of the Alberta Bone and Joint Health Institute and served as director of the Alberta Provincial CIHR training program in bone and joint health, a combined graduate program of the University of Calgary and University of Alberta.

Zernicke has taught courses in biomechanics and injury mechanisms at the university level for more than 30 years. He received the UCLA Distinguished Teaching Award as well as the City of Calgary Community Achievement Award in Education. He has authored more than 545 peer-reviewed research publications and two books, including the first edition of this book, which received the Preeminent Scholarly Publication Award from CSUN in 2002.

Zernicke has served on the editorial boards of *Journal of Motor Behavior, Exercise and Sport Sciences Reviews, Journal of Biomechanics,* and *Clinical Journal of Sport Medicine.* He is a fellow of the American College of Sports Medicine and the Canadian Society of Biomechanics and an international fellow of the American Academy of Kinesiology and Physical Education.

Zernicke has also served as president of the International Society of Biomechanics (ISB), the American Society of Biomechanics (ASB), and the Canadian Society of Biomechanics (CSB). He continues to be an active member of the ISB, ASB, CSB, and the Canadian Orthopedic Research Society, Orthopedic Research Society (USA), American Society for Bone and Mineral Research, and Biomedical Engineering Society (USA).

Zernicke is the recipient of numerous achievement awards, most notably the NASA Cosmos Achievement Award awarded by the United States National Aeronautics and Space Administration, CIHR Partnership Award, Founders Award (Best Research) given by the Canadian Orthopaedic Research Society, and Alumnus of the Year from Concordia University Chicago.

In his leisure time, Zernicke enjoys reading, cross-country skiing, and hiking. He lives in Ann Arbor with his wife, Kathy.